CARTOGRAPHY
Thematic Map Design

CARTOGRAPHY
Thematic Map Design

Fifth Edition

CARTOGRAPHY
Thematic Map Design

Borden D. Dent
Georgia State University

Boston   Burr Ridge, IL   Dubuque, IA   Madison, WI   New York   San Francisco   St. Louis
Bangkok   Bogotá   Caracas   Lisbon   London   Madrid
Mexico City   Milan   New Delhi   Seoul   Singapore   Sydney   Taipei   Toronto

WCB/McGraw-Hill

A Division of The **McGraw·Hill** *Companies*

CARTOGRAPHY THEMATIC MAP DESIGN, FIFTH EDITION

Copyright ©1999 by The McGraw-Hill Companies, Inc. All rights reserved. Previous edition © 1996 by Times Mirror Higher Education Group, Inc. All rights reserved. Printed in the United States of America. Except as permitted under the United States Copyright Act of 1976, no part of this publication may be reproduced or distributed in any form or by any means, or stored in a data base or retrieval system, without the prior written permission of the publisher.

This book is printed on acid-free paper.

2 3 4 5 6 7 8 9 0 KGP/KGP 9 3 2 1 0 9

ISBN 0-679-38495-0

Vice president and editorial director: *Kevin T. Kane*
Publisher: *Edward E. Bartell*
Sponsoring editor: *Daryl Bruflodt*
Marketing manager: *Lisa L. Gottschalk*
Senior project manager: *Kay J. Brimeyer*
Production supervisor: *Deborah Donner*
Designer: *K. Wayne Harms*
Compositor: *Shepherd, Inc.*
Typeface: *10/12 Times Roman*
Printer: *Quebecor Printing Book Group/Kingsport*

Interior design: *Rokusek Design*
Cover design: *Lisa Gravunder*

Library of Congress Cataloging-in-Publication Data

Dent, Borden D.
    Cartography thematic map design / Borden D. Dent. — 5th ed.
        p.   cm.
    Includes bibliographical references and index.
    ISBN 0-697-38495-0
    1. Cartography. I. Title.
GA105.3.D45   1999
526—dc21   98-25683
                                                            CIP

www.mhhe.com

FOR JEANNE, ANDREW, JEFFREY, JENNIFER, LAUREN, AND ANNA

CONTENTS

PREFACE

Cartography continues to change, and in this edition of *Cartography: Thematic Map Design,* I attempt to incorporate the major changes, yet retain the central themes that have made the book popular among so many colleges and teachers. As computer visualization has crept into nearly all aspects of modern academic life, including cartography, map makers have suggested new paradigms to better account for and explain our activities. Philosophically, many cartographers have abandoned the idea that *communication* is a major overarching goal in mapping. Although some cling to it as an overriding paradigm, some have placed it in the realm of *public cartography* only, and visualization someplace that can be considered *private cartography.* This model makes a certain amount of intuitive sense, and one feature implicit in this new approach is that as long as we communicate at all, we need some form of map *design* to guide us. The central themes of this text are therefore retained, and still find a place in the education of the thematic cartographer.

An area of thematic cartography, which I believe has not changed as rapidly as others, especially compared to the technological changes, is with symbolization. There have been several strides made in some areas such as with animation and multi-media applications, but fundamental exploration into ways that data are inscripted into graphic language has not. The work of Dorling (discussed in the Cartogram chapter) may be an exception here, but there are precious few other major contributions as to new ways of taking non-map data and have them function in graphic ways. Displays in thematic atlases still rely on choropleth, dot, proportional symbol, isoplethic, flow maps, or perhaps cartogrammetric techniques. It is difficult to believe that the rich and innovative talent that thematic cartographers have would not by this time have led to many other forms. Perhaps we are seeing a natural limit.

The first edition of this book appeared just as the microcomputer revolution in our society was beginning, and no one knew precisely what that meant for cartography. The first edition covered materials on manual drafting that most college students today would find tiring and difficult. Gradually most material in this book that dealt with manual drafting have disappeared (although even today some feel that manual methods still need to be included). By the last edition, just three years ago, the World Wide Web allowed students to capture data and maps with this electronic means. The World Wide Web is even more embedded in the fabric of map making today, and shows no signs of relinquishing its hold on our activities.

Retained in this edition is the overall chapter presentation for the different kinds of thematic maps. Having them separated into organized "chunks" works well for most students, and I have been reluctant to change from this venue.

Several reviewers over the years have commented that the chapters on map projections (Chapters 2 and 3) are not necessary, or are too technical, for a book about thematic mapping. I simply cannot disagree more. To me, the foundation of any map, the structural basis, is the projection on which the remainder of the map rests. The proper selection of the projection is *fundamental* to the design of the map. The map's message hangs on its projection, and the final display of data is altered by the map's projection. To design a thematic map without any attention given to its projection would be like designing a building without giving any thought to the structural framework. These chapters remain, in my mind, central to this book.

This edition of *Cartography: Thematic Map Design* once again includes the Mercator poster of different map projections, from the United States Geological Survey, that has proved so useful to my students.

Some reviewers, too, have commented that the chapters on Printing Fundamentals (especially) and Desktop Mapping are not warranted in this text. While this may be true in some college classrooms, I feel that *map production* continues to be central to the total map design activity. In a great many cases cartography students go on to become professional cartographers, and map production techniques require basic notions of these processes.

There are several changes to this edition. I have endeavored to eliminate all errors from the previous edition, including mistakes that crept into the maps and graphics. Several graphics have been changed to make them more design-compatible with others throughout the book, a comment correctly made by several reviewers. There continues to be a strong emphasis on having all map and graphic examples be highly integrated with the text. Throughout the book, materials have been removed that were anachronistic relative to the practice of cartography today.

The major change is the addition of a chapter on geographic information systems (GIS). Several comments from reviewers at various times have indicated their desire to have such a chapter for their students. A chapter has thus been added at the end of the first section. Elaine Hallisey Hendrix provided this material, and I wish to thank her for this contribution, and her willingness to make the chapter integrate with others.

Another change occurs with the blending of the two chapters on The Map Design Process (formerly Chapter 12), and the one on Total Map Organization (formerly chapter 13). There was some overlap among these two, and putting them together made pedagogical, reasonable, and economic sense.

This edition contains relevant new references as they relate to map design. Again, as before for this text, the chief sources for bibliographic entries have been those books, articles, and other manuscripts published in the key English professional marketplaces. Texts referenced as this one required up-to-date literature to keep it fresh and resourceful for students. One apparent trend with map design literature is that much of the newer literature is not focused on the *appearance* of maps. Much of the literature tells me that cartographers have dwelled on either philosophy or theory and on technological achievements, but not on *design per se*. I have made every attempt to incorporate the major design trends whenever possible, but much has remained unchanged. In a way, revealing this absence should foster, I hope, young researchers into doing more with the look of maps. I believe I have captured the majority of the principal entries that have made a difference to map design.

Major contributions to this edition are two items found in the back jacket of the book. Thanks to the efforts of Christian Harder of Environmental Systems Research Institute (ESRI), and my editor Daryl Bruflodt at WCB/McGraw-Hill, a fully functioning version of ArcView 3.0 is provided with this edition. Map exercises keyed to various chapters, developed by Elaine Hallisey Hendrix, and tested by graduate student Betsy Hermann and me, appear on the disk for instructors and students. These exercises use the fully functioning ArcView program included, and will provide a systematic introduction to this useful desktop GIS program and to the practice of good thematic map design.

Appendix material remains largely unchanged except for the appendix of map sources, which has been dropped in favor of a listing of the most useful World Wide Web sites for cartographers. The included web sites are those that will not likely change in the near future. The ones listed are, of course, not exhaustive, but should provide a major starting point for students to get an idea of the wealth of material at these sources.

In each new edition of this book I try to develop at least one new map in which I explore some new design/symbolization plan. In the second edition, I examined a linearly scaled cartogram, *Airlines' View of the United States*. For the third edition, on the cover, I looked at showing thematic symbols shown on a satellite image. This edition includes a map called *Oklahoma: Reported Tornadoes, 1950–1991* (Chapter 9). This map symbolizes tornado numbers with symbols that appear as three-dimensional spheres,

but are range-graded into numeric classes. The reader will not need to attempt to scale these spheres with magnitude estimation, which results in considerable underestimation (see Chapter 9). The symbols were scaled to their areas, and then range graded, but have the added feature of three-dimensionality, which provides a fluid-like quality to the map.

I have been fortunate in having excellent manuscript and user reviewers over the years for this text, and this edition is no different. I have read these reviews carefully and have incorporated their comments wherever possible. Their excellent suggestions and comments have made the book a much better product and students using the text have greatly benefitted. I hope that I convey my sincere thanks to them for their guidance in this process. The reviewers of this edition include:

- Jeffery D. Colby
  East Carolina University
- Nicole Devine
  University of Washington—Seattle
- Dennis Fitzsimons
  Southwest Texas State University
- Erick Howenstine
  Northeastern Illinois University
- Robert B. Kent
  University of Akron
- Dr. James Saku
  Frostburg State University
- Eugene Turner
  California State University—Northridge

Ultimately, of course, I take the responsibility for the text. Finally, the editorial team at WCB/McGraw-Hill, and the production team led by Kay Brimeyer, deserve much credit for making the book a handsome and readable volume.

I wish to again express my thanks to Georgia State University for its support over the years. Georgia State has provided me a wonderful climate for teaching cartography to a great many students. My students are always critical, forgiving, and helpful. And the administrative support of the dean's office is always appreciated. Of course, I could not pursue these extra activities without the support of my family. They too have been wonderful to me during the many hours spent revising this book.

Borden D. Dent
Atlanta, Georgia
August, 1998

PART I

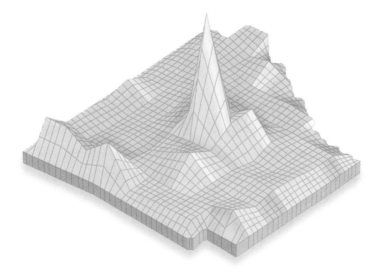

THEMATIC MAPPING ESSENTIALS

The first chapter in this part contains an introduction to the world of the thematic map and presents a backdrop to this increasingly used form of map. Various map examples highlight the rich variety. A discussion is included about map design and the role the cartographer plays in the design activity. After the introductory chapter, this part provides material necessary to begin the thematic mapping task. Thematic maps contain two chief components, the base map and the thematic overlay. The next two chapters provide background and techniques to organize and develop the base-map portion of the complete thematic map, and include a discussion of the variety of map projections useful for the thematic designer. A knowledge of map projections and their characteristics is required for successful map design.

Before the thematic cartographer can begin mapping, a fundamental awareness of the nature of geographical inquiry and the ways geographic data present themselves must be grasped. Ability to recognize data forms assists the map designer in the important activity of symbol design logic. Another important part of the cartographer's training is a knowledge of census or enumeration data, notably that provided by the United States Census Bureau. A brief background is provided here, as well as other data sources. For most cartographic design tasks the cartographer needs to develop and transform original data into forms that can be mapped. To do this, a knowledge of at least descriptive statistics is essential, and these are discussed in Chapter 5. The last chapter in this part presents the fundamental ideas and relationships of cartography and geographic information systems (GIS).

CHAPTER

1

INTRODUCTION TO THEMATIC MAPPING

CHAPTER PREVIEW

Maps are graphic representations of the cultural and physical environment. Two subclasses of maps exist: general-purpose (reference) maps and thematic maps. This text concerns the design of the thematic map, which shows the spatial distribution of a particular geographic phenomenon. Map scale, or the amount of reduction of the real world, is critical to the cartographer, because it determines selection and generalization. Mapmaking may be viewed in a larger context of cartographic thinking and cartographic communication. The designer plays an im- *portant role in this context as one who deals with the appropriate and creative symbolization of the image selected for communication after visualization and experimentation has taken place. Generalization takes place in the process, which includes selection, classification, simplification, and symbolization. Thematic map design is the aggregate of all the mental processes that lead to solutions in the abstraction phase of cartographic communication. Ethics in cartography is a subject of concern and plays a role in the way one views maps.*

The purpose of this book is to introduce several principles of thematic map design. The thematic map, only one of many map forms, will be more precisely defined later in this chapter. For now, some general comments about all maps will help set the stage. Maps seem to be everywhere we look: in daily newspapers and weekly newsmagazines, in books, on television, on trains, in kiosks, and even on table place mats. There is also great variety in types of maps. Some are greatly detailed and look like engineering drawings; others appear as freehand sketches or simple way-finding diagrams. Some show the whole world, others an area no larger than your backyard.

As our society has become more complex, the needs and uses for maps of all kinds have increased. Local governments and planning agencies use them for plotting environmental and resource data. Soil, geology, and water resource professionals use them daily in their work and planning. Public utility and engineering firms consult technical maps in order to complete their tasks. Land-use maps are utilized by planners; detailed cadastral maps are indispensable to city tax recorders. (See Figure 1.1.) Astronauts have used maps to help them land on the moon. The list of uses and users is virtually endless. In response to the rapidly increasing uses for maps, the United States government established the National Mapping Program in 1973 to coordinate the efforts of the many federal agencies involved in map production.[1] As a result, these agencies have become more responsive to the needs of map users, both public and private, across the country.

Since the 1960s, we have also seen a proliferation of state atlases. These have particular relevance to our discussion because they consist mainly of thematic maps, collected to assist educators and other professionals in displaying historical and geographic information about the state.

All of cartography, including thematic mapping, is undergoing rapid changes today. These changes are reflected in the publication of several books on cartography that now use *visualization* in the title, relatively easy map production from computer software, cartographic interaction on the World Wide Web, the fuzzy boundary between cartography and geographic information systems, and others. And the rapid changes do not seem to be slowing. Today maps can be made easily by those not trained in cartography, which has led to many maps of dubious quality. Poorly selected symbol scaling, inappropriate symbolization for the data used, colors without plans, and ineffective design have muddied the scene. But, maps *empower* through their very use, says Denis Wood,[2] notwithstanding their lack of accuracy.

Mapmaking is an interesting subject to study and an activity that has enjoyed a long history that is closely tied to the history of civilization itself. Maps date as far back as the fifth or sixth century B.C.[3] Mapmaking has come to be respected as a disciplined field of study in its own right. In this country, mapmaking has been closely associated with geography curricula since the early decades of this century. Although still linked in many ways to the study of geography, and rightfully so, cartography is recognized more and

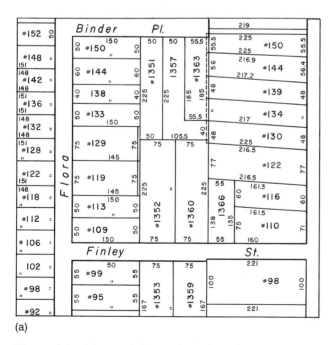

(a)

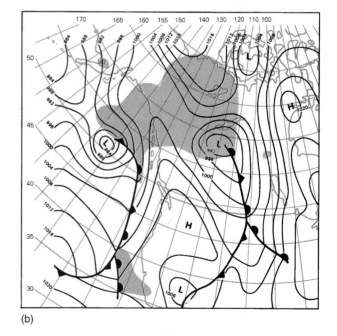

(b)

Figure 1.1   Maps satisfy a variety of needs.
A record of city land lots is kept on cadastral maps (a), and the National Weather Service produces a daily weather map, as in (b). These are but two of the hundreds of uses of maps.

more as a distinct field. With the growth of mapmaking has come a partitioning of the field into several component parts, each with its own scope, educational requirements, technology, and philosophical underpinnings. The practice of mapping today is far different from that used by our ancestors, although there are some common bonds.

In this chapter, we will look at the foundations of the **thematic map.** Thematic maps came late in the development of cartography; they were not widely introduced until the early nineteenth century. Today, such maps are produced quickly and cheaply, not only because the difficult process of base mapping has already been done by others, but also because of the benefits of computer technology. The last 30 years have been referred to as the "era of thematic mapping," and this trend is expected to continue in the future. The ultimate impact on the development and production of thematic maps by the easy access to both data and mapping software on the World Wide Web is not yet certain, although the impact thus far has been dramatic. Thematic maps make it easier for professional geographers, planners, and other scientists and academicians to view the spatial distribution of phenomena.

THE REALM OF MAPS

What is astounding in written language is that new, often provocative, concepts and meanings can be generated by the combination of just a few simple words ("the universe began with a big bang"). The same can be said for a map with just a few graphic elements, and this is the essence of the cartographic instruction presented here. Professional cartographers think of maps as vehicles for the transmission of knowledge (and for analysis). The communication idea is central to this book and will serve as our point of departure for introducing thematic cartography.

THE MAP DEFINED

Although there are many kinds of maps, one description can be adopted that defines all maps: "A **map** is a graphic representation of the milieu."[4] In this context, *milieu* is used broadly to include all aspects of the cultural and physical environment. It is important to note that this definition includes *mental abstractions* that are not physically present on the geographical landscape. It is possible, for example, to map people's attitudes, although these do not occupy physical space. In this context, maps also have been described as "models of reality"—although one's reality may be different than another's.

It is also assumed here that a map is a tangible product. There are such things as **mental maps,** generally described as mental images that have spatial attributes. Mental maps are developed in our minds over time by the accumulation of many sensory inputs, including tangible maps. Figure 1.2, for example, is an image of an interesting composite

mental map. The definition used throughout this book, however, assumes that the map is a *physical object* that can be touched.

WHAT IS CARTOGRAPHY?

The newcomer will probably be confused by the array of terms, names, and descriptions associated with mapmaking. It is unlikely that complete agreement on terminology will ever be reached by all those involved. For present purposes, we will adopt the following definitions. First, **mapmaking,** or mapping, refers to the production of a tangible map and is defined as "the aggregate of those individual and largely technical processes of data collection, cartographic design and construction (drafting, scribing, display), reproduction, et cetera, normally associated with the actual production of maps."[5] Mapping, then, is the process of "designing, compiling, and producing maps."[6] The mapmaker may also be called a *cartographer.*[7]

A second problem is the proper definition of **cartography.** As the discipline has matured and become broader in scope, many professional cartographers have come to make a distinction between mapmaking and cartography. In general, cartography is viewed as broader than mapmaking, for it requires the study of the philosophical and theoretical bases of the rules for mapmaking, including the study of map communication.[8] It is often thought to be the study of the artistic and scientific foundations of mapmaking. The International Cartographic Association defines cartography as follows:

> *The art, science, and technology of making maps, together with their study as scientific documents and works of art. In this context may be regarded as including all types of maps, plans, charts, and sections, three-dimensional models and globes representing the Earth or any celestial body at any scale.*[9]

This definition is broad enough in scope to be acceptable to most practitioners. It certainly will serve us adequately in this introduction.

GEOGRAPHIC CARTOGRAPHY

Geographic cartography, although certainly a part of all cartography, should be defined a bit further. *Geographic cartography* is distinct from other branches of cartography in that it alone is the tool and product of the geographer. The geographic cartographer understands the spatial perspective of the physical environment and has the skills to abstract and symbolize this environment. The cartographer specializing in this branch of cartography is skillful in map projection selection and the mapping and understanding of areal relationships, and has a thorough knowledge of the importance of scale to the final presentation of spatial

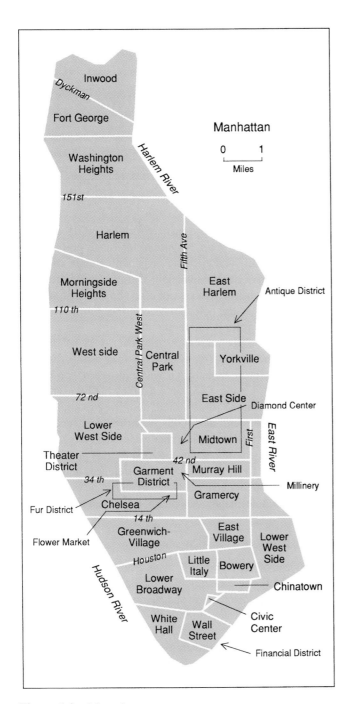

Figure 1.2   Mental map.

In this case, a composite or shared view of the neighborhoods of Manhattan is shown. Mental maps of this sort develop stereotypes and provide additional meaning to people not familiar with the mapped areas. *Source: Redrawn from John F. Rooney, Jr. et al. eds.* This Remarkable Continent: An Atlas of the United States and Canadian Society and Cultures *(College Station, Texas: Texas A&M University Press, 1982), p. 99.*

data.[10] Furthermore, geographic cartography, involving an intimacy with the abstraction of geographical reality and its symbolization to the printed page, is capable of "unrav-

eling" or revising the process, that is, geographic cartographers are very adept at map reading.[11] For the most part, geographic cartographers are involved in producing thematic maps, whether quantitative or qualitative, and are not usually associated with the production of highly detailed, large-scale reference (topographic) maps, photogrammetric products, surveying methods, or remotely sensed images.

On the other hand, geographic cartographers are associated with the production of thematic maps, are versed in the reading of photomaps and other remotely sensed images, and use these latter products in the preparation of their special-purpose (including atlas) maps. Geographic cartography is a branch of the broader science, and geographic cartographers understand spatial methodology.

Geographic cartography is seen as having a new strength as the notion of visualization has been more or less adopted by professionals as having a major role to play in cartography.[12] The kinds of expertise described above are those that we employ as we use the new powers of computer visualization in cartography.

Geographers have for many years embraced the study of maps and often suggested that maps are at the heart of their discipline. For example, the famous Berkeley geographer Carl O. Sauer writes:

> *Show me a geographer who does not need them [maps] constantly and want them about him, and I shall have my doubts as to whether he has made the right choice in life. The map speaks across the barriers of language.*[13]

More recently British geographer Peter Haggett writes, while exploring new ways of mapping accessibility in the Pacific Basin:

> *Although the new map is unfamiliar, it shows the locational forces at work in the Pacific in a dramatic way. In the physical world the earth's crust is reshaped by the massive slow forces of plate tectonics. So also technological changes of great speed are grinding and tearing the world map into new shapes. Capturing those spatial shifts [on maps] is at the heart of modern geography.*[14]

And American geographer John Borchert comments on the importance of maps to geography:

> *In short, maps and other graphics comprise one of three major modes of communication, together with words and numbers. Because of the distinctive subject matter of geography, the language of maps is the distinctive language of geography. Hence sophistication in map reading and composition, and ability to translate between the languages of maps, words, and numbers are fundamental to the study and practice of geography.*[15]

Maps provide us with a structure for storing geographic knowledge and experience. Without them, we would find it difficult, if not impossible, to orient ourselves in larger environments. We would be dependent upon the close, familiar world of personal experience and would be hesitant—since many of us lack the explorer's intrepid sense of adventure—to strike out into unknown, uncharted terrain. Moreover, maps give us a means not only for storing information, but for ana-

lyzing it, comparing it, generalizing or abstracting from it. From thousands of separate experiences of places, we create larger spatial clusters that become neighborhoods, districts, routes, regions, countries, all in relation to one another.

Source: Michael Southworth and Susan Southworth, *Maps: A Visual Survey and Design Guide.* A New York Graphic Society Book (Boston: Little, Brown, 1982), p. 11.

Atlas Mapping

Many cartographers believe that atlas production provides the best opportunities for thematic mapping [although some see the future strength of thematic mapping in geographic information systems (GIS)]. Certainly the proliferation of state and national atlases supports this view. The *Atlas of Pennsylvania,* for example, contains over one thousand separate maps, most of them thematic.[16] Another intriguing atlas is *This Remarkable Continent: An Atlas of United States and Canadian Society and Cultures,* edited by John F. Rooney, Jr., and colleagues. This latter atlas contains 390 maps taken from many different geographical studies done over a number of years. The traditional view of the atlas is changing, though. In the future, electronic atlases, with maps displayed on computer monitors and with the user capable of interacting with the map, may take over the now common ink-paper format. Regardless of the medium, however, the thematic cartographer continues to need sound design principles.

CARTOGRAPHY AND GEOGRAPHIC INFORMATION SYSTEMS

The distinction between cartography, especially thematic cartography, as a discipline and a geographic information system is increasingly blurred. Further compounding the boundary is the dramatic change and range of possibilities in scientific visualization. There are many definitions of GIS. Accepting one view, that a "geographical information system is a computer-based system that processes geographical information,"[17] does not easily clarify the role of thematic mapping in GIS. It is not uncommon today to see full functioning desktop GIS programs with several thematic mapping alternatives. The ease with which socioeconomic data may be incorporated into the programs also makes thematic mapping all that much easier. Notwithstanding any connections between the two, the proper methods of processing data, symbolization, and display require continued specialized study and examination.

Increasingly, display components of GIS assumes no knowledge of cartographic methods nor any fundamental

ideas about good design principles. Many GIS software packages contain map "wizards," or those prepackaged *macros* that produce thematic maps with simple and minimal alternative design selection. Unfortunately, these embedded map instructions do not offer the best that thematic cartography has the potential of offering. Further study of thematic principles is still required.

Among novice map users, however, GIS has become the *standard* to ask for when what they really need is a *map.* Many professionals, for whatever reasons, often say they want to get into GIS, but what they really mean is that they want a way to *display* data, not a way to *analyze* data as can be done with most GIS. Why there is this confusion between the two is not altogether clear.

A discussion about these two activities appears to be the best way to shed light on their differences and similarities so that the boundary between them can be better understood. A whole chapter, Chapter 6, is devoted to this topic.

KINDS OF MAPS

A review of the variety of maps leads to a better understanding of where quantitative thematic mapping fits into the larger realm of maps and mapping. At one time, thematic mapping was estimated to be only 10 percent of the field.[18] In general, all maps may be classified as either general-purpose or thematic types. However, it is possible to develop the classification further to include a greater variety, and one such scheme is depicted here. (See Figure 1.3.) These map types are discussed below.

General-Purpose Maps

Another name commonly applied to the **general-purpose map** is *reference map.*[19] Such maps customarily display objects (both natural and man-made) from the geographical environment. The emphasis is on location, and the purpose is to show a variety of features of the world or a portion of it.[20] Examples of such maps are topographic maps and atlas maps. The kinds of features found on these maps include coastlines, lakes, ponds, rivers, canals, political boundaries, roads, houses, and similar objects. (See Figure 1.4.) In the

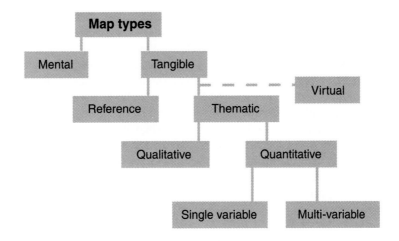

Figure 1.3   A classification of kinds of maps.
This organization is helpful in understanding the different types of maps. Only tangible maps are dealt with in this text.

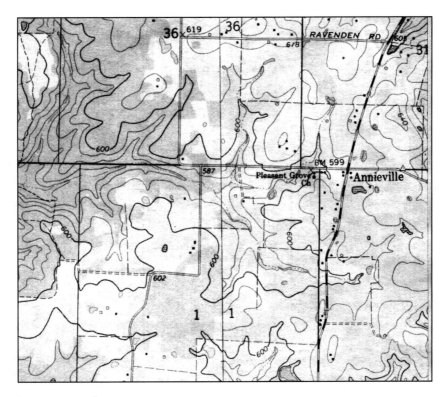

Figure 1.4   A general-purpose or reference map.
This is a portion of the Ravenden and the Imboden, Arkansas, 7½-minute USGS topographic map sheet and represents a typical reference map at large scale, showing both physical and cultural features.

United States, topographic maps are produced by the United States Geological Survey (USGS).

Historically, the general-purpose or reference map was the prevalent form until the middle of the eighteenth century. Geographers, explorers, and cartographers were pre-occupied with "filling in" the world map. Because knowledge about the world was still accumulating, emphasis was placed on this form. It was not until later, when scientists began to seize the opportunity to express the spatial attributes of social and scientific data, that thematic maps began to appear. Such subjects as climate, vegetation, geology, and trade, to mention a few, were mapped.

Thematic Maps

The other major class of map is the thematic map, also called a *special-purpose, single-topic,* or *statistical* map. The International Cartographic Association defines the thematic map this way: "A map designed to demonstrate particular features or concepts. In conventional use this term excludes topographic maps."[21]

The purpose of all thematic maps is to illustrate the "structural characteristics of some particular geographical distribution."[22] This involves the mapping of physical and cultural phenomena or abstract ideas about them. Structural features include distance and directional relationships, patterns of location, or spatial attributes of magnitude change. (See Figure 1.5.)

A thematic map, as its name implies, presents a *graphic theme* about a subject. It must be remembered that a single theme is chosen for such a map; this is what distinguishes it from a reference map.

Thematic maps may be subdivided into two groups, **qualitative** and **quantitative.** The principal purpose of a qualitative thematic map is to show the spatial distribution or location of kind (nominal data). For example, the mapping of the places and their boundaries used by the United States Weather Service in the south central area of the country is pictured in Figure 1.6. This map depicts a portion of the Geographical Area Designator Map used by the U.S. Weather Service in their aviation weather forecasts. This map does not show any quantities at all, but purely qualitative information, and is not precise but rather generalized in its record. On this form of map, the reader cannot determine quantity, except as shown by relative areal extent.

Quantitative thematic maps, on the other hand, display the spatial aspects of numerical data. In most instances, a single variable, such as corn, people, or income, is chosen, and the map focuses on the variation of the feature from

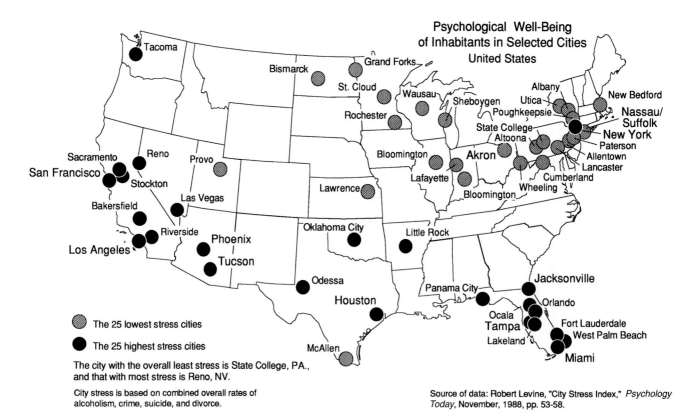

Figure 1.5 Thematic maps portray a variety of topics.
Thematic maps often show topical subjects to elicit questions and develop working hypotheses. In this case one might ask why so many so-called Sunbelt cities are in the high-stress category. Sociologist Levine suggests that new arrivals and retirees to these predominately southern cities brought their problems with them.

Figure 1.6   A qualitative thematic map.
The purpose here is to show location and approximate boundaries of places identified by the U.S. Weather Service in its aviation weather forecasts. *(Source: National Weather Service.)*

place to place. These maps may illustrate numerical data on the ordinal (less than/greater than) scale or the interval/ratio (how much different) scale. (See Figure 1.7.) These measurement scales will be treated in depth in a later chapter.

Quantitative mapping, as already pointed out, functions to show *how much* of something is present in the mapped area. The principal operation in quantitative thematic mapping is in the transformation of tabular data (an aspatial format) into the spatial format of the map.[23] The qualities that the map format provides (distance, direction, shape, and location) are not easily obtainable from the aspatial tabular listing. In fact, this is the quantitative map's *raison d'être*. If the transformation does not add any spatial understanding, the map should not be considered an alternative form for the reader; the table will suffice.

Furthermore, if exact amounts are required by the reader, a quantitative thematic map is not the answer. The results of the transformations in mapping are *generalized* pictures of the original data. The map is therefore an inefficient and inaccurate form of the original data. "A statistical map is a symbolized generalization of the information contained in a table."[24] Yet the map is the only graphic means we have of showing the spatial attributes of quantitative geographic phenomena. The special process of abstracting,

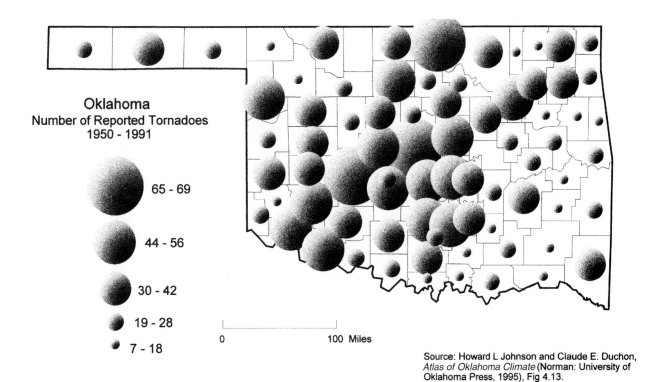

Oklahoma
Number of Reported Tornadoes
1950 - 1991

65 - 69

44 - 56

30 - 42

19 - 28

7 - 18

0   100 Miles

Source: Howard L Johnson and Claude E. Duchon, *Atlas of Oklahoma Climate* (Norman: University of Oklahoma Press, 1995), Fig 4.13.

Figure 1.7   Quantitative thematic map.
The clustering of tornadoes in the central part of Oklahoma is apparent on this map. This map provides an experimental approach to symbolization. The reader is not being asked to respond to the volume of the spheres, but to the areas beneath them. Symbolizing by spheres is dealt with later in this book.

generalizing, and mapping data (even at the expense of losing detail) is therefore justified when the spatial dimension is to be communicated.

In a book on map appreciation, the author identifies at least 10 different types or groups of maps, several of which are thematic in nature: photomaps, maps of the landscape and atmosphere, population maps, political maps, maps of the municipality, journalistic maps, advertising and persuasive maps, and computer maps.[25] Journalistic maps have been investigated recently, for example, and are defined as ". . . maps which are published in the mass news media . . . either accompanying and relating to news stories or other features, or as a primary news vehicle."[26] The types of maps are quite varied!

As further examples of the different forms that quantitative thematic maps may take, the reader is directed to Figures 1.8, and 1.9. In the first, a cartogrammatic, or very abstract, view of the *accessibility of France* in Europe is mapped, albeit unconventionally. This map illustrates how Europe would look in the year 2015 if the distances between cities are drawn proportional to their time (duration) of proposed high-speed rail. It is an intriguing map and portrays a concept in an interesting and forceful way. This map *points out that map content is a very important aspect of thematic map design.*

Figure 1.9 presents a thematic map illustrating how thematic quantitative data can be portrayed "over" a shaded relief base map.

Mental Maps

Mental maps, as mentioned earlier, are entirely different in that they are not tangible "paper" maps, but maps that reside in our heads. They are images we have stored away that contain features about our environment. Most intriguing is that these images are rarely accurate, at least in the traditional geometric way. These images often distort areas and distances, usually in ways that have meaning only to us. Mental maps are not simply stored images of paper maps (but can be), but are abstract constructs that allow us to operate on them, to problem-solve, perhaps to way-find. So, indeed, our mental maps often affect our spatial behavior.

Components of the Thematic Map   Every thematic map is composed of two important elements: a

Figure 1.8   A remarkable cartogram.
This map is a linear transformation of space in Europe. As such, it shows the "centrality" of Paris to the remainder of Europe.
Source: Copied with permission from the map's author, Colette Cauvin-Reymond, and the Delegation of l' Amenagement du Territoire er a Action Regionale, Visages de la France *(Paris: Montpellier, 1993), p. 7.*

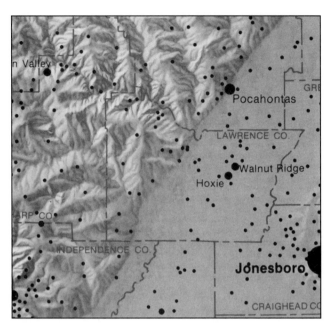

Figure 1.9   Arkansas population distribution.
On this thematic map, proportional circles and dots representing population have been printed on a shaded relief base map. *(Map detail part of map entitled* Arkansas Population Distribution with Shaded Relief features of the Physical Landscape. *Copyright Borden D. Dent, 1984.)*

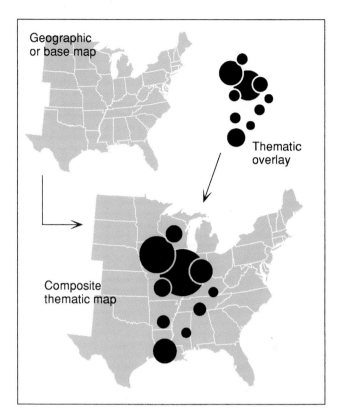

Figure 1.10   The major components of the thematic map. The composite or completed thematic map is made up of two important components: the geographic or base map and the thematic overlay.

geographic or base map and a **thematic overlay.** (See Figure 1.10.) The user of a thematic map must integrate these two, visually and intellectually, during map reading. The purpose of the geographic base map is to provide locational information to which the thematic overlay can be related. It must be well-designed *and include only the amount of information thought necessary to convey the map's message.* Simplicity and clarity are important design features of the thematic overlay. Design strategies for each of these components are dealt with in later chapters.

MAP SCALE

When cartographers decide on graphic representation of the environment or a portion of it, an early choice to be made is that of **map scale.** Scale is the amount of reduction that takes place in going from real-world dimensions to the new mapped area on the map plane. Technically, map scale is defined as a ratio of map distance to earth distance, with each distance expressed in the same units of measurement and customarily reduced so that unity appears in the numerator (e.g., 1:25,000). A more detailed discussion of scale appears in the next chapter. For now, our discussion is focused on simply the idea of scale and its relationship to such design considerations as generalization and symbolization.

Scale selection has important consequences for the map's appearance and its potential as a communication device. Scale operates along a continuum from large scale to small scale. (See Figure 1.11 and Plate 13.) Large-scale maps show small portions of the earth's surface; *detailed information* may therefore be shown. Small-scale maps show large areas, so only *limited detail* can be carried on the map. Which final scale is selected for a given map design problem will depend on the map's purpose and physical size. The amount of geographical detail necessary to satisfy the purpose of the map will also act as a constraint in scale selection. Generally, the scale used will be a compromise between these two controlling factors.

Another important consequence of scale selection is its impact on *symbolization.* In changing from large scale to small scale, map objects must increasingly be represented with symbols that are no longer true to scale and thus are more *generalized.* At large scales, the outline and area of a city may be shown in proportion to its actual size—that is, occupy areas on the map proportional to the city's area. At smaller scales, whole cities may be represented by a single dot having no size relation to the city's real size.

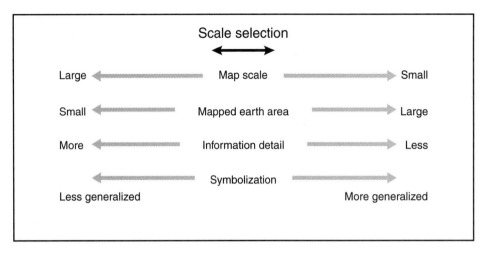

Figure 1.11 Map scale and its effect on mapped earth area, map information, and symbolization.
The selection of a map scale has definite consequences for map design. For example, small-scale maps contain large earth areas and less specific detail and must use symbols that are more generalized. Selection of map scale is a very important design consideration because it will affect other map elements.

Scale varies over the map depending on the projection used, as will be explained in Chapter 2. Scale, symbolization, and map projection are thus interdependent, and the selection of each will have considerable effect on the final map. *The selection of scale is perhaps the most important decision a cartographer makes about any map.*

In general, there is an inverse relationship between reference maps and thematic maps regarding scale. In other words, most thematic maps are made at small scales and reference maps at larger scales. As thematic cartographers generally work at small map scales, this requires them to be especially attentive to the operations of cartographic generalization. This is especially true for geographic cartographers.

MODERN VIEWS OF MAP COMMUNICATION

Views of how maps may function today in the contexts of visualization and communication are changing. The relatively simple *linear* interpretation of map communication as data field to map to map reader has given way to new concepts. With reference to new models, several ideas important to design are presented here.

Map Communication and Visualization

There is considerable discussion, debate, and even disagreement today about the roles of *communication* and *visualization* within the context of modern cartography. Scientific visualization, in general, can be thought of as a method that incorporates computers (especially graphic) that can transform data into visual models that could not have been seen ordinarily. Scientific visualization (SVIS) in cartography and geography, in general, has led to other

> Making a road relatively wider than it is on the Earth makes it visible; distorting distances on a projection enables the map user to see the whole Earth at once; separating features by greater than Earth distances allows representation of relative positions. Distortion is necessary in order that the map reader be permitted to comprehend the meaning of the map.
>
> Source: Mark S. Monmonier, *Maps, Distortion, and Meaning.* Resource Paper No. 75–4 (Washington, DC: Association of American Geographers, 1977), p. 7.

terms, such as cartographic visualization (CVIS) and geographic visualization (GVIS). Within the discipline over the past 15 years, there generally has been a sort of demise of *cartographic communication* as the appropriate model for cartography (especially thematic). Many cartographers have not completely given up this view, and some have accommodated it within their new constructs of what cartography, GIS, and cartographic visualization are.

David DiBiase's view of visualization in scientific research includes visual communication as in the "public realm" portion of his model. His model suggests that visualization takes place along a continuum, with exploration and confirmation in the private realm, and synthesis and presentation in the public realm. For him the private realm constitutes "visual thinking" and the public realm is "visual communication."[27] According to cartographic researcher Alan MacEachren this view is using the term visualization in a new way to describe cartography as a research tool.

MacEachren himself places cartographic communication within a "cartography cube" anchored at one vertex, with visualization at the other.[28] (See Figure 1.12). An interesting aspect of the discussion of cartography and visualization is that cartographic communication is not dead, but is incorporated into more complex descriptions of cartography, indeed, as an important component. As MacEachren himself says, "All authors, however, seem to agree that visualization includes both an analysis/visual thinking component and a communication/presentation component and suggest (or at least imply) that communication is a subcomponent of visualization."[29]

In reciting some of the history of cartography and communication, Hilary Hearnshaw and David Unwin state that the distinction between cartographic communication and cartographic visualization is, respectively, that the former deals with an "optimal map" whose purpose is to communicate a specific message, and the latter concerns a message that is unknown and for which there is no optimal map.[30] In a sense, this idea follows much of the thinking that is imbedded in the difference between deterministic thinking and probabalistic thinking, and this characterizes much of scientific thinking of this century. It also mirrors the difference between linear processes and parallel processes. The significance, though, is that the newer views of visualization in cartography and communication recognize the importance of the *map user* in the communication process, who was in the past often forgotten.

So, within these new contexts, where does map communication fit? Do we expect the map user (or map reader) to interact with the map in a cognitive way? One contribution that cartographers have made in the last 20 years is recognizing that map readers are different, and not simple mechanical unthinking parts of the process, that they bring to the map reading activity their own experiences and cognition. *The view held in this book in general supports the idea* that *map communication is the component of thematic mapping whose purpose is to present one of many possible results of a geographical inquiry, and that while the cartographer may want to convey a specific message, he/she cannot with certainty do this.* Maps, however, are useful in ways other than to convey (communicate) findings to others. They are also useful as tools to ". . . prompt insight, reveal patterns in data, and highlight anomalies."[31] In this context, maps are seen as tools for the researcher in *finding patterns and relationships among mapped data,* not simply for the communication of ideas to others.

Cartographic design can be practiced, then, within a communication system framework. The kinds of questions asked by the designer and the steps he or she will take are better controlled by this approach as they best focus the design activity. In practice, maps are always made with a purpose; the map reader's abilities and needs and the limitations of the graphic media all influence design decisions.[32] This is what makes cartographic design so interesting.

A very general graphic depiction of thematic map communication would be that represented in Figure 1.13. Two broad components recognizing the contributions of others are, first, "cartographic thinking," those activities in the private realm including visualization of data looking for patterns and relationships and using unencumbered and unstructured symbolization. This idea is not new, however, as mentioned previously. In this context, maps are seen as tools for the researcher in *finding patterns and relationships among mapped data,* not simply for the communication of ideas to others. The second major portion of the model is "cartographic communication," which resides in the public realm and involves the making of a final or optimal map or maps (within the context of the problem) with structured symbolization. Making the map includes those activities usually associated with map design—scale, symbolization, color choice, lettering, and so forth. Going from the private realm of unstructured symbolization to structured symbolization involves the cartographic processes of generalization and abstraction (discussed later).

In this picture of cartography and communication, *mapmaking* involves the conversion of the visualized data into a set of graphic marks (symbols), that are applied in rigorous ways and are placed on the map. Cartographic communication requires that the cartographer know something about the map reader, so the message is transferred to the reader with as much possibility of success as possible. Communication error develops when there is a discrepancy between the message intended by the map author and the one gained by the map reader. The cartographer may never be certain that the intended message is conveyed.

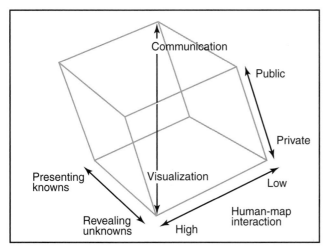

Figure 1.12 A model of map visualization and communication.
See text for explanation. *Source: Redrawn from Alan MacEachren, "Visualization in Modern Cartography: Setting the Agenda," in A.M. MacEachren and D.R.F. Taylor (eds.),* Visualization in Modern Cartography. *Oxford, England: Elsevier, 1994, p. 3.*

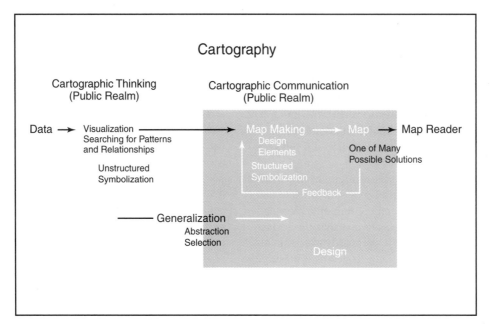

Figure 1.13   A model of map communication. See text for explanation.

A **map author** is someone who wishes to convey a spatial message. He or she may or may not be a mapmaker (or cartographer in the broadest use of the term).[33] When a map author wishes to structure a spatial message most effectively, he or she may employ a trained cartographer to design the map. A cartographer who originates a message is his or her own designer. Going from the private world of visualization and exploration to a final map used for communication requires *cartographic generalization,* which involves *selection, classification, simplification,* and *symbolization.*[34] Going from unmapped data to map form is sometimes referred to as the **cartographic process.** These topics will be more fully discussed later in this chapter.

Several constraints are imposed on the map author or cartographer in developing the message. These include, especially, the purpose of the map, map format, scale, symbolization, graphic and printing limitations, and economic considerations.[35] These may be considered to be *design constraints* in cartographic communication; they are the main subject matter of this book.

Within the context of mapmaking, initiation of a map can come from a variety of sources. (See Table 1.1.) Maps are made for aims other than pure communication, as the last item in Table 1.1 suggests. Some are constructed as decorative pieces of art, purely for visual appreciation. In such cases, map content plays a lesser role in the map's design.

In cartographic communication and design a fundamental idea is that the mapmaker and map percipient are not

The map provides an information source which can offer up the makings of many unexpected patterns from which insights may emerge. As even the most complex maps contain limited input data, when compared to the real world and do not have simply defined messages, observed patterns must be the result of an interactive cognitive dialogue between the map and the viewer's mind. The perfect visual working environment for this process is data-rich and well designed.

Source: Hilary M. Hearnshaw and David J. Unwin, eds., *Visualization in Geographical Information Systems* (New York: Wiley, 1994), p. 15.

independent of each other.[36] A **map percipient** is one who gains a spatial knowledge by looking at a map;[37] most thematic maps are designed for percipients.

Cartographic researchers recognize that the map reading activity is very complex and involves the human nervous system. Professor Phillip Muerhcke states that **map use** comprises reading, analysis, and interpretation.[38] In **map reading,** the viewer looks at a map and determines what is displayed and how the mapmaker did it. On closer inspection, the map user begins to see different patterns; this initiates thoughtful **map analysis.** Finally, a desire to explain these patterns leads to **map interpretation.** In this last case, causal explanations (probably not displayed on the map) are sought by the map user.

Table 1.1   Sources for the Initiation of a Map Message

1. A decision by intending map users or their representatives.
2. A decision by mapmakers in anticipation of a need.
3. A decision by a map author as a consequence of his or her own need for explanation.
4. A decision by a scientific body to contribute to the total information in a specialist field.
5. A decision by an individual to express himself or herself through making a map.

Source: Adapted with changes for J. S. Keates, Understanding Maps *(New York: Wiley, 1982), pp. 103–4.*

Table 1.2   Map-Reading Tasks

1. Pre-Map-Reading Tasks
   Obtaining, unfolding, etc.
   Orienting
2. Detection, Discrimination, and Recognition Tasks
   Search
   Locate
   Identify
   Delimit
   Verify
3. Estimation Tasks
   Count
   Compare or contrast
   Measure
   a. Direct estimation
   b. Indirect estimation
4. Attitudes on Map Style
   Pleasantness
   Preference

Source: Joel L. Morrison, "Towards a Functional Definition of the Science of Cartography with Emphasis on Map Reading," American Cartographer *5 (1978): 106.*

Many complex reading tasks have been identified. (See Table 1.2.) It is no wonder that map communication is so difficult! Maps become especially difficult to read when the map author or cartographer incorporates several tasks in one design. Simplicity in design is a goal and can be achieved in part by reducing the number of map-reading tasks on a single map. Reader training has been shown to improve the efficiency of map communication.[39]

Older cartographic communication models (including the one used in previous editions of this book) usually contained an element called *feedback*. In the linear sense, feedback is helpful to the map author in that it allows him or her to alter the design by incorporating positive changes suggested by the map user. Unfortunately, most cartographic designers are separated from users by time and space, so it is difficult to make use of feedback. The wise designer will make use of it when possible to improve designs.

The Importance of Meaning

Discussion among professional cartographers has emphasized the importance of *meaning* in cartographic communication. The *needs* of the map user are very important in the transfer of knowledge between map author and user. It is essential to include only those objects of specific relevance to the context of the map's message. For example, John Dornbach incorporated the users' requirements in his map design strategy during the development of aeronautical charts some years ago.[40] The pilots of modern jet aircraft need maps with specific information that can be easily perceived. Not just any map will do.

In a given design task, then, *only information that is potentially meaningful to the context should be included on the map.*[41] Because a map becomes meaningful only in relation to the previous knowledge the user brings to it,[42] the probable knowledge level of users should be taken into account before the final design of a thematic map is specified.

In sum, success in map communication depends on how well the cartographic designer has been able to interpret the requirements of the user.[43] Cartography has been called "the science of communicating information between individuals by the use of maps."[44] This statement implies that both map authors and map users are part of the process; the cartographer must pay particular attention to the needs of the user in designing each map.

CARTOGRAPHIC ABSTRACTION AND GENERALIZATION

Cartographic abstraction is that part of the mapping activity wherein the map author or cartographer *transforms* unmapped data into map form and *selects* and *organizes* the information necessary to develop the user's understanding of the concepts. "When we accept the idea that not all the available information needs to be presented, that instead information must be selected for particular purposes, then the mapping task becomes the identification of relevant elements."[45] Of course, the selection is guided by the purpose of the map.

Selection, classification, simplification, and symbolization are each part of cartographic abstraction and are generalizing operations. Each results in a reduction of the amount of specific detail carried on the map, yet the end result presents the map reader with enough information to

The possibility of placing data in inappropriate or misleading contexts is a real danger. Just as the meaning of statistics can be radically altered by judicious manipulation of their context, so too can the meaning of map data be changed according to the contextual information included. It is, therefore, essential that a cartographer have a broad background in a wide range of subjects to enable him or her to select appropriate contextual information against which the data mapped can be more readily comprehended. This ability implies that the cartographer has a knowledge of factors behind distributional patterns and some understanding of their causal interconnections.

Source: Leonard Guelke, "Cartographic Communication and Geographic Understanding," *Canadian Cartographer* 13 (1976): 107–22.

grasp the conceptual meaning of the map. Generalization takes place in the context of designing a map to meet user needs. The generalization processes lead to simple visual images, which are more apt to remain in the map user's memory.[46] Unless simplicity is achieved, the map will likely be cluttered with unnecessary detail. Appropriate generalization will result in a spatial message that is efficiently structured for the reader. On the other hand, excessive generalization may cause map images to contain so little useful information that there is no transfer of knowledge. A balance must be struck by the cartographer.

SELECTION

The **selection** process in the generalization operation of cartographic design begins the mapmaking activity. Selection involves early decisions regarding the geographic space to be mapped, map scale, map projection and aspect, which data variables are appropriate for the map's purpose, and any data gathering or sampling methods that must be employed. Selection is critical and may involve working very closely with the map author or map client.[47] The selection activity, finally, requires the cartographer to be at least somewhat familiar with the map's content—what is being mapped—and this cannot be overly emphasized.

CLASSIFICATION

Classification is a process in which objects are placed in groups having identical or similar features. The individuality and detail of each element is lost. Information is conveyed through identification of the boundaries of the group. Classification *reduces the complexity* of the map image, helps to *organize* the mapped information, and thus enhances communication.

In thematic mapping, classification can be carried out with qualitative or quantitative information. Qualitative data might include the identification of geographical regions; for example, the wheat belt, corn belt, or bible belt. It may be far simpler to communicate the concept of region by one large area than to show individual elements. Quantitative classification is normally in numerical data applications. Generally, an entire data array is divided into numerical classes, and each value is placed in its proper class. Only class boundaries are shown on the map. This process reduces overall information, but usually results in a map that is more meaningful.

SIMPLIFICATION

Selection and classification are examples of **simplification,** but simplification may take other forms as well. An example might be the *smoothing* of natural or man-made lines on the map to eliminate unnecessary detail. In the selection process, the cartographer may choose to include in the map's base material a road classed as all-weather. Simplification would be the process in which its path is *straightened* between two points so that it no longer retains exact planimetric location (even though this might be possible at the given scale). It can be straightened because the purpose of this map is simply to show *connectivity* between two points, not to illustrate the road's precise locational features. (See Figure 1.14.)

SYMBOLIZATION

Perhaps the most complex of the mapping abstractions is **symbolization.** Developing a map requires symbolization, because it is not possible to create a reduced image of the real world without devising a set of marks (symbols) that stand for real-world things. In thematic mapping, if an element is mapped, it is usually said to be *symbolized*.

Two major classes of symbols are used for thematic maps: **replicative** and **abstract.** Replicative symbols are those that are designed to look like their real-world counterparts; they are used only to stand for tangible objects. Coastlines, trees, railroads, houses, and cars are examples. Base-map symbols are replicative in nature, whereas thematic-overlay symbols may be either replicative or abstract. Abstract symbols generally take the form of geometric shapes, such as circles, squares, and triangles. They are traditionally used to represent amounts that vary from place to place; they can represent anything and require sophistication of the map user. A detailed legend is required.

By its very nature, the symbolization process is a generalizing activity shaped by the influence of *scale*. At smaller scales, it is virtually impossible to represent geographical features at true-to-scale likeness. Distortions are necessary.

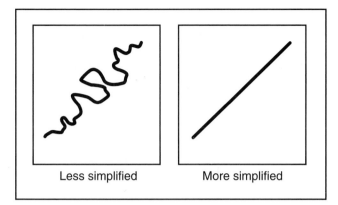

Figure 1.14 Classification and simplification in cartographic communication. More effective cartographic communication may result if data are first classified and simplified.

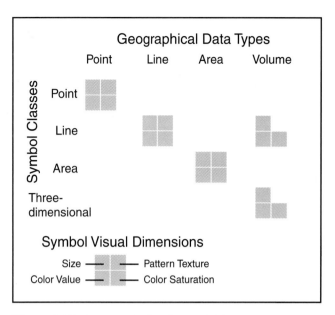

Figure 1.15 Map symbolization variables.
The selection of thematic symbolization is compounded because of the complex nature of the world of symbols. Geographical data are found in one of four classes but must be symbolized in only one of three symbol classes. Furthermore, each symbol may or may not be varied, depending on four visual dimensions of symbols.

For example, rivers on base maps are widened, and cities that have irregular boundaries in reality are represented by squares or dots.

The cartographer's choices of symbols are not so automatic that every possible mapping problem will yield similar symbol solutions. The symbolization process in thematic mapping is more complex than that, owing to a variety of factors. (See Figure 1.15.) This topic will be discussed in fuller detail in Chapter 4. It will suffice to say here that the selection of map symbols should be based on the logical association between the level of measurement of the map's data and certain graphic variables (sometimes called "primitives"), conventions or standards, appropriateness, readers' abilities to use the symbols, ease of construction, and similar considerations. Although various attempts have been made, very little standardization exists in the realm of thematic map symbolization. The student is encouraged to browse through the many thematic atlases in the research room in the library to see many of the symbolizations used today. This is a good laboratory exercise and often leads to very interesting classroom discussions.

The Art in Cartography

Professional cartographers have long discussed whether cartography should be regarded as an art or as a science. Most would probably agree that it is not art in the traditional view of that activity—that is, an expressive medium. Most would certainly agree that it is a science in the sense that the methods used to study its nature are scientific. The idea that there is an art to cartography suggests that its practice as a profession involves more than the mere learning of a set of well-established rules and conventions. *The art in cartography is the cartographer's ability to synthesize the various ingredients involved in the abstraction process into an organized whole that facilitates the communication of ideas.*

One prominent British cartographer very much believes that art is a part of cartography. In his design process, there is a critical point that is "a series of design decisions embodied in the symbol specifications, the point at which design moves from any general ideas to the particular."[48] What is important here is that at that point the designer is not primarily interested in communication (the science of) but *representation,* and a desire for the map to be not only informationally effective, but also aesthetically pleasing.

Each mapping problem is unique; its solutions cannot be predetermined by rigid formulas. How well the cartographic designer can orchestrate all the variables in the abstraction processes is the measure of how much of an artist he or she is. The reader must be considered, the final map solution must reduce complexity, and the map must heighten the reader's interest.[49] These considerations require an artist's skill, which can be acquired only through learning and experience.

THEMATIC MAP DESIGN

Now that cartography within the context of cartographic thinking and communication has been described, it is possible to develop the concepts of thematic map design. Cartographic design has been described as the "most fundamental, challenging, and creative aspect of the cartographic process."[50] But what is design? Is it a thing or an activity?

It is evident that cartography is not merely a technical art. It is for the greater part an applied art, an art governed and determined by scientific laws. But how can cartography avoid the rigid rules of mathematical precision? The decisive turning-point, according to my opinion, lies in the transition from the topographic to the general map. As long as the scale allows the objects in nature to be represented in their true proportion on the map, technical skill alone is necessary. Where this possibility ends the art of the cartographer begins. With generalization art enters into the making of maps.

Source: Max Eckert, "On the Nature of Maps and Map Logic," *Cartographica* 19 (1977): 1–7.

The cartographer is essentially an engineer attempting to construct a visual device which will effectively communicate geographical information to a percipient. . . . As an engineer, the map designer cannot solve his design problems unless he has carefully settled on the purpose of the map he is going to design.

Source: Arthur H. Robinson, "Map Design," *Proceedings* (International Symposium on Computer-Assisted Cartography, 1975): 9–14.

We can say, "That map has a good design," or, "Don't bother me while I'm designing." Is there a way to relate the noun to the verb? Indeed, there is.

WHAT IS MAP DESIGN?

What is design? There are almost as many definitions as there are designers. Design is the optimum use of tools for the creation of better solutions to the problems that confront us. The end result of design "is the initiation of change in man-made things."[51] Things are not just products, such as automobiles or toasters—they can be laws, institutions, processes, opinions, and the like—but they do include physical objects, including maps. Furthermore, design is a *dynamic activity*. "To design is to conceive, to innovate, to create."[52] It is a process, a sequential (and repetitive) ordering of events. To include the fashioning of everything from a child's soapbox racer to an artificial human heart, we can accept this definition: "The planning and patterning of any act towards a desired foreseeable end constitutes the design process."[53]

Design operates in at least three broad categories: product design (things), environmental design (places), and communications design (messages).[54] Cartographic design fits this classification most aptly in the category of communications, although it performs its functions with things (maps).

Map design is the aggregate of all the thought processes that cartographers go through during the abstraction phase of the cartographic process. It "involves all major decision-making having to do with specification of scale, projection, symbology, typography, color, and so on."[55] Designers speak of the *principle of synthesis:* "All features of a product must combine to satisfy all the characteristics we expect it to possess with an acceptable relative importance for as long as we wish, bearing in mind the resources available to make and use it."[56] The carto-

graphic designer seeks an organized whole for the graphic elements on the map, in order to achieve an efficient and accurate transfer of knowledge between map author and map user. Map design is the *functional relationship* between these two.

To reach the communication goals set forth for a given map design problem, the cartographer looks carefully at the intended audience and then defines the map's purpose with precision.[57] This involves an investigation into the functional context of the map, its intended uses, and its communication objectives. Every manipulation of the marks on the map is planned so that the end result will yield a structured visual whole that serves the map's purpose.

Thus, cartographic design is a complex activity involving both intellectual and visual aspects. It is intellectual in the sense that the cartographer relies on the foundations of sciences such as communication, geography, and psychology during the creation of the map. Map design is visual in the sense that the cartographer is striving to reach goals of communication through a visual medium. When we speak of a map as having a design, we are actually looking at the product as the result of design *activity*. (See Figure 1.16.)

A more detailed examination of the *design process* occurs in Chapter 13. There, such topics as design evaluation, creativity and ideation, visualization, and map aesthetics are dealt with.

A rich topic of the 1990s is map design and GIS. As more and more the boundary between cartography and GIS becomes blurred, and as it is becoming evident that design is wanting in that arena, it is inevitable that cartographic researchers will turn their design talents in that direction. In the meantime, what you will learn in this book about design and technique can be applied to maps in geographic information systems.

ETHICS IN CARTOGRAPHY

Ethical questions regarding mapmaking are of increasing concern to professional cartographers—so much so that several sessions have been devoted to this topic at cartographic conferences. In fact, a published book titled *How to*

Figure 1.16　Ernst Hoffman at work on Hammond's terrascope globe.
Despite the trend toward widespread computerization of cartography, manual techniques by talented cartographers remain and contribute to our enjoyment of maps. *(Photo courtesy of Hammond, Incorporated, Maplewood, N.J.)*

Lie with Maps addresses this topic. Professor Mark Monmonier, the author of the book, states in his introduction:

> *The purpose of this book is to promote a healthy skepticism about maps, not to foster either cynicism or deliberate dishonesty. In showing how to lie with maps, I want to make readers aware that maps, like speeches and paintings, are authored collections of information and also are subject to distortions arising from ignorance, greed, ideological blindness, or malice.*[58]

The important point is that because maps are made by humans they may as a consequence contain purposeful errors (lies) or errors of oversight and poor judgment, or both.

As you proceed through this book you will discover that there are no absolutely correct thematic maps, but only ones that are the best among several alternatives. Which to use? What should guide me? Professor Monmonier addresses this problem as well, and I am inclined to agree with his philosophy when he remarks (on the subject of one-map solutions):

> *Any single map is but one of many cartographic views of a variable or a set of data. Because the statistical map is a rhetorical device as well as an analytical tool, ethics requires that a single map not impose a deceptively erroneous or carelessly incomplete cartographic view of the data. Scholars must look carefully at their data, experiment with different representation, weigh both the requirements of the analysis and the likely perceptions of the reader, and consider presenting complementary views with multiple maps.*[59]

My cartography students, both graduate and undergraduate, have looked at issues related to cartographic ethics for a number of years. The culmination of many discussions have led to several "codes of ethics" for the thematic cartographer, and includes these, in various forms:

1. Always have a straightforward agenda, and have a defining purpose or goal for each map.
2. Always strive to know your audience (the map reader).
3. Do not intentionally lie with data.
4. Always show all relevant data whenever possible.
5. Data should not be discarded simply because they are contrary to the position held by the cartographer.
6. At a given scale, strive for an accurate portrayal of the data.
7. The cartographer should avoid plagiarizing; report all data sources.
8. Symbolization should not be selected to bias the interpretation of the map.
9. The mapped result should be able to be repeated by other cartographers.
10. Attention should be given to differing cultural values and principles.

A good place to initiate any discussion of cartographic ethics may begin with these. It makes a great classroom exercise.

The possibilities today for maps without ethics are compounded by the proliferation of "off-the-shelf" computer programs allowing non-cartography-trained persons to produce maps that may *look* good, but are not consistent with any established professional standards or conventions.

So, the question is what do we do about this? At present, no licensing or certification agencies oversee the cartography profession (as they do surveyors, for example). Certainly there are no governing bodies charged with guiding the ethical practices of thematic mappers. Thus the intent, honesty, and credentials of each cartographer must be dealt with on an individual basis, and his or her reputation must be considered. At the very least, cartographers should make it a practice to provide all base-map and data sources, methods of compilation, and appropriate dates on each product they design.

NOTES

1. Morris M. Thompson, *Maps for America* (U.S. Department of the Interior, Geological Survey—Washington, DC: USGPO, 1982), p. 13.

2. Denis Wood, *The Power of Maps* (New York: Guilford, 1992).

3. Leo Bagrow, *History of Cartography,* rev. ed. (Cambridge: Harvard University Press, 1966), p. 25.

4. Arthur H. Robinson and Barbara Bartz Petchenik, *The Nature of Maps: Essays toward Understanding Maps and Meaning* (Chicago: University of Chicago Press, 1976), pp. 16–17.

5. Phillip C. Muehrcke, *Thematic Cartography* (Commission on College Geography, Resource Paper No. 19—Washington, DC: Association of American Geographers, 1972), p. 1.

6. Mark S. Monmonier, *Maps, Distortion, and Meaning* (Resource Paper No. 75–4—Washington, DC: Association of American Geographers, 1977), p. 9.

7. Robinson and Petchenik, *The Nature of Maps,* p. 19.

8. Muehrcke, *Thematic Cartography,* p. 1; and Monmonier, *Maps, Distortion, and Meaning,* p. 9.

9. E. Meynen, ed., *Multilingual Dictionary of Technical Terms in Cartography* (International Cartographic Association, Commission II—Wiesbaden: Franz Steiner Verlag, 1973), p. 1.

10. Arthur H. Robinson, "Geographical Cartography," in *American Geography, Inventory and Prospect,* eds. Preston E. James and Clarence F. Jones (Syracuse, NY: Syracuse University Press, 1954), pp. 553–77.

11. Phillip C. Muehrcke, "Maps in Geography," *Cartographica* 18 (1981): 1–37.

12. Alan M. MacEachren, "Visualization in Modern Cartography," in *Visualization in Modern Cartography,* eds. Alan M. MacEachren and D. R. Fraser Taylor (Oxford, England: Pergamon, 1994), p. 8.

13. Carl O. Sauer, "The Education of a Geographer," *Annals* (Association of American Geographers) 46 (1956): 287–99.

14. Peter Haggett, *The Geographer's Art* (Oxford: Basil Blackwell, 1990), p. 54.

15. John R. Borchert, "Maps, Geography, and Geographers," *The Professional Geographer* 39 (1987): 387–89.

16. David J. Cuff, William J. Young, Edward K. Muller, Wilbur Zelinsky, and Ronald F. Alber, eds., *The Atlas of Pennsylvania* (Philadelphia: Temple University Press, 1990).

17. Michael N. DeMers, *Fundamentals of Geographic Information Systems* (New York: Wiley, 1997), p. 7.

18. D. P. Bickmore, "The Relevance of Cartography," in *Display and Analysis of Spatial Data,* eds. John C. Davis and Michael J. McCullagh (New York: Wiley, 1975), p. 330.

19. Arthur H. Robinson, "Map Design," *Proceedings* (International Symposium on Computer-Assisted Cartography, 1975), pp. 9–14.

20. Robinson and Petchenik, *The Nature of Maps,* pp. 116–17.

21. Meynen, *Multilingual Dictionary,* p. 291.

22. Robinson, "Map Design," pp. 9–14.

23. George F. Jenks, "Contemporary Statistical Maps—Evidence of Spatial and Graphic Ignorance," *American Cartographer* 3 (1976): 11–19.

24. Ibid.

25. Mark S. Monmonier and George A. Schnell, *Map Appreciation* (Englewood Cliffs, NJ: Prentice Hall, 1988).

26. Patricia Gilmartin, "The Design of Journalistic Maps: Purposes, Parameters, and Prospects," *Cartographica* 22 (1985): 1–18; see also Jeffrey S. Murray, "The Map Is the Message," *The Geographical Magazine* 59 (1987): 237–41.

27. David DiBiase, "Visualization in the Earth Sciences," *Earth and Mineral Sciences* (Bulletin of the College of Earth and Mineral Sciences, Pennsylvania State University), 59, no. 2 (1990) pp. 13–18.

28. MacEachren, "Visualization in Modern Cartography," pp. 5–8.

29. Ibid., p. 5.

30. Hilary M. Hearnshaw and David J. Unwin, eds., *Visualization in Geographical Information Systems* (New York: Wiley, 1994), p. 13.

31. Alan M. MacEachren, *Some Truth with Maps: A Primer on Symbolization and Design* (Washington, DC: Association of American Geographers, 1994), p. 2; see also David DiBiase, "Visualization in the Earth Sciences," *Earth and Mineral Sciences,* (Bulletin of the College of Earth and Mineral Sciences, Pennsylvania State University) 59, no. 2 (1990): 13–18.

32. These ideas were first expressed by Arthur H. Robinson in *The Look of Maps* (Madison: University of Wisconsin Press, 1966), p. 13.

33. Monmonier, *Maps, Distortion, and Meaning,* pp. 9–10.

34. Phillip C. Muehrcke, *Map Use, Reading, Analysis, and Interpretation* (Madison, WI: JP Publications, 1978), p. 19; see also Joel L. Morrison, "Map Generalization: Theory, Practice, and Economics," *Proceedings* (International Symposium on Computer-Assisted Cartography, 1975): 99–112.

35. Ibid., pp. 99–112.

36. Arthur H. Robinson and Barbara Bartz Petchenik, "The Map as a Communication System," *Cartographic Journal* 12 (1975): 7–15.

37. Robinson and Petchenik, *The Nature of Maps,* p. 20.

38. Muehrcke, *Map Use, Reading, Analysis, and Interpretation,* p. 8.

39. Judy M. Olson, "Experience and the Improvement of Cartographic Communication," *Cartographic Journal* 12 (1975): 94–108.

40. John E. Dornbach, *An Analysis of the Map as an Information Display System* (unpublished Ph.D. dissertation, Department of Geography, Clark University, Worcester, MA, 1967).

41. Leonard Guelke, "Cartographic Communication and Geographic Understanding," *Canadian Cartographer* 13 (1976): 107–22.

42. Barbara Bartz Petchenik, "Cognition in Cartography," *Proceedings* (International Symposium on Computer-Assisted Cartography, 1975), pp. 183–93.

43. Guelke, "Cartographic Communication," pp. 107–22.

44. Joel L. Morrison, "Towards a Functional Definition of the Science of Cartography with Emphasis on Map Reading," *American Cartographer* 5 (1978): 97–110.

45. Gershon Weltman, ed., *Maps: A Guide to Innovative Design* (Woodland Hills, CA: Perceptronics, 1979), p. 25.

46. Rudolf Arnheim, "The Perception of Maps," *American Cartographer* 3 (1976): 5–10.

47. Rosemary E. Ommer and Clifford H. Wood, "Data, Concept and the Translation to Graphics," *Cartographica* 22 (1985): 44–62.

48. J. S. Keates, "Cartographic Art," *Cartographica* 21 (1984): 37–43.

49. Mark Monmonier, *Maps, Distortion, and Meaning,* p. 14; see also George F. McCleary, "In Pursuit of the Map User," *Proceedings* (International Symposium on Computer-Assisted Cartography, 1975): 238–49.

50. Alan DeLucia, "Design: The Fundamental Cartographic Process," *Proceedings* (Association of American Geographers) 6 (1974): 83–86.

51. Christopher J. Jones, *Design Methods: Seeds of Human Futures* (New York: Wiley, 1981), p. 6.

52. Warren J. Luzadder, *Innovative Design, with an Introduction to Graphic Design* (Englewood Cliffs, NJ: Prentice Hall, 1975), p. 13.

53. Victor Papenek, *Design for the Real World* (New York: Pantheon Books, 1971), p. 3.

54. Norman Potter, *What Is a Designer: Education and Practice* (London: Studio-Vista, 1969), pp. 9–10.

55. Robinson and Petchenik, *The Nature of Maps,* p. 19.

56. W. H. Mayall, *Principles of Design* (New York: Van Nostrand Reinhold, 1979), p. 90.

57. Robinson, "Map Design," pp. 9–14.

58. Mark Monmonier, *How to Lie with Maps* (Chicago: University of Chicago Press, 1991), p. 2; see also Patrick McHaffie et al., "Ethical Problems in Cartography a Round Table Commentary," *Cartographic Perspectives* 7 (Fall 1990): 3–13; and Mark Monmonier, "Ethics and Map Design: Six Strategies for Confronting the Traditional One-Map Solution," *Cartographic Perspectives* 10 (Summer 1991): 3–7.

59. Mark Monmonier, *Mapping It Out: Expository Cartography for the Humanities and Social Sciences* (Chicago: University of Chicago Press, 1993), p. 185.

GLOSSARY

abstract symbols   may represent anything; usually take the form of geometrical shapes such as circles, squares, or triangles, p. 17

cartographic abstraction   transformational process in which map author or cartographer selects and organizes material to be mapped; identification of relevant elements, p. 16

cartographic generalization   transformation process of abstraction involving selection, classification, simplification, and symbolization, p. 16

cartographic process   the transformation of unmapped data into map form involving symbolization and the map-reading activities whereby map users gain information, p. 14

cartography   somewhat broader in scope than mapmaking; the study of all subjects that are concurrent with maps including their history, reading and use, and collection, p. 4

classification   placing objects in groups having identical or similar features; reduces complexity and helps organize materials for communication, p. 17

design   a dynamic activity that deals with the optimum use of tools for the creation of better solutions to life; initiation of change in man-made things, p. 18

general-purpose maps   also called reference maps; show natural and man-made features of general interest for widespread public use, p. 7

geographic map   base-map component of the thematic map; used to provide locational information for the map user, p. 12

map   a graphic representation of the milieu, p. 4

map analysis   map-use activity in which the map user begins to see patterns, p. 15

map author   anyone wishing to convey a spatial message; may or may not be a cartographer, p. 14

map design   aggregate of all thought processes that the map author or cartographer goes through during the abstraction phase of the cartographic process; activity seeking graphic solutions, p. 19

map interpretation   map-use activity in which the map user attempts to explain mapped patterns, p. 15

mapmaking   also called mapping; all processes associated with the production of maps; the designing, compiling, and producing of maps, p. 4

map meaning   supplied to the map by the user; instrumental in the communication process; not evident from the physical marks on the map, p. 16

map percipient   anyone who gains spatial knowledge by looking at a map, p. 15

map reading   map-use activity in which the user simply determines what is displayed and how the mapmaker did it, p. 15

map scale   amount of reduction that takes place in going from real-world to map plane; ratio of map distance to earth distance, p. 12

map use   comprises map reading, map analysis, and map interpretation, p. 15

mental maps   mental images having spatial attributes, p. 4

qualitative map   thematic map whose main purpose is to show locational features of nominal data, p. 8

quantitative map   thematic map whose main purpose is to show spatial aspects of numerical data-spatial variation of amount, p. 8

replicative symbols   designed to appear like their real-world counterparts, p. 17

selection   a generalization process in which selections of items to be mapped are made, all within the context of map purpose and map scale, pp. 16–17

simplification   reducing amount of information by selection, classification, or smoothing, p. 17

symbolization   devising a set of graphic marks that stand for real-world things, p. 17

thematic map   special-purpose, single-topic map; designed to demonstrate particular features or concepts, p. 4

thematic overlay   that part of the thematic map that contains the specific information (map subject), p. 12

READINGS FOR FURTHER UNDERSTANDING

Arnheim, Rudolf. "The Perception of Maps." *American Cartographer* 3 (1976): 5–10.

Bagrow, Leo. *History of Cartography.* Revised and enlarged by R. A. Skelton. Cambridge: Harvard University Press, 1966.

Bertin, Jacque. "Visual Perception and Cartographic Transcription." *World Cartography* 15 (1979): 17–27.

Bickmore, D. P. "The Relevance of Cartography." In *Display and Analysis of Spatial Data,* eds. John C. Davis and Michael J. McCullagh. New York: Wiley, 1975, pp. 328–67.

Board, Christopher. "Cartographic Communication." In *Maps in Modern Geography: Geographical Perspectives on the New Cartography,* ed. Leonard Guelke. *Cartographica,* Monograph 27. Toronto: Univ. of Toronto, 1981, pp. 42–78.

Borchert, John R. "Maps, Geography, and Geographers." *The Professional Geographer* 39 (1987): 387–89.

Bos, E. S. "Another Approach to the Identity of Cartography." *ITC Journal* 2 (1982): 104–8.

DeLucia, Alan. "Design: The Fundamental Cartographic Process." *Proceedings* (Association of American Geographers) 6 (1974): 83–86.

DeMers, Michael N. *Fundamentals of Geographic Information Systems.* New York: Wiley, 1997.

Dent, Borden D. Visual Organization and Thematic Map Communication." *Annals* (Association of American Geographers) 61 (1972): 79–92.

———. "Postulates on the Nature of Map Reading." Paper presented at the Georgia Academy of Science, Annual Meeting, 1976.

Dornbach, John E. *An Analysis of the Map as an Information Display System.* Unpublished Ph.D. dissertation, Department of Geography, Clark University, Worcester, MA, 1967.

Eckert, Max. "On the Nature of Map and Map Logic." *Cartographica* 19 (1977): 1–7.

Edwards, Betty. *Drawing on the Right Side of the Brain.* Los Angeles: J. P. Tarcher, 1979.

Farrell, Barbara. "Map Evaluation." In *World Mapping Today,* eds. R. B. Parry and C. R. Perkins. London: Butterworths, 1987, pp. 27–34.

Gersmehl, Philip J. "The Data, the Reader, and the Innocent Bystander—A Parable for Map Users." *The Professional Geographer* 37 (1985): 329–34.

Gilmartin, Patricia. "The Design of Journalistic Maps: Purposes, Parameters, and Prospects." *Cartographica* 22 (1985): 1–18.

Gould, Peter, and Rodney White. *Mental Maps.* 2d ed. Boston: Allen and Unwin, 1986.

Guelke, Leonard. "Cartographic Communication and Geographic Understanding." *Canadian Cartographer* 13 (1976): 107–22.

Haggett, Peter. *The Geographer's Art.* Oxford: Basil Blackwell, 1990.

Institute of Community and Area Development. *Interactive Atlas of Georgia.* Athens: Institute of Community and Area Development, University of Georgia, 1994.

Jenks, George F. "Contemporary Statistical Maps—Evidence of Spatial and Graphic Ignorance." *American Cartographer* 3 (1976): 11–19.

Jones, Christopher J. *Design Methods: Seeds of Human Futures.* New York: Wiley, 1981.

Keates, J. S. *Understanding Maps.* New York: Wiley, 1982.

———. "Cartographic Art," *Cartographica* 21 (1984): 37–43.

Kolacny, A. "Cartographic Information—A Fundamental Concept and Term in Modern Cartography." *Cartographic Journal* 6 (1969): 47–49.

Kraak, M. J., and F. J. Ormeling. *Cartography: Visualization of Spatial Data.* Essex, England: Addison-Wesley Longman, 1996.

Luzadder, Warren J. *Innovative Design, with an Introduction to Design Graphics.* Englewood Cliffs, NJ: Prentice Hall, 1975.

MacEachren, Alan M. "Map Use and Map Making Education: Attention to Sources of Geographic Information." *Cartographic Journal* 23 (1986): 115–22.

———. *Some Truth with Maps: A Primer on Symbolization and Design.* Washington, DC: Association of American Geographers, 1994.

———. **and D. R. Fraser Taylor, eds.** *Visualization on Modern Cartography.* Oxford, England: Pergamon, 1994.

Mayall, W. H. *Principles in Design.* New York: Van Nostrand Reinhold, 1979.

McCleary, George F. "In Pursuit of the Map User." *Proceedings* (International Symposium on Computer-Assisted Cartography, 1975): 238–49.

Meine, Karl-Heintz. "Thematic Mapping: Present and Future Capabilities." *World Cartography* 15 (1979): 1–16.

Meynen, E., ed. *Multilingual Dictionary of Technical Terms in Cartography.* International Cartographic Association, Commission II. Wiesbaden: Franz Steiner Verlag, 1973.

Monmonier, Mark S. *Maps, Distortion, and Meaning.* Resource Paper No. 75–4. Washington, DC: Association of American Geographers, 1977.

———. *How to Lie with Maps.* Chicago: University of Chicago Press, 1991.

———. *Mapping It Out.* Chicago: University of Chicago Press, 1993.

———, **and George A. Schnell.** *Map Appreciation.* Englewood Cliffs, NJ: Prentice-Hall, 1988.

Morrison, Joel L. "Map Generalization: Theory, Practice and Economics." *Proceedings* (International Symposium on Computer-Assisted Cartography, 1975): 99–112.

———. "Towards a Functional Definition of the Science of Cartography with Emphasis on Map Reading." *American Cartographer* 5 (1978): 97–110.

Muehrcke, Phillip C. *Thematic Cartography.* Commission on College Geography, Resource Paper No. 19.

Washington, DC: Association of American Geographers, 1972.

———. *Map Use, Reading, Analysis, and Interpretation.* 4th edition. Madison, WI: JP Publications, 1997.

———. "Maps in Geography." In *Maps in Modern Geography: Geographical Perspectives on the New Cartography,* ed. Leonard Guelke. Cartographica, Monograph 27. Toronto: Univ. of Toronto, 1981, pp.1–41.

———. "Whatever Happened to Geographic Cartography?" *The Professional Geographer* 33 (1981): 397–405.

———. "An Integrated Approach to Map Design and Production." *American Cartographer* 9 (1982): 109–22.

Olson, Judy M. "Experience and the Improvement of Cartographic Communication." *Cartographic Journal* 12 (1975): 94–108.

Ommer, Rosemary E., and Clifford H. Wood. "Data, Concept and the Translation to Graphics." *Cartographica* 22 (1985): 44–62.

Papenek, Victor. *Design for the Real World.* New York: Pantheon Books, 1971.

Petchenik, Barbara Bartz. "Cognition in Cartography." *Proceedings* (International Symposium on Computer-Assisted Cartography, 1975): 183–93.

———. "From Place to Space: The Psychological Achievement of Thematic Mapping." *American Cartographer* 6 (1979): 5–12.

Potter, Norman. *What Is a Designer: Education and Practice.* London: Studio-Vista, 1969.

Robinson, Arthur H. "Geographical Cartography." In *American Geography, Inventory and Prospect,* eds. Preston E. James and Clarence F. Jones. Syracuse: Syracuse University Press, 1954.

———. *The Look of Maps.* Madison: University of Wisconsin Press, 1966.

———. "An International Standard Symbolism for Thematic Maps: Approaches and Problems." *International Yearbook of Cartography* 13 (1973): 19–26.

———. "Map Design." *Proceedings* (International Symposium on Computer-Assisted Cartography, 1975): 9–14.

———. *The Nature of Maps: Essays toward Understanding Maps and Mapping.* Chicago: University of Chicago Press, 1976.

———. "Cartography as an Art." In *Cartography Past, Present, and Future,* eds. D. W. Rhind and D. R. F. Taylor. London: Elsevier, 1989.

———, **and Barbara Bartz Petchenik.** "The Map as a Communication System." *Cartographic Journal* 12 (1975): 7–15.

Salichtchev, K. A. "Some Reflections on the Subject and Method of Cartography after the Sixth International

Cartographic Conference." *Cartographica* 19 (1977): 111–16.

Southworth, Michael, and Susan Southworth. *Maps: A Visual Survey and Design Guide.* A New York Graphic Society Book. Boston: Little, Brown, 1982.

Tufte, Edward R. *The Visual Display of Quantitative Information.* Cheshire, CT: Graphics Press, 1983.

———. *Envisioning Information.* Cheshire, CT: Graphics Press, 1990.

Weltman, Gershon, ed. *Maps: A Guide to Innovative Design.* Woodland Hills, CA: Perceptronics, 1979.

Wolter, John A. "Cartography—An Emerging Discipline." *Canadian Cartographer* 12 (1975): 210–16.

Wood, Denis. *The Power of Maps.* New York: Guilford, 1992.

Wood, Michael. "Human Factors in Cartographic Communication." *Cartographic Journal* 9 (1972): 123–32.

Wright, John K. "Map Makers Are Human." *Geographical Review* 32 (1944): 527–44.

CHAPTER

THE ROUND EARTH TO FLAT MAP: MAP PROJECTIONS FOR DESIGNERS

CHAPTER PREVIEW

The earth is an irregularly shaped body, roughly approximating a sphere but more precisely defined by a reference ellipsoid. Although the earth can be accurately described this way, it is assumed to be perfectly spherical in most applications of thematic cartography. The slight imperfections of the earth's sphere simply cannot be accommodated on page-size maps because of the great reduction. The cartographer must have a fundamental knowledge of the earth's geographic grid, as it appears on a perfect sphere, for the preparation and selection of map projections. Map projections provide geometric control over the locational accuracy and appearance of the final thematic map; the cartographer must thoroughly understand the distortion brought on by the projection process. Certain map projections are better than others for representing different parts of the earth's land areas. The Albers conical equal-area projection, for example, is uniquely suited for portraying the United States.

Thematic maps are composed of two main structural elements: the base map and the thematic overlay. The appearance of the base map and its appropriateness to the whole map depend upon the projection on which it is compiled. In choosing an appropriate projection, cartographers must know at least the fundamental concepts of plane and spherical geometry.

Map projection is one element in the total design effort. This chapter introduces the main aspects of the earth's size and shape as a foundation for the description of its spherical or geographic coordinate system. These topics will lead to an explanation of map projections.

THE SIZE AND SHAPE OF THE EARTH

It is not known exactly when the earth was first thought to be round or spherical in form, but Pythagoras (sixth century B.C.) and Aristotle (384–322 B.C.) are known to have decided that the earth was round. Aristotle based his conclusions partly on the idea, then widely held among Greek philosophers, that the sphere was a perfect shape and that the earth must therefore be spherical. Celestial observations, notably lunar eclipses, also helped him reach this important conclusion. The idea of a spherical earth soon became adopted by most philosophers and mathematicians.

Greek scholars turned their attention to measurement of the earth. In fact, the earth's size was measured quite accurately by a Greek scholar living in Alexandria. **Eratosthenes** (276–194 B.C.) calculated the equatorial circumference to be 40,233 km (25,000 mi), remarkably close to today's measurement of 40,072 km (24,900 mi).[1] Another Greek mathematician, Poseidonius (c. 130–51 B.C.), measured the earth's equatorial circumference to be 38,622 km (24,000 mi).

Eratosthenes' ingenious method of measuring the earth employed simple geometrical calculations. (See Figure 2.1.) In fact, the method is still used today. Eratosthenes noticed on the day of the summer solstice that the noon sun shone directly down a well at Syene, near present-day Aswan in southern Egypt. However, the sun was not directly overhead at Alexandria but rather cast a shadow that was 7°12′ off the vertical. Applying geometrical principles, he knew that the deviation of the sun's rays from the vertical would subtend an angle of 7°12′ at the center of the earth. This angle is 1/50 the whole circumference (360). The only remaining measurement needed to complete the calculations was the distance between Alexandria and Syene; this was estimated at 804.65 km (500 mi). Multiplying this figure by 50, he calculated the total circumference to be 40,233 km (25,000 mi).

Three assumptions on Eratosthenes' part led to error in the results, although these partially compensated for each other. Alexandria and Syene are not on the same meridian. Syene's latitude is 24°5′30″—it is not at the Tropic of Cancer, where the sun's rays are perfectly vertical at the sum-

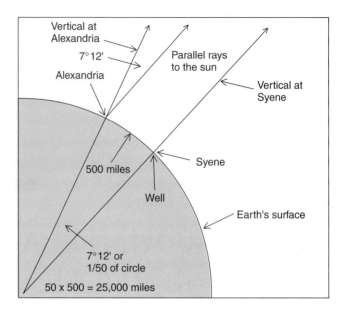

Figure 2.1 Eratosthenes' method of measuring the size of the earth.
See the text for a complete explanation.

mer solstice. Finally, the actual distance between Alexandria and Syene is 729 km (453 mi). Regardless of these sources of error, Eratosthenes made remarkably accurate calculations.

Not until the end of the seventeenth century was the notion of an imperfectly shaped earth introduced. By that time, accurate measurement of gravitational pull was possible. Most notably, Newton in England and Huygens in Holland put forward the theory that the earth was flattened at the poles and extended (bulged) at the equator. This idea was later tested by field observation in Peru and Lapland by the prestigious French Academy of Sciences. The earth was indeed flattened at the poles! It is interesting to note that the first indication of this flattening came from seamen who noticed that their chronometers were not keeping consistent time as they sailed great latitudinal distances. The unequal pull of gravity, caused by the imperfectly shaped earth, created different gravitational effects on the pendulums of their clocks.

Today geodesists describe the shape of the earth as a **geoid** (earth-shaped). Satellite measurements of the earth since 1958 suggest that, in addition to being flattened at the poles and extended at the equator, the earth also contains great areas of depressions and bulges.[2] (See Figures 2.2 and 2.3.) Of course, these irregularities are not noticeable to us on the earth's surface. They do, however, have significance for precise survey work and geodetic measurement. **Geodesy** is the science of earth measurement.

For the purposes of geodesy, and for some branches of cartography, the earth's flattened shape can be best described by reference to an **ellipsoid.** This geometrical solid is generated by rotating an ellipse around its minor axis and choosing the lengths of the major and minor axes that best

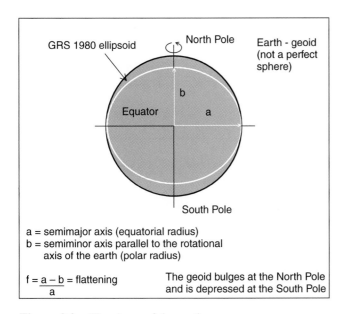

Figure 2.2   The shape of the earth.
Recent satellite measurements show that there are minor bulges and depressions away from the Fischer reference ellipsoid. We cannot detect these on the surface of the earth. The Fischer ellipsoid is the International Astronomical Union figure for the earth (1968) and has a flattening of 1:298.25.

fit those of the real earth. (See Figure 2.2.) Various reference ellipsoids have been adopted by different countries throughout the world for their official mapping programs, based on the local precision of the ellipsoid in describing their part of the earth's surface. (See Table 2.1.)

Until recently, the United States used the **Clarke ellipsoid of 1866** as its reference ellipsoid in establishing a *datum* (starting point) for horizontal control in large-scale mapping. Because of recent advances in satellite geodesy with higher degrees of accuracy, and because many surveying errors have built-up since the last datum was adopted (North American Datum—NAD27), a new datum was selected in 1983. The NAD83 uses the **Geodetic Reference System (GRS80)** reference ellipsoid. This ellipsoid is earth-centered (center of mass), and does not have a single origin (a place where the ellipsoid theoretically touches the geoid), as does the Clarke ellipsoid of 1866.[3]

As a result of adopting the new datum, almost all places in North America will have a slightly different latitude and longitude. These shifts (some may be as much as 300 m) are noted at the margins of newly printed large-scale topographic maps. Table 2.2 shows the dimensions of the earth as described by the GRS80 reference ellipsoid.

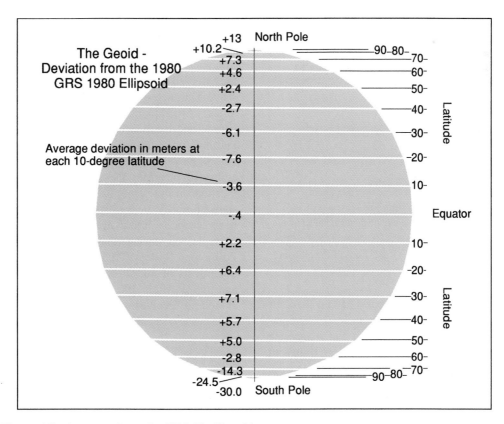

Figure 2.3   The geoid's departure from the GRS 80 ellipsoid.
This diagram represents how the geoid varies from the GRS-80 reference ellipsoid. This picture illustrates that the earth is indeed a complex figure. (*Source: Drawn from data supplied by the United States Coast and Geodetic Survey.*)

Table 2.1   Reference Ellipsoids

| Ellipsoid | Equatorial Radius (a) | | Polar Radius (b) | |
	Nautical Miles	Statute Miles	Nautical Miles	Statute Miles
Airy (1830)	3,443.609	3,962.56	3,432.104	3,949.32
Austrian Nat'l–South Am. (1969)	3,443.931	3,962.93	3,432.384	3,949.64
Bessel (1841)	3,443.520	3,962.46	3,432.010	3,949.21
Clarke (1866)	3,443.957	3,962.96	3,432.281	3,949.53
Clarke (1880)	3,443.980	3,962.99	3,432.245	3,949.48
Everest (1830)	3,443.454	3,962.38	3,432.006	3,949.21
Geodetic Reference System (1980)	3,443.939	3,962.94	3,432.392	3,949.65
International (1924)	3,444.054	3,963.07	3,432.459	3,949.73
Krasovskiy (1940)	3,443.977	3,962.98	3,432.430	3,949.70
World Geodetic System (1972)	3,443.917	3,962.92	3,432.370	3,949.62

Note: The nautical mile as adopted by the International Hydrographic Bureau is 6,076.12 feet, and is approximately 1.1507 statute miles. The statute mile is 5,280 feet.

Table 2.2   Earth Measurements (After GRS80 ellipsoid)

	Statute Miles (U.S.)	Kilometers
Equatorial diameter	7,926.59	12,756.27
Polar diameter	7,900.01	12,713.50
Equatorial circumference	24,902.13	40,075.00
Polar circumference	24,818.64	39,940.64
Area*	197,000,000 sq mi	509,000,000 sq km

*Obtained by using the main radius (3,958.51 mi) of the ellipsoid.

Although any discussion of the true size and shape of the earth is interesting and useful to the thematic cartographer, we need not dwell on it at length; we can assume the earth to be a perfect sphere for purposes of practically all cartography. The development of the base map begins from a small model of the real earth. If we assume this model to be, for example, an ellipsoid whose semimajor axis is 30.48 cm (12 in), then its semiminor axis will be 30.37 cm (11.5 in)—a difference of only .10 cm (.04 in). For most projects in thematic cartography, this introduces precision not ordinarily obtainable, or even necessarily desirable.

COORDINATE GEOMETRY FOR THE CARTOGRAPHER

Location was the key idea behind the historical development of the earth's coordinate geometry. Pre-Christian astronomers were naturally concerned with this question as they delved into debates related to the earth's size and shape. During the fifteenth, sixteenth, and seventeenth centuries, when exploration flourished, exactness in ocean navigation and location became critical. Death often awaited mariners who did not know their way along treacherous coasts. Naval military operations in the seventeenth and eighteenth centuries also required precise determination of location on the globe, as is the case today. Considerable sums of money are now spent on orbiting satellites that can beam locational information to earth. The United States, for example, has launched several satellites of a new Global Positioning System (GPS) which serves the locational requirements of the military. Precise location within 10 m and timing within 100 billionths of a second are possible. Handheld receiving units are operational now. We will have more to say about GPS later in this chapter.

PLANE COORDINATE GEOMETRY

Perhaps the best way to introduce the earth's spherical coordinate system is to examine *plane coordinate geometry.* **Descartes,** a French mathematician of the seventeenth century, devised a system for geometric interpretation of algebraic relationships. This eventually led to the branch of mathematics called analytic geometry. It is from his

Mean Radius	Ellipticity	
(2a + b)/3	(Flattening)	
Statute Miles	$f = \dfrac{a-b}{a}$	Where Used
3,958.15	1/299.32	Great Britain
3,958.50	1/298.25	Australia, South America
3,958.04	1/299.15	China, Korea, Japan
3,958.48	1/294.98	North America, Central America, Greenland
3,958.49	1/293.46	Much of Africa, some countries of the Middle East
3,957.99	1/300.80	India, Southeast Asia, Indonesia
3,958.51	1/298.26	Newly adopted for North America
3,958.63	1/297.00	Europe, Individual States in South America
3,958.56	1/298.30	Russia (and former socialist states)
3,958.49	1/298.26	NASA; U.S. Department of Defense; oil companies, Russia

Source: Defense Mapping Agency, Hydrographic Center, American Practical Navigator, *vol. 1 (Washington, DC: Defense Mapping Agency, 1977), pp. 117–120; John P. Snyder,* Map Projections—A Working Manual. *U.S. Geological Survey Professional Paper 1395 (Washington DC: USGPO, 1987), p. 12.*

Until 1929, the *nautical mile* was equivalent to the length of one minute of arc along a great circle on a sphere having a surface area equal to that of the Clarke reference ellipsoid of 1866. This equaled 6,082.2 feet. Actually, on the Clarke spheroid of 1866, one minute of latitude varies from 6,046 feet at the equator to 6.108 feet at the poles. This nautical standard was changed by international agreement in 1929 to 1852 meters (6,076.12 ft) and became known as the *international nautical mile.* For most purposes, however, one minute of latitude along any parallel or along any great circle arc is customarily referred to as a nautical mile. The *land* or *statute mile,* in the United States, is 5,280 feet. This mile is also used for navigation on inland waters and lakes. One nautical mile is equal to 1.1507 statute miles. Because there are 1,852 meters in a nautical mile, there are 1,610.4 meters in a statute mile. Therefore, one statute mile equals 1.6 kilometers.

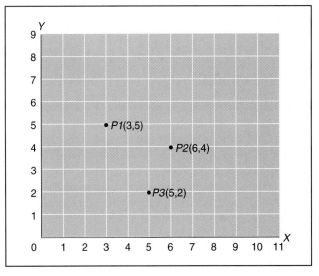

Figure 2.4   The Cartesian coordinate system.
Points (P1, P2, P3, and so on) can have their exact locations defined by reference to the grid. Absolute and relative locations are therefore easy to determine.

contributions that we have **Cartesian coordinate geometry.** This system of intersecting perpendicular lines on a plane contains two principal axes, called the *x*- and *y*-axes. (See Figure 2.4.) The vertical axis is usually referred to as the *y*-axis and the horizontal as the *x*-axis.

The plane of Cartesian space is marked at intervals by equally spaced lines. The position of any point (*Pxy*) can be specified by simply indicating the values of *x* and *y* and plotting its location with respect to the values of the Cartesian plane. In this manner, each point can have its own unique, unambiguous location. Relative location can easily be shown by plotting several points in the space. Cartography uses Cartesian geometry in a variety of ways. In addition to being a good method of introducing the earth's spherical coordinate system (because of similarities), Cartesian geometry is also of use in the development of computer maps, a process called **digitizing.** Digitizing involves

Converting conventional angular measure of the geographic coordinate system into decimal degrees is a fairly simple task. Decimal degrees are often useful for navigation and other map or location work. Since each degree contains 60 minutes and each minute contains 60 seconds, there are 3,600 seconds in a degree. If 52°17′32″ is to be converted, first multiply 17 × 60 (= 1,020), then add 32 to 1,020 (= 1,052). A ratio is then computed, 1,052/3,600 (= .2922). The answer is 52.2922°. For practice, convert 17°52′08″ into decimal degrees. Convert 18.52° into conventional notation.

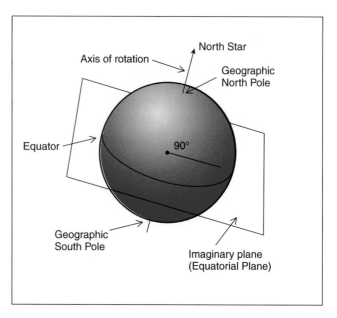

Figure 2.5   Determination of the earth's poles and the equator.

specifying the locations of geographical points on a map in plane Cartesian space, usually by electronic means.[4] The Cartesian system is also used in hand plotting map projections. In fact, the art and drafting workstation that commonly employs the T-square and triangle has the Cartesian system as a structural foundation. Finally, the thematic cartographer often deals with statistical concepts that are best portrayed in Cartesian space.

Cartographers often must change from one grid system to another. This is true especially in the work with geographic information systems. When cartographers specify Cartesian coordinates into their geographic reference points, or when other source documents (air photos, remotely sensed images) are referenced to a common coordinate system, it is called *georeferencing*.

EARTH COORDINATE GEOMETRY

Concepts similar to those used in plane or Cartesian coordinate geometry are incorporated in the earth's coordinate system. The earth's geometry is somewhat more complex because of its spherical shape. Nonetheless, it can be easily learned. To specify location on the earth (or any spherical body), angular measurement must be used in addition to the elements of the ordinary plane system. Angular measurement is based on a **sexagesimal** scale: division of a circle into 360 degrees, each degree into 60 minutes, and each minute into 60 seconds.

Our planet rotates about an imaginary axis, called the **axis of rotation.** (See Figure 2.5.) If extended, one of the axes points to a fixed star, the North Star. (This is not exactly accurate because this extended axis is actually about 0.5° off at the present time.) The place on earth where this axis of rotation emerges is referred to as **geographic north** (the North Pole). The opposite, or **antipodal point,** is called **geographic south** or the South Pole. These points are very important because the entire coordinate geometry of the earth is keyed to them.

If we were to pass through the earth an imaginary plane that bisected the axis of rotation and was perpendicular to it, the intersection of the plane with the surface of the earth would form a complete circle (assuming that the earth is

perfectly spherical). This imaginary circle is referred to as the earth's **equator.** (See Figure 2.5.) The North and South Poles and the equator are the most important elements of the earth's coordinate system.

Latitude Determination

Latitude is simply the location on the earth's surface between the equator and either the North or the South Pole. Latitude determination is easily accomplished: It is a function of the angle between the horizon and the North Star (or some other fixed star). (See Figure 2.6.) As one travels closer to the pole, this angle increases. It can be demonstrated that the angle subtended at the center of the earth between a radius to any point on the earth's surface and the equator is identical in magnitude to the angle made between the horizon and the North Star at that location. If an imaginary plane is passed through this point parallel to the equatorial plane, it will intersect the earth's surface, forming a **small circle,** or a **parallel of latitude.** (See Figure 2.6.) There are, of course, an infinite number of these parallels, and every place on the earth can have a parallel. Conceptually, the parallels of latitude are *x*-axes similar to those of plane Cartesian space.

Latitude designation is in angular degrees, from 0° at the equator to 90° at the poles. It is customary to label with a capital N or S the position north or south of the equator. Thus common latitude designations would be 82° N, 16° S, 47° N, and so on. Angular degrees are subdivided into minutes and seconds; exact latitudinal positions on the earth require minute and second determination.

Longitude Determination

Longitude on the earth's surface has always been more difficult to determine than latitude. It baffled early astronomers

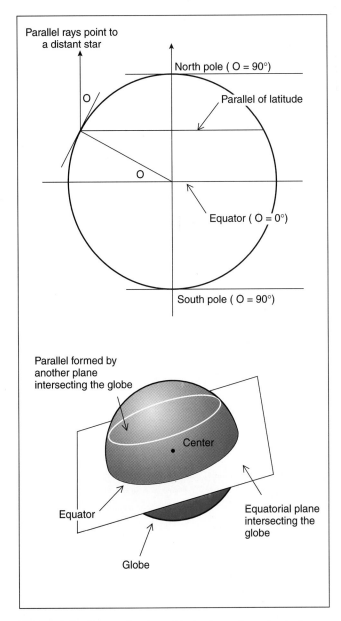

Figure 2.6 Determination of latitude on the spherical earth.

The concept was simple enough, but the technology of measuring time was slow in coming. In 1714, the British Parliament and its newly formed Board of Longitude announced a competition to build and test such an accurate timepiece. After several unsuccessful attempts, John Harrison built his famous **marine chronometer** in 1761. It was accurate to within 1.25 nautical miles. The puzzle of longitude was solved.[5]

At first, each country specified some place within its boundaries as the fixed reference point for reckoning longitude. By international agreement, most countries now recognize the meridian passing through the British Royal Observatory at Greenwich as the *fixed* or *reference* line. Some countries still retain local reference lines.

If an imaginary plane is passed through the earth so that it intersects the axis of rotation in a line, it will intersect the surface of the earth as a complete circle. One half of this circle, from pole to pole, is called a **meridian.** (See Figure 2.7.) The meridian passing through Greenwich is referred to as the **prime meridian** and has the angular designation 0°. Although there are 360 angular degrees in a circle, longitude position is designated as 0° to 180° east or west of the prime meridian. Separating east and west longitude on the side of the earth opposite the prime meridian is the meridian of the International Dateline, along which the date changes. Meridians can be thought of as corresponding to the *y*-axis of the plane coordinate system.

Until the advent of radio after World War I and radar after World War II, navigation and reckoning one's position on the earth were accomplished by "shooting the stars" (celestial navigation), by marine chronometer, and by compass. These still form the basis for navigation, but radio, radar, and satellite telemetering are more often used today. The precision made possible by satellite-beamed signals is remarkable.

With the system of parallels and meridians just described, also called the **geographic grid,** the earth's spherical coordinate geometry is complete. (See Figure 2.8.) Its similarities to the plane Cartesian system should be apparent. By studying Figure 2.8 it should be clear that (1) *latitude is measured by counting angular degrees along a meridian* and (2) *longitude is measured by counting degrees along a parallel.* The earth's coordinate system and especially these last two points must be understood by the thematic cartographer for a fundamental knowledge of projections and to select them wisely. Knowledge of how these elements are arranged with respect to each other is essential, and we will look at these next.

Global Positioning System

The **Global Positioning System (GPS)** is a "constellation" of 21 high-altitude satellites (called NAVSTARS) that beam radio signals to earth-based receivers. These receivers provide location by giving latitude and longitude positions, and also give elevation and precise time from the received signals. Originally conceived by the United States Navy to

and seamen, not so much for its concept, but for the instrumentation required to record it. Because the earth rotates on its axis, there is no fixed point at which to begin counting position. Navigators, cartographers, and others from the fourteenth to the seventeenth centuries knew that in practice they would need a fixed reference point. They also knew that the earth rotated on its axis approximately every 24 hours. Any point on the earth would thus move through 360 angular degrees in a day's time, or 15 degrees in each hour. If a navigator could keep a record of the time at some agreed-upon fixed point and determine the difference in time between the local time and the point of reference, this could be converted into angular degrees and hence position.

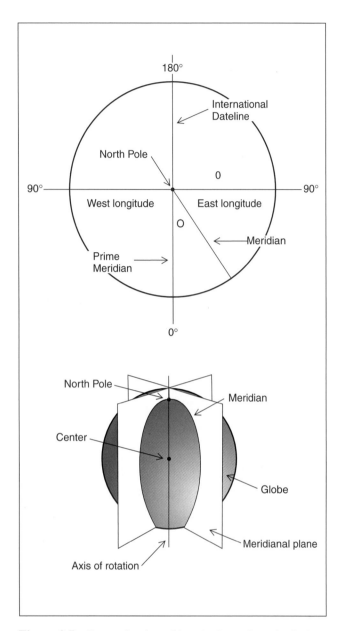

Figure 2.7 Determination of longitude on the spherical earth.

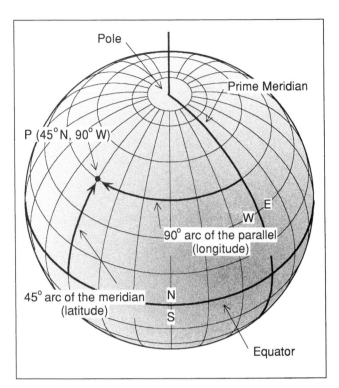

Figure 2.8 The complete geographic grid. The similarity of the geographic grid to the Cartesian grid is evident.

provide navigational information to ships at sea, the system has become important for other branches of the military (especially the Army, which used it successfully in the 1991 conflict with Iraq). In addition, and more important for us in cartography, GPS has revolutionized surveying and is fast becoming a necessary ingredient in geographic information systems (GIS) technology to help rectify satellite data to map bases.

A GPS receiver can be a stationary or mobile unit, some of the smallest weighing just several ounces. (See Figure 2.9.) An inexpensive receiver can be accurate to within 15 meters and more sophisticated ones can achieve greater accuracy. Many surveyors today receive signals from larger ground-based units that retransmit GPS signals providing accuracies to within a few centimeters.

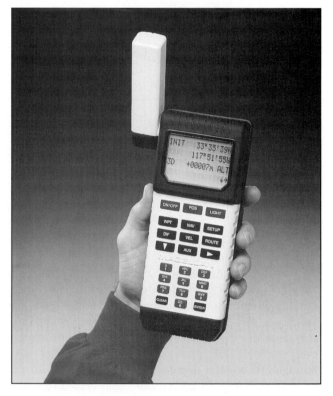

Figure 2.9 A small handheld GPS receiver. Small portable receivers such as this one are revolutionizing survey and other ground-location work. *(Photo courtesy of Magellan Systems Corporation.)*

As part of the PAGIS [Pittsburgh–Allegheny Geographic Information System] project, the Planning Department discovered that the city's survey monumentation was insufficient; it could not provide accurate control points for aerial photos. The key strategy at this point was to use a new technology called Global Positioning System (GPS) control, which used satellite signals to compute the precise positions of control points at a fraction of the time and cost of traditional methods.

Source: Alex G. Sciull, "Pittsburgh's AM/FM Strategies," *Geo Info Systems*, March 1991, pp. 38–42.

A GPS receiver works by receiving signals from at least three satellites at the same time. Surveyors like to work during "windows" when strong signals from four satellites are possible because they can achieve greater accuracy. Surveying problems in urban areas are compounded somewhat because the receivers rely on "line of sight" and often tall buildings interfere with direct observation. This is less a problem when the surveyor is using signals from ground-based stations. Adjustments are made to the postprocessing software of the receiver to accommodate for different ellipsoids used in the survey operation. Basic rules of trigonometry are applied to then determine precise location on the earth.

A wide spectrum of uses of GPS technology are here: aircraft avoidance (collision) systems, aircraft landing systems (which will make older instrument landing systems obsolete), and such consumer uses as electronic maps in personal automobiles. We will be able to determine precise locations for such things as restaurants, shops, and so forth, and maps in our cars will tell us exactly where we are at all times. Commercial airlines and private aircraft are using GPS as backup navigational aids already.

PRINCIPAL GEOMETRIC AND MATHEMATICAL RELATIONSHIPS OF THE EARTH'S COORDINATE GRID

The thematic cartographer is interested in the geographic grid and its portrayal as a map projection because it must be understood in order to render the best possible thematic map for the purpose at hand. The appropriateness of the final map depends in large measure on how well the cartographer knows the relationships of the elements of the grid. The present discussion is intended to provide the student with a basic understanding.

Linear

Lengths of lines of the spherical grid have fixed relations to each other. The most important length is the magnitude of the radius (*r*); from this and easily learned formulas, most other line lengths can be calculated. For example, the diameter is 2*r*. For the perfect sphere, the polar radius is identi-

Familiarity with the spherical grid and the characteristics of the arrangement of meridians and parallels is important in estimating graticule distortion on the flat map. The student is encouraged to inspect a globe and its grid. These important generalizations should be noted:

1. Scale is the same everywhere on the globe; all great circles have equal lengths; all meridians are of equal length and equal to the equator; the poles are points.
2. Meridians are spaced evenly on parallels; meridians converge toward the poles and diverge toward the equator.
3. Parallels are parallel and are spaced equally on the meridians.
4. Meridians and parallels intersect at right angles.
5. Quadrilaterals that are formed between any two parallels and that have equal longitudinal extent have equal areas.
6. The areas of quadrilaterals between any two meridians and between similarly spaced parallels decrease poleward and increase equatorward.

cal to the equatorial radius. There is only one circumference, and it is equal to $2\pi r$ (π = 3.1416). On the real earth, of course, the equatorial circumference does not equal the meridional circumferences because of flattening at the poles.

Meridional lengths are the easiest to handle. On the perfect sphere, a meridian is one-half the circumference of the globe. Normally, we want to deal only with the length of the degree along a meridian. On the perfect sphere, this length is simply the circumference divided by 360; every degree is equal to every other degree. On the real earth, however, polar flattening causes the radius of the arc of the meridian to change, fitting it to an ellipse. Consequently, the length of the degree along the meridian is not constant. (Appendix A describes these lengths based on the GRS80 ellipsoid of 1980.) The meridian is equal to one-half the circumference—or half the length of the equator. Knowing this is critical to the evaluation of projection properties.

Parallels of latitude also have fixed linear relationships. No parallel in one hemisphere is equal in length to any other. Parallels decrease in length at high latitudes. This relationship has a very definite mathematical expression, namely,

length of parallel at latitude λ = (cosine of λ) × (length of the equator)

The length of the degree of the parallel is determined by dividing its whole extent by 360. The cosine of 60° is .5. Thus the length of the degree along the 60° parallel is but one-half that at the equator. (See Table 2.3.) This information can be used intelligently in assessing map projections and consequent distortions. The lengths of the degree along the parallels on the real earth are listed in Appendix A.

Table 2.3 Length of Parallels on a Perfect Sphere
($r = 1.0$) Circumference (Equator) = πd = 6.28318

Latitude (°)	Length of Parallel	Percent of Equatorial Length
0 (equator)	6.2831	100.00
5	6.2592	99.61
10	6.1877	98.48
15	6.0690	96.59
20	5.9042	93.96
25	5.6945	90.63
30	5.4414	86.60
35	5.1468	81.91
40	4.8132	76.60
45	4.4428	70.71
50	4.0387	64.27
55	3.6039	57.35
60	3.1416	50.00
65	2.6554	42.26
70	2.1490	34.20
75	1.6262	25.88
80	1.0911	17.36
85	.5476	8.71
90	.0000	.00

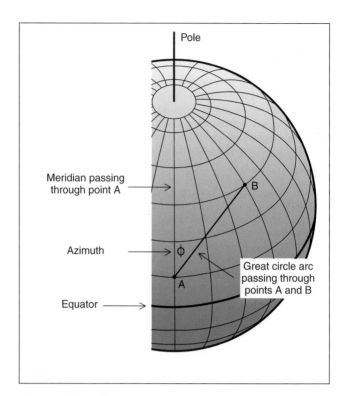

Figure 2.10 The determination of azimuth.

Angular

Inspection of a globe printed with a geographic grid will reveal important angular characteristics of the grid's elements. Most notably, it is easy to see that meridians converge poleward and diverge equatorward. Parallels, by definition, are parallel. What is more subtle is that meridians and parallels intersect at right angles. This is an important characteristic of the grid which can help in the evaluation of projection properties.

A special line on the globe is a **loxodrome,** which has a constant compass bearing. The equator, all meridians, and all parallels are loxodromes. Other loxodromes, called oblique loxodromes, are special, and they too maintain constant compass bearings, because they intersect all meridians at equal angles. Because meridians converge, special loxodromes tend to spiral toward the pole, theoretically never reaching it. Throughout history the loxodrome has always been important in sailing; mariners often wish to maintain the same heading throughout much of a journey. Unfortunately, loxodromes do not follow the course of the shortest distance between points on the earth, which is a **great circle arc.** Navigators would usually approximate a great circle arc by subdividing it into loxodrome segments in order to reduce the number of heading changes during travel. Great circle arcs are followed very closely today, especially in airplane navigation.

Azimuth   Azimuth has a very specific definition in cartography. (See Figure 2.10.) Azimuth is always defined in reference to two points, for example A and B. *The azimuth from A to B is the angle made between the meridian passing through A and the great circle arc passing from A to B.* It is customary to specify azimuth as an angle counting clockwise from geographic north through 360 degrees. Most azimuths are from geographic north, although possible from the south. The angular measure is followed by either N or S, for example 49 N.

Area

The areas of quadrilaterals found between bounding meridians and parallels are important in understanding the areal aspects of the earth's spherical coordinate system. (See Figure 2.11.) Between two bounding meridians, for identical latitudinal extents, the quadrilateral areas decrease poleward. Any misrepresentation of this feature during projection will have a profound effect on the appearance of the final land/water areas of the map. The relationship of these areas on a perfect sphere, based on change of latitude, is represented in Table 2.4. The decrease in area relative to the lowest quadrilateral is more dramatic at the higher latitudes because of rapid meridional convergence. At lower latitudes, meridians converge slowly; as a result, the change in area is less marked.

Points

Perfect spheres are considered to be *allside surfaces,* on which there are no differences from point to point. Every

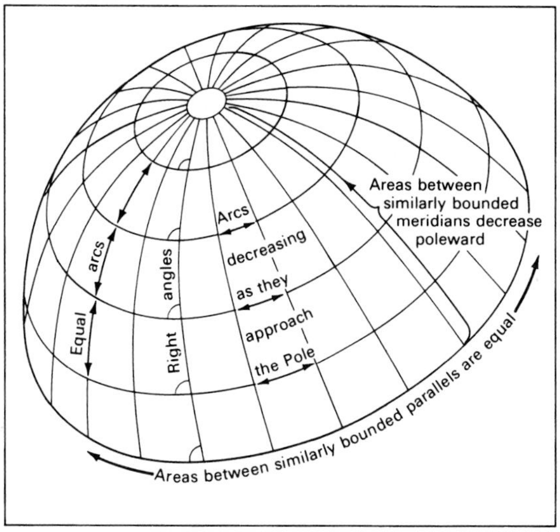

Figure 2.11   The principal geometrical relationships of the earth's coordinate system.

Table 2.4   Areal Relationships of Quadrilaterals Using the Geographic Grid on a Perfect Sphere ($r = 1.0$) 5° Longitudinal Extent

Latitude (°)	Area	Percent of Lowest Quadrilateral
0–5	.007605	100.00
5–10	.007547	99.23
10–15	.007432	97.72
15–20	.007260	95.46
20–25	.007033	92.47
25–30	.006752	88.78
30–35	.006420	84.41
35–40	.006039	79.41
40–45	.005612	73.79
45–50	.005143	67.62
50–55	.004634	60.93
55–60	.004090	53.78
60–65	.003515	46.21
65–70	.002913	38.30
70–75	.002289	30.09
75–80	.001647	21.66
80–85	.000993	13.06
85–90	.000332	4.36

point is like every other point, with the surface falling away from each point in a similar manner everywhere. In dealing with spheres and with the projection of the spherical grid onto a plane surface, we may think of points as having dimensional qualities. That is, they are commensurate figures, though infinitesimally small.

Circles on the Grid

Two special circles appear on the spherical grid. A *great circle* is formed by passing a plane through the center of the sphere. This plane forms a perfect circle where it intersects with the sphere's surface. Certain qualities about the circle are worth knowing: (1) The plane forming the great circle bisects the spherical surface. (2) Great circles always bisect other great circles. (3) An arc segment of a great circle is the shortest distance between two points on the spherical surface. Some of the elements of the geographic grid are great circles All meridians are great circles; the equator is also a great circle.

Circles on the grid that are not great circles are called *small circles.* Parallels of latitude are small circles. On the earth, to travel along a meridian (N–S) is to go the shortest distance. Traveling along a parallel (E–W) is *not* the shortest distance! Following the path of the equator, however, is an efficient way of travel.

AN INTRODUCTION TO MAP PROJECTIONS FOR THE DESIGNER

Map projections are some of the more interesting aspects of thematic mapping. They have occupied the thinking of mathematicians, cartographers, and geographers for many centuries, and new ones appear frequently. Projections are important for the designer, for they are the locational framework of the thematic map. Much as steel girders make up the skeleton of a modern skyscraper, the map projection provides the framework on which the remainder of the map rests. Projections can serve to focus the map reader's attention; they amplify and provide selective detail for the map's message. Every thematic map has a projection, and understanding this should be part of the education of a map designer.

THE MAP PROJECTION PROCESS

In thematic mapping, as in all cartography, we consider the production of a map a process of representing the earth (or a part of it) as a *reduced model of reality.* This involves at least four alterations. (See Figure 2.12.) The earth's real shape (geoid) is represented by an ellipsoid of reference through the adoption of the magnitudes of the major and minor semi-axes that best fit this real shape. For most thematic mapping, cartographers reduce this ellipsoid model to a **reference globe** (also called a *nominal* or *generating globe*), from which a map projection is generated.

Map projection is the transformation of the spherical surface to a plane surface; it occurs at the last step in the series of alterations from the earth to the flat map. *Any map projection is the systematic arrangement of the earth's (or generating globe's) meridians and parallels onto a plane surface.* Together, these spherical elements, the meridians and parallels, are called the projection **graticule.**

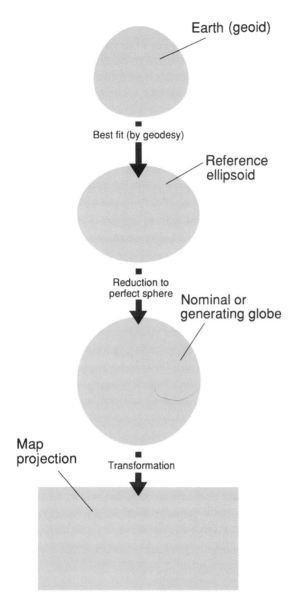

Figure 2.12   The map projection process.
The final development of a map projection is the result of several best-fit and reduction alterations of the actual size and shape of the earth.

The map projection process also includes the transformation of other map features in addition to the graticule. For example, coastlines and boundaries are also transformed and, as a result, such map objects as landmasses and bodies of water may be distorted.

Nominal Scale

In Chapter 1, we introduced the concept of map scale. These additional comments will also prove useful to an understanding of scale. The generating globe chosen has a nominal, or defined, **scale.** In cartography, scale is represented by a fraction:

$$\text{map scale} = \frac{\text{map distance}}{\text{earth distance}}$$

It is generally expressed as a representative fraction (RF), and will always contain unity in the numerator. Fractions such as 1:25,000, 1:50,000, or 1:50,000,000 are examples of RF scales. They are to be read, "One unit on the map *represents* so many of the *same* units on the earth." The number in the fraction may be in any units, but both numerator and denominator will be in the *same* units. The cartographer should never say, "One unit on the map *equals* so many of the same units on the earth." This is incorrect and logically inconsistent.

Generating globe sizes will be chosen with a nominal scale in mind, usually one that describes the radius of the globe. A generating globe that has a 15-in radius will have a nominal scale of 1:16,740,000:

$$\frac{\text{map distance}}{\text{earth distance}} = \frac{\text{globe radius}}{\text{earth radius}}$$
$$= \frac{15 \text{ in}}{3,963 \text{ mi} \times 63,360 \text{ in}}$$
$$= \frac{1}{16,740,000}$$
$$= 1:16,740,000$$

Every line and every point on the generating globe will have this same scale. Thus, all meridians and parallels, the equator, or other great circle arcs will be the same scale. This is a fundamental characteristic of the spherical or all-side geometrical figure.

There are three customary ways of expressing scale on a map: representative fraction, graphic, and verbal scale. *Representative fraction* (RF) is a ratio expressing the relationship of the number of units on the map to the number of the same units on the real earth. An example would be 1:500,000. The RF usually refers to the scale of a standard line and in fact changes over the map, depending on the selected projection.

On many maps, a *graphic* bar scale is included. This bar is usually divided into equally spaced segments and labeled with familiar linear units. This scale is read the same as an RF scale. The distance between any two divisions can be measured with a drafting scale (inches or centimeters); this distance represents the earth distance as labeled on the bar. Usually, the labeled units must be converted to those on the map and the ratio reduced to unity in the numerator. This form of scale is very useful when the map is to be reduced during reproduction because it changes in correct proportion to the amount of reduction.

Another common scale is the simple *verbal* scale. This is a simple expression on the face of the map stating the linear relationship. For example, "five miles to the inch" is an example of a verbal scale. This scale form is easily converted to an RF scale between map and earth distances.

Regardless of the scale form chosen, *almost all maps should contain a scale.*

In cartography, it has become conventional to refer to map scale as large, intermediate, or small. These terms are relative; rigorous numerical boundaries do not exist. Small scales are those that map large earth areas; a typical scale might be 1:30,000,000. An example would be a world map on an 8½ × 11-inch piece of paper. A large scale is 1:24,000 (or larger) and shows only small earth areas. The United States Geological Survey's Topographic Map series at 1:24,000 is an example. Such maps can show numerous physical and cultural features. Intermediate scales would be RFs between these two extremes.

Two sample scale problems for practice follow:

1. Determine the fractional scale, which is expressed verbally as:
   (A) 1 inch to 4 miles.
   Remember, the basic scale ratio is map distance to earth distance (always expressed in the same units). The problem here is that we have inches and miles. *Solution:* Convert 4 miles into inches (4 × 63,360 = 253,440), and the representative fraction becomes 1:253,440.
   (B) 4 inches to the mile.
   This is a bit more complicated. *Solution:* First, convert all units into the same measure. To change miles to inches: 1 mile × 63,360 = 63,360 inches. Now, we have 4 inches to 63,360 inches. This is not correct, though, because we need the numerator expressed as unity. Verbally, we say that if there are 4 inches to 63,360 inches, how many would there be for one? Let us simply cast two ratios (quotients) as equal, like this:

$$\underset{63,360}{\overset{4}{\text{ratio 1}}} = \underset{\text{unknown}}{\overset{1}{\text{ratio 2}}}$$

We can do this because the quotient yielded in ratio 2 must be the same as in ratio 1 (we are not changing the *relationship,* only the numbers with which we are working). Then (by applying the cross-multiplication rule of proportions), we get

$$\text{unknown} = \frac{(1) \cdot (63,360)}{4}$$
$$\text{unknown} = 15,840$$

The answer, then, is 1:15,840.

2. Typical map scale problem:
   (A)
   The distance between two known points on a map is 5 miles. What is the scale of a map on which the points are 3.168 inches apart?
   Solution: At first glance, this may dismay you. Remember the first rule: cast the scale as a ratio of map distance to earth distance:

$$\frac{\text{map distance}}{\text{earth distance}}$$

The earth distance given is 5 miles, but change this to inches right away, 5 × 63,360 = 316,800 inches. The map distance given is 3.168 inches, so our scale is

$$\frac{3.168}{316,800}$$

but this is not the correct form (numerator should be expressed as unity). So, then, do what we did in (1B), that is,:

$$\underset{316,800}{\overset{3.168}{\text{ratio 1}}} = \underset{\text{unknown}}{\overset{1}{\text{ratio 2}}}$$

then:

$$(3.168) \cdot (\text{unknown}) = (1) \cdot (316,800)$$
$$\text{unknown} = \frac{(1) \cdot (316,800)}{3.168}$$
$$\text{unknown} = 100,000$$

The answer is: 1:100,000.
(Note: See the *simplicity* of all of this? Just remember to cast the scale as the ratio of map distance to earth distance, note your given quantities, change the units, and solve.)

SURFACE TRANSFORMATION AND MAP DISTORTION

In the transformation process from the spherical surface to a plane, some distortion occurs that cannot be completely eliminated. Although designers strive to develop the perfect map, free of error, *all* maps contain errors because of the transformation process. It is impossible to render the spherical surface of the generating globe as a flat map without distortion error caused by *tearing, shearing,* or *compression* of the surface. (See Figure 2.13.) The designer's task is to select the most appropriate projection so that there is a measure of control over the unwanted error.

These distortions and their consequences for the appearance of the map vary with scale. One can think of the globe as being made up of very small quadrilaterals. If each quadrilateral were extremely small, it would not differ significantly from a plane surface. For mapping small earth areas (large-scale mapping), distortion is not a major design problem. It can be ignored because it cannot be dealt with in the preparation of the final map.

As the mapped area increases to subcontinental or continental proportions, distortion becomes a significant design problem for the cartographer. In designing maps to portray the whole earth, we can no longer ignore this surface distortion. At such scales, the map designer must contend with

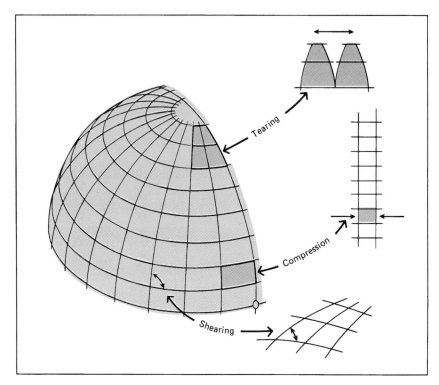

Figure 2.13 Distortion caused by the projection process.
Transformation from sphere to plane may cause distortion brought about by tearing, shearing, or compression.

alterations of area, shape, distance, and direction. No projection of the globe's graticule can maintain all of these properties simultaneously.

Equal-Area Mapping

Map projections on which area relationships of all parts of the globe are maintained are called **equal-area** (or equivalent) **map projections.** Linear or distance distortion often occurs in such projections. It is impossible for one projection to maintain both equivalency and conformality (shape preservation). On equal-area projections, therefore, shape is often quite skewed. Also, the intersections of meridians and parallels are not at right angles. Nonetheless, areas of similarly bounded quadrilaterals maintain correct area properties. The cartographer must select the most appropriate features for the design purpose.

Equivalent projections are very important for general quantitative thematic map work. It is usually desirable to retain area properties, because area is often part of the data being mapped. Population density is an example. Unless the map calls for unique projection attributes, equivalent projections have the greater all-around utility, regardless of the extent of the earth area mapped. Marschner wrote, several decades ago:

It is hardly necessary to recapitulate here the arguments used in favor of equal-area representation. There is no doubt but that the preservation of a true areal expression of maps, used as a basis for land economic investigation and research, is more important than a theoretical reten-

tion of angular values. Of the three geometrical elements that can be considered in mapping, the linear, angular, and areal, the last is the one around which land economic questions usually revolve. They do so for obvious reasons. Man does not inhabit a line, but occupies the area; he does not cultivate an angle of land, but cultivates and utilizes the land area. One of the principal functions of lines and angles is to define the boundary of the area. . . . They provide only the framework for controlling the relative position of features in the area, and with it a means for controlling the areal expression of the map itself.[6]

Conformal Mapping

Conformal (or orthomorphic) mapping of the sphere means that angles are preserved around points and that the shapes of small areas are preserved. The quality of orthomorphism applies to small areas (theoretically only to points). On **conformal projections,** meridians intersect parallels at right angles, and the scale is the same in all directions about a point. Scale may change from point to point, however. It is misleading to think that shapes over large areas can be held true. Although shapes for small areas are maintained, the shapes of larger regions, such as continents, may be severely distorted. Areas are also distorted on orthographic projections.

The shape quality of mapped areas is an elusive element. If we view a continent on a globe so that our eyes are perpendicular to the globe at a point near the center of the continent, we see a shape of that continent. However, the shape of the continent is distorted because the globe's surface is falling away from the center point of our vision. We

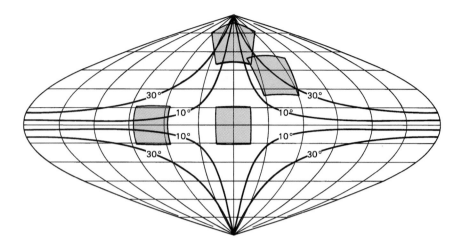

Figure 2.14   Shape distortion on an equal-area projection.
On the globe, the squares in the figure are all the same size. On this sinusoidal projection, the angular relations (and hence shape) are not preserved. Notice that distortion occurs most severely at the edges of the projection, as shown by the plot of the 2w values. (See text for explanation.) *(Source: Squares have been drawn from information obtainable from Wellman Chamberlin,* The Round Earth on Flat Paper *[Washington, D.C.: National Geographic Society, 1947], p. 97.)*

can view but one point orthographically at a time. If we select another point, the view changes, and so does our perception of the continent's shape. It becomes difficult for us to compare the shapes we see on a map to those on the globe, and it is safe to say that shapes of large areas on conformal projections are to be believed with caution.[7] For large-scale mapping of small earth areas, distortion is not significant. Indeed, the choice between an equivalent or an orthographic projection becomes virtually a moot exercise. At intermediate to small scales, the selection of the projection is more critical in the design process. Even at these scales, however, it is seldom necessary to specify a conformal projection, except in rare circumstances. Mapping phenomena with circular radiational patterns may warrant such a choice. Radio broadcast areas, seismic wave patterns, or average wind directions are examples.

Equidistance Mapping
The property of equidistance on projections refers to the preservation of great circle distances. There are certain limitations: distance can be held true from one to all other points, or from a few points to others, but not from all points to all other points. The distance property is never global. Scale will be uniform along the lines whose distances are true. Projections that contain these properties are called **equidistant projections.**

Azimuthal Mapping
On **azimuthal projections,** true directions are shown from one central point to all other points. Directions or azimuths from points other than the central point to other points are not accurate. The quality of azimuthality is not an exclusive projection quality. It can occur with equivalency, conformality, and equidistance.

Over the years, some cartographers have stressed the idea that **minimum error projections** are best suited for general geographic cartography. These projections contain all of the four previously mentioned errors, to a greater or lesser degree, but as a whole provide us with a good picture of the globe or parts of it. These projections are chosen by the designer on much the same basis as a projection having a uniquely preserved property: by selection of those total qualities that best suit the mapping task.[8]

DETERMINING DEFORMATION AND ITS DISTRIBUTION OVER THE PROJECTION

Two chief methods are available for determining projection distortion and its distribution over the map. One is to depict a geometrical figure (square, triangle, or circle) or familiar object (such as a person's head) and plot it at several locations on the projection graticule. (See Figure 2.14.) Distortion on the projection is readily apparent. This method is very effective and quite sufficient for most general cartographic analyses; its weakness is the lack of a general quantitative index of distortion.

Another method, conceptually and mathematically more complex, uses **Tissot's indicatrix.** Tissot, a French mathematician working in the latter part of the nineteenth century, developed a way to show distortion at points on the projection graticule.[9] Following his work, others have computed these indices and mapped them on several projections to show the patterns of distortion. The weakness of Tissot's method is that it is somewhat more complex mathematically than plotting simple geometrical shapes. Its strength lies in its quantitative ability to describe distortion.

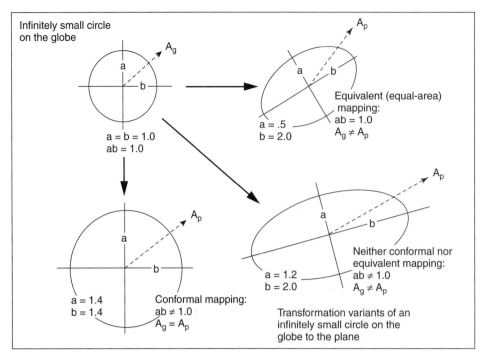

Figure 2.15   Variations of the standard line and the determination of scale factor.

The construct of Tissot's indicatrix consists of a very small circle, whose scale is unity (1.0), on the globe's surface. This small circle and two perpendicular radii appear on the plane map surface during transformation as a circle of the same size, a circle of different size, or an ellipse. For equal-area mapping, the circle is transformed into an ellipse with the ratio of the new semimajor and semiminor axes such that its area remains as unity. Angular properties are not preserved. (See Figure 2.15.) On conformal projections, the small circle is transformed on the projection as a circle, although its size (area) varies over the map. Because the small circle is accurately projected as a circle, angular relationships are maintained. On some projections, the small circle becomes an ellipse that preserves neither equivalency nor conformality. Such projections are not classified as either equivalent or conformal.

For purposes of computation and explanation, the indicatrix has these qualities:

- Maximum angular distortion = 2ω
- Scale along the ellipse major axis = a
- Scale along the ellipse minor axis = b
- Maximum areal distortion = S

Every point on the new surface (map) has values computed for 2ω, a, b, and S. These conditions inform us of the distortion characteristics:

- If S = 1.0, there is no areal distortion, and the projection is equal-area. If S = 3.0, for example, *local* features are three times their proper areal size.

- If $a = b$ at all points on the map, it is a conformal projection. S varies, of course.
- If $a \neq b$, it is not conformal, and the amount of angular distortion is represented by 2ω.

It should be pointed out that no distortion occurs on standard lines, and values of the indicatrix need not be computed along these lines.

Values of S or 2ω are usually mapped on projection graticules to illustrate the patterns of distortion. These will be discussed more fully below. Value of 2ω have been plotted on the sinusoidal projection in Figure 2.14.

Standard Lines and Points, Scale Factor

Standard lines are lines (usually meridians or parallels, especially the equator) on a projection that have identical dimensions to their corresponding lines on the nominal or reference globe. For example, if the circumference of the generating globe is 15 inches, the equator on the projection is drawn 15 inches long if it is a standard line. Scale can be thought to exist at **points,** so every point along a standard line has an unchanging or true scale when compared to the scale of the generating globe. On standard lines, scales are true. At all points on a standard line, using the indicatrix as a guide, $a = b$, S = 1.0, and $2\omega = 0°$. (See Figure 2.16.)

In practice, the lines on the projection can be compared to their corresponding lines on the reference globe by a ratio called **scale factor** (S.F.):

$$\frac{\textbf{projection scale fraction}}{\textbf{nominal scale fraction}}$$

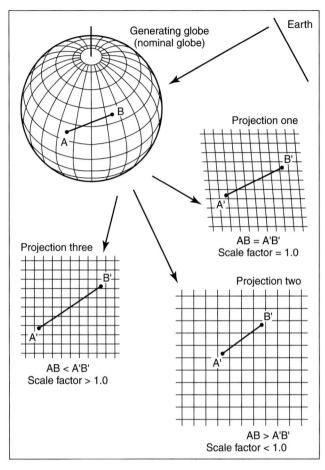

Figure 2.16 Variations of the standard line and the determination of scale factor.

If the scale of a generating globe is 1:3,000,000 and the equator on the projection is drafted as a standard line, then

$$\frac{\text{projection scale fraction}}{\text{nominal scale fraction}} = \frac{\dfrac{1}{3,000,000}}{\dfrac{1}{3,000,000}}$$

$$= \frac{3,000,000}{3,000,000}$$

$$= 1.0 \text{ (S.F.)}$$

The scale factor of the equator will have a value of 1.0. For another example, suppose the nominal scale is 1:3,000,000 and a line on the projection has a linear scale of 1:1,500,000:

$$\frac{\text{projection scale fraction}}{\text{nominal scale fraction}} = \frac{\dfrac{1}{1,500,000}}{\dfrac{1}{3,000,000}}$$

$$= \frac{3,000,000}{1,500,000}$$

$$= 2.0 \text{ (S.F.)}$$

In this example, the line is two times *longer* than it should be, because of stretching. In our final example, the nominal scale is 1:3,000,000 and a line on the projection is drafted at 1:6,000,000:

$$\frac{\text{projection scale fraction}}{\text{nominal scale fraction}} = \frac{\dfrac{1}{6,000,000}}{\dfrac{1}{3,000,000}}$$

$$= \frac{3,000,000}{6,000,000}$$

$$= 0.5 \text{ (S.F.)}$$

Here we find the line to be one-half as long as its corresponding line on the globe. *Compression* has taken place in the transformation process.

Standard lines and scale factors are important to the overall understanding of projection distortion. The idea of scale factor can also be used in the assessment of areal scales. In this instance, small quadrilateral areas on the projection can be compared to the corresponding areas on the globe to determine the amount of *areal exaggeration* occurring on the projection.

PATTERNS OF DEFORMATION

Patterns of deformation on projections can best be approached by first looking at the different *projection families.* There are literally hundreds of individual projections. Some are quite old, such as the Mercator—dating back hundreds of years. Some have been devised within the last several years, most notably the Robinson pseudocylindrical projection.[10] This projection was selected in 1988 by the National Geographic Society as its base projection for world political maps. Although there is great diversity among projections, similarities in construction and appearance yield enough common elements to classify them into a few groups. The approach presented here is a conventional one.

Some map projections can be produced by descriptive, projective geometry. They can be developed with T-square and triangle on a drafting table. These **developable projections** form the basis for the classification presented here. There are three basic forms of developable surfaces: plane, cylinder, and cone. These yield the azimuthal, cylindrical, and conic projection families. (See Figure 2.17.) The purely mathematical projections (those that cannot be developed by projective geometry) are classified into these families on the basis of their appearances. A few projections bear striking resemblance to the developable ones, but are different enough to be classed as pseudocylindrical, pseudoconic, and pseudoazimuthal.

Azimuthal Family

In the *plane* or *azimuthal* class, the spherical grid is projected onto a plane. This plane can be tangential to the

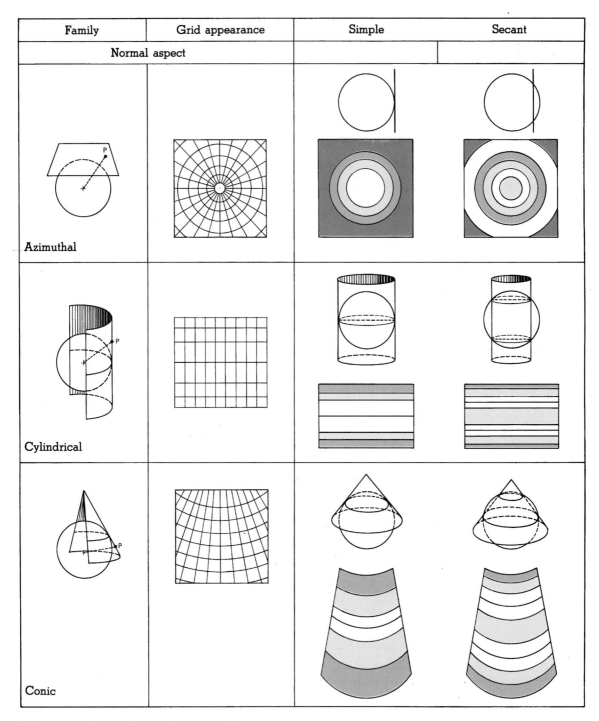

Family	Grid appearance	Simple	Secant
Normal aspect			
Azimuthal			
Cylindrical			
Conic			

Figure 2.17 Projection families and patterns of deformation.
The patterns are clearly related to the way the projection is devised. Darker areas represent greater distances.

sphere at a point (simple form), or pass through the sphere, making it tangent along a small circle (secant form). There are numerous versions of this group, distinguished from one another by the location of the assumed light source for the projection. In Figure 2.17, the light is shown emanating from the center of the globe; this is a gnomonic projection. If the light source is at the point opposite the point of tan-

gency, the projection is stereographic; if it is at infinity (outside the generating globe), an orthographic projection results.

The plane may be tangent, of course, at any point on the spherical grid, depending on the **projection aspect.**[11] Tangency at the pole is a *polar aspect;* at mid-latitude, an *oblique aspect;* and at the equator, an *equatorial aspect.*

The *normal aspect* is the position that produces the simplest graticule. Normal aspect for this family is the polar position when the plane is tangent at one of the poles. In this case, the meridians are straight lines intersecting the pole, and parallels are concentric circles having the pole as their centers. Directions to any point from the point of tangency (pole) are held true. All lines drawn to the center are great circles, as is also the case for equatorial and oblique aspects. Normally, only one hemisphere is shown on these projections. Some versions, for example the **azimuthal equidistant map of the world,** do indeed map the entire globe, but the grid departs radically from what we are accustomed to seeing. The point of tangency is a standard point; in the secant case, the standard line is the line of tangency. Scale deformation is nonexistent at these locations. Azimuthal (also called zenithal) projections became quite popular during World War II, when there was considerable circumpolar air navigation; they have remained so today.

It is possible to compute the values of *a, b,* S, and 2ω, plot these on a projection, and thereby see the pattern of deformation. Repetition of this procedure for many projections shows that the pattern of deformation is closely associated with the manner in which each was developed. Patterns of deformation begin to emerge for the azimuthal class. (See Figure 2.17.) As is the case for all projections, deformation increases with distance from either the standard point or the standard line. In the present case, the deformation increases outward in concentric bands. On conformal projections, areal exaggeration is extreme at greater distances from tangency. Likewise, for equivalent azimuthal projections, shape distortion reaches extremes at the outer edges.

Cylindrical Family

Cylindrical or rectangular projections are common forms, frequently seen in atlases and other maps portraying the whole world. They are developed (graphically or mathematically) by wrapping a flat plane (sheet) into a cylinder and making it tangent along a line or lines on the sphere. Points on the spherical grid can be transferred to this cylinder, which is then unrolled into a flat map. The normal aspect for these projections is the equatorial aspect, with the equator as the standard line. In this case, the standard line is also a great circle. Oblique cases of this projection, notably the oblique Mercator, also have the tangent line as a great circle.

This class of projection is often used to show the worldwide distribution of a variety of geographic phenomena. A great deal of effort has therefore gone into the development of equivalent cylindrical projections. Most of the common ones used today are, in fact, pseudocylindrical versions. The popular ones include the Eckert family, Mollweide, Boggs Eumorphic, and Goode's Interrupted Homolosine. Two recent developments are the Robinson and Tobler pseudocylindricals.

The patterns of deformation for all rectangular projections depend on their method of development. (See Figure 2.17.) Areas of least distortion are bands parallel to the line(s) of tangency, with increasing exaggeration toward the outer edges of the map plane. Here, as with the azimuthal class, distortion may appear in area, angle, distance, or direction.

Conic Family

Conic projections are constructed by transferring points from the generating globe grid to a cone enveloped around the sphere. This cone is then unrolled into a flat plane. In the normal aspect, the axis of the cone coincides with the axis of the sphere. This aspect yields either straight or curved meridians that converge on the near pole and parallels that are arcs of circles. In the simple conic projection (normal aspect), the cone is tangent along a chosen parallel, along which there is no distortion. In the secant case, the cone intersects the sphere along two parallels. This reduces distortion.

One special form of the conic family is the polyconic projection. This projection uses several cones of development and consequently has several standard lines. Theoretically, each parallel is the base of a tangent cone. This form of polyconic projection is not conformal. It is often used for mapping areas of great latitudinal extent; the polyconic projection was used by the United States Geological Survey (USGS) in its topographic mapping program until it was replaced in the 1950s.

The pattern of deformation includes concentric bands parallel to the standard parallels of the projection. (See Figure 2.18.) Secant conics tend to compress scale in areas between the standard lines and to exaggerate scale elsewhere. There are a number of very useful conic projections, one of which will be discussed at length below. Conic projections, simple or secant, are best for mapping earth areas having greater east–west extent than north–south.

Analysis of a variety of conic projections suggests that the **Albers conical equal-area projection,** with standard parallels at 29 1/2° and 45 1/2°, is a good choice for mapping the United States. (See Figure 2.18.)

This projection, at scales of 1:7,500,000, 1:17,000,000, and 1:34,000,000, was used exclusively for mapping the United States in the *National Atlas of the United States of America.*[12] The reasons for its use, as stated by the atlas design team, were as follows:

1. Equivalent projection.
2. Simple to construct.
3. Easy to segment and reassemble.
4. Contains small errors in scale.
5. Long been popular in mapping the United States, with its large east-west expanse.

Indicatrix values for this projection suggest very little distortion in an area the size of the United States. Values of 2ω nowhere exceed 2° , and linear values (*a,b*) are very nearly 1.0. The index value for areal change, S, is 0.0, as

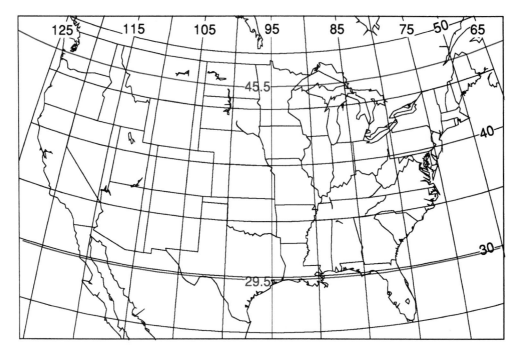

Figure 2.18   The United States, plotted on an Albers conical equal-area projection.
Angular distortion is minor, nowhere exceeding 2°. This projection is frequently used for mapping the United States. The standard parallels are at 29½° and 45½° north.

this is an equivalent projection. Thematic cartographers seeking a projection suitable for mapping the continental United States will find a most useful projection in the Albers equal-area.

Students interested in projections, and especially the history of projections, are indeed fortunate in having a splendid published work, *Flattening the Earth: Two Thousand Years of Map Projections* by John P. Synder. Not only an interesting narrative, the book also contains many projection illustrations, some very unique.[13]

NOTES

1. Lloyd A. Brown, *The Story of Maps* (New York: Bonanza Books, 1959), pp. 28–29.
2. Desmond King-Hele, "The Shape of the Earth," *Science* 183 (1967): 67–76; and Desmond King-Hele, "The Shape of the Earth," *Science* 192 (1976): 1293–1300.
3. John P. Snyder, *Map Projections—A Working Manual* (U.S. Geological Survey, Professional Paper 1395—Washington, DC: USGPO, 1987), p. 13.
4. Mark S. Monmonier, *Computer-Assisted Cartography, Principles and Prospects* (Englewood Cliffs, NJ: Prentice Hall, 1982), p. 195.
5. Brown, *The Story of Maps,* pp. 209–40; see also Dava Sobel, *Longitude* (New York: Penguin Books, 1995).
6. F. J. Marschner, "Structural Properties of Medium- and Small-Scale Maps," *Annals* (Association of American Geographers) 34 (1944): 44.
7. Borden D. Dent, "Continental Shapes on World Projections: The Design of a Poly-Centered Oblique Orthographic World Projection," *Cartographic Journal* 24 (1987): 117–24.
8. Frank Canters, "New Projections for World Maps: A Quantitative-Perceptive Approach," *Cartographica* 26 (1989): 53–71.
9. A. Tissot, *Memoire sur la Representation des Surfaces et les Projections des Cartes Geographiques* (Paris, 1881).
10. Arthur H. Robinson, "A New Map Projection: Its Development and Characteristics," *International Yearbook of Cartography* 14 (1974): 145–55; see also Robert T. Richardson, "Area Deformation on the Robinson Projection," Technical Note, *American Cartographer* 16 (1989): 294–96.
11. James A. Hilliard, Umit Bosoglu, and Phillip C. Muehrcke, *A Projection Handbook* (Madison: University of Wisconsin: Cartographic Laboratory, 1978), pp. 3–5.
12. U.S. Department of the Interior, Geological Survey, *National Atlas of the United States of America* (Washington, DC: USGPO, 1970).
13. John P. Snyder, *Flattening the Earth: Two Thousand Years of Map Projections* (Chicago: University of Chicago Press, 1993).

GLOSSARY

Albers conical equal-area projection   customarily used for small-scale mapping of the United States; a good selection for mapping any earth area at mid-latitude with considerable east–west extent, p. 43

antipodal point   point opposite; on the earth, the North Pole is antipodal to the South Pole; 20° S, 60° E is antipodal to 20° N, 120° W, p. 30

axis of rotation   imaginary line around which the earth rotates, p. 30

azimuthal projection   directions from the projection's center to all points are correct; also called a zenithal projection, p. 39

Cartesian coordinate geometry   system of intersecting perpendicular lines in plane space, useful in analytic geometry and the precise specification of location, p. 29

Clarke ellipsoid of 1866   a reference ellipsoid used by the United States and other countries of North America, p. 27

conformal projection   preserves angular relationships at points during the transformation process; cannot be equal-area; also called an orthomorphic projection, p. 38

Descartes   French mathematician whose early studies of algebra and geometry led to analytic geometry, p. 28

developable projection   can be constructed using the ordinary means of the draftsperson and the principles of projective geometry, p. 41

digitizing   transforming spatial elements of a map—or other two dimensional images—into x- and y-coordinates of Cartesian space, p. 29-30

ellipsoid   a geometrical solid developed by the rotation of a plane ellipse about its minor axis, p. 26-27

equal-area map projection   no areal deformation; cannot be conformal; also called an equivalent projection, p. 38

equator   imaginary line of the earth's coordinate system that is formed by passing a plane through the center of the earth perpendicular to the axis of rotation, midway between the poles, p. 30

equidistant projection   preserves correct linear relationships between a point and several other points, or between two points; cannot show correct linear distance between all points to all other points, p. 39

Eratosthenes   (276–194 B.C.) Greek scholar living in Alexandria who first accurately measured the size of the earth, p. 26

geodesy   the science that measures the size and shape of the earth; often involves the measurement of the external gravitational field of the earth, p. 26

Geodetic Reference System (GRS80)   reference ellipsoid used by the United States in the NAD83 datum, p. 27

geographic grid   spherical coordinate system used for the determination of location on the earth's surface, p. 31

geographic north and south   the imaginary line forming the earth's axis of rotation intersects the earth's surface at two locations, the North and South Poles, referred to as geographic north or south, p. 30

geoid   term used to describe the shape of the earth; means "earth-shaped" and does not refer to a mathematical model, p. 26

Global Positioning System (GPS)   a system of 21 orbiting satellites that transmit locational information to ground-based receivers, p. 31-32

graticule   meridians and parallels on a map projection, p. 35

great circle arc   segment of a great circle that is the shortest distance between two points on the spherical surface, p. 34

latitude   position north or south of the earth's equator; designation is by identifying the parallel passing through the position; determined by the angle subtended at the center of a sphere by a radius drawn to a point on the surface, p. 30

longitude   position east or west of the prime meridian; designation is by identifying the meridian passing through the position; determined by angular degrees subtended at the center of a sphere by a radius drawn to the meridian and the position in question, p. 30-31

loxodrome   a line on the earth that intersects every meridian at the same angle; because of meridional convergence, a loxodrome theoretically never reaches the pole; also called a rhumb line, p. 34

map projection   the systematic arrangement of the earth's spherical or geographic coordinate system onto a plane; a transformation process, p. 35

marine chronometer   extremely accurate timepiece used to determine longitude; perfected by John Harrison in 1761, p. 31

meridian   great circle of the earth's geographic coordinate system formed by passing a plane through the axis of rotation; meridional number designation ranges from 0° to 180° E or W of the prime meridian, p. 31

minimum error projection   no equivalency, conformality, azimuthality, or equidistance; chosen for its overall utility and distinctive characteristics, p. 39

nominal scale   the scale of the reference globe, expressed as a representative fraction; also called the defined scale, p. 36

parallel of latitude   small circle of the earth's geographic coordinate system formed by passing a plane through the earth parallel to the equator; parallel number designation ranges from 0° at the equator (a great circle) to 90° at the Pole (either North or South), p. 30

pattern of deformation   distribution of distortion over a projection; customarily increases away from point or line(s) of tangency of plane to sphere, p. 41

prime meridian meridian adopted by most countries as the point of origin (0°) for determination of east or west longitude; passes through the British Royal Observatory at Greenwich, England, p. 31

projection aspect the position of the projected graticule relative to the ordinary position of the geographic grid on the earth, p. 42-43

reference globe the reduced model of the spherical earth from which projections are constructed; also called a nominal or generating globe, p. 35

scale factor ratio of the scale of the projection to the scale of the reference globe; 1.0 on standard lines, at standard points, and at other places, depending on the system of projection, p. 40-41

sexagesimal system of numbering that proceeds in increments of 60; for example, the division of a circle into 360 degrees, a degree into 60 minutes, and a minute into 60 seconds, p. 30

small circles any circles on the spherical surface that are not great circles; parallels are small circles, p. 30

standard lines and points transformed from the spherical surface to the plane surface without distortion; scale factors on these lines or points are 1.0, p. 40-41

Tissot's indicatrix mathematical construct that yields quantitative indices of distortion at points on map projections, p. 39

READINGS FOR FURTHER UNDERSTANDING

Brown, Lloyd A. *The Story of Maps.* New York: Bonanza Books, 1959.

Burkard, R. K. *Geodesy for the Layman.* St. Louis: Geophysical and Space Sciences Branch, Aeronautical Chart and Information Center, 1964.

Cotter, Charles H. *The Astronomical and Mathematical Foundations of Geography.* New York: American Elsevier, 1966.

Deetz, Charles H., and Oscar S. Adams. *Elements of Map Projection.* U.S. Department of Commerce, Special Publication No. 68. Washington, DC: USGPO, 1945.

Defense Mapping Agency, Hydrographic Center. *American Practical Navigator.* Washington, DC: Defense Mapping Agency, 1977.

Dent, Borden D. "Continental Shapes on World Projections: The Design of a Poly-Centered Oblique Orthographic World Projection." *The Cartographic Journal* 24 (1987): 117–24.

Hilliard, James A., Umit Bosoglu, and Phillip S. Muehrcke, comp. *A Projection Handbook.*

Cartographic Laboratory, Paper No. 2. Madison: University of Wisconsin, 1978.

Hsu, Mei-Ling. "The Role of Projections in Modern Map Design." *Cartographica* 18 (1981): 151–86.

Institute of Navigation. *Global Positioning System.* Washington, DC, 1986.

Jackson, J. E. *Sphere, Spheroid and Projections for Surveyors.* New York: Halsted Press, 1980.

King-Hele, Desmond. "The Shape of the Earth." *Science* 183 (1967): 67–76.

———. "The Shape of the Earth." *Science* 192 (1976): 1293–1300.

Leick, A. *GPS Satellite Surveying.* New York: Wiley Interscience, 1990.

Maling, D. H. *Coordinate Systems and Map Projections.* London: George Philip, 1973. A new edition of this very informative book has been announced by Pergamon Press.

———. *Measurements from Maps: Principles and Methods of Cartometry.* Oxford, England: Pergamon, 1989.

Marschner, F. J. "Structural Properties of Medium- and Small-Scale Maps." *Annals* (Association of American Geographers) 34 (1944): 1–46.

Morgan, J. G. "The North American Datum of 1983." *Geophysics: The Leading Edge of Exploration,* January, 1987, pp. 27–32.

Richardus, Peter, and Ron K. Adler. *Map Projections.* New York: American Elsevier, 1972.

Robinson, Arthur H. "An Analytical Approach to Map Projections." *Annals* (Association of American Geographers) 39 (1949): 283–90.

———. "A New Map Projection: Its Development and Characteristics." *International Yearbook of Cartography* 14 (1974): 145–55.

Snyder, John P. "The Perspective Map Projection of the Earth." *American Cartographer* 8 (1981): 149–60.

———. *Flattening the Earth: Two Thousand Years of Map Projections.* Chicago: University of Chicago Press, 1993.

Sobel, Dava. *Longitude.* New York: Penguin Books, 1995; *See also* Dava Sobel, "Longitude?" *Harvard Magazine* (March/April, 1998): 45–52.

Steers, J. A. *An Introduction to the Study of Map Projections.* London: University of London Press, 1962.

———. *Map Projections—A Working Manual.* U.S. Geological Survey, Professional Paper 1395. Washington, DC: USGPO, 1987.

Thompson, Morris M. *Maps for America.* 2d ed. U.S. Department of the Interior, Geological Survey. Washington, DC: USGPO, 1982.

Vogel, Steven A. "Network for the 21st Century." *ACSM Bulletin* (August 1988): 21–23.

CHAPTER

3

EMPLOYMENT OF PROJECTIONS AND THEMATIC BASE-MAP COMPILATION

CHAPTER PREVIEW

Thematic base-map compilation begins by selecting an appropriate projection. This may involve developing the projection from the beginning, or selecting it from digital files. Base map compilation continues by addressing such questions as appropriateness of scale, level of generalization of coastlines and other lines on the map, and feature selection. Designers of thematic maps have a choice of hundreds of projections, but the range is reduced by differences in the suitability of projections for mapping the desired part of the globe. Because the equal-area property is important in mapping most themes, the choice is further restricted. For world mapping on one sheet, only a
handful of projections have become widely used. Only three are discussed in detail: Mollweide, Hammer, and Boggs. Projections suitable for continent- and country-scale mapping are generally different from those employed at world scales. Unique solutions to projections may also need to be explored.

A compiler's chief concern is the evaluation of the accuracy and reliability of source maps and materials. Good judgment comes with care and experience. Compilation may include the development of a draft map to be scanned for later digital production. The designer needs to be cautious when using other source maps and materials to avoid infringement of others' legal rights.

Chapter 2 introduced the earth's spherical grid, the development of projections, and patterns of deformation in projections. This chapter addresses two additional topics of importance: employment of projections and base-map compilation. A thematic map is comprised of a base map and a thematic overlay. Only topics central to the preparation of the base map are discussed in this chapter; subsequent chapters deal with the operations that go into the production of various thematic overlays.

THEMATIC MAP COMPILATION

Compilation is composition using materials from other documents. **Map compilation** has been given considerable treatment in the cartographic literature. The International Cartographic Association, in its *Multilingual Dictionary,* defines it as follows:

> *The selection, assembly and graphic presentation of all relevant information required for the preparation of a map. Such information may be derived from other maps and from other sources.*[1]

Another view is that compilation includes all "aspects of collection, evaluation, interpretation and technical assembly of cartographic material."[2]

Thematic map compilation refers to the construction of a special-purpose map from a variety of previously existing sources: base maps, other thematic maps, or both, and involves numerous steps. (See Figure 3.1.) In compilation,

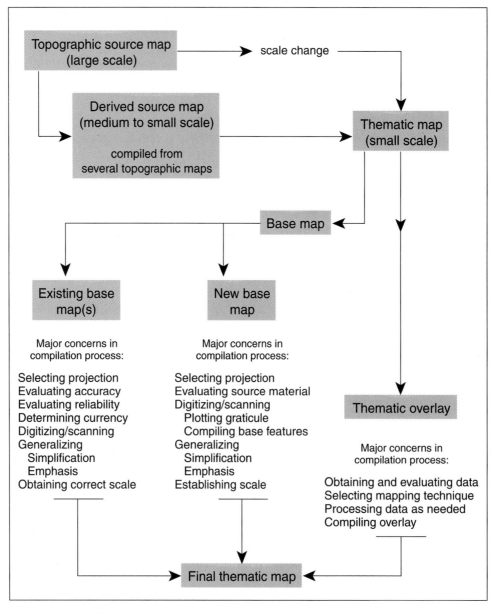

Figure 3.1   Thematic map compilation.
See the text for an explanation.

it is important to use only good sources.[3] Compilers of small-scale thematic maps can use either large-scale maps, usually topographic survey maps, or derived medium- to small-scale maps as source maps. All **derived maps** are made from large-scale maps. United States Geological Survey maps at large scales are used as source maps for the production of smaller-scale maps in this country. Compilers of thematic maps use derived maps more frequently than maps at topographic scales. The availability, accuracy, reliability, and currency of derived maps are of great concern to the designer. Generalization, scale changing, and plotting methods are also important aspects of compilation.

If a thematic map is to be compiled without using a previously existing base map, a new base map must be constructed, or selected from carefully chosen digital base-map files. A projection and its features must be selected, the coastline selected, and major cities, political boundaries, and physical features located. Developing a new base map requires reference to previously produced maps, from which locational information about the mapped features can be taken. Consideration must be given to the accuracy and reliability of all previously produced source maps, and this includes digital files. There will be questions about reliability in these cases, too.

THE EMPLOYMENT OF MAP PROJECTIONS

The thematic cartographer must select the proper map projection for a given map design problem. Fortunately, most map projections today are available in applications software, freeing the cartographer from the laborious task of drawing them by manual methods. These programs are relatively simple to use and provide the cartographers with many options and features. For the first time, geographers and cartographers can spend more useful time in the proper *selection* of the projection, without concerning themselves with the routine of plotting it.

Several essential elements must be carefully considered in the selection of a map projection:

1. *Projection properties.* Are the properties of a particular projection suited to the design problem at hand? Are equivalency, conformality, equidistance, or azimuthality needed?
2. *Deformational patterns.* Are the deformational aspects of the projection acceptable for the mapped area? Is linear scale and its variation over the projection within the limits specified in the design goals? Do the characteristics of linear scale over the projection benefit the *shape* of the mapped area?
3. *Projection center.* Can the projection be centered easily for the design problem? Can the design accommodate experimentation with the recentering of the projection?
4. *Familiarity.* Will the projection and the appearance of its meridians and parallels be familiar to most readers?

Will the form of the graticule detract from the main purpose of the map?
5. *Cost.* Is using a particular projection economical? Is there one already at hand, or will it need to be freshly plotted, by hand or by computer? Have land areas and political boundaries already been plotted on the graticule?

Although there are literally hundreds of projections from which the cartographer may choose, certain ones have proven more useful in mapping particular places. (See Table 3.1.) The ones included in our discussion are offered only as a guide.

WORLD PROJECTIONS

Many world projections may be selected for thematic mapping. Equivalency is of overriding concern. Three projections, all equal-area, are presented here as good choices when mapping at the world scale on a single sheet: the Mollweide (or homolographic), the Boggs eumorphic, and the Hammer equal-area. None of these and no world equal-area projection on a single sheet can avoid considerable shape distortion, especially along the peripheries of the map. These disadvantages must be accepted in order to represent the whole earth on one sheet.

The **Mollweide projection,** named after Carl B. Mollweide, who developed it in 1805, has become widely used for mapping world distributions.[4] The equator is a standard line, equally divided. (See Figure 3.2a.) The central meridian is one-half the length of the equator and drawn perpendicular to it. Parallels are straight lines parallel to the equator but are not drawn with lengths true to scale, except for the parallel at 40°40′. Each parallel, however, is divided equally along its length. The parallels are spaced along the central meridian to achieve equivalency.

Meridians on the Mollweide are curved, and the ones at 90° from the central meridian form a complete circle. Thus one hemisphere is illustrated by a whole circle. Other meridians are elliptical arcs, with the shape of the whole projection an ellipse.[5]

Shape distortion on the Mollweide projection is consistent with the overall pattern of deformation for world pseudocylindrical equal-area projections. Maximum shape deformation occurs at the extremes, where the intersections of the meridians and parallels are the most oblique. The principal use for the Mollweide is the plotting of world thematic distributions. Its overall shape is rather pleasing, and if the distortion of shape at the peripheries is not at odds with the design purpose, this projection is an acceptable choice.

Very similar to the Mollweide are the Hammer equal-area and Boggs eumorphic projections. The **Hammer projection,** developed in Germany in 1892, was for many years erroneously called the Aitoff.[6] The principal difference between this projection and the Mollweide is that the Hammer has curved parallels. (See Figure 3.2b.) This curvature results in less oblique intersections of meridians and

Table 3.1   Guide to the Employment of Projections for World-, Continental-, and Country-Scale Thematic Maps

Principal Use	Suitable Projections	Principal Features		
		Parallels	**Meridians**	**Other**
1. Maps of the world in one sheet				
Equal-area	Sinusoidal (Sanson-Flamstead)	Horizontal, spaced equally at true distances	Sine curves, spaced equally, true on each parallel	Awkward shape
Equal-area	Eumorphic (Boggs)	Horizontal, spaced equally	Curved	Arithmetic mean between sinusoidal and Mollweide
Equal-area	Mollweide	Horizontal, spaced closer near poles	Ellipses, spaced equally	Pleasing shape
Equal-area	Hammer	Curved, spaced equally	Curved, spaced closer on central meridian	
Equal-area	Eckert (esp. IV and VI)	Horizontal, spaced closer near poles; poles are lines half length of equator	Ellipses, spaced equally. Sine curves, spaced equally	Poles represented as lines; considerable distortion in poleward areas
Conformal	Mercator	Horizontal, spaced closer near equator	Vertical, spaced equally, true on equator	All bearings correct
2. Continental areas				
A. Asia and North America				
Equal-area	Bonne	Concentric circles, spaced equally at true distances	Curves, spaced equally, true on each parallel	Considerable distortion in NE and NW corners
Equal-area	Lambert azimuthal	Curves, spaced closer near poles on central meridian	Curves, spaced closer near sides of equator	Bearings true from center
B. Europe and Australia				
Equal-area	Lambert azimuthal, Bonne			
Equal-area	Albers with two standard parallels	Concentric circles, spaced closer at N and S ends	Radiating straight lines, spaced equally, true on one or two standard parallels	Ideal for United States

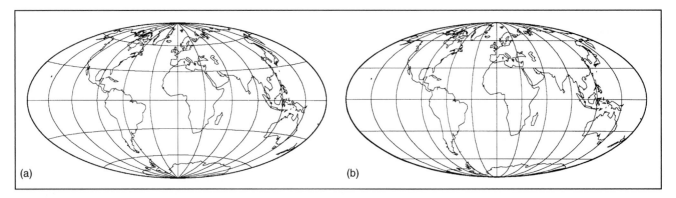

Figure 3.2   Two world projections.
The Hammer projection is shown at the left, and the Mollweide right. Both of these projections are equal-area.

Principal Use	Suitable Projections	Principal Features		
		Parallels	**Meridians**	**Other**
C. Africa and South America				
Equal-area	Mollweide			
Equal-area	Bonne			
Equal-area	Sinusoidal			
Equal-area	Lambert azimuthal			
3. Large countries in mid-latitudes				
A. United States, Russia, China				
Equal-area	Lambert azimuthal			
Equal-area	Albers equal-area			
Equal-area	Bonne			
Conformal	Lambert conformal conic	Concentric circles, spaced wider at N and S ends	Radiating straight lines spaced equally true on one or two standard parallels	
4. Small countries in mid-latitudes				
Equal-area	Albers equal-area			
Equal-area	Bonne			
5. Polar regions				
Equal-area	Lambert azimuthal			
6. Hemispheres and continents				
Visual	Orthographic	Ellipses, spaced closer near periphery	Ellipses, spaced closer near periphery	View of earth as if from space; neither equal-area nor conformal

Source: Compiled from a variety of sources; see especially Erwin Raisz, Principles of Cartography (New York: McGraw-Hill, 1962), p. 189; J. A. Steers, An Introduction to the Study of Map Projections (London: University of London Press, 1962), pp. 222–25; and John P. Snyder, Map Projections—A Working Manual. U.S. Geological Survey Professional Paper 1395 (Washington, DC: USGPO, 1987), table "The Properties and Uses of Selected Map Projections," in pocket.

parallels at the extremities, and thus reduces shape distortions in these areas. The outline (i.e., the ellipses forming the outermost meridians) is identical to the Mollweide.

Construction of the Hammer projection is somewhat more difficult than the Mollweide because of the curved parallels. This is a minor inconvenience, especially if machine plotting is available. Because the parallels are unequally curved, they are not true to scale.

Hammer's projection also is quite acceptable for mapping world distributions. A comparison of it with the Mollweide shows little difference. Because the parallels are curved, east-west exaggeration at the poles is less on the Hammer than on the Mollweide. This is most notable when comparing the Antarctica land masses. Africa is less stretched along the north-south axis on the Hammer. Overall, however, these projections are very similar in appearance and attributes.

The **Boggs eumorphic projection** was developed by Whittemore Boggs in 1929.[7] In contrast to the Mollweide and the Hammer projections, the poles on the Boggs projection are accentuated by the converging meridians. This projection is a combination of the sinusoidal (Sanson-Flamstead) and the Mollweide. It is considered an arithmetic average between the two. Above 62° latitude, angular distortion is less than on sinusoidal projection. A chief advantage of the Boggs projection over the

RESOLUTION REGARDING THE USE OF RECTANGULAR WORLD MAPS

WHEREAS, the earth is round with a coordinate system composed entirely of circles, and

WHEREAS, flat world maps are more useful than globe maps, but flattening the globe surface necessarily greatly changes the appearance of the earth's features and coordinate system, and

WHEREAS, world maps have a powerful and lasting effect on peoples' impressions of the shapes and sizes of lands and seas, their arrangement, and the nature of the coordinate system, and

WHEREAS, frequently seeing a greatly distorted map tends to make it "look right,"

THEREFORE, we strongly urge book and map publishers, the media, and government agencies to cease using rectangular world maps for general purposes or artistic displays. Such maps promote serious, erroneous conceptions by severely distorting large sections of the world, by showing the round earth as having straight edges and sharp corners, by representing most distances and direct routes incorrectly, and by portraying the circular coordinate system as a squared grid. The most widely displayed rectangular world map is the Mercator (in fact a navigational diagram devised for nautical charts), but other rectangular world maps proposed as replacements for the Mercator also display a greatly distorted image of the spherical earth.

American Cartographic Association
American Geographical Society
Association of American Geographers
Canadian Cartographic Association
National Council for Geographic Education
National Geographic Society
Special Libraries Association, Geography and Map
    Division

Source: Committee on Map Projections, "Technical Notes," *American Cartographer* 16 (1989): 223.

Mollweide is better shape preservation along the equator, brought about by greater equality of linear scales in both east–west and north–south directions. There is greater north–south stretching at the equator (e.g., Africa) on the Mollweide.

The number of other projections that can be chosen for world thematic mapping on one sheet is quite large. Numerous publications treat this subject in more depth than space allows here. A particularly interesting book on projections, entitled *Matching the Map Projection to the Need,* offers some provocative choices. This book and others are listed in the readings at the end of the chapter.

One final comment may be made regarding the selection of projections for world mapping. Concern about the use of rectangular projections for world presentations recently reached the attention of the Committee on Map Projections of the American Cartographic Association, and the resolution it passed has the endorsement of most professional geography and cartography associations. (See boxed article.) They do not endorse the use of any rectangular world projection. None of the world projections mentioned previously are of the rectangular type, as they have rounded margins.

PROJECTIONS FOR MAPPING CONTINENTS

The projections that are suitable for world maps are generally not the best for mapping continental areas. Either equivalency or conformality can be better preserved at the larger scales by selecting other projections. As indicated in Table 3.1, the Bonne, Lambert azimuthal, Albers equal-area, and sinusoidal projections, in addition to the Mollweide, are wise choices when mapping continent-size areas. We will look at the Bonne projection in detail.

The **Bonne projection** is named after its inventor, Rigobert Bonne (1727–1795).[8] It is an equal-area conical projection, with a central meridian and the cone assumed tangent to a standard parallel. (See Figure 3.3.) All parallels are concentric circles, with the center of the standard parallel the apex of the cone. The central meridian is divided true to scale. All parallels are drawn with their lengths true to scale, and each is divided truly. The meridians are drawn through the points of division along the parallels.[9] If the standard parallel selected is the equator, the projection becomes identical to the sinusoidal. Map designers select the Bonne projection for a variety of continental mapping cases. It is commonly used to map Asia, North America, South America, Australia, and other large areas. Europe may also be adequately mapped with the Bonne projection. Caution should be exercised, however; although equivalency is maintained throughout, shape distortion is particularly evident at the northeast and northwest corners. Because of this, the Bonne projection is really best suited for mapping compact regions lying on only one side of the equator.[10] Because shape is best along the central meridian, the distortion becoming objectionable at greater distances from it, the selection of the central meridian relative to the important mapped area is critical.

MAPPING LARGE AND SMALL COUNTRIES AT MID-LATITUDES

Mapping large countries at mid-latitudes can be handled in a variety of ways. The Bonne, the Lambert azimuthal equal-area, or the Albers equal-area projections may be used. If conformality is desired, the Lambert conformal conic can be selected. In general, a conic projection is usually adequate for mapping rather large countries lying in the mid-latitudes.

New World Map Cuts Nations Down to Size

From Wire Reports

WASHINGTON — Pentagon officials are in for a pleasant surprise: The Soviet Union finally has been cut down to size — reduced by 63 percent, in fact.

But Canadians probably will be furious to see their country trimmed by two-thirds. And real estate agents all over the world will be alarmed to find the Earth's land surface has shrunk dramatically.

Blame it on the National Geographic Society.

"It's not every day that you get to change the world," National Geographic spokesman Robert B. Sims said Thursday as he unveiled the society's new official world map.

"The Robinson Projection more accurately portrays round Earth on a flat surface," said Gilbert M. Grosvenor, president and chairman of the society.

The map is named for Arthur H. Robinson, professor emeritus of geography and cartography at the University of Wisconsin at Madison, who came up with the design in 1963. National Geographic also honored him with a medal Thursday for his contributions to the field.

The new map, like the old — the Van der Grinten projection — has the inherent problem of trying to capture a three-dimensional world on paper.

It differs from the old primarily in the polar regions, which are less exaggerated. Greenland, for example, was 5½ times its actual size on the old map, which made it look about the size of South America. On the new map

it is a mere 60 percent larger than life. Antarctica no longer is as big as Asia.

On Van der Grinten maps, the Soviet Union appears 223 percent too large and the United States seems 68 percent bigger than it actually is. On the Robinson maps, the Soviet Union is 18 percent larger than it should be, and the United States is 3 percent smaller.

By reducing the sizes of land masses relative to oceans, the Robinson Projection shows the Earth's surface more like what it is — 71 percent water.

Finally, because most people live in temperate and tropical zones, the new map includes more place names in those areas.

National Geographic's move to the new map coincides with the organization's centennial. The society, which has 11 million members, estimates 40 million people will use the new map.

Although the society fully endorses the map, it — and the rest of the world — still will use the Mercator Projection, a map designed in the 16th century, to show areas smaller than the whole world. The Mercator greatly distorts the size of the world's land masses

Even the creator of the new map admits it is not perfect.

"I hesitate to say this, but it is pretty much an artistic process rather than a scientific one. You try to get the most realistic view of the world," Mr. Robinson said. "I worked with the variables until it got to the point where, if I changed one of them, it didn't get any better. And so I stopped."

A New Look at the World

After more than 50 years of using the Van der Grinten projection for its world maps, the National Geographic Society has adopted the Robinson projection.

Percentages show the distortion of land areas.

ROBINSON (new map)
(Robinson map shows the United States as 3% smaller than it appears on a globe. Van der Grinten shows it 68% larger.)

VAN DER GRINTEN (old map)

Source: National Geographic Society

Note: In early 1998, the National Geographic Society has dropped the Robinson projection for many of its world maps in favor of the Winkel Tripel projection. Shape distortion at higher latitudes is less than on the Robinson.

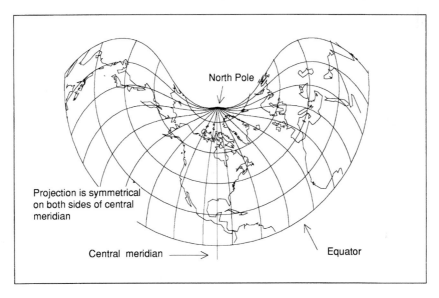

Figure 3.3   The Bonne projection.

This projection is suitable for mapping continents, but should never be used for mapping complete hemispheres. Notice severe shape distortion, brought about by shearing, at the northeast and northwest corners of the projection.

All simple azimuthal projections are developed onto a plane that is tangent to the generating globe at a point. The **Lambert equal-area azimuthal projections** follow this pattern. They may be tangent at any point, although the polar case is considered the normal aspect. It is re-

ferred to as the Lambert equal-area meridional (or central equivalent) projection when the point of tangency is the equator and some meridian.[11] It may also be centered at some latitude between the equator and the pole; this is called the *oblique* case. (See Figure 3.4.) This projection

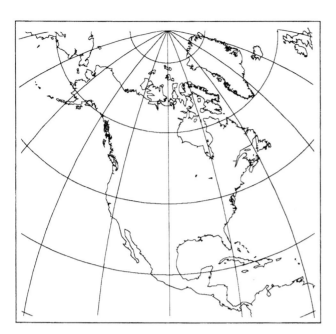

Figure 3.4 An oblique Lambert equal-area projection. This projection is suitable for mapping both large and small countries at mid-latitudes.

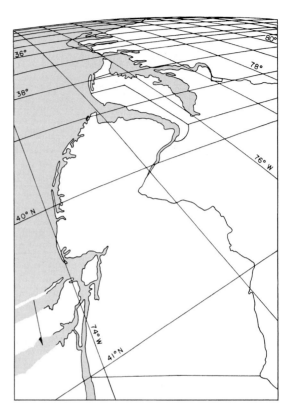

Figure 3.5 A tilted perspective or "space photo" projection.
The view is of the eastern seaboard of the United States, viewed from about 160 km above Newburgh, N.Y. Cartographers now have greater opportunity to select unique projections because of the proliferation of computer programs to plot them. *(Redrawn by permission from John P. Snyder, "The Perspective Map Projection of the Earth,"* American Cartographer *8 [1981]: 155).*

is quite adequate for showing countries that have symmetrical shapes; little shape distortion is evident. In addition to the equivalency feature, because the Lambert is an azimuthal projection, the azimuth of any point on the map (as measured from the center of the projection) is correct. This makes this projection especially useful when mapping phenomena having an important directional relationship to the central point chosen for the map.

The **Albers equal-area projection** is a conic projection having two standard parallels. Its utility is not limited to mapping the United States; any land area having considerable east-west extent is well represented by the Albers. Overall scale distortion is nearly the lowest possible for an area the size of the United States. The desirable properties of this projection are as follows:[12]

1. It is an equal-area projection.
2. Maximum-scale error is approximately 1.25 percent over an area the size of the United States.
3. Meridians are straight lines that intersect parallels at right angles.
4. Parallels are concentric circles, making construction relatively easy.
5. Because it is a single-cone conic projection, its properties do not deteriorate in an east-west direction.
6. It is suitable for a series of "section" maps, because adjacent sections of this map will fit together exactly.

Equal-area mapping of small countries at mid-latitudes (e.g., France, Spain) can be accomplished by using such projections as the Bonne or the Albers. As mapped areas become smaller in extent, the selection of the projection becomes less critical; potential scale errors begin to drop off considerably.

MAPPING AT LOW LATITUDES

Many of the projections already discussed are suitable for mapping countries, large and small, on or near the equator. Most of the equal-area projections suited for world mapping (e.g., Mollweide, Hammer, sinusoidal) are cylindrical projections having relatively little angular or linear scale distortion at low latitudes. Coupled with a larger scale (showing smaller earth areas), these projections are quite satisfactory for most thematic map applications in equatorial regions.

UNIQUE SOLUTIONS TO THE EMPLOYMENT OF PROJECTIONS

Thematic cartographers should spend considerable time on the selection of a projection in furthering the communication effort of the map.[13] They should therefore be willing to investigate the impact of recentering the projection and to see the effects of new orientations on the map's message. Two such maps are reproduced here as Figures 3.5 and 3.6.

Unique solutions to the employment of projections have become more commonplace with the increasing use of

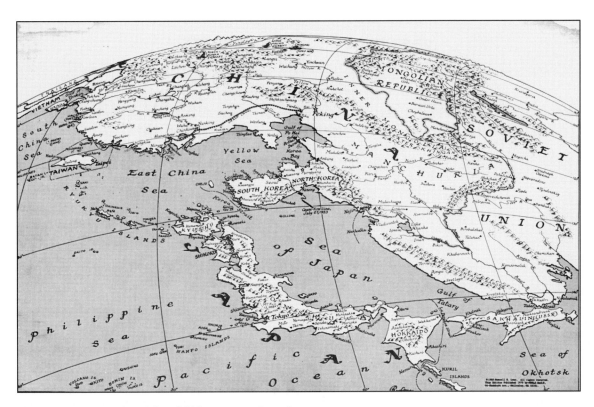

Figure 3.6   The People's Republic of China as seen from Japan.
The orientation of projection, which was developed by photographing the graticule on a globe, gives this map its unique quality. This map is one of a sries done by cartographer Russell Lenz. (© *Russell Lenz. All rights reserved. Reproduced by permission.*)

computer mapping. Cartographers were previously discouraged from experimentation by the drudgery of computing and plotting projections. Now the emphasis can be on selection.

In sum, the projection provides the geometrical framework for all the points and areas compiled on the final map. Projection is thus a key element in the overall design. It controls the locational accuracy and final appearance of the mapped area and can be used to provide a focus for the map's message. The designer must have a firm grasp of the earth's spherical coordinate system in order to understand projections and to use them effectively in the design effort. And finally, the cartographer should provide the map reader with information about the essential projection properties. These include, where applicable: scale, central meridian, standard parallel(s), central parallel, scale factor on the central meridian, and any other unique attributes.[14]

Mapping the earth in a continuous, uninterrupted way has become common today, but was not always the case. In many older atlases the world was shown in two halves, most often the eastern and western hemispheres. There was no pretense in trying to show it as a one-sided globe. After all, we can see only one-half at a time. A version of this form of world mapping is presented here, in Figure 3.7. This view is based on the Hammer projection.

A poster titled *Map Projections,* produced by the National Mapping Program of the Department of the Interior, United States Geological Survey, accompanies this book and illustrates many features of 17 different map projec-

tions. Class of projection, properties (such as deformation), information regarding sizes and shapes of continental areas, characteristics of the grids, and special features of each projection are included. The most appropriate uses for each projection (that is, for mapping world, hemispheric, continent/ocean, and regional areas, and at what scales) are listed also in a convenient table. Principal definitions (for example, aspect, conformality, equivalency, linear scale, and several others) are provided also. Accompanying each projection is a representation of its grid, and shapes of continental areas.

Careful examination of this poster and the many definitions on it can provide you with a solid introduction to these projections. The poster, along with this chapter, will serve as a useful study and reference guide.

GALL-PETERS PROJECTION

Special consideration is devoted here to what has become known as the **Gall-Peters projection,** primarily to stimulate the cartographic designer to look further into the literature, and partly because of the controversy surrounding its use. In recent times, probably no other map projection has received as much attention in both the scientific *and* popular literature.

Dr. Arno Peters, of Germany, in 1972 published what he called a new map projection—the Peters projection.[15] In fact, this projection had been devised earlier by Gall in the

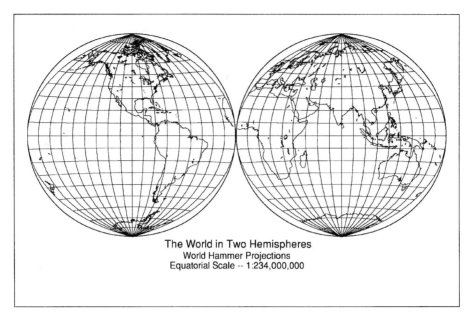

The World in Two Hemispheres
World Hammer Projections
Equatorial Scale -- 1:234,000,000

Figure 3.7 The world in two hemispheres.
This is not an ordinary use of the Hammer projection, but shows how a projection may be customized for a particular use.

mid-1880s, who called it the orthographic—a form of cylindrical projection (but it may be speculated that Peters had no knowledge of Gall's previous work).[16] In the conventional aspect, the Peters projection is an equal-area rectangular projection with standard parallels at 45° north and south. (See Figure 3.8.) Peters was reacting to the areal misrepresentation on the Mercator projection, which he believed showed European dominance over the Third World.[17] Shapes are terribly distorted on the Mercator projection. Because the projection is not new, it is referred to as the Gall-Peters projection.

The purpose behind the invention, as Peters argues, is that the Mercator projection, often used for mapping the world, is inadequate for that purpose because it so grossly distorts area, which is true. Peters, therefore, stressed that his projection be used exclusively, maintaining that it portrays distances accurately (which is false). The real objection to Peters is with the misconceptions he renders about the projection. It has been widely adopted by three United Nations organizations (UNESCO, UNDP, and UNICEF), the National Council of Churches, and Lutheran and Methodist organizations.[18]

One good thing has come from the controversy, namely, that the inadequacies of using the Mercator projection for world thematic mapping have finally been noted to the general population. Snyder, for example, has said:

Nevertheless, Peters and Kaiser (his agent) appear to have successfully accomplished a feat that most cartographers only dream of achieving. Professional mapmakers have been wringing their hands for decades about the misuse of the Mercator Projection, but, as Peters stresses,

the Mercator is still widely misused by school teachers, television news broadcasters and others. At least Peters' supporters are rightly communicating the fact that the Mercator should not be used for geographical purposes, and numerous cartographers agree. The Gall-Peters Projection does show many people that there is another way of depicting the world.[19]

THE SELECTION OF PROJECTIONS FOR INDIVIDUAL STATES AND THE STATE PLANE COORDINATE SYSTEM

For most large-scale thematic mapping, the selection of a projection on which to plot a state outline and its geographical features does not present any significant problem. In general, the projection should be equal-area, and other desired qualities can be chosen as appropriate (azimuthality from the center, for example). It is important to select a projection having the properties that are suited for the state. For example, a good choice for mapping Tennessee would be an equal-area conic projection whose standard parallel runs through the main east-west axis of the state.

Thematic cartographers are finding that much of their work involves using remotely sensed, digital images that are best processed in geographic information system (GIS) environments. These environments are really nothing other than large computer programs that assist in spatial data handling and spatial problem solving:

Technically defined, GIS is the integration of administrative data bases, spatial modeling, and cartographic display technology. . . . [it] provides us with the unique

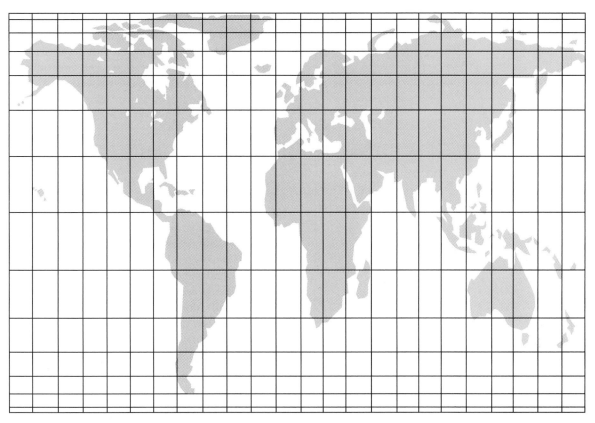

Figure 3.8   Gall-Peters projection.
Projection advanced by Arno Peters in 1972 to overcome area distortion of the Mercator projection. May be based on the Gall cylindrical orthographic.

ability to catalog, analyze, and present a wide variety of spatial information in a clear and relevant manner so that existing conditions can be characterized, trends can be identified, and the future can be predicted.[20]

The subject of GIS and cartography is dealt with in much more detail in this book in Chapter 6.

What is important for us here is that most GIS systems offer the user a variety of ellipsoids and grid reference systems to which they may match their imported data. So the thematic mapper at least must have a fundamental knowledge of these ellipsoids and grid systems. The ellipsoids were introduced in the last chapter. We will look briefly at the **state plane coordinate system** (SPCS) here.

The original state plane coordinate system was developed in the early 1930s by the then United States Coast and Geodetic Survey (USCGS), now the National Ocean Service.[21] It was devised so that local engineers, surveyors, and others could tie their work into the reference then used, the Clarke ellipsoid of 1866, which was used in the North American Datum (NAD) of 1927. What they desired was a simple rectangular coordinate system on which easy plane geometry and trigonometry could be applied for surveying, because working with spherical coordinates was cumbersome.

In the last chapter we mentioned that if the area of the earth being mapped is small enough, virtually no distortion exists. This is the principle behind the SPCS. The USCGS thus decided to use three conformal projections to map the states—the **Lambert conformal conic** for states with long east-west dimensions, the **transverse Mercator** for states with long north-south dimensions, and the oblique Mercator for portions of Alaska. In each case, over small areas, these projections essentially project as rectangular grids, with little or no areal and distance distortion (Because they are conformal, no angular distortion is present).

The transverse Mercator is developed by "rotating" the normal aspect of the Mercator where the cylinder is tangent to the equator to a position where it is tangent to a meridian. The transverse Mercator projections used for the SPCS are secant in form, thus they are tangent to two meridians, reducing overall distortion. Along these meridians (standard lines) there is no distortion. The meridian lying midway between these two is referred to as the central meridian. Between the central meridian and the two standard meridians, scale distortion is reduced; beyond the standard meridians, scale distortion (what little there is) is enlarged.

In a sense, then, a rectangular grid is superimposed over the projected graticule. Mathematical formulas are computed so that accurate conversions at points can be made

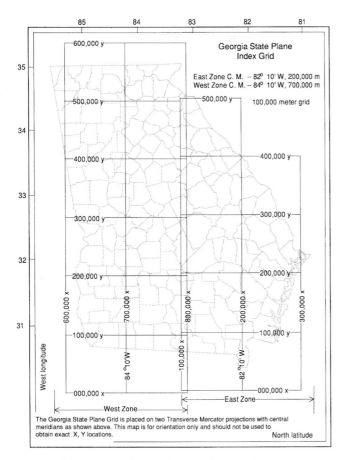

Figure 3.9 Orientation grid of the Georgia State Plane Coordinate System (SPCS).

from the SPCS grid to the projection grid, and to act as guides in large-scale map compilation. To ensure stated accuracies of less than 1 part in 10,000, some states (such as Arizona, New Mexico, Alabama, and many others) have several SPCS grids applied to them. (See Appendix B.) The original grids formed in the early 1930s were specified in feet. Since the adoption of the North American Datum of 1983 (using the GRS80 reference ellipsoid), all new computations (and newer topographic maps) and grids are specified in meters. The central meridian of each state's grid has an "easting" value assigned to it (the X value), and each grid begins with a specified false "northing" (the Y value). In those states with two (or more) grids, each zone has its own central meridian. (See Figure 3.9.) The projections, central meridians, and grid origin designations for all states are included in Appendix B for ready reference.

Today and for many years, local civil engineers (especially transportation), surveyors, utility companies, and others have used the state plane grid to map and specify locations. With the coming of GIS, where so many spatial data are applied to ground networks of information, we see nothing but continued and expanded use of this handy rectangular grid.

The United States Geological Survey's state base maps are produced on either the Lambert conformal conic or the transverse Mercator. Because earth areas are relatively small and the mapping scale so large, areal distortion of these conformal projections, even at state scales, is negligible. Therefore, the thematic map designer has a convenient source for state base maps on acceptable projections.

BASE-MAP COMPILATION

The remainder of this chapter treats base-map compilation, previously defined in broad terms. More likely than not, today base-map compilation will take place at the computer. A projection must be selected, its aspect and limits chosen, its scale determined, and a level of generalization chosen. These operations must be made with a good knowledge of several subjects. So, we now turn to some specific aspects of the subject: issues related to generalization in the computer map world, compilation sources, preparing a base map for scanning, and copyright issues. Not so many years ago, these topics included both manual and computer compilation, but today and in this text, they are applied to computer compilation.

COMPILATION AND GENERALIZATION

Compiling thematic maps involves a knowledge of **cartographic generalization,** which ordinarily means simplification.[22] Simplification requires the cartographer to use good judgment in the selection of detail as he/she begins to assemble the base map, which is usually taken from preexisting computer map files. Selection and emphasis are also important in generalization, which is considered one of the principal functions of the cartographer.[23] In thematic map design, poorly performed generalization can cause the whole map effort to fail. Unfortunately, generalization and simplification in thematic mapping are not subject to quantification or strict rules, so the cartographer must become proficient in these areas through experience and good judgment.

In the days of manual base-map production, guidelines were suggested to help deal with problems in generalization and compilation, and the ones that continue to bear on computer map compilation are these:[24]

1. *Map purpose.* Keep this clearly in mind.
2. *Use objective evaluation:* Utilize many sources in the selection process.
3. *Avoid personal bias.* Become acquainted with details of the whole region, so that areas about which you have greater knowledge do not appear with inordinate precision and detail.
4. Determine what elements really typify the *character* of an area, and retain only these during the selection process.

> What's the map for? To direct visitors how to get to your house from the station? Then subordinate everything else to that, even scale. The point is to simplify your guest's way to your residence. The simplification may reduce your project from a map to a mere cartogram. No matter: it will still be cartography. If it works, that's something. A map's first business is to function. Utter simplification need not inhibit interest, and it may even improve the design. The simplest map can be fascinating, though obvious.
>
> Source: David Greenhood, *Mapping* (Chicago: University of Chicago Press, 1964), p. 175.

5. Strive for *uniformity* of treatment over the new map. Uniformity applies to the *level* of generalization, as well as to the amount of detail chosen for the map.

These guidelines, although quite general, will serve you well and are worth learning.

What Is Compiled?

The answer must be in terms of the purpose of the map, because we are concerned with the compilation of the base-map portion of the thematic map. Map purpose will also dictate the level of generalization believed important for a particular message. First, let us look at the selection of features.

Base-map information serves to help map readers orient the thematic information to a spatial or geographical frame of reference. The designer's task is thus to select only those base-map features that will help the reader in the context of the map's purpose. For example, a map that shows urban employment in aircraft manufacturing will not need rivers on the base map. State boundaries, and perhaps cities not having employment, should be shown. A map showing major cattle drives on the Great Plains in the 1870s, however, should probably include the major rivers. The designer should ask, "Does this information assist in orienting the reader? Is this information helpful to the map purpose?"

The kinds of base-map information that often prove useful are the inclusion of coastlines, lakes, rivers and streams, major or minor physical features (e.g., mountains), and other special physical features. Other base data helpful for reader orientation are political boundaries and the locations of cities and towns. Of particular importance is the geographic grid (the graticule), especially if the mapped subject is dependent on its location on the earth. For example, a map showing annual temperature should include the graticule, or at least small *tick marks* along the map border.

Many beginning cartographers are reluctant to reduce the amount of base-map information for fear of producing

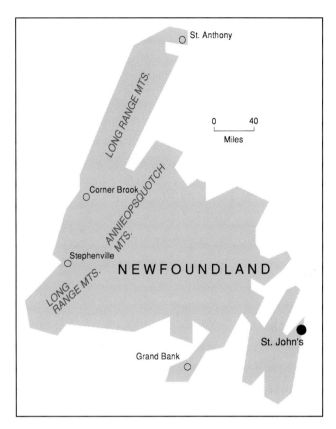

Figure 3.10   Map with a high coastline generalization. Coastline recognition can take place even with extreme generalization.

an "unscientific" map. Even many experienced cartographers are reluctant to adopt a high level of generalization. This is unfortunate because ideas can often be communicated with simple solutions. In Figure 3.10, for example, the overall message intended by the cartographer is effectively conveyed even though the outline of the island is highly generalized. The map's purpose and the reader's preexisting knowledge about shapes and features will determine the level of generalization and the selection of base-map information.[25]

Line Generalization in Computer Files

Normally, as you visually "focus in" in on cartographic lines drawn by hand you discover more detail in a particular line than when you stand back from it. The eye has a tendency to generalize (simplify) that line the farther back from it you are. Actually, the same effect applies to a computer-drawn line on the screen. If you stand back, it loses detail. However, many computer programs allow you to "zoom in" on a line, or feature. The effect is not the same as you get by moving closer to the line. As you zoom in further you actually see that the line has less detail. (See Figure 3.11.)

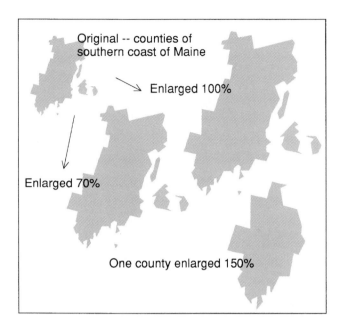

Original -- counties of southern coast of Maine

Enlarged 100%

Enlarged 70%

One county enlarged 150%

Figure 3.11 The effect on coastline generalization by computer software "zooming in."
Looking closer at a coastline by zooming in reveals that details are not part of the computer file.

In the compilation process when you are using computer line files, it is very easy to change scales of the map by using the zoom function. In previous times the cartographer had to optically change scales or use other manual means, which took considerable time. *In that process the cartographer also changed the amount of detail* in the line to suit the map purpose. In the computer process, this does not take place. Consequently, extreme care must be taken when changing scales with computer files to make certain the amount of line detail matches the map purpose.

In a very general way, a map that contains lines with too little detail for the map scale will look inappropriate, unscientific, or even "childish." And conversely, a map at a certain scale with too much detail looks equally awkward. Just what amount of detail at what scale often is a function of comparative convention, and the map purpose.

COMPILATION SOURCES

Today, thematic base maps are often compiled from digital files. A fundamental concern is with the selection of the projection, its scale, aspect, and limits. Certain features, notably coastlines, islands, rivers, and lakes, will have to be selected to go along with the projection. Likewise, cultural features such as roads, railroads, towns, and cities, may have to be obtained. If digital files are not available, cartographers can scan paper maps into computer software and develop their own files. Likewise, it is possible to scan air photos and make digital base maps from these. Knowing

about map sources that can be used in compilation is most helpful to the cartographer.

Fortunately, many high-quality source maps are available. Many public and private agencies produce very useful base maps and other thematic map materials. The principal government source for maps in this country is the United States Geological Survey, which produces maps ranging in scale from 1:24,000 (topographic scale) to 1:1,000,000. (See Table 3.2.) The large-scale topographic maps include detailed physical and cultural features. The smaller-scale (1:250,000 or smaller) maps show a limited number of features, such as boundaries, state parks, airports, major roads, and railroads. Government sources also produce many digital products from which base maps can be derived.

Over 30 federal agencies, states, counties, and municipalities are involved in producing maps of all kinds. It is estimated that more than 25,000 new maps are produced each year. These range from topographic maps to census maps to maps of the moon. Information concerning the products, sources, and addresses of government mapping activities is provided by this coordinating agency:

Earth Science Information Center
United States Geological Survey
507 National Center
Reston, Virginia 22092
(703) 860-6045 or 1-800-USA-MAPS

Appendix E provides important Internet sites for cartographers, including data and map sources, newsgroups, and mailing lists.

Of special interest to the thematic map compiler are the Geographic Names Information System (GNIS) files produced by the Branch of Geographic Names, Office of Geographic Research of the USGS. These files contain nearly two million name entries, with information about each feature, its category (stream, town, etc.), and its location by geographic coordinates, county, and topographic map. The files are organized by state and are available in printed form or are on CD-ROM. They are extremely helpful in checking the location of small places.

Cartographic designers often require other sources for mapping foreign lands. Especially useful here are two sources: the International Map of the World (IMW) series, produced by the USGS at a scale of 1:1,000,000; and the Digital Chart of the World, now called Vector Map Level 0 (VMap), produced by NIMA (National Image and Mapping Agency, formerly the Defense Mapping Agency). NIMA can be reached at www.nima.mil, and the United States Geological Survey at www.usgs.gov. The United States Air Force Navigation Charts are also quite useful. All these source maps have good reliability. Governmental agencies are not the only places to find map source materials. Private map and atlas producers are extremely valuable resources. The thematic map designer should gather a personal inventory of atlases and single-sheet maps for reference.[26] It is

Table 3.2   Scales of National Topographic Maps Produced by the U.S. Geological Survey

Series	Scale	1 Inch Represents	1 Centimeter Represents	Standard Quadrangle Size (Latitude-Longitude)	Quadrangle Area (Sq Mi)
7.5 min	1:24,000	2,000 ft	240 m	7.5 × 7.5 min	49–70
7.5 × 15 min	1:25,000	2,083 ft	250 m	7.5 × 15 min	98–140
Puerto Rico					
7.5 min	1:20,000	1,667 ft	200 m	7.5 × 7.5 min	71
15 min	1:62,500	1 mi	625 m	15 × 15 min	197–282
Alaska	1:63,360	1 mi	634 m	15 × 20 to 36 min	207–281 (by country)
Intermediate	1:50,000	3.2 mi	2 km		
Intermediate	1:100,000	1.6 mi	1 km	30 × 60 min	1,568–2,240
United States	1:1,000,000	4 mi	2.5 km	1° × 2° or 3°	4,580–8,669
State maps	1:500,000	8 mi	5 km		
United States	1:1,000,000	16 mi	10 km	4° × 6°	73,734–102,759
Antarctica	1:250,000	4 mi	2.5 km	1° × 3° to 15°	4,089–8,336
Antarctica	1:500,000	8 mi	5 km	2° × 7.5°	28,174–30,462

A lthough the objective of the map must be kept constantly in view, freedom of expression must be permitted. . . . Practical experience combined with common sense and a flair for the subject is the essential requirement for successful generalization of information.

Source: O. M. Miller and Robert J. Voskuill, "Thematic Map Generalization," *Geographical Review* 54 (1964): 13–19.

W e are all aware that to produce a successful thematic map the summarizing or generalizing process rightly starts at the planning stage and continues on through the steps of design, compilation, and drafting. But before planning can even begin, the source material must be sought out and appraised for its reliability and significance.

Source: O. M. Miller and Robert J. Voskuill, "Thematic Map Generalization," *Geographical Review* 54 (1964): 13–19.

especially important to collect a variety of maps having different scales, projections, and content. A good gazetteer is useful, as well as government-produced national atlases.

There are many map sources located on the World Wide Web, and many maps may be downloaded from these sites. Many digital maps are also available from private sources as well as from governmental agencies. Appendix E in the back of this text lists some of the most useful Web sites. Most of the private sources for digital maps have copyright protection, so the cartographer must be extremely careful not to infringe on their rights. (See page 63.) If you are not certain if these materials are in the public domain, it is best to first secure permission to use them.

Accuracy and Reliability

Having obtained a source map or maps, the designer must judge their accuracy and reliability. The newly compiled base map can be no more accurate than the one on which it is based. The locational accuracy of features on USGS topographic maps is quite good; occasional errors, mostly because of omission, are found among names, symbols, and

spellings, so they must be checked. Generally, however, federal standards of mapping accuracy are maintained.

The word *accuracy* can be applied only to base-map features: the location of features or the graticule. **Map accuracy** is the degree of conformance to an established standard. Because thematic maps are not devised relative to standards, their accuracy cannot be properly evaluated. Thus, when other thematic maps are used as sources for a new compilation, there is no accuracy standard by which to judge them. Only their base-map component can be so evaluated; even locational accuracy is often irrelevant, because some thematic maps do not require planimetric precision. Because of this, thematic map compilers need to familiarize themselves with all aspects of the mapped area so that the accuracy and reliability of a source map can be better judged.[27] The key to source evaluation is comparison and an accumulated sense of the reliability of the map-producing agency. There are few substitutes for experience and knowledge.

A source map should be checked to determine if it is a *basic* or a **derived** map. Basic maps result from original

UNITED STATES NATIONAL MAP ACCURACY STANDARDS

With a view to the utmost economy and expedition in producing maps which fulfill not only the broad needs for standard or principal maps, but also the reasonable particular needs of individual agencies, standards of accuracy for published maps are defined as follows:

1.   Horizontal accuracy. For maps on publication scales larger than 1:20,000, not more than 10 percent of the points tested shall be in error by more than 1/30 inch, measured on the publication scale; for maps on publication scales of 1:20,000 or smaller, 1/50 inch. These limits of accuracy shall apply in all cases to positions of well-defined points only. Well-defined points are those that are easily visible or recoverable on the ground, such as the following: monuments or markers, such as bench marks, property boundary monuments; intersections of roads, railroads, etc.; corners of large buildings or structures (or center points of small buildings); etc. In general, what is well-defined will also be determined by what is plottable on the scale of the map within 1/100 inch. Thus while the intersection of two road or property lines meeting at right angles, would come within a sensible interpretation, identification of the intersection of such lines meeting at an acute angle would obviously not be practicable within 1/100 inch. Similarly, features not identifiable upon the ground within close limits are not to be considered as test points within the limits quoted, even though their positions may be scaled closely upon the map. In this class would come timber lines, soil boundaries, etc.

2.   Vertical accuracy, as applied to contour maps on all publication scales, shall be such that not more than 10 percent of the elevations tested shall be in error more than one-half the contour interval. In checking elevations taken from the map, the apparent vertical error may be decreased by assuming a horizontal displacement within the permissible horizontal error for a map of that scale.

3.   The accuracy of any map may be tested by comparing the positions of points whose locations or elevations are shown upon it with corresponding positions as determined by surveys of a higher accuracy. Tests shall be made by the producing agency, which shall also determine which of its maps are to be tested, and the extent of such testing.

4.   Published maps meeting these accuracy requirements shall note this fact in their legends, as follows: "This map compiles with National Map Accuracy Standards."

5.   Published maps whose errors exceed those aforestated shall omit from their legends all mention of standard accuracy.

6.   When a published map is a considerable enlargement of a map drawing (manuscript) or of a published map, that fact shall be stated in the legend. For example, "This map is an enlargement of a 1:20,000-scale map drawing," or "This map is an enlargement of a 1:24,000-scale published map."

7.   To facilitate ready interchange and use of basic information for map construction among all Federal mapmaking agencies, manuscript maps and published maps, wherever economically feasible and consistent with the use to which the map is to be put, shall conform to latitude and longitude boundaries, being 15 minutes of latitude and longitude, or 7 1/2 minutes, or 3 3/4 minutes in size.

Source: Morris M. Thompson, *Maps for America*. U.S. Geological Survey (Washington, DC: USGPO, 1982), p. 104.

survey work; derived maps have been compiled from basic maps. The potential for error is greater on derived maps.

PREPARING A BASE MAP FOR SCANNING

Many base maps are developed by scanning them into computer software to make them useable for later digital thematic map compilation. Often this includes preparing a "draft" pencil map for scanning. Preparing such a base map involves selecting a projection proper for the final map, selecting and plotting relevant coastlines, rivers, lakes, and other physical features, and providing any man-made features called for on the final map. Sometimes the production of this initial base map calls for the cartographer to use one of several compilation methods: the reduction method, the common-scale method, and the one-to-one method.

Reduction methods involve "lifting" or tracing the selected objects from one source map onto the worksheet of a draft map. The worksheet is then reduced, by one of several means discussed later in this chapter, to the desired manuscript or artwork map size to be scanned into the computer. Locational accuracy is the chief advantage of the reduction method. The compiler makes selection and generalization decisions while tracing. False information should not be introduced, and only relevant information is traced.[28] The reduction method is *not* a simple tracing operation.

Common-scale methods involve compilation from various sources onto one worksheet. A political base outline may be traced from one source map and another feature (e.g., railroads) traced from another. The different source maps are likely to be at different scales—perhaps even on different projections—because they have been produced for different purposes. Through reduction methods, generalization, and selection, the compiler develops a single-scale draft map that is homogeneous in detail.[29] Extreme caution must be exercised at this stage because of the danger that variations in scale and level of generalization may be introduced into the new map. Especially careful treatment is needed in order to assure the locational accuracy of features compiled from maps having different projections.

In **one-to-one methods,** the cartographer simply traces, at the same scale, the features from the source map onto the new worksheet. Locational accuracy is very good. As with other compilation methods, generalization and selection are performed during the tracing process.

In some mapping programs used in desktop mapping compilation, decisions are made by selecting from a menu in the applications program. In others, the cartographer continues to do compilation and generalization manually, albeit at the computer screen. The cartographer will still be required to know the best level of generalization applicable to the map at hand so that, at a given scale, the amount of information provided is appropriate. Unfortunately, there are no rigid guidelines to assist the cartographer, other than experience and good judgment.

Plotting Projections

One of the most time-consuming of compilation activities is the plotting of projections. If the compiler has not been fortunate enough to obtain a graticule of the desired projection, one must be drawn. Computer plotting is the easiest and most flexible, if the facilities are available. (See Chapter 17 for a more complete discussion of this topic.) Drawing the graticule from projection tables is next best, and developing projections from scratch is the least desirable. With each of these three methods, the projection can be custom designed and developed at the desired draft map size. The graticule of the projection should include only the number of meridians and parallels suitable for the map purpose. Too dense a graticule detracts from the thematic map.

Changing Scales

Changing map scale is a compilation feature even for the production of draft maps for the scanner. Usually the purpose of changing the scale of the draft map is to accommodate the size of the scanner. Scale changing may be accomplished by these methods: *photography, electrostatic (photocopy), or computer methods (scanning and digitizing).* Extreme caution is recommended with this latter method (as with the others) because there is a tendency to enlarge images by computers, which usually introduces unwanted generalization: careful reduction is acceptable, however. The most convenient method today is by electrostatic copier. It is very convenient simply to select a precise percentage reduction and get results quickly. Copies are not produced precisely at the size specified, however, so one must be careful. The copies may be off by about 1 percent.

If possible, it is best to use only the center of the image area, where accuracy is greatest.

The rule is to go from larger to smaller scale for any of these methods of scale changing. In this way, the compiler

can select for simplicity, whereas in the opposite direction, information must be added, which may result in error.

COPYRIGHT

Map copyright in the United States, and all copyright, " . . . is a form of protection provided by the laws of the United States (Title 17, U.S. Code) to the authors of 'original works of authorship.' . . ."[30] The category of cartography includes maps, globes, and relief models. Copyright infringement is any violation of the rights of the copyright holder. The rights most applicable to the map case are as follows:

1. To reproduce the copyrighted work in copies.
2. To prepare derivative works based on the original copyrighted work.
3. To distribute copies of the copyrighted work to the public by sale or other transfer of ownership or by rental, lease, or lending.

The rules governing the copyright of maps should be followed as carefully as possible. In general, a map must display an "appreciable" amount of original cartographic material to be copyrightable. A reprint of a map that is already in the public domain cannot be copyrighted. Generally, outline or base maps are not copyrightable. For example,

A revised version of a previously published map may be copyrighted as a "new work" if the additions and changes in the new version are copyrightable in themselves. The preparation of most present-day maps involves the use of previously published source material to a significant degree, and the copyrightable authorship therefore is generally based upon elements such as compilation and drawing rather than upon original surveying, aerial photography, and field work alone. Where any substantial portion of the work submitted for registration is based upon previously published sources, a statement of the nature of the new, copyrightable authorship should be given at the appropriate space on the application submitted to the Copyright Office.[31]

Compilers of thematic maps must exercise caution when using the source materials of others. Care and good judgment are required so that others' rights are not violated. In the United States, maps produced by the federal government are in the public domain and are therefore free of potential copyright problems. The use of privately produced maps, on the other hand, may require permission from the copyright holder.

Many commercial mapmakers inbed *copyright traps* or *hooks* in obscure places on their maps to protect them from unscrupulous copying. These may be misspelled street names, streets that do not exist, or other incorrect features. These provide proof that a map has been copied, as the illegal copier is probably not aware of the trap.

The complexities regarding copyright are compounded today because in many instances the source materials are in electronic media form. If you are using map and data files that are owned by another (company, corporation, or other) in the production of a thematic map, *it is imperative that you read all licensing agreements regarding the duplication and distribution of its materials.* Failure to do so may result in copyright infringement.

In some instances, the owner of the source materials may require you to acknowledge ownership on any hard copy rendition of their files. In some instances, duplication within a company is permitted, and in other cases not permitted. Some map and database owners allow you to make up to eight copies for distribution, if the purpose is noncommercial, and if you belong to a government agency or accredited academic institution. For example, Strategic Mapping, Inc., provides this statement regarding duplication:

> *If you have received the 5-Digit Zip Code Files ("ZIP Code File"), you shall not use the ZIP Code File to publish products in the form of atlases, maps or map series, tabular compendiums of information, or routing services. However, delivery of eight (8) or fewer printed copies as a customized service will not constitute such publication.*[32]

Caution is the key word when using source materials in the production and distribution of a new thematic map when there are any concerns regarding copyright.

NOTES

1. E. Meynen, ed., *Multilingual Dictionary of Technical Terms in Cartography* (International Cartographic Association, Commission II—Wiesbaden: Franz Steiner Verlag, 1973), p. 137.
2. H. V. Steward, *Cartographic Generalization* (Cartographica, Monograph No. 10—Toronto: University of Toronto Press, 1974), p. 20.
3. David Greenhood, *Mapping* (Chicago: University of Chicago Press, 1964), p. 176.
4. F. McBryde and Paul D. Thomas, *Equal Area Projections for World Statistical Maps* (U.S. Department of Commerce, Coast, and Geodetic Survey, Special Publication No. 245—Washington, DC: USGPO, 1949), p. 7.
5. Charles H. Deetz and Oscar S. Adams, *Elements of Map Projections* (U.S. Department of Commerce, Coast, and Geodetic Survey, Special Publication No. 68—Washington, DC: USGPO, 1944), pp. 163–64.
6. J. A. Steers, *An Introduction to the Study of Map Projections* (London: University of London Press, 1962), pp. 161–63.
7. Ibid., p. 182.
8. Erwin Raisz, *Principles of Cartography* (New York: McGraw-Hill, 1962), p. 177.
9. Deetz and Adams, *Elements of Map Projections,* p. 67.
10. Porter W. McDonnell, Jr., *An Introduction to Map Projections* (New York: Marcel Dekker, 1979), p. 61.
11. Deetz and Adams, *Elements of Map Projections,* p. 75.
12. Ibid., pp. 94–96.
13. Mei-Ling Hsu, "The Role of Projections in Modern Map Design," *Cartographica* 18 (1981): 151–86. This informative contribution to the literature of thematic mapping is highly recommended.
14. John P. Snyder, "Labeling Projections on Published Maps," *American Cartographer* 14 (1987): 21–27.
15. Arthur H. Robinson, "Arno Peters and His New Cartography," Views and Opinions, *American Cartographer* 12 (1985): 103–11; John P. Snyder, "Social Consciousness and World Maps," *The Christian Century,* February 24, 1988, pp. 190–92.
16. John Loxton, "The Peters Phenomenon," *The Cartographic Journal* 22 (1985): 106–8.
17. Marshall Faintick, "The Politics of Geometry," *Computer Graphics World,* May 1986, pp. 101–4.
18. Loxton, "The Peters Phenomenon," pp. 106–8; A multicolored, 35- × 52-inch version may be obtained from the Friendship Press in Cincinnati, OH.
19. Snyder, "Social Consciousness," p. 192.
20. Stephen Kinzy, "GIS Professionals: Keepers of the Flame," *Geo Info Systems,* February 1991, pp. 10–12. No single definition has emerged that defines adequately the entire field of GIS. The student is encouraged to look at Jeffrey Star and John Estes, *Geographic Information Systems* (Englewood Cliffs, NJ: Prentice Hall, 1990), pp. 2–3, for an additional definition.
21. National Oceanic and Atmospheric Administration, *State Plane Coordinate System of 1983,* Manual NOS NGS 5 (Rockville, MD: National Geodetic Information Center, 1989) p. 2.
22. O. M. Miller and Robert J. Voskuil, "Thematic Map Generalization," *Geographical Review* 54 (1964): 13–19.
23. A. J. Pannekoek, "Generalization of Coastlines and Contours," *International Yearbook of Cartography* 2 (1962): 55–73.
24. Gösta Lundquist, "Generalization—A Preliminary Survey of an Important Subject," *Canadian Survey,* 1959, pp. 466–70.
25. Steward, *Cartographic Generalization,* p. 30.
26. Alan G. Hodgkiss, *Maps for Books and Theses* (New York: Pica Press, 1970), pp. 29–32.
27. Barbara A. Bond, "Cartographic Source Material and Its Evaluation," *Cartographic Journal* 10 (1973): 54–58.
28. David Cuff and Mark T. Mattson, *Thematic Maps, Their Design and Production* (New York: Methuen, 1982), p. 88.
29. J. S. Keates, *Cartographic Design and Production* (New York: Wiley, 1973), p. 170.
30. James W. Cerney, "Awareness of Maps as Objects of Copyright," *American Cartographer* 5 (1978): 45–56.

31. Copyright Office, *Copyright Basics,* Circular 1 (Washington, DC: USGPO, 1993), p. 2; and Copyright Office, *Copyright Registration for Works of the Visual Arts,* Circular 40 (Washington, DC: USGPO, 1993), p. 2.

32. Strategic Mapping, Inc., Licensing Agreement, Atlas GIS, 1994.

GLOSSARY

Albers equal-area projection   secant conical projection having equal-area properties; useful for mapping continent-size areas on the earth; used most frequently for medium-scale maps of the United States, p. 54

Boggs eumorphic projection   equal-area projection useful for mapping the world on one sheet; adequate for small-scale thematic maps containing a world theme, pp. 51–52

Bonne projection   simple conical equal-area projection, useful for mapping continent-size areas of the earth; should not be used for areas of considerable east-west extent, p. 52

cartographic generalization   a central function in cartographic compilation; involves simplification, selection, and emphasis, pp. 58–59

common-scale method   compilation of a new manuscript map from more than one source map, often at different scales, each containing unique information useful to the new map, pp. 62–63

derived maps   maps, usually at intermediate to small scale, that have been compiled from topographic scale maps, pp. 61–62

Earth Science Information Center   a centralized source for obtaining information about national cartographic products and ordering them, p. 60

Gall-Peters projection   based on the earlier work of James Gall in 1855 and adopted by several UN organizations, pp. 55–56

Hammer projection   equal-area projection useful for mapping the world on one sheet; adequate for small-scale thematic maps containing a world theme, pp. 49–51

Lambert equal-area azimuthal projection   has equal-area properties useful for mapping continent-size areas on the earth, pp. 53–54

Lambert conformal conic projection   has conformal properties; used as the standard reference projection for the plane grid system for many states, p. 57

map accuracy   degree of conformance to established standards; has particular significance with reference to maps at topographic scales but loses relevance when applied to thematic maps, p. 61

map compilation   includes all aspects of collection, evaluation, interpretation, and technical assembly of cartographic material, p. 48

map copyright   legal protection against the infringement of the rights of the copyright holder of a map, pp. 63–64

Mollweide (or homolographic) projection   equal-area projection useful for mapping the world on one sheet; adequate for small-scale thematic maps containing a world theme, p. 49

one-to-one method   compilation by simple tracing of information from a source map at the required scale onto the new manuscript map, p. 63

reduction method   compilation by tracing of information from one source map onto the new manuscript map being compiled; the new map is then reduced to final size, p. 62

state plane coordinate system   rectangular plane coordinate system applied to states; used especially by highway engineers, utility companies, and planners; grids labelled in feet and meters, p. 57

thematic map compilation   construction of a special-purpose map from a variety of previously existing sources, including maps and nonmap materials, pp. 48–49

transverse Mercator projection   has conformal properties; used as the standard reference projection for the plane grid system for many states, p. 57

READINGS FOR FURTHER UNDERSTANDING

Birch, T. W. *Maps, Topographical and Statistical.* London: Oxford University Press, 1964.

Bond, Barbara A. "Cartographic Source Material and Its Evaluation." *Cartographic Journal* 10 (1973): 54–58.

Cerny, James W. "Awareness of Maps as Objects of Copyright." *American Cartographer* 5 (1978): 45–56.

Committee on Map Projections. *Which Map Is Best?* Falls Church, VA: American Congress on Surveying and Mapping, 1986.

———. *Choosing a World Map.* Falls Church, VA: American Congress on Surveying and Mapping, 1988.

———. *Matching the Map Projection to the Need.* Falls Church, VA: American Congress on Surveying and Mapping, 1991.

Copyright Office. *Copyright for Maps.* Circular 40F. Washington, DC: USGPO, 1974.

Cuff, David, and Mark T. Mattson. *Thematic Maps, Their Design and Production.* New York: Methuen, 1982.

Deetz, Charles H., and Oscar S. Adams. *Elements of Map Projections.* U.S. Department of Commerce, Coast, and Geodetic Survey, Special Publication No. 68. Washington, DC: USGPO, 1944.

Dent, Borden D. "Continental Shapes on World Projections: The Design of a Poly-Centered Oblique Orthographic World Projection." *Cartographic Journal* 24 (1987): 21–27.

Dickinson, G. C. *Maps and Air Photographs.* New York: Wiley, 1979.

Dracup, Joseph F. "Plane Coordinate Systems," *ACSM Bulletin* 59 (1977): 27.

Espenshade, Edward B., ed. *Goode's World Atlas.* 16th ed. Chicago: Rand McNally, 1983.

Faintich, Marshall. "The Politics of Geometry." *Computer Graphics World,* May 1986, pp. 101–4.

Gilmantin, Patricia P. "Aesthetic Preferences for the Proportions and Forms of Graticules." *The Cartographic Journal* 20 (1983): 95–100.

Greenhood, David. *Mapping.* Chicago: University of Chicago Press, 1964.

Hazelwood, L. K. "Semantic Capabilities of Thematic Maps." *Cartography* 7 (1970): 69–75, 87.

Hodgkiss, Alan G. *Maps for Books and Theses.* New York: Pica Press, 1970.

Hsu, Mei-Ling. "The Role of Projections in Modern Map Design." *Cartographica* 18 (1981): 151–86.

Keates, J. S. *Cartographic Design and Production.* New York: Wiley, 1973.

Kelley, Philip S. *Information and Generalization in Cartographic Communication.* Unpublished Ph.D. dissertation. Seattle: University of Washington, Department of Geography, 1977.

Lawrence, G. R. P. *Cartographic Methods.* London: Methuen, 1971.

Loxton, John. "The Peters Phenomenon." *The Cartographic Journal* 22 (1985): 106–8.

Lundquist, Gösta. "Generalization—A Preliminary Survey of an Important Subject." *Canadian Survey,* 1959, pp. 466–70.

Mainwaring, James. *An Introduction to the Study of Map Projection.* London: Macmillan, 1942.

McBryde, F., and Paul O. Thomas. *Equal Area Projections for World Statistical Maps.* U.S. Department of Commerce, Coast, and Geodetic Survey, Special Publication No. 245. Washington, DC: USGPO, 1949.

McDonnell, Porter W., Jr. *Introduction to Map Projections.* New York: Marcel Dekker, 1979.

Meynen, E., ed. *Multilingual Dictionary of Technical Terms in Cartography.* International Cartographic Association, Commission II. Wiesbaden: Franz Steiner Verlag, 1973.

Miller, O. M., and Robert J. Voskuill. "Thematic Map Generalization." *Geographical Review* 54 (1964): 13–19.

National Oceanic and Atmospheric Administration. *State Plane Coordinate System of 1983.* Manual NOS NGS 5. Rockville, MD: National Geodetic Information Center, 1989.

Pannekoek, A. J. "Generalization of Coastlines and Contours." *International Yearbook of Cartography* 2 (1962): 55–73.

Pearson, Frederick, II. *Map Projection Methods.* Blacksburg, VA: Sigma Scientific, 1984.

———. *Map Projection Software.* Blacksburg, VA: Sigma Scientific, 1984.

Raisz, Erwin. *Principles of Cartography.* New York: McGraw-Hill, 1962.

Robinson, Arthur H. "Arno Peters and His New Cartography." Views and Opinions. *American Cartographer* 12 (1985): 103–11.

———, **Randall Sale, and Joel Morrison.** *Elements of Cartography.* 4th ed. New York: Wiley, 1978.

Snyder, John P. *Map Projections Used by the U.S. Geological Survey.* 2d ed. U.S. Geological Survey, Geological Survey Bulletin 1532. Washington, DC: USGPO, 1983.

———. "Labeling Projections on Published Maps." *American Cartographer* 14 (1987): 21–27.

———. *Map Projections—A Working Manual.* U.S. Geological Survey, Professional Paper 1395. Washington, DC: USGPO, 1987.

———. *Map Projections Used for Large-Scale Quadrangles by the U.S. Geological Survey.* U.S. Geological Survey, Circular 982. Denver: U.S. Geological Survey, 1987.

———. "Social Consciousness and World Maps." *The Christian Century,* February 24, 1988, pp. 190–92.

Steers, J. A. *An Introduction to the Study of Map Projections.* London: University of London Press, 1962.

Steward, H. J. *Cartographic Generalization.* Cartographica, Monograph No. 10. Toronto: University of Toronto Press, 1974.

Thompson, Morris M. *Maps for America.* U.S. Geological Survey. Washington, DC: USGPO, 1982.

Thrower, Norman J. W. *Maps and Man.* Englewood Cliffs, NJ: Prentice Hall, 1972.

CHAPTER

THE NATURE OF GEOGRAPHIC PHENOMENA AND THE SELECTION OF THEMATIC MAP SYMBOLS

CHAPTER PREVIEW

Cartographers are often faced with decisions about the presentation and symbolization of geographic phenomena and data. They must therefore learn the fundamental concepts of geography: direction, distance, scale, location, distribution, localization, functional association, spatial interaction, and the concept of region. Geographic phenomena are measured on four levels: nominal, ordinal, interval, and ratio. It is entirely possible to symbolize geographic data at each of these levels of measurement. Cartographic symbol schemes are examined by looking at map types (quantitative or qualitative), symbol dimensions, symbols at the different measurement levels, and the functions of cartographic symbols. The designer must know the concepts of geography so that appropriate symbols can be chosen to best function with their referents. Knowledge of data sources, especially enumeration data provided by the United States Census Bureau and Statistics Canada, is very helpful for the thematic cartographer.

Cartography involves the graphic presentation of some aspect of the real world. A mapmaker usually works in one of two ways: as content author of the map's presentation of reality or on commission to render someone else's version of reality in graphic form. In either case, the cartographer deals with a wide variety of geographic phenomena. Understanding and interpretation of geographical analyses is often necessary to the selection of the appropriate mapping solution.

GEOGRAPHY AND GEOGRAPHIC PHENOMENA

Designers are increasingly faced with perplexing graphic design problems as geographers become more sophisticated in their analytic methods. Indeed, the types of presentation problems brought about by complex modeling and quantitative approaches used today are forcing cartographers to be more innovative than ever before. Familiarity with basic geographical concepts, trends, and data forms is important for the student designer.

GEOGRAPHY DEFINED

Our present concern is with modern geography only, not the many changes and philosophical viewpoints in the history of the discipline. Even for present and future geographers, the following integrating materials are useful in illuminating geography's relationship to modern thematic cartography.

In the words of one prominent scholar on the philosophy of geography, Professor Richard Hartshorne, "The intrinsic characteristics of geography are the product of man's effort to know and understand the combinations of phenomena as they exist in areal interrelation in his world."[1] Hartshorne also pointed out that geography does not contain its own particular set of phenomena, but studies the integration of heterogeneous phenomena over area, and that geography is both a social and a physical science. The discipline is so diverse that geographers are equally at home studying the distribution of climates throughout the world as they are analyzing transport between large urban centers. A wide range of research interests is one of the hallmarks of geography.

Geography is the study of areal interrelation; it can be called the *science of spatial analysis.* The distinctive thrust of the discipline is its attempt to answer the question, "Why are spatial distributions structured the way they are?"[2] Geographers attempt to reveal the underlying processes that are causally related to spatial patterns or structures. Its central focus on the spatial dimension sets the field of geography apart from other sciences.

In asking questions about the spatial world, geographers look at a wide variety of phenomena: drugstore locations in urban commercial districts, the distribution of rat bites in slum areas, patterns of land ownership in Third World countries, and worldwide language regions. Any phenomenon that has or can be thought to have a spatial attribute is subject to geographical inquiry.

Certain limits, however, appear to restrict the field of geography. Geographers do not normally pay attention to spaces smaller than architectural or larger than terrestrial. As a matter of fact, room geography (the architectural scale) has become the subject of some geographic researchers. The earth and its varying phenomena, at the other extreme, are often studied in geographic research. Worldwide distribu-

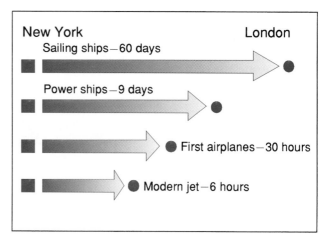

Figure 4.1 The Shrinking Earth.
As travel times decrease, relative distances diminish. What will it be like in 2025? What effect do shrinking distances have on the way we perceive different people?

tional patterns (e.g., climates, soils, or religions) have been of interest to geographers for many years. Within the range between the architectural and terrestrial scales, geographical inquiry has been free to operate.

What? Where? Why? These are the typical questions geographers ask. Answers to "what" questions have traditionally been descriptive. For much of the history of geography, such questions have attracted the most attention, as is natural for an emerging science. For most of our history, we did not know "what" because most of the world was unknown. "Where" questions have also concerned geographers for a long time, and the answers have been largely descriptive until recent years. **Absolute location** with reference to a fixed coordinate system, such as the earth's geographic reference system, is characteristic of the answers to "where" questions. Answering questions in terms of absolute location allowed geographers and cartographers to "fill in" the world map.

As geographical inquiry has become more sophisticated, "where" questions have taken on more dynamic qualities as **relative location** has gained prominence.[3] The traditional view of location is based on a rigid grid system—an arbitrary Cartesian one or the geographic grid. Regardless of the reference grid adopted, location was viewed as absolute and unchanging. In the new view, location is a relative matter, dynamic and ever-changing; the distance between any two places may be stretching, or more likely shrinking. For example, London is closer to New York now than it was 100 years ago because travel time has decreased. (See Figure 4.1.) Remote places on the earth have become more accessible because transportation technology has led to more rapid travel and telecommunications has led to a more rapid exchange of ideas.

Geographical exploration of these new spatial dimensions has resulted in different ways of mapping. In the con-

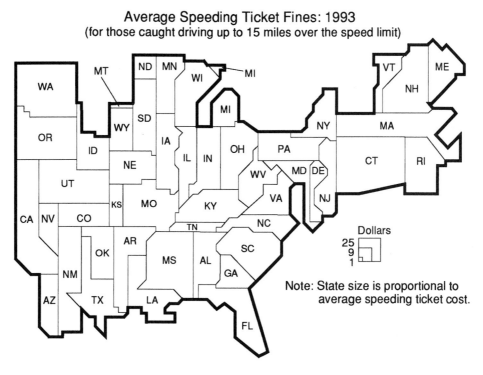

Average Speeding Ticket Fines: 1993
(for those caught driving up to 15 miles over the speed limit)

Source: Money Magazine, June 1994

Figure 4.2   Value-by-area cartogram of speeding ticket fines in the United States, 1993.
While the majority of the southern, middle, and western states appear to yield a rather uniform distribution, the sizes of the New England states show that they are disproportionately large. *(Map compiled by John Freeman, Georgia State University. Used by permission.)*

T here are many, many maps of each fragment of the earth's surface and they change as the information changes. The last definitive map of any area can never be drawn.

Source: Peter Haggett, *The Geographer's Art* (Oxford: Basil Blackwell, 1990), p. 52.

ventional geography of absolute location, the mapping activity is based on Euclidean geometry. True and predictable distances and angles are the main features of Euclidean space. In order to map spaces that reflect relative perspectives (e.g., social, psychological, cost, or time), new forms of maps have emerged. (Review Figure 1.7 of Chapter 1.) For example, special transformations (cartograms) show content spaces and illustrate relative perspectives better than traditional maps and are often useful in showing the underlying spatial structure of dynamic distributions. (See Figure 4.2.) Nonconventional mapping often employs logarithmic projections (Figure 4.3), which tend to enlarge places closer to the center and diminish peripheral spaces. These projections are useful in showing geographical distributions or processes whose magnitudes decrease outward

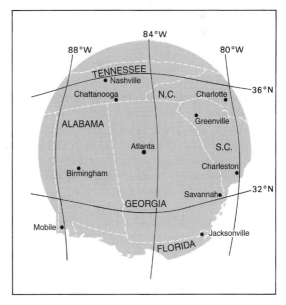

Figure 4.3   Logarithmic projection is suitable for showing activities in relative space.
Here we see that the nature of the projection enlarges places closer to the projection center and reduces places farther away.

with distance. For example, migration patterns often show numerous short moves closer to home and fewer long-distance moves.

MAJOR THEMES IN GEOGRAPHIC RESEARCH

Certain major themes have dominated geographical inquiry over the past several decades.[4]

Areal or Spatial Association

The geographer studies the spatial location of a variable with respect to another or others functionally or causally related to it. An example at the microscale would be the locational linkage of florist shops to urban high-rise office buildings.

Forms and Processes

In this approach, the researcher examines the processes that result in random, clustered, or regular spatial distributions. For example, some studies show the location, extent, and spacing of central places (towns); others attempt to describe the tectonic processes that yield landform types and their distributions.

Spatial Interaction

Investigations of this type seek explanations for the flow (interaction) of people or goods from place to place. Forces of attraction and friction over distance are integrating ideas. An example of this theme would be the development of road networks in response to regional traffic patterns.

Distance Decay

Distance-decay investigations examine the decreasing occurrence of certain events from a central point. Nodal area development (the influence of a city in its region) or "action spaces" (places where daily activities take place) can be determined by distance-decay studies.

A geographer may conduct research in many of these theme areas, and sometimes a particular study may overlap several areas. Geography is an eclectic field of study—one reason for its richness and attraction.

KEY CONCEPTS IN GEOGRAPHY

Regardless of the approach taken or the nature of the content studied, certain key concepts bind geography together. The cartographer deals with geographical items and must therefore be thoroughly familiar with the concepts of geography.

Direction

Direction is a geometric property that can be used to describe relative location. True direction (azimuth) is measured by reference to a meridian. Direction can also be used to describe the location of a point in other spaces (such as

---

Geography is a very old field of study. Writers of geography were among the earliest scholars of antiquity, and spoken geography must have been widely practiced long before the invention of writing. During the long history of man's effort to gain further understanding of the forces and objects of his environment, the essential nature of geography remained unchanged. Today, as in the past, geography is concerned with the arrangement of things on the face of the earth, and with the association of things that give character to particular places. Those who face problems involving the factor of location, or involving the examination of conditions peculiar to specific locations, are concerned with geography, just as those who must be informed about a sequence of events in the past are concerned with history.

Source: Preston E. James, "Introduction: The Field of Geography," in *American Geography, Inventory and Prospect,* eds. Preston E. James and Clarence F. Jones (Washington DC: Association of American Geographers, 1954), p. 4.

---

on cartograms), which are more difficult because a different system of reference must be established.

Distance

The property of **distance** has long been instrumental in geographical inquiry. Distance is conventionally expressed in *units of length,* based on Euclidean geometry, but it can be expressed in terms of time, cost, stress, or any number of other units. As geographical inquiry becomes more and more behaviorally oriented, a greater variety of units is likely to be used to measure distance.

Geographic Scale

Scale relates to the size of the area being studied and determines the level of precision and generalization applied in the study. Microscale studies are done over small earth areas; macroscale studies deal with larger areas. The nature of the inquiry sets the scale, and the scale in turn determines the degree of generalization. The study of the distribution of manufacturing employment in United States cities (macroscale) usually dictates that cities be perceived as points; therefore, the intracity variation in employment areas becomes unimportant. Studying this same phenomenon at the intracity level (microscale), however, leads the researcher to focus on the distribution of employment areas within the city. Also, human interaction varies with scale, as shown in Table 4.1. Voice, touch, taste, and smell can link people at the personal-space scale but are impossible at greater distances. At the global-space scale, whole new networks are formed to facilitate interaction. This is especially true today with the ever increasing World Wide Web telecommuncations/computer networks. Today people correspond worldwide quickly and easily making the world seem smaller still.

Table 4.1   Human Interaction at Varying Scales

Type of Space	Spatial Range; Radius from Person (Representative Fraction)	Some Interaction Characteristics	
		Primary Modes of Interaction	Number of People Involved
Personal	Arm's length (1:1)	Voice, touch, taste, smell	1–3
Living or working (private space)	10–50 ft (1:50)	Audio and visual (sharp focus on facial expression, slight movement, or tonal changes)	50–400
House and neighborhood	100–1000 ft: line of sight (1:500)	Audiovisual (sound amplification)	100–1000
Limit of Proxemic Interaction			
City-hinterland	4–40 mi: 60-minute one-way commute (1:50,000)	Local news media, TV, urban institutions, commuter systems	50,000–10,000,000
Regional-national	200–3000 mi (1:500,000)	National network of news media, national organizations (common language desirable)	200,000,000+
Global	12,000 mi (1:50,000,000)	International communication networks, translation services, world organizations, international blocs	3,000,000,000+

Source: Modified from John F. Kolars and John D. Nystuen, Geography: The Study of Location, Culture, and Environment *(New York: McGraw-Hill, 1974), p. 29.*

Location

A geographical feature can be evaluated in terms of either absolute or relative location. Absolute location is restricted to a reference or fixed-grid system. An object's location relative to another (or to all others) involves dynamic aspects of the elements and reflects cultural and technological change.

Distribution

Distribution is the spatial property of areal *pattern.* The chief pattern types are *dispersed* and *agglomerated.* A densely settled area is an example of agglomerated pattern, as is clustering of manufacturers in a given area.

Localization

Localization applies to spatial clustering; it measures the concentration of an activity in a limited area. The concentration of petrochemical plants along the Gulf Coast is an example of localization of industry.

Functional Association

The concept of **functional association** involves the *occurrence* of objects or activities in space as a response to the presence of other elements. One feature is located as it is because another is close by. Medical offices tend to locate near

hospitals. *Connectivity* is a special case of functional association; it implies actual ties.[5] Functional association can also be called *areal coherence.* Much of geographical explanation results from the discovery of functional associations.

Spatial Interaction

Chief among the geographical concepts is **spatial interaction.** It involves the movement of people, ideas, or goods from place to place over the surface of the earth. A special case of interaction is *spatial diffusion*—the spread of people (migration), things (e.g., agricultural tools), or ideas.

The Regional Concept

A **region** is defined by the researcher to study the likenesses and differences of areas on the earth. Regions may be cultural or physical. A region has an *internal homogeneity* that sets it apart from others. Defining regions often provides insight into geographical distributions and the explanations for them.

Concept of Change

Last but not least is the **concept of change.**[6] There is really no such thing as equilibrium in geography—only temporary phases in the process of transformation. If change is

Table 4.2   The Six Essential Elements and Eighteen Geography Standards of the National Geography Standards Project

I. The World In Spatial Terms
1. Maps and other geographic tools for information in a spatial perspective.
2. Mental maps and spatial context.
3. Spatial organization of Earth.

II. Places and Regions
4. Physical and human characteristics of places.
5. Regions interpret Earth's complexity.
6. Culture and experience influence perception of places and regions.

III. Physical Systems
7. Physical processes shape patterns of Earth's surface.
8. Characteristics and distribution of Earth's ecosystems.

IV. Human Systems
9. Human populations.
10. The nature and complexity of Earth's cultural mosaics.
11. Patterns and networks of economic interdependence.
12. Human settlement.
13. Forces of cooperation and conflict that shape Earth's surface.

V. Environment and Society
14. Human actions modify physical environment.
15. Physical systems affect human systems.
16. Meaning, distribution, and importance of resources.

VI. The Uses of Geography
17. How to apply geography to interpret the past.
18. How to apply geography to interpret the present and plan for the future.

Source:Geography Education Standards Project, Key to the National Standards *(Washington, DC: National Geographic Society, 1995), p. 2.*

By the year 2000, planet Earth will be more crowded, the physical environment more threatened, natural resources more depleted, the global economy more competitive, and world events more interconnected. Dealing with these challenges requires an understanding of geography.

Geography is the science of space and place on Earth's surface. Its subject matter is the physical and human phenomena that make up the world's environments and places. Geography asks us to look at the world as a whole, to understand the connections between places, to recognize that the local affects the global and vice versa.

The power and beauty of geography lie in seeing, understanding, and appreciating the web of relationships among people, places, and environments.

Source: Geography Education Standards Project, *Geography for Life.* (Washington, DC: National Geographic Research and Exploration, 1994), p. 3.

---

accepted as a general principle, then rate, regularity, and direction of change become important. For example, one of the most dynamic elements of a city is its land-use map. It should never be viewed as a static picture of conditions, but rather an inventory of conditions for a small slice of time. Tomorrow it will be different. To what degree the land-use map changes, how fast, and with what regularity and direction, are important in explaining the economic forces in urban land-use theory.

GEOGRAPHY FOR LIFE

In response to the federal act *1994 Goals: Educate America Act,* section 102, stating that students in three grade levels, including the twelfth, should have basic competency in ge-

ography (among other core areas), a project funded by three federal agencies, and supervised by the National Council of Geographic Education, produced a book entitled *Geography for Life.*[7] The committee of diverse geographers identified *standards and skills,* working from previously published **five themes of geography.**[8] These themes, and the subsequent standards and skills, have been very largely adopted by many public schools across the country, often working in concert with Geographic Alliances, partially funded by the National Geographic Society. By the mid-1990s these themes have become part of the rallying cry for national geographic education.

The five themes are **location, place, human/environment interaction, movement, and regions.** To a very large degree, these themes bring together most of the ideas expressed above about the definition of geography and its major research themes. Within these thematic areas come the **six essential elements** and from these, eighteen **geography standards.** (See Table 4.2.)

It is very interesting to note that *maps, mental maps, spatial context,* and *spatial organization* form the first three of the eighteen geography standards.

The **location** theme refers to both absolute and relative location, and is defined in terms that were discussed above. The **place** theme deals with the *distinguishability* of a place by looking at its physical and human characteristics. **Human/environment interaction** deals with the study of how people relate to their environments *within* places. **Movement** deals with the study of the mobility of people, goods, and ideas. For example, what is the structure of getting goods and products to markets? The study of **regions,** as noted earlier, is a bulwark of geography. Regions, of course, may be defined in terms of their physical or cultural characteristics.

MEASUREMENT IN GEOGRAPHY

Geographers do not study a unique set of things that they can call their own. Geography is eclectic; the geographer uses techniques of observation in seeking to explain images of reality.[9] Geographers can study any phenomenon that has variability of spatial dimension, be it from the cultural world (people oriented) or the physical (nature oriented). Geography can be applied to such diverse topics as human poverty and drifting sand on a beach.

A distinction must be made between **geographic phenomena** and **geographic data.** Data are facts from which conclusions can be drawn. *Information* is another word that can be used in place of *data.* Geographic data are selected features (usually numerical) that geographers use to describe or measure, directly or indirectly, phenomena that have a spatial quality. For example, the phenomenon of climate can be described in part by looking at precipitation data. From the cultural world, the phenomenon of spatial interaction can be described and measured by collecting and analyzing data on telephone calls. In a recent study ranking cities, the phenomenon of stress brought about by psychosocial pathology was recorded with rates of alcoholism, suicide, divorce, and crime *data.*[10] Cartographers often map data when they think they are mapping phenomena.

SPATIAL DIMENSIONS OF GEOGRAPHIC PHENOMENA

Various **forms of geographic phenomena** can be considered to exist:

1. *Point* (zero-dimensional)
2. *Line* (one-dimensional)
3. *Area* (two-dimensional)
4. *Volume* (three-dimensional)
5. *Space-time* (four-dimensional)[11]

Most geographical study involves explanation of phenomena of the first four kinds, and to some degree the fifth.

An interesting feature of phenomena is that their form is intricately related to scale and may change with the level of inquiry. A city, for example, may be a point phenomenon at the macroscale but can be considered to have two-dimensional qualities when examined at micro levels. At one level of investigation, a road may be a link between points (one-dimensional), but at micro levels the road can have the two-dimensional, areal qualities of length and width.

Examples of volume phenomena include landforms, oceans, and the atmosphere. Other geographic phenomena that are conventionally treated as three-dimensional because of their similarity to volumes are rainfall, temperature, growing-season days, and such derived ratios as population density and disease-related deaths. Rainfall collects in a glass and is volumetric; people piled close together make up a volume. Space-time phenomena are best exem-

Figure 4.4   Continuous and discontinuous scales. Simple diagrams often help in understanding the difference between these scales. The upper series suggests a continuous scale and the lower a discontinuous scale.

plified by succession (e.g., human settlement over time) and migration/diffusion.

Thematic cartographers are faced with a dilemma: they do not have a unique way of symbolizing each different class or form of geographic phenomena.

Discrete, Sequential, and Continuous Phenomena

Besides the above dimensional classification of geographic phenomena, they can be classed as *discrete* (or *discontinuous*), *sequential,* or *continuous.* **Discrete geographic phenomena** are those that do not occur between spatial observations, that form a separate entity, and that are individually distinguishable from place to place. (See Figure 4.4.) Cows on the landscape are discrete phenomena because there are no other cows between those observed. The number of crimes in Baltimore, for example, is another; there are only whole crimes, not fractional crimes. Family size is yet another, as people are not fractional. Discrete phenomena can be either dispersed or concentrated; this is largely determined by scale. The important point is that the *counting unit* is the distinguishing characteristic of discrete phenomena.[12] Discrete scales used by geographers involve whole numbers (numbers of people, rivers, cities, cases of disease, and so forth), and in general are the results of *counting.* The most important distinguishing characteristic of discrete variables is that they *do not allow for fractional parts.*

In special cases, lines, or linear phenomena, can be referred to as *sequential.* A series of points or discrete elements makes up a sequential phenomenon. Roads, telephone lines, and boundaries are examples; they can be redefined as points appearing close together.

Continuous geographic phenomena, either two- or three-dimensional, exist spatially between observations. For example, landform elevation is a continuous phenomenon, because everywhere between observations there are other elevations. Temperature is another example: Between spatially arranged weather stations where observations are taken, temperature exists. Weight, time, stream velocity, industrialization, income, and rate of crime are other examples. With continuous phenomena any number of possible values may occur between observations, and division of whole numbers is likely (45.22 m for example). Values used in continuous scales are the result, usually, of *measurement,* and may contain decimals. The distinguishing characteristic of continuous variables is that *fractional parts are allowed.*

The cartographic process of symbolization has traditionally mapped a variety of geographic phenomena by symbol schemes that are based on the concept of discrete and continuous forms. For example, although population is discrete (people do not exist between people), population density is usually considered to be continuous. The use of an areal unit of land whose ratio is continuous (people/ sq km) to express density assumes continuity from place to place. To confuse the issue even more, population density is often mapped using data aggregated by discrete real units (e.g., states or counties). Although these areal units are discrete, the data are assumed to be uniform or continuous within units and change only between units.

MEASUREMENT SCALES

S. S. Stevens, a noted scientist and psychologist, has said that measurement is the "assignment of numerals to things so as to represent facts and conventions about them."[13] It has been customary in recent years for geographers to classify the ways they measure events into **measurement scales.** Measurement is an attempt to structure observations about reality. Ways of doing this can be grouped into four levels, depending on the mathematical attributes of the observed facts. A given measurement system can be assigned to one of these four levels: nominal, ordinal, interval, and ratio, listed in increasing order of sophistication of measurement. Methods of cartographic symbolization are chosen specially for representing geographic phenomena or data at these levels of measurement.

Nominal Scaling

Nominal scaling is the simplest level of measurement scale,[14] sometimes considered to be no more than a zero or 1 classification. In this system, objects are classed into groups that are often labeled with letters or numbers. At this level of measurement, we cannot perform any mathematical operations between classes. We can only ascertain equality or inequality between groups (region A = region B; region A ≠ region C). *Identification* is the key word in the use of this measurement scale. An example would be the identification of wheat regions, corn regions, and soybean regions. Each crop region is distinct; arithmetic operations between the regions are not possible at this level. Political party affiliation (Democrat, Independent, Republican), sex (male, female), and response (yes, no) are other examples.

Ordinal Scaling

The underlying structure of ordinal scales is a hierarchy of rank. Objects or events are arranged from least to most or vice versa, and the information obtainable is of the "greater than" or "less than" variety. Ordinal scaling provides no way of determining how much distance separates the items in the array.

There are several types of ordinal measurement. In *complete* ordering, every element in the array has its own position, and no other element can share this position. This kind of ordinal scale is considered relatively strong because *statistical* observations about the ranking are possible. This is not the case with the second main type, *weak* ordering, in which elements can share positions (called *paired ranks*) along the ordinal continuum.

One interesting feature of ordinal scales is that we can attach any numerical scale to the ranking without violating the underlying structure. If we know that the order is *K L M,* we can have either $K = 5$, $K = 3$, $M = 1$, or $K = 500$, $L = 300$, $M = 1$, while retaining the original scale. Remember that in ordinal scaling we do not know how much difference separates the events in the array. Several geography students spend a summer touring 50 large cities throughout North America. After they return, their professor asks them to rank the cities based on their appeal, from best liked to least liked. In this array, a city's position is known in the overall ranking, but it is not possible to discern how much it differs from those ranking above and below it.

Other examples at the ordinal scale are social class, social power (more, less), and agreement (strongly agree, strongly disagree). Again, the characteristic that sets ordinal scales apart from others is that we cannot discern magnitude differences between the observations.

Interval Scaling

At this measurement scale, we can array the events in order of rank and know the distance between ranks. Observations with numerical scores at the interval scale are important in geographical analysis because data at this level are needed to perform fundamental statistical tests having predictive power. Units on an interval scale are equal throughout; that is, a degree on the Fahrenheit scale is assumed to be the same regardless of whether it's at 22–23 or 78–79 degrees.

Magnitude scales at the interval level have no natural origin; any beginning point may be used. The classic example is the Fahrenheit temperature scale. There are no absolute values associated with interval scales; they are *relative.* In interval scaling, units are agreed upon by researchers and are assumed to be standard from one set of conditions to another. Variables at the interval scale *do not have zero as a starting point.*

Try this experiment for an example. Place a yardstick next to a tabletop, but not touching the floor. You can slide the yardstick up and down *relative* to the tabletop, but as it is not touching the floor, there is no absolute height for the table, only a relative height as shown by the sliding yardstick.

Table 4.3   Measurement Scales and Methods of Studying Them

| | | Appropriate Statistics | | |
		Central Tendency	Dispersion	Tests
Scale	**Relations**			
Nominal	Equivalence	Mole	Variation ratio	Chi-square contingency coefficient
Ordinal	Equal to; greater than	Median	Percentiles	Spearman's rho; Kendall's tau
Interval	Equal to; greater than; difference of interval	Mean	Standard deviation	F-test; t-test; Pearson's V
Ratio	Equal to; greater than; difference of interval; difference of ratio	Geometric mean	Coefficient of variation	F-test; t-test; Pearson's V

Source: Adapted from Peter J. Taylor, Quantitative Methods in Geography: An Introduction to Spatial Analysis *(Boston: Houghton Mifflin, 1977), p. 41; and David Harvey,* Explanations in Geography *(New York: St. Martin's Press, 1969), p. 312, originally published in S. S. Stevens, "On the Theory of Scales of Measurement,"* Science *103 (1946): 677–80.*

Ratio Scaling

Like interval scaling, ratio scaling involves ordering events with known distances separating the events. The difference is that ratio-scale magnitudes are absolute, having a known starting point. This scale of measurement has a zero (absence of a magnitude) as its starting point. The Kelvin absolute temperature scale is an example here. Other examples include weight (20 lbs is twice as large as 10 lbs), and 100 miles is twice as far as 50 miles. Elevation above sea level is another ratio scale. Ratio scaling is important to geography because more sophisticated statistical tests can be performed using this level of measurement.

In the yardstick/tabletop example, if the yardstick is placed on the floor, then the height of the tabletop can be measured relative to a zero starting point (the floor). If we chop off its legs, we know exactly how much shorter it has become.

The different levels of measurement are summarized in Table 4.3 and discussed in more detail in Chapter 5. The amount of information obtainable, statistical confidence, and predictive power increase as one progresses from nominal to ratio scaling.

MEASUREMENT ERROR

It is not usually possible to measure, without some error, one or all of the following types: observer error, instrument error, environmental error, or error that is due to the observed material. Cartographers are faced with these types of error to a lesser or greater degree and should be sufficiently aware of their dangers to develop methods of minimizing them.

Observer error relates to the performance of the investigator, resulting in biased measurement. This type of error

usually stems from the inability of observers to remove themselves from the measurement process. For example, an interviewer may consistently word questions in such a way as to favor a particular kind of response; an instrument reader may misread a dial because of visual parallax.

Errors caused by poorly designed or malfunctioning recording instruments belong to a class called *instrument error.* These errors may be biased (i.e., in one direction) or random. The classic example of the former is a thermometer that consistently reads one-half degree above the true temperature. If at all possible, the researcher should state the degree of instrument error in order to alert the reader to the measurement's reliability.

Conditions extraneous to the measurement can lead to *environmental error.* Errors in instrument recording result if the conditions deviate excessively from those in which the instrument was designed to operate. Observer error occurs, for example, when excessively cold temperatures cause the observer to lose concentration. Errors in the observed material could result from such environmental factors as high winds causing freak variations in the object of the measurement. The researcher must provide adequate environmental control in the context of the nature of the study. Environmental conditions may need to be stated as part of the study.

Errors that are due to changes in the *observed material* can lead to imperfect results. People interviewed for a psychological survey may interact with the researcher, causing the observer to measure the wrong behavior. These forms of error may be random or persistent (noncompensating). Observers need to watch out for such situations and inform readers of any potential sources of error they recognize.

Of special concern today are errors that may lurk in computerized data bases. MacEachren, for example, warns us:

Because of the precision that computer systems are capable of, there is a disturbing tendency to accept solutions arrived at by these systems as being accurate as well. In order for students to become successful makers or users of maps, they must learn to critically evaluate the potential for error inherent in mapping systems and geographic data bases that are increasingly being used to generate maps.[15]

In cartography, error sources range from false images of reality to tricky reproduction techniques. Not all errors can be eliminated, but every conceivable means must be employed to reduce error to tolerable levels. Each mapping method discussed in this book has its own error sources, based on its unique way of mapping and symbolization. The thematic map designer must learn to take these error sources into account.

DATA SOURCES

Geographic data are of three kinds: those obtained from direct field observation, those obtained from archival sources, and those generated from theoretical work.[16] Field observations may be either quantitative or qualitative in nature. Archival sources may be areal (maps, air photographs), or essentially nonareal. Theoretical work can involve modeling in a variety of forms. Data are becoming increasingly available on-line through the World Wide Web, making acquisition relatively easy (and caution necessary). Cartographers are called on from time to time to treat data from each of these sources. Of particular interest, because of their extent of coverage and easy accessibility, are enumeration (census) data aggregated for areal units. Later in this chapter we will talk about remote sensing.

THE GEOGRAPHICAL AREAL UNIT

Geographical areal units may be of two kinds, natural and artificial.[17] *Natural areal units* are exemplified by such phenomena as lakes, countries, farms, cities, and other geographic elements having two dimensions. The boundaries are intimately tied to the feature. *Artificial areal units* are imposed by researchers in their attempt to organize reality and facilitate data collection. Examples are regions, Metropolitan Statistical Areas (MSAs), and census tracts.

It is likewise possible to work with either singular or aggregate areal units. *Singular areal units* correspond to only themselves, whereas one *aggregate unit* is composed of several singular units. The important distinction is that data collected at the singular level can be used in combination to make inferences about aggregate units. However, it is difficult to make inferences about singular units from aggregate unit data. Suppose you have collected data at the

singular level. It is possible to compute a mean and use it to describe the nature of the region composed of the individual units. By contrast, if the mean is all you have, you cannot describe the nature of each singular unit—the process of data aggregation can actually yield a *loss* of information.

THE MULTIDIMENSIONAL CHARACTERISTICS OF THEMATIC MAP SYMBOLS

The selection of symbols is one of the more interesting tasks for the cartographic designer. The choice is wide, and no firm rules prevail. Symbol selection, however, is increasingly based on a compelling system of logic tied to both the type of geographic phenomenon mapped and certain graphic primitives or variables.

Symbols are the graphic marks used to encode the thematic distribution onto the map. From a vast array of symbols having different dimensions, the designer selects the symbol that best represents the geographic phenomena. Fortunately, the task is reduced somewhat by controlling factors, such as cartographic convention and the inability of most map readers to easily understand the more complex symbols that might be chosen.

SYMBOL TYPES, SYMBOL VISUAL DIMENSIONS, QUALITATIVE AND QUANTITATIVE MAPPING

The three generally recognized cartographic **symbol types** are *point, line,* and *area.*[18] To this list might be added the three-dimensional diagram so commonly used to describe volumetric distributions (sometimes referred to as a volume or *surface* symbol. (See Figure 4.5.) The first three symbol types have been virtually standard in thematic mapping.

The design of cartographic symbols requires attention to their visual qualities. These **symbol dimensions** are *shape, size, color hue, color value, color saturation, pattern orientation, pattern arrangement,* and *pattern texture.*[19] To this list we might add *location.* Location as a graphic variable is intrinsic to the symbol itself and relates to it in the *x–y* system of the map. The cartographic designer makes selections on the basis of visual dimensions, such that the symbol is tied functionally to the level of measurement of the data. For example, we can draw circles sized proportionally to magnitude in a data set because size variations correspond to intervals in the array. Size, however, cannot be the visual dimension when we scale symbols to nominal data because we do not know the numerical distance between nominal classes.

Thematic maps can be either qualitative or quantitative. The chief purpose of qualitative mapping is to show the locational characteristics of geographic phenomena. Amounts, other than the "eyeballing" possible by comparing map areas, are not the content communicated by such mapping. The location graphic primitive functions chiefly

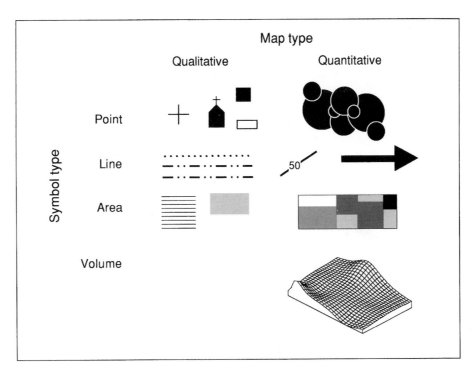

Figure 4.5   Various map symbols and map types.

Some common symbols used in both map types, qualitative and quantitative, are illustrated here. These are only a few of the large number of individual symbols that may be chosen. Point, line, and area symbols have been used for many years. To this group we may add the three-dimensional diagram made popular by the use of computer graphics (for quantitative mapping).

on these maps. Quantitative maps, on the other hand, stress how much of something is present at various locations. Not all symbol dimensions can be used in both qualitative and quantitative maps. (See Table 4.4.) The dimensional qualities of each symbol type (point, line, and area) have a unique graphic expression for each map type. A *point* symbol, for example, can be used in either qualitative or quantitative mapping. In qualitative maps, *shape* is the quality of visual dimension, whereas *size* is the dimensional quality in quantitative maps.

The design of thematic map symbols takes place within the limits of these visual dimensions and is governed by the type of thematic map. Literally hundreds of kinds of symbols are devised, even within these constraints, although certain ones are standard. Figure 4.6 presents a number of unique symbols that fit each of the categories listed by symbol and map type. Symbol design requires considerably more research.

CHARACTERISTICS OF GEOGRAPHIC PHENOMENA AND SYMBOL SELECTION

There is a logical (and traditional) correspondence between geographic phenomena (point, line, area, and volume) and the employment of symbol types (point, line, area). Of course, the match is not a convenient one-to-one correspondence. Point data (e.g., cities at the appropriate

Table 4.4   Symbol Dimensions and Map Type (Applicable to Point, Line, and Area Symbols)

	Map Type	
Symbol Dimension	**Qualitative**	**Quantitative**
Shape	X	
Size		X
Color		
Hue	X	
Value		X
Saturation		X
Pattern		
Orientation	X	
Arrangement	X	
Texture		X

Source: Adapted from Phillip C. Muehrcke, Map Use, Reading, Analysis and Interpretation *(Madison, WI: JP Publications, 1978), pp .62–63.*

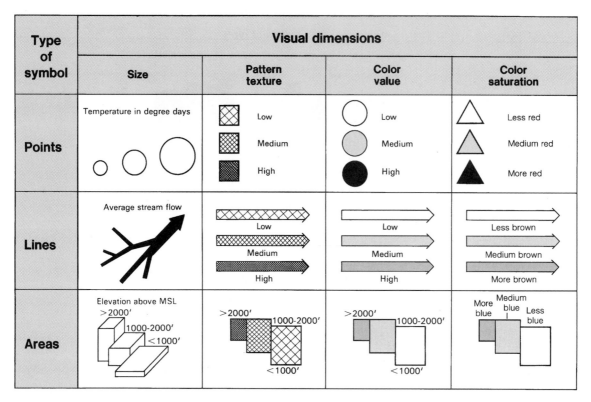

Figure 4.6   Types and symbols useful in quantitative mapping and their visual dimensions.
(Courtesy JP Publications. Modified by author.)

	Symbol type		
Level of measurement	Point	Line	Area
Nominal	House types, barns, churches,	Roads and railroads	Different crop areas
Ordinal	Small, medium, large of the same feature	Types of roads, county, state, other	Rank of crop yield areas
Interval	Temperatures at points	Isolines showing year of westward expansion	Areas showing average SAT scores
Ratio	Graduated circles of urban populations	Graduated flow lines of commodity exchange	Value-by area cartogram of population density

Figure 4.7   Typology of map symbols.
Classification of thematic map symbols id possible by examining the school type and the level of measurement being used.

scale) are customarily mapped by point symbols such as dots, or scaled circles. Roads, which are linear phenomena, can be mapped by line symbols; geographic phenomena having areal extent (lakes, countries, nations) can be mapped with areal symbolization—patterns or tints. Geographical landforms have been mapped by a form of area symbolization called *hypsometric tinting:* areas between selected elevation boundaries are rendered in various color shades. Elevation is also mapped by *contours of elevation,* which are line symbols. The use of three-dimensional *dia-*

grams may become a standard form of symbol to represent volume phenomena.

The three symbol types may be matched to the four measurement scales to produce a valuable **typology of map symbols.** (See Figure 4.7.) Cartographic presentations are thus possible even at the highest (interval/ratio) levels of measurement.[20] Choosing to symbolize the data at a level below that possible for their level of precision is unwise and results in loss of information.[21] Symbolization above the level of precision of the data is questionable practice because it forces the cartographer to make assumptions not warranted by the data.

Many cartographers refer to map symbols as the *language* or *language elements* of maps. As such, they do perform as the most important functional parts of the map. One very prolific cartographer, Mark Monmonier, has said:

> As elements in a constructed view of reality, cartographic symbols must communicate differences among geographic features, functional relationships, and relative significance.[22]

Continuing in this context, Monmonier lists several ways that the visual variables may be used to satisfy several functions of map symbols:[23]

1. *Portraying a location and its neighbors.* An example here would be a simple locator map (parks, or national monuments).
2. *Portraying causal relationships among features.* Adding causally relevant features to a simple locator map can provide explanation.
3. *Portraying routes.* These itinerary maps can provide a substitute for mundane written descriptions.
4. *Portraying a phenomenon with multiple instances.* These maps include the familiar one-to-one dot maps (see Figure 8.1 later in the textbook); for example, a map showing the locations of all automobile assembly plants in the United States.
5. *Portraying variation in count or magnitude.* These maps may be those that use proportional symbols at points, such as those represented in Chapter 9.
6. *Portraying variation in density.* Common maps are those that show the density of population, agricultural production per acre, and others.

In practice, a large portion of the thematic map designer's role is to select a symbolization plan that best portrays the appropriate concept of geography. Regardless of the nature or level of sophistication of the geographical study, the cartographic communication of the findings can usually be reduced to one of the key concepts. It is therefore important for cartographers to be familiar with the nature of geography. Matching the symbolization plan to the geographical concept is part of the challenge and enjoyment of cartographic design.

ENUMERATION DATA, GEOGRAPHICAL UNITS, AND CENSUS DEFINITIONS

Geographers and cartographers frequently use enumeration data provided by federal government census. Enumeration data are aggregated data, tabulated by specific geographic areas, and are often the only information available. The geographical units for which these data are published are important background for the thematic cartographer.

> I n short, maps and other graphics comprise one of three major modes of communication, together with words and numbers. Because of the distinctive subject matter of geography, the language of maps is the distinctive language of geography. Hence sophistication in map reading and composition, and ability to translate between the languages of maps, words, and numbers are fundamental to the study and practice of geography.
>
> Source: John R. Borchert, "Maps, Geography, and Geographers," *The Professional Geographer* 39 (1987): 387–89.

THE UNITED STATES CENSUS

One of the principal sources of geographic data (information associated with a particular areal unit) is the United States decennial *Census of Population and Housing.* Census data have been collected for the United States since 1790. The purpose of the census of population is to take a "head count" for determining representation in the House of Representatives. Its scope, thoroughness, and reliability have increased over the years; it now constitutes the single best source of broad-based demographic and socioeconomic data available to geographic researchers.

Census Geography

Of particular interest to the cartographer is the nature of the **census geography** used in the decennial censuses. The Census Bureau tabulates data into more than 40 types of geographic areas. (See Table 4.5 for some types.) Other censuses are also conducted by the Census Bureau, including the Census of Governments, six Economic Censuses, and the Census of Agriculture. As Table 4.5 shows, more geographic detail is provided by the Census of Population and Housing.

The larger the geographic area, the greater the data detail available. At the block level, for example, the only 1990 population data published are total population; number of blacks, of Asian/Pacific Islanders, and those of Spanish origin; number under 18 years old; and number 65 years or older. A few items of housing are also available at the block level. At the census tract level, numerous socioeconomic data are published, in addition to expanded population characteristics. This is partly because of disclosure rules. The Census Bureau does not publish data that permit violation of confidentiality; this could be a problem in geographic areas having very small populations. The boundaries of census geographic areas are determined by a variety of agencies. Governmental authorities provide the Census Bureau with information on the boundaries of such units as states, congressional districts, Indian reservations, election districts, counties, minor civil divisions, incorporated places, and city wards. Metropolitan areas are statistical divisions defined by the Office of Management and

Table 4.5   Major Geographic Areas and Census Bureau Programs

	Population and Housing Censuses		Economic Censuses							
	Area Reports	Subject Reports	Census of Govern- ments	Retail Trade	Whole- sale Trade	Selected Services	Manu- factures	Mineral Indus- tries	Construc- tion Indus- tries	Agriculture
United States	a	a	a	a	a	a	a	a	a	a
Census regions	a	s	a	a	a	a	a	a	a	a
Census divisions	a	s	a	a	a	a	a	a	a	a
States	a	s	a	a	a	a	a	a	a	a
Metropolitan areas	a	s	a	a	a	a	a			a
Urbanized areas	a									
Counties	a		a	a	a	a	a	a		a
Incorporated places	a		a	s	s	s	s			
Census designated places (CDPs)	a									
Minor civil divisions (MCDs)	a		a							
Census county divisions (CCDs)	a									
Census tracts	a									
Enumeration districts (EDs) and block groups*	a									
ZIP code areas*	a									
Blocks*	a									
Central business districts**	c			a						
Major retail centers**				a						

Note: Other areas unique to the population and housing census are urbanized areas, urban/rural, and congressional districts.
a All areas.
c All, by addition of components.
s Selected areas—larger or with more activity.
*Not in printed reports.

**These areas are dropped beginning with the 1987 Census of Retail Trade. *Source: U.S. Bureau of the Census,* Census Geography. *Data Access Description No. 33 (Washington DC: USGPO, 1979), p. 2; U.S. Bureau of the Census,* Census Geography—Concepts and Products. *Factfinder No. 8 rev. (August 1985): 6.*

Table 4.6   Governmental and Statistical Geographical Units Used by the U.S. Bureau of the Census

Governmental Areas	Statistical Areas
United States	Geographic regions of the United States
States (and outlying areas), American Indian	Geographic divisions of the United States
reservations, Native Villages	Metropolitan Statistical Areas (MSAs)
Counties (and county equivalents)	Urbanized areas (UAs)
Minor civil divisions (MCDs)	Urban and rural areas
Incorporated places	Census designated places (CDPs)
Congressional districts	Census county divisions (CCDs)
Election precincts (in selected areas)	Census tracts
ZIP code areas (a governmental administrative area)	Block numbering areas (BNAs)
	Block groups (BGs)
	Census blocks
	County groups (in the public—use samples)

Source: *U.S. Bureau of the Census,* Census Geography. *Data Access Description No. 33 (Washington, DC: USGPO, 1979), pp. 2, 6; and U.S. Bureau of the Census,* Census Geography—Concepts and Products. *Factfinder No. 8 rev. (August 1985): 2.*

Budget. Local government agencies or public committees and the Census Bureau determine other statistical geographic areas, including urbanized areas (UAs), census county divisions (CCDs), census designated places (CDPs), census tracts, enumeration districts (EDs), blocks, and block groups (BGs).

There is a major distinction between *governmental area units* and *statistical area units.* The boundaries of governmental units are established by law; they really exist and are often marked on the landscape. Census statistical units, on the other hand, are established merely for convenience in enumeration and tabulation. They have been devised to satisfy the needs of governmental agencies and public users for analyzing population characteristics or change. For example, urbanized area boundaries are used to measure the areal extent of larger American cities, but they do not exist on the landscape—only on maps provided by the Census Bureau.

Table 4.6 lists the governmental and statistical area units for which the census, currently or in the past, has provided data, and the organizational and hierarchical arrangements of these are illustrated in Figure 4.8. Figure 4.9 presents the geographical relationships of the principal metropolitan areal units. The Athens, Georgia, urbanized area map is reproduced as an example in Figure 4.10. Definitions of many of the geographical units are included in Appendix C.

Caution is necessary when dealing with census data taken from decennial censuses. Definitions often change from one census to the next, so they must be checked before using these data sources. Comparability is doubtful in

many instances. Furthermore, census geography may likewise change from time to time; urbanized area boundaries, for example, are usually different from one census to the next. In fact, the urbanized area concept was not even used before 1950. Census tract boundaries may change because boundary features may disappear (perhaps because of urban renewal). A thorough examination of all data definitions and geographical descriptions is recommended before using these data.

The 1990 Census

Products and definitions of the 1990 Census of Population and Housing are essentially the same as for the 1980 census. However, important changes in media formats of the products will make it easier to use census materials. Since the census of 1790, there has always been an ink/paper version, and this has continued for the 1990 census. In 1965, the Bureau of the Census began to distribute its files in computer tape form, a practice that continued through the 1980 census. This tape program is extended for the 1990 census.

Limited data files of the 1960 and 1970 censuses were made available on microfiche, but were broadly used for the 1980 census. Microfiche offers much data for little cost, especially for the user who has no computer capabilities. However, because of rapid advances and lower costs in microcomputer technology, fewer products of the 1990 census will be offered on microfiche and, instead, more will be available on CD-ROM. (See Chapter 17.) Many other censuses (for example the 1987 Census of Retail Trade and the Census of Manufactures) are presently available on these

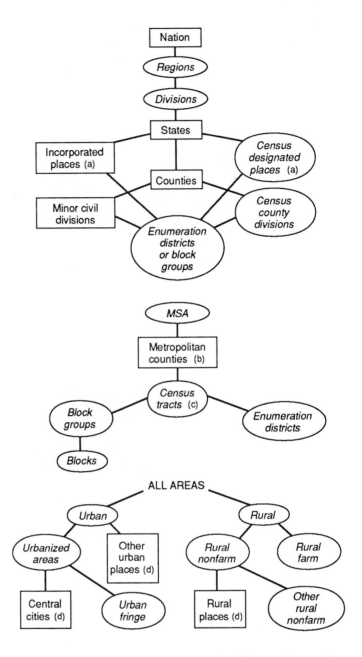

Key:

Notes:

(a) Note that places (incorporated and census designated) are not shown within the county and county subdivision hierarchy, since places may cross the boundries of these areas.

(b) In New England, MSAs are defined in terms of towns and cities, rather than counties (as in the rest of the country).

(c) Census tracts subdivide most MSA counties as well as about 200 other counties. As tracts may cross MCDs and place boundries, MCDs and places are not shown in the hiearachy.

(d) Includes both governmental and statistical units.

Figure 4.8 Hierarchial arrangement of census geographical units.
Note that some hierarchies overlap. For example, counties may be subdivided into minor civil divisions (MCDs) or census county divisions (CCDs). *(Source: Redrawn from the United States Bureau of the Census, "Census Geography-Concepts and Products," Factfinder, No. 8 rev. [August 1985]: 2.)*

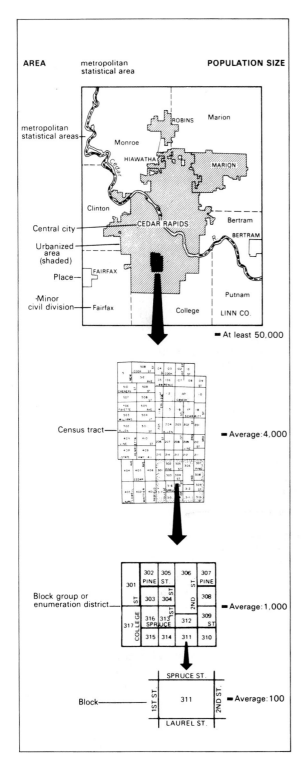

Figure 4.9   Metropolitan census geographic areas.
(Source: United States Bureau of the Census, Census Geography, *Data Access Description No. 33 [Washington, DC: USGPO, 1979], p 4.)*

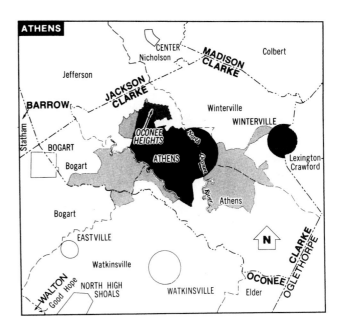

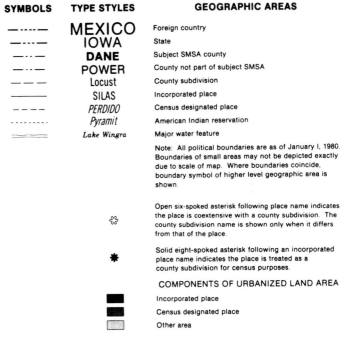

Figure 4.10   Athens, Georgia, urbanized area (UA) map.
Note the areas that go to make up the urbanized area. *(Source: United States Bureau of the Census, 1980 Census of Housing. General Housing Characteristics: Georgia [Washington, DC:USPGO, 1982], p. 298.)*

Table 4.7   1987 Economic Census on CD-ROM

Census	Publication Vol. 1	Vol. 2
Census of Retail Trade		
Geographic areas	•	
Establishment/firm size	•	
Merchandise line sales (selected 1982 and 1977 MLS)	•	
Nonemployer statistics	•	
Special report	•	
ZIP code		•
Census of Wholesale Trade		
Commodity line sales	•	
Establishment/firm size	•	
Geographic areas (selected 1982 and 1977 MLS)	•	
Census of Service Industries		
Establishment/firm size	•	
Nonemployer statistics	•	
ZIP code statistics	•	
Census of Manufactures		
Preliminary U.S. statistics**	•	
Industry series (final data)	•	
General summary	•	
Geographic areas (selected 1982 and 1977)	•	
Location of manufacturing plants		•
Census of Mineral Industries		
Geographic areas	•	
Industry series	•	
Census of Construction Industries		
Geographic areas	•	
Industry series	•	
Census of Transportation		
Geographic areas	•	
Survey of Minority-Owned Business	•	
Survey of Women-Owned Business	•	

*Volume 1 when completely published will contain five discs, and Volume 2 will contain two discs of information at the ZIP code level.
**Preliminary manufactures data will not be included on the final Volume 1e CD-ROM.

laser discs. (See Table 4.7.) In addition, several products are available on floppy diskettes.

For the 1990 census, 100-percent data (i.e., not sample data) of population and housing counts and characteristics for states and subareas down to the block group level are provided on CD-ROM. These will become available, how-

The Census Bureau has developed an automated geographic data base, known as the TIGER (Topologically Integrated Geographic Encoding and Referencing) System, that will allow the production of various geographic products to support the 1990 census. It provides coordinate-based digital map information for the entire United States, Puerto Rico, the Virgin Islands, and for areas over which the United States has jurisdiction.

The TIGER System will significantly improve 1990 census maps and geographic reference products and will permit users to generate, by computer, maps at different scales for any geographic area of the country.

The final list of TIGER products has not been determined, but the Census Bureau expects that extracts from the TIGER System will be released in several formats. One extract of selected geographic and cartographic information is called the TIGER/Line files. TIGER/Line files contain basic data for the segments of each boundary or feature (e.g., roads, railroads, and rivers), including adjacent census geographic area codes, latitude/longitude coordinates of segment end points, the name and type of the feature, and the relevant census feature class code identifying the feature segment by category. TIGER/Line files also furnish address ranges and associated ZIP Codes for each side of street segments for areas approximating the urbanized areas.

The TIGER/Line files are organized by county and are available to the public now in a precensus version and later in a final 1990 census version. The files are released on computer tape and CD-ROM.

Source: U.S. Department of Commerce, Bureau of the Census, *Census '90 Basics* (Washington, DC: U.S. Bureau of the Census, January 1990), pp. 16–17.

ever, only after the paper and computer tape files have been released. Sample data for social, economic, and housing characteristics down to the block group level also will be available on CD-ROM. More information on these products is included in Appendix C.

Two publications that are useful references for geographers and cartographers are the *County and City Data Book* and the *State and Metropolitan Area Data Book,* published every five years (on the years ending in 2 and 7 in each decade). These are available on computer tape, CD-ROM, and floppy diskette. An interesting and helpful publication on census geography entitled *Maps and More: Your Guide to Census Bureau Geography* is available from the U.S. Census Bureau Customer Services (301) 763–4100. There is one final publication of the Census Bureau worthy of note here, *1992 TIGER/Line Files: Helping You Map Things Out.* This publication provides a brief history of the TIGER mapping program and the types of geographic elements found in the files.

Census Bureau on the World Wide Web

The United States Census Bureau now provides virtually all its material on its World Wide Web site,

For over 120 years the *Canada Year Book* has recorded the economic, social and political life of Canada.

Like its predecessors, the *Canada Year Book 1990* brings together in a single volume a wealth of information from various sources to provide a composite portrait of Canada in all its diversity and richness. Over the years, it has become the standard statistical reference source on Canada, widely consulted by librarians, parliamentarians, teachers, diplomats, journalists and others.

To mark the opening of the last decade of this century, the 1990 edition features a variety of computer-generated maps, illustrating key social and economic trends.

Readers with an interest in following up in more detail on a particular topic will be pleased with another innovation to the 1990 edition. A selected list of related publications has been added to the end of each of the 23 chapters. In addition, the index has been significantly expanded and improved to provide more ready access to the wealth of information contained herein.

The content for the *Canada Year Book* is drawn from over 300 contributors, making it impossible to acknowledge each individually. Nevertheless, our gratitude to each remains, as does our gratitude to the Canadian public for responding to the surveys and providing the data that are the foundation of this nation's statistical system.

As were the publishers of the first edition of the *Year Book,* we are

witnesses of the extreme care taken to ensure accuracy, and believing the contents of the Year-Book *to be of general usefulness, feel sanguine that the work will meet with public favour.*

Source: *1990 Canada Year Book* (Ottawa: Statistics Canada, 1990), Preface.

http://www.census.gov. Not only can data tables be found, but also thematic maps, TIGER files, and area profiles, to name a few.

THE CENSUS OF CANADA

The Census of Canada is conducted decennially and published in the first year of the decade; the current census is from 1981, the twelfth since Canadian Confederation in 1867. It is taken by law for the purpose of establishing representation in the Federal House of Commons. Like the United States *Census of Population and Housing,* the *Decennial Census of Canada* has become more complex. It is used by federal, provincial, and municipal governments for planning and decision making. Private business and industry also use the information it provides.

A quinquennial (every five years) census was added in 1956 providing mid-decennial demographic and socioeconomic statistics. The most recent was published in 1986. Like the *United States Statistical Abstract,* Statistics Canada publishes a very useful yearly national summary called the *Canada Year Book.* (See boxed article.)

Geographical units for which data are tabulated are as follows: Canada, provinces, census divisions, census subdivisions, unorganized townships, federal electoral districts, census metropolitan areas, census agglomerations, and census tracts.[24] As is the case in the United States census, the geographic divisions of the census of Canada are hierarchical and can be divided into both governmental and statistical area units. A perusal of a current census of Canada will demonstrate that using its documents is similar to using those of the United States census.

THEMATIC MAPPING AND REMOTELY SENSED DATA

Geographers have been involved in remote sensing for approximately 150 years, first by direct observation, then by camera from low-altitude balloons. This evolved into using cameras from fixed-wing aircraft, and more recently by scanners on earth-orbiting satellites. The technical aspects of this kind of data gathering are best left to other books that treat the subject in detail.[25] Here we will present only some organizing ideas and illustrate how many of these products are used by modern geographers and cartographers.

In addition to the customary photographic products obtained by cameras aboard both low- and high-altitude aircraft, today many of the products from fixed-orbit satellites come in the form of *digital* information that is then reassembled into ordinary-looking, photographic images. Postprocessing of the digital information is done using electronic, photographic, and other techniques which lead to a printing negative that produces a final image. The creation of most images by remote sensing is a complex, tedious, and custom operation. (See boxed article.) Three satellite-imaging platforms are most useful for the thematic mapper: (1) **Landsat,** first launched in 1972 by the United States. In 1984 Landsat 5 was launched and Landsat 6 was placed into orbit in mid-1992; (2) **SPOT,** launched by France in 1986, and the first commercially produced imaging satellite; and (3) **METSAT,** or meteorological satellites operated by the National Oceanographic and Atmospheric Administration (NOAA).

Landsat operations have been turned over to a quasi-governmental agency called the Earth Observation Satellite Company (EOSAT). Landsat 5 orbits the earth about every 99 minutes, in a near north-south orbit at a 450-mile altitude. It contains sensors that record electromagnetic energy in as many as seven bands of the energy spectrum. These are stored in the satellite and periodically sent to ground receivers in digital format. Image-processing computers on the ground recreate the images.

Although considerable mapping capability exists for cartography by using the information and image-producing techniques associated with satellite data gathering, for the most part only qualitative thematic maps are possible (or ordinal data on quantitative maps). The only exceptions are the relative areal amounts of elements on the map. For

METHODS USED

After Landsat scenes are acquired and evaluated, they are processed either photographically or digitally, to produce halftone separates for printing. Digital image data are registered to a map base using a series of control points and calculated transformation coefficients. Images are digitally mosaicked using the points and carefully chosen cut lines. Radiometric differences are balanced and images are abutted as part of the mosaicking step. Photographically, images of each spectral band are enlarged, registered to a control base at publication scale, tone-matched, mosaicked, and exposed through a halftone screen. For digitally mosaicked Landsat data in quadrangle format, each band is electronically scanned, enlarged, and halftoned using specified density aimpoints for lithography. Final-scale image separates are registered with line and lettering separates. Color proofs are prepared and evaluated against initial press sheets for register, completeness of linework, and color balance prior to approval for final printing.

Source: J. Denegre, *Thematic Mapping from Satellite Imagery* (London: Elsevier, 1988), p. 32.

Table 4.8 Recent Applications of Landsat Imagery

Archaeological exploration
Flood assessment
Forest–agriculture environmental mapping
General (unspecific) cartographic images
Impacts of surface mining
Land cover (land use)
Mineral exploration
Oil (hydrocarbon) exploration
Soil mapping
Terrain (landscape) mapping
Water quality
Wetland mapping

example, it is impossible to show how many people live in urban areas, although urban images show the extent of built-up urban areas exceptionally well. Qualitative classification is possible, but quantitative is not. What is done quantitatively is added to the images in postprocessing operations.

An examination of recent literature attests to the wide applications of remote sensing to cartography.[26] For example, landscape (terrain) mapping, wetlands and water-quality mapping, mapping of ice and snow fields, and vegetation mapping have been done successfully. (See Table 4.8.) It

appears that satellite images are to be used as *base-map* components in final, quantitative thematic mapping, but they have a much wider use in geographic information systems (GIS) applications (see Chapter 6).

Digital data from sources other than satellites also are useful for thematic cartographers. The United States Department of the Interior provides information on its digital data—elevation data, planimetric data, land-use and land-cover data, and geographic names—through its publication, *US GeoData,* available from that department. This publication describes the information available and points to some potential uses. Use of remotely gathered digital data is discussed more in Chapter 17.

NOTES

1. Richard Hartshorne, "The Concept of Geography as a Science of Space, from Kant and Humboldt to Hettner," in *Introduction to Geography: Selected Readings,* eds. Fred E. Dohrs and Lawrence M. Sommers. (New York: Thomas Y. Crowell, 1967), pp. 89–90.
2. Ronald Abler, John S. Adams, and Peter Gould, *Spatial Organization: The Geographer's View of the World* (Englewood Cliffs, NJ: Prentice Hall, 1971), p. 56.
3. Ibid., p. 59.
4. Douglas Amedea and Reginald G. Golledge, *An Introduction to Scientific Reasoning in Geography* (New York: Wiley, 1973), pp. 4–6. Students are encouraged to read also *Guidelines for Geographic Education: Elementary and Secondary Schools,* a booklet prepared by the Association of American Geographers and the National Geographic Society, 1986. Although written to a different audience, the five themes mentioned in this publication—location, place, human-environment interactions, movement, and regions—capture the spirit of much of modern geography.
5. John F. Kolars and John D. Nystuen, *Geography: The Study of Location, Culture, and Environment* (New York: McGraw-Hill, 1974), pp. 14–16.
6. Jan O. M. Broek, *Geography: Its Scope and Spirit* (Columbus, OH: Merrill, 1965), pp. 72–76.
7. Geography Education Standards Project, *Geography for Life* (Washington, DC: National Geographic Research and Exploration, 1994).
8. Geography Education Standards Project, *Key to the National Geography Standards* (Washington, DC: National Geographic Society, 1995, pp. 3–4.
9. David Harvey, *Explanation in Geography* (New York: St. Martin's Press, 1969), pp. 293–98.
10. Robert Levine, "City Stress Index," *Psychology Today,* November 1988, pp. 53–58.
11. Harvey, *Explanation in Geography,* p. 351.
12. Kirk W. Elifson, *Fundamentals of Social Statistics* (Reading, MA: Addison-Wesley, 1982), p. 31.

13. Harvey, *Explanation in Geography,* p. 306.

14. Peter J. Taylor, *Quantitative Methods in Geography: An Introduction to Spatial Analysis* (Boston: Houghton Mifflin, 1977), p. 39. See also J. Chapman McGrew, Jr., and Charles B. Monroe, *An Introduction to Statistical Problem Solving in Geography* (Dubuque: Wm. C. Brown, 1993), pp. 18–20.

15. Alan M. MacEachren, "Map Use and Map Making Education: Attention to Sources of Geographic Education," *Cartographic Journal* 23 (1986): 115–22.

16. Peter Haggett, *Locational Analysis in Human Geography* (New York: St. Martin's Press, 1966), p. 186.

17. Harvey, *Explanation in Geography,* pp. 351–52.

18. Kang-tsung Chang, "Data Differentiation and Cartographic Symbolization," *Canadian Cartographer* 13 (1976): 60–68.

19. Joel Morrison, "A Theoretical Framework for Cartographic Symbolization," *International Yearbook of Cartography* 14 (1974): 115–27; see also Jacques Bertin, *Semiology of Graphics: Diagrams, Networks, Maps,* trans. William J. Berg (Madison: University of Wisconsin Press, 1983).

20. Alan M. MacEachren, *Some Truth with Maps: A Primer on Symbolization and Design* (Washington, DC: Association of American Geographers, 1994), pp. 15–19.

21. Phillip Muehrcke, *Thematic Cartography* (Commission on College Geography, Resource Paper No. 19— Washington, DC: Association of American Geographers, 1972), p. 15.

22. Mark Monmonier, *Mapping It Out: Expository Cartography for the Humanities and Social Sciences* (Chicago: University of Chicago Press, 1993), p. 76.

23. Ibid, pp. 76–89.

24. Statistics Canada, *1971 Census of Canada,* vol. 1, pt. 1, *Population* (Ottawa: Ministry of Industry, Trade, and Commerce, 1974).

25. A number of good books on remote sensing are available. For a beginning treatment, see Thomas D. Rabenhorst and Paul D. McDermott, *Applied Cartography: Introduction to Remote Sensing* (Columbus, OH: Merrill, 1989).

26. J. Denègre, ed., *Thematic Mapping from Satellite Imagery* (London: Elsevier, 1988).

GLOSSARY

absolute location   specifies position relative to a fixed grid system, p. 68

census geography   the divisions and descriptions of areal entities, both political and statistical; used to tabulate census (enumerated) data, p. 79–81

concept of change   fundamental idea in geography; both space and time events are merely brief states of equilibrium separating periods of change, p. 71–72

continuous geographic phenomena   extend unbroken and without interruptions, p. 73–74

direction   a line going from one place to another; azimuth is a formal way of specifying direction, p. 70

discrete geographic phenomena   do not occur between spatial observations; distinguishable individual entities, p. 73

distance   the number of linear units separating two objects; in relative location, distance can be measured in units of time, cost, psychological factors (e.g., stress), or other units, p. 70

distance decay   key concept in geography; decreasing occurrence of events with distance from a central point, p. 70

distribution   characteristic of the spatial pattern of geographic phenomena; dispersed and agglomerated are two kinds, p. 71

five themes of geography   from *Geography for Life;* includes location; place, human/environment interaction; movement, and regions, p. 72

forms of geographic phenomena   point, linear, areal, volumetric, and space-time; sometimes called the *spatial dimensions* of geographic phenomena, p. 73

functional association   one object is located relative to another because of a link (e.g., physical or economic) between the two; also called *areal coherence* or *connectivity,* p. 71

geographical areal units   entities either natural (e.g., lakes, countries, oceans) or artificial (e.g., winter wheat region, urbanized areas); often used to tabulate aggregated data, p. 76

geographic data   facts about which conclusions can be drawn; chosen to describe geographic phenomena; associated with a spatial dimension, p. 73

geographical phenomena   elements of reality that have spatial attributes; any spatial phenomena can be the subject of geographical analysis within the limits of scale, p. 73

geographic scale   relates to the size of the area being studied; determines the types of inquiry questions that may be asked, p. 70

geography   science that deals with the analysis of spatially distributed phenomena, p. 68

human/environment interaction   a major theme in geography that deals with how people relate to their environment, p. 72

Landsat   United Sates satellite that provides digital data from which images of the earth's surface can be created, p. 85

location   a major research theme in geography; can be absolute or relative, p. 72

localization   relates to spatial clustering; measure of concentration, p. 71

measurement scales   definite rules established for each of four levels describing numerical events; the nominal, ordinal, interval, and ratio scales; precision increases from nominal to interval/ratio, p. 74–75

METSAT NOAA meteorological satellite(s) useful in producing a variety of earth images, p. 85

movement a major theme in geography study of the mobility of people, p. 72

place a major theme in geography; deals with distinguishability, p. 72

region major research theme in geography; mental constructs formed for the purpose of understanding spatial patterns of geographic phenomena; regions are based on the idea of internal homogeneity, p. 71

relative location position of an object relative to another object; dynamic in nature, p. 68

scale level of inquiry, ranging from micro (small earth areas) to macro (large earth areas); scale determines precision and generalization, p. 70

spatial association study of functional associations in a spatial context, p. 71

spatial interaction movement of people, goods, or ideas between places, p. 71

SPOT French-owned satellite useful in creating images of the earth's surface from a variety of on-board sensors, p. 85

symbol dimensions shape, size, color hue, color value, color intensity, pattern orientation, pattern arrangement, and pattern texture, p. 76

symbol types cartographic symbols are classed as point, line, or areal symbols, p. 76

typology of map symbols description of thematic map symbols based on a cross-tabulation of measurement scales and symbol types, p. 78–79

READINGS FOR FURTHER UNDERSTANDING

Abler, Ronald, John S. Adams, and Peter Gould. *Spatial Organization: The Geographer's View of the World.* New York: Prentice Hall, 1971.

Ackerman, Edward A. *Geography as a Fundamental Research Discipline.* Department of Geography, Research Paper No. 53. Chicago: University of Chicago, 1958.

Amedeo, Douglas, and Reginald G. Golledge. *An Introduction to Scientific Reasoning in Geography.* New York: Wiley, 1975.

Borchert, John R. "Maps, Geography, and Geographers." *The Professional Geographer* 39 (1987): 387–89.

Broek, Jan O. M. *Geography: Its Scope and Spirit.* Columbus, OH: Merrill, 1965.

Chang, Kang-tsung. "Data Differentiation and Cartographic Symbolization." *Canadian Geographer* 13 (1976): 60–68.

Dent, Borden D., ed. *Census Data: Geographic Significance and Classroom Utility.* NCGE Pacesetter Series. Chicago: National Council for Geographic Education, 1976.

Elifson, Kirk W. *Fundamentals of Social Statistics.* Reading, MA: Addison-Wesley, 1982.

Haggett, Peter. *Locational Analysis in Human Geography.* New York: St. Martin's Press, 1966.

Hartshorne, Richard. "The Concept of Geography as a Science of Space, from Kant and Humboldt to Hettner." In *Introduction to Geography: Selected Readings,* eds. Fred E. Dohrs and Lawrence M. Sommers. New York: Thomas Y. Crowell, 1967.

Harvey, David. *Explanation in Geography.* New York: St. Martin's Press, 1969.

Hsu, Mei-Ling. "The Cartographer's Conceptual Process and Thematic Symbolization." *American Cartographer* 6 (1979): 117–27.

Jenks, George. "Conceptual and Perceptual Error in Thematic Mapping." *Proceedings* (American Congress on Surveying and Mapping, March 1970): 174–88.

Kolars, John F., and John D. Nystauen. *Geography: The Study of Location, Culture and Environment.* New York: McGraw-Hill, 1974.

Levine, Robert. "City Stress Index." *Psychology Today,* November 1988, pp. 53–58.

MacEachren, Alan M. "Map Use and Map Making Education: Attention to Sources of Geographic Education." *Cartographic Journal* 23 (1986): 115–22.

McGrew, J. Chapman, Jr., and Charles B. Monroe. *An Introduction to Statistical Problem Solving in Geography.* Dubuque: Wm. C. Brown, 1993.

Monmonier, Mark S. *Maps, Distortion, and Meaning.* Association of American Geographers, Resource Paper No. 75–4. Washington, DC: Association of American Geographers, 1977.

———. *Mapping It Out: Expository Cartography for the Humanities and Social Sciences.* Chicago: University of Chicago Press, 1993.

Morrison, Joel. "A Theoretical Framework for Cartographic Generalization, with Emphasis on the Process of Symbolization." *International Yearbook of Cartography* 14 (1974): 115–27.

Muehrcke, Phillip. *Thematic Cartography.* Commission on College Geography, Resource Paper No. 19. Washington, DC: Association of American Geographers, 1972.

Statistics Canada. *1971 Census of Canada.* Vol. 1, pt. 1, *Population.* Ottawa: Ministry of Industry, Trade and Commerce, 1974. See also later years.

Stevens, S. S. "On the Theory of Scales of Measurement." *Science* 103 (1946): 677–80.

Taaffe, Edward J. "The Spatial View in Context." *Annals* (Association of American Geographers) 64 (1974): 1–16.

Taylor, Peter J. *Quantitative Methods in Geography: An Introduction to Spatial Analysis.* Boston: Houghton Mifflin, 1977.

Unwin, David. *Introductory Spatial Analysis.* London: Methuen, 1981.

CHAPTER

5

PROCESSING GEOGRAPHIC DATA: COMMON MEASURES USEFUL IN THEMATIC MAPPING

CHAPTER PREVIEW

Processing involves both mathematical and statistical measures designed to develop single-value statistics that describe nominal, ordinal, or interval/ratio data distributions: notably the mode, median, and mean. Dispersion characteristics are also part of the description of distributions: the variation ratio, percentiles, and the standard deviation. Skewness and kurtosis are descriptors of distribution shapes and prove useful in classifying activities. The useful measures of areal concentration are the coefficient of areal correspondence, areal means, and the location quotient. The mathematical measurement of association between spatial distributions is another important processing activity. Regression, correlation, and mapping residuals from regression are very useful data-processing techniques. Data classification, especially numerical classification, is a most important processing task in which groups are developed, based on numerical similarity. Graphic methods and methods of one-way analysis of variance are borrowed from statistics for use in numerical classification.

Data processing—analyzing and preparing geographic data for mapping—has become increasingly important in the broader contexts of thematic mapping. Cartographers often work with professionals, other than geographers, who customarily deal with statistics or who need data summarized before portraying them in graphic form. This chapter introduces the more common measures used in processing data for mapping.

THE NEED FOR DATA PROCESSING

The goal of cartographic design is to convey thought in graphic form as simply as possible. The contents may be simple, based on routinely derived measures, or complex, based on lengthy and sophisticated data compilations. With the widespread use of computers for data storage and processing, cartographers are finding it increasingly difficult to portray the data simply. Processing data is one way to reduce them into forms more suitable for straightforward communication. It provides mechanisms for data reduction, enhancement, retention of key features, and simplification of spatial patterns.[1]

MATHEMATICAL AND STATISTICAL METHODS

The principal methods of processing data for mapping are mathematical and statistical. Mathematical measures are usually quite simple and involve the production of common measures such as ratios, proportions, and percentages, including such activities as determining area and shape, and calculating trends or indices of comparison. For our purposes, we will consider *mathematics* to mean arithmetic, geometry, algebra, and other methods dealing with magnitudes and relationships that can be expressed in terms of numbers and symbols.

Statistics is a separate discipline and has been defined as follows:

> *Statistics is a body of* concepts *and* methods *used to collect and interpret data concerning a particular area of investigation and to draw conclusions in situations where uncertainty and variation are present.*[2]

Statistics is *not* the displaying of large numerical arrays such as census enumeration tables, although statisticians have historically spent much time compiling such data.

Three goals of statistics are[3]

1. To summarize observations.
2. To describe relationships between two variables.
3. To make inferences, both estimations and tests of significance.

RATIO, PROPORTION, AND PERCENT

Three of the simplest *derived* indices used by geographers and cartographers are ratios, proportions, and percents. The first two are often used interchangeably, but in fact they differ. **Ratio** is a good way of expressing the *relationship* between data. It is expressed as

$$\frac{f_a}{f_b}$$

where f_a is the number of items in one class and f_b the number in another class. The number of items is referred to as the *frequency,* a term used in most statistical work.

As an example, consider the ratio of the sexes in the population of a small town. If we discover that there are 3,000 males, this is one thing, but if we discover that there are also 1,500 females, we learn something else. The ratio is expressed this way:

$$\frac{3,000}{1,500} = 2$$

The 2 really means

$$\frac{2}{1}$$

Thus, there are two males in the town for every female.

A familiar ratio in geography is *population density,* defined as the number of people per square kilometer or other areal unit. Thus, if the same small town has a total area of 10 square kilometers, the density would be

$$\frac{4,500}{10}$$

We must reduce this ratio so that unity (1 sq km) is in the denominator:

$$\frac{4,500}{10} = \frac{x}{1}; \quad 10x = 4,500; \quad x = 450$$

$$\frac{4,500}{10} = \frac{450}{1}$$

or, simply, the population density is 450.

Proportion is the ratio of the number of items in one class to the total of all items. It is written

$$\frac{f_a}{N}$$

where f_a is the number of items (frequency) in a class and N is the total number of items or total frequency. In our small town, the proportion of males would be

$$\frac{3,000}{4,500} = \frac{30}{45} = \frac{2}{3} = .66$$

Typically, proportions are multiplied by 100, yielding **percentage.** In this case,

$$.66 \times 100 = 66\%$$

It is easier to describe the proportion of males in the small town as 66 percent than as .66.

Percentage is a familiar and useful index in many geographical analyses. It is simply derived and, when coupled with location, can yield information about geographical concentration.

VARIABLES, VALUES, AND ARRAYS

A distinction must be made between variables and values. **Variables** are the subjects used in statistical analyses: for example, height, rainfall, elevation, and income. Individual observations of the variable are called **values.** Values of the rainfall variable could be 16, 18, 19, 25, 34, and so on. It is typical in statistics to designate *variables* with a capital letter (e.g., *X* or *Y*) and *values* with subscripts of the lowercase form of the letters: x_1, x_2, x_3, x_4, x_5, and so on. To complete the symbology, the subscript is usually referred to simply as *i*. Thus x_i refers to any unspecified value of the variable *X*.

After an examiner completes the process of collecting measures, the values are arranged in an order called an **array.** Typically, an array arranges the values in ascending or descending order of magnitude or else lists nominal observations in similar groups. The following is a nominal array:

> **apple apple apple orange orange**
> **banana banana banana**
> **peach peach peach peach**

Arrays may also be developed for ordinal data:

> **city city city town town town town**
> **town village village village village village**

Here is an example of an interval array:

> **1,500 1,450 1,400 1,395 1,200 1,195 1,100**
> **1,050 1,025 955 905 859 825 795 750**

FREQUENCY DISTRIBUTIONS AND HISTOGRAMS

A *frequency distribution* is an ordered array that shows the frequency of occurrence of each value. The distribution is simply a method of portraying the data in a way that allows for easy inspection. The simple array can be placed in a table, as in Table 5.1.

When the number of observations is large, a simple table showing the frequency of each class or value becomes quite cumbersome. A common solution is to develop a frequency table from which a histogram can easily be drawn. The frequency table is constructed by dividing the total range of the data (the difference between the lowest and highest values in the array) into nonoverlapping subgroups called *classes* (or *cells*). The number of values belonging to each class is then counted. (The number of values in each cell is called the *class frequency.*) Usually, the relative frequency in each cell is determined. **Relative frequency** is that proportion of the observations belonging to a particular class:

$$\text{relative frequency of a class} = \frac{\text{class frequency}}{\text{total number of observations}}$$

When added together, all relative frequencies should total 1.00. (See Table 5.2.)

A note must be made about how statisticians group data to portray relative frequency. The classes shown in Table 5.2 are referred to as *overlapping* classes because a given value may be placed in either of two adjoining classes. For example, an actual value of 30,000 can be in the third or fourth class. This is a standard method of presenting data by the statistician, and appears from a long tradition. But this practice is not used by the cartographer, as will be discussed in Chapter 7. Cartographers do not like the ambiguity imposed by this strategy. **Histograms** are special graphs drawn using the information from a relative frequency table. Most often, a *relative frequency histogram* is drawn by plotting the class boundaries on the horizontal axis and

Table 5.1   Typical Frequency Distributions in Nominal, Ordinal, and Interval Scales

Nominal		Ordinal		Interval	
Crop	Frequency	Hardiness	Frequency	Yields	Frequency
Wheat	10	Most	12	400	18
Barley	8	Intermediate	15	300	22
Corn	6	Least	8	250	23
				175	15
				125	10

Table 5.2 Classed Frequency Distribution
and Relative Frequency Table

Class Interval	Frequency	Relative Frequency
0–10,000	39	.204
10,000–20,000	51	.267
20,000–30,000	33	.172
30,000–40,000	25	.130
40,000–50,000	20	.104
50,000–60,000	18	.094
60,000–70,000	5	0.26
Total	191	1.000

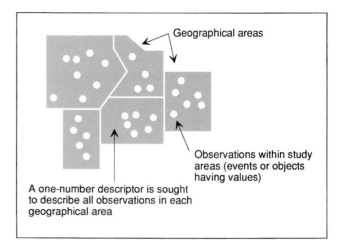

Figure 5.2 Summarizing geographic data.
Summarizing geographic data usually involves developing a one-number descriptor from several numerical observations occurring in geographical study areas.

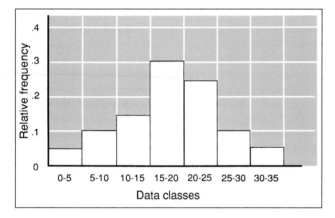

Figure 5.1 Relative frequency histogram.
Heights of bars are drawn proportional to their relative frequencies. The sum of the bar heights equals one.

erecting rectangles over each class as a base. The vertical axis is scaled in divisions of relative frequency. If the horizontal axis is laid out with the widths of the rectangle bases equal, their heights are simply scaled to the relative frequencies shown on the vertical axis. (See Figure 5.1.)

Relative frequency histograms with proportional areas are more useful than simple histograms that have their vertical axes scaled to frequency. Because the total height of all rectangles equals one, a quick visual inspection will reveal how much of the total each class represents.

SUMMARIZING DATA DISTRIBUTIONS

The principal purpose of most descriptive statistics is to develop one best numerical description for a data set. (See Figure 5.2.) In many mapping problems, events or observations are found at specified locations within a bounded area. The area may be a specially created study region, a

political subdivision, or a statistical division defined by the Census Bureau. It is advantageous to work with the latter areas because they function as enumeration areas for which there are numerous data available.

Each of the events or observations conceptualized in Figure 5.2 has a value associated with it. Because mapping the values at each location is usually impracticable, one *best* numerical descriptor is often computed for all observed values within the bounded area. The values are *aggregated*—this has already been done for us in many census enumerations. For example, the median family income in the Little Rock, Arkansas, Metropolitan Statistical Area (MSA) in 1969 was $8,285; this is an aggregated and summarized descriptor for all the individual incomes reported in the MSA.

Several methods exist for determining the one best descriptor for a data set, depending on the scale of measurement. It is often desirable to know how values are arrayed around the descriptor, because this can show how effective the descriptor is. Are values clustered closely around the descriptor or spread uniformly throughout the data range?

NOMINAL SCALE: THE MODE AND THE VARIATION RATIO

Nominal scale data are those that are the result of *counting*. For example, there are 10 hats, 5 coats, 22 pants, and 12 pairs of shoes. Generally, nominal scale data are considered the least sophisticated because only simple arithmetic operations can be performed on such data.

For nominal scale measurement, the *mode* is the one best numerical descriptor, or summary statistic, and the *variation ratio* is the index that tells us how the values are arrayed in the distribution and defines a distribution in terms of its dominant class. The **mode** is the name or

Table 5.3   Results of Marketing Opinion Research on Types of Goods Sold at Women's Apparel Store
(m = Modal Class)

Zip code area 1	Frequency	ZIP code area 6	Frequency
A	10	A	0
B	147 m	B	172 m
C	32	C	26
D	11	D	2
ZIP code area 2	Frequency	ZIP code are 7	Frequency
A	125 m	A	38
B	12	B	70 m
C	33	C	42
D	30	D	50
ZIP code area 3	Frequency	ZIP code area 8	Frequency
A	137 m	A	24
B	10	B	20
C	25	C	24
D	28	D	132 m
ZIP code area 4	Frequency	ZIP code area 9	Frequency
A	153 m	A	36
B	5	B	40
C	5	C	46
D	37	D	78 m
ZIP code area 5	Frequency	ZIP code area 10	Frequency
A	5	A	32
B	135 m	B	17
C	50	C	9
D	10	D	142 m

ZIP Code Area	Variation Ratio
1	.265
2	.375
3	.315
4	.235
5	.325
6	.140
7	.650
8	.340
9	.610
10	.290

number of the class in a nominal distribution having the greatest frequency.

A market researcher conducted a study to determine shoppers' opinions of the types of apparel sold in a women's apparel store. Ten ZIP code areas immediately surrounding the store were identified as the study area, and 200 people from each ZIP code area were asked to choose among the following alternatives:

A. Prefer the better-quality merchandise, and am willing to pay for it.
B. Always seek lower-quality merchandise, will usually not pay for high-priced goods.

C. Prefer the better-quality merchandise but will buy it only if on sale.
D. Like the current mix of merchandise and will buy both the better-quality and lower-quality merchandise.

Table 5.3 lists the final results of the study. The modal class is identified for each ZIP code area,

In this example, people in ZIP code areas 2, 3, and 4 clearly prefer the better-quality merchandise and are apparently willing to pay for it; people in ZIP code areas 8, 9, and 10 have no clear-cut preferences. The results of this study could be used by the store manager to direct advertising to certain geographical locations.

It is clear from these results that the frequencies tallied in ZIP code areas 7 and 9 do not indicate modes as strong as for the others. The variation ratio (v) is a quantification of how well the mode describes the distribution. The **variation ratio** is the proportion of occurrences not in the modal class. It is computed this way:

$$v = 1 - \frac{f_{modal}}{N}$$

where f_{modal} is the frequency of the modal class and N equals the total number of occurrences in all classes. The smaller the v is, the better it summarizes the distribution; the higher the v, the less satisfactory it is.

The variation ratio has been computed for each ZIP code area in our example. It is clear that areas 7 and 9 contain less definite modal classes than the others ($v = .65$ and .61, respectively). This indicates to the advertiser that a different approach might be used in these areas.

Modes and variation ratios are the most important indices of central tendency and dispersion for nominally scaled data. They can also be used for interval/ratio scales, although other statistics are preferable at such scales.

ORDINAL SCALE: THE MEDIAN AND PERCENTILES

If data have been collected that permit *ranking,* they are said to be at the **ordinal** level of measurement. With this kind of data, numerical distance between values is ordinarily not desirable, only that a position in the array is more than or less than another.

When data are collected at the ordinal scale, the statistic that describes central tendency is the *median,* and one of several *decile ranges* describes the dispersion of observations around the median. The **median** can be defined as the point that neither exceeds nor is exceeded in rank by more than 50 percent of the observations in the distribution. Its determination is straightforward when the number of values is an odd number of ranks; special rules apply when the number of values is an even number.

Just as the mode can be determined for interval/ratio observations, the median can be calculated for such higher-level data. It is probably more common for the median and decile ranges to be used in this way.

Table 5.4 illustrates an array of 1980 birth rates for an odd number of Arkansas counties. The median birth rank is 10; it is the typical rank. Its value of 14.3 is the middle value of the array. For this geographic area, it might be said that 14.3 births per 1,000 people is the typical birth rate. We judge the other ranks, or counties, as being high or low *relative to the median value.*

Special problems arise when the number of ranks is even. When dealing solely with ordinal data, we can say only that the median *position* is halfway between the two middle ranks. We *cannot* say that it is *exactly* halfway in terms of value—with ordinal data, we have no idea of val-

Table 5.4 Arkansas: Health System Area I

Counties Ranked by Birth Rates (per 1,000 Population)

County	Rank		Birth Rate
Boone	1		26.2
Perry	2		16.1
Washington	2		16.1
Newton	3		16.0
Crawford	4		15.8
Polk	5		15.6
Sebastian	6		15.5
Madison	6		15.5
Hot Spring	7		15.0
Conway	8		14.7
Johnson	9		14.5
Montgomery	9		14.5
Pope	10	Median	14.3
Logan	11		13.7
Franklin	11		13.7
Clark	12		13.3
Pike	13		13.2
Carroll	14		12.3
Searcy	14		12.3
Garland	15		12.2
Benton	16		11.5
Yell	16		11.5
Marion	17		11.2
Scott	18		10.9
Baxter	19		9.8

Source: Arkansas Division of Health Statistics, Arkansas Vital Statistics, *1980, p. 61.*

ues. However, when dealing with ratio data with an even number of ranks, the median is the average of the values of the two middle ranks. (See Table 5.5.)

Because the concept of the median is related to position, a median value may be the same for two distributions but may fail to provide adequate comparison between the two data sets. (See Table 5.6.) When comparing distributions, it is best not to attach too much significance to the median values unless additional measures of dispersion are computed.

One method of looking at the dispersion of ranks or scores for ranked observations is to compute **quartile values,** which are obtained in a manner similar to finding the median. The lower (first) quartile separates the lower 25 percent of the observations from the upper 75 percent and is called the 25th **percentile.** The second quartile is the median, and the upper (third) quartile marks the division between the lower 75 percent of the observations from the upper 25 percent and is referred to as the 75th percentile. Any percentile can be computed, but it is not customarily done unless there are at least 25 observations in the array.[4]

Table 5.5   Arkansas: Health Systems Area III

Counties Ranked by Birth Rates (per 1,000 Population)

County	Rank		Birth Rate
Monroe	1		19.4
Pulaski	1		19.4
Lonoke	2	Median	16.8
Saline	3		13.8
Faulkner	4		13.6
Prairie	5		12.1

Median rank = 2-1/2

Median value = $\dfrac{16.8 + 13.8}{2}$ = 15.3

Source: Arkansas Division of Health Statistics, Arkansas Vital Statistics, *1980, p. 73.*

Table 5.6   Two Hypothetical Ranked Distributions with Identical Medians

Distribution A		Distribution B
69		722
67		341
66		250
52		99
49		98
37	Median	37
36		30
32		28
32		14
32		1
30		1

The *quartile deviation* is defined as

upper quartile – lower quartile

$$\frac{\textbf{(interquartile range)}}{2}$$

and is the average expectation of how the occurrences will be distributed on either side of the median.[5] The employment of the quartile measure or the quartile deviation is most useful in comparing distributions.

The interquartile range identifies the 50 percent of all values that are closest to the median value. This is useful in processing data before mapping, because mapping those geographic areas within the interquartile range reveals the geographical pattern of values nearest the typical (median) value. (See Table 5.7 and Figure 5.3.)

Geographical patterns of abortion rates by states in the United States show only a moderate amount of clustering. Those states in the upper quartile (*Q*1) are in the Northeast, the extreme Southeast in Florida, and along the Pacific Coast. Those states near the median (the interquartile range) appear to be spread out throughout the country with no particular concentration or clustering, leading one to speculate that there are no regional cultural operants governing the practice of child abortion.

INTERVAL AND RATIO SCALES: THE ARITHMETIC MEAN AND THE STANDARD DEVIATION

Numerical data that are the result of *measuring* are usually referred to as being at the **interval** or **ratio** level of measurement. Interval scales have arbitrary starting points (such as the Fahrenheit temperature scale), and ratio scales have zero as a starting point (the Kelvin temperature scale).

Data that exist at the interval or ratio scales are sometimes more useful in geographical analysis because more statistical measures can be applied to them. When we summarize a distribution at these scales, the goal is to obtain a one-number index to describe the array of values. The statistic most frequently used for this purpose is the *arithmetic mean,* and the statistic to describe dispersion around the mean is the *standard deviation.*

The **arithmetic mean** is derived by adding all the values and dividing by the number of values:

$$\overline{X} = \frac{\Sigma x}{N}$$

where $\overline{X}$ is the mean; Σ is a summation sign; x is a value in the array (Σx means to sum all x-values), and N is the total number of x-values.

It may be easier to see the idea of the mean, and later the standard deviation, by looking at Figure 5.4, a distribution of values on a number line. The mean is a point on the number line around which all values can be thought to balance. The arithmetic mean is considered the best one-number description of all values in a data array.

Like the other measures of central tendency (the mode and median), the mean does not show *how* the values are arranged around the mean. Two quite different numerical arrays can have identical arithmetic means. In Table 5.8, for example, the range of the values (difference between highest and lowest values) for variable *X* is 48, whereas it is only 10 for variable *Y.* The identical means do not reflect these quite different deviation patterns. We may also call this the *variation* around the mean.

One might think that a good way to express this deviation is to calculate the average deviation. Doing this, however, yields no measure at all! If we calculate each *x*-value's deviation from the mean and sum these (as we must do in computing an average), they would total 0. The positive deviations will be offset by the negative deviations.

Table 5.7   Abortion Rates in the United States, 1993

Alabama	18.2	Montana	18.2
Alaska	16.5	Nebraska	15.7
Arizona	24.1	Nevada	44.2
Arkansas	13.5	New Hampshire	14.6
California	42.1	New Jersey	31.0
Colorado	23.6	New Mexico	17.7
Connecticut	26.2	New York	46.2
Delaware	35.2	North Carolina	22.4
District of Columbia	138.4	North Dakota	10.7
Florida	30.0	Ohio	19.2
Georgia	24.0	Oklahoma	12.5
Hawaii	46.0	Oregon	23.9
Idaho	7.2	Pennsylvania	18.6
Illinois	25.4	Rhode Island	30.0
Indiana	12.0	South Carolina	14.2
Iowa	11.4	South Dakota	6.8
Kansas	22.4	Tennessee	16.2
Kentucky	11.4	Texas	23.1
Louisiana	13.4	Utah	9.3
Maine	14.7	Vermont	21.2
Maryland	26.4	Virginia	22.7
Massachusetts	28.4	Washington	27.7
Michigan	25.2	West Virginia	7.7
Minnesota	15.6	Wisconsin	13.6
Mississippi	12.4	Wyoming	4.3
Missouri	11.6		

Median ($Q2$) = 18.9; $Q1$ = 13.45; $Q3$ = 26.3;
Interquartile range = 26.3 – 13.45 = 12.85
Quartile deviation = 12.85/2 = 6.425
Note: Abortion rates per 1,000 women aged 15–44.

Source: Stanley K. Henshaw and Jennifer Van Vort, "Abortion Services in the United States," Family Planning Perspectives *26 (1994). pp. 100–106.*

Figure 5.3   Mapping the interquartile range.
States whose values are in the interquartile range are mapped, as are those above Q_3 and below Q_1, with distinct area patterns. There appears to be an absence of spatial clustering of states in the quartile ranges. This map presents the data of Table 5.7.

The large Greek sigma Σ is used in statistical notation to denote summation. It may be $\sum_{i=1}^{n}$. This means that all values from the first ($i = 1$) to the nth are summed in the operation. Hence, $\sum_{i=1}^{n}$ directs you to add all x-values from the first to the nth.

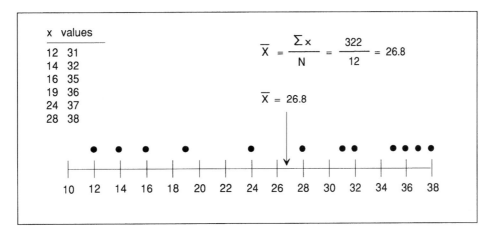

Figure 5.4   Plotting the arithmetic mean on a number line.
The arithmetic mean can be thought of as a balance point for the values on the line.

Table 5.8   Two Data Sets with Identical Means (Units = Tons)

Variable X	Variable Y
6	28
39	24
11	26
22	29
23	23
44	27
49	21
1	20
31	21
19	29
28	30
27	22

$\Sigma x = 300$
$N = 12$
$\overline{X} = 25$
Range $= |\,49 - 1\,| = 48$
$\overline{X}^2 = 625$
$\Sigma x^2 = 9{,}864$
σx^2 (variance) $= 196.84$
σ (standard deviation)
  $= 14.03$ tons

$\Sigma y = 300$
$N = 12$
$\overline{Y} = 25$
Range $= |\,30 - 20\,| = 10$
$\overline{Y}^2 = 625$
$\Sigma y^2 = 7{,}642$
σy^2 (variance) $= 11.76$
σ (standard deviation)
  $= 3.43$ tons

The standard deviation is the solution to this dilemma. We begin by first squaring each deviation. This eliminates the negative signs (minus times minus equals a plus). Now we can sum them, divide by N, and compute the mean squared deviation. This is also called the **variance** and is symbolically written

$$\sigma^2 = \frac{\Sigma(x - \overline{X})^2}{N}$$

It is sometimes necessary to compute the mean of grouped data. In these cases, we use this formula:

$$\overline{X} = \frac{\Sigma fx}{\Sigma f}$$

The xs are now class midpoint values, and the fs are frequencies in each class. The denominator is the sum of all frequencies.

where Σ^2 is the variance; $(x - \overline{X})^2$ means to compute each x's deviation from the mean $\overline{X}$, square this deviation and sum them all; N is the number of values.

The formula also may be written this way:

$$\sigma^2 = \frac{\Sigma x^2 - N\overline{X}^2}{N}$$

This version of the formula is preferred because it is easy to use, especially when using electronic calculators. First, each x is squared and these are summed. The average is squared, multiplied by N, and this product is subtracted from the sum of x-values squared. Finally, this is divided by the number (N) of observations.

There are two main advantages to using this method of determining variation. First, it is relatively simple; many handheld calculators perform these computations automatically. Second, the units of the variance are identical to the original units of the variable. If variable X in Table 5.8 is in tons, the variance is expressed in (tons)^2. We take the square root of the variance to obtain the measure of deviation in the same units as the data, and this is called the **standard deviation:**

$$\sigma = \sqrt{\text{variance}} = \sqrt{\frac{\Sigma x^2 - N\overline{X}^2}{N}}$$

The standard deviation is especially useful when comparing two variables. The smaller of two standard deviations indicates values more closely packed around the mean than a larger standard deviation. An examination of Table 5.8 reveals this. Variable X's Σ of 14.03 indicates a wider dispersion than variable Y's 3.43. Their ranges (48 and 10 respectively) can also be compared. Of course, comparison of standard deviations is only meaningful with variables in the same units.

Using the standard deviation as a measure of dispersion in one distribution is more subtle, but there are certain relationships that exist between X and the standard deviation that assist in this analysis. Using Chebyshev's rule, these summaries may be drawn *for all data sets:*[6]

1. The space between $\overline{X} - 2\sigma$ and $\overline{X} + 2\sigma$ contains *at least* 75 percent of the observations.
2. The space between $\overline{X} - 3\sigma$ and $\overline{X} + 3\sigma$ contains *at least* 88 percent of the observations.
3. In general cases, for any multiplier K (>1), the space between $\overline{X} - K\sigma$ and $\overline{X} + K\sigma$ contains at least $1 - \dfrac{1}{k^2}$ fraction of the observations.

Frequency or relative frequency distributions having certain specified shapes represented by bell-shaped curves are known as **normal distributions.** It has long been thought that a great many naturally occurring phenomena display such a distributional shape. This is not necessarily true, but the normal distribution remains at the heart of much statistical analysis. If data are distributed normally, *specific* percentages of observations occurring within spaces between $\overline{X}$ and σ can be specified. (See Figure 5.5.)

Central to the use of normal distributions is the calculation of the *probability* that certain observed values will occur. For data distributed normally, events that occur at further distances from the mean are less likely to occur.

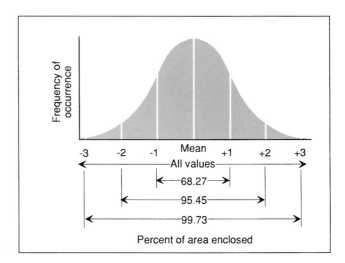

Figure 5.5   The normal distribution.
This figure illustrates the percentage of observations that fall within the standard deviations ±1, ± 2, and ± 3. Note that values above or below three standard deviations are very rare (less than 1 percent).

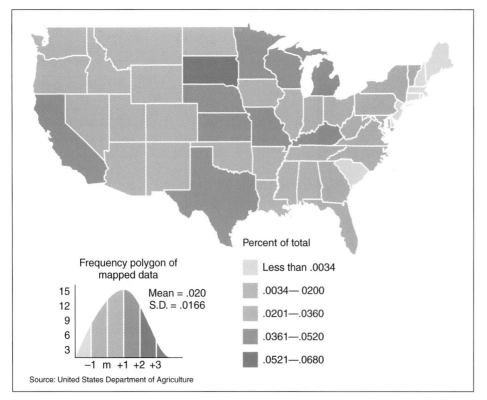

Figure 5.6   Mapping data using means and standard deviations.
Processing data can involve using means and standard deviations and may be used to present final geographical analyses or simply as work-maps. In final presentations, it is helpful to include as part of the legend material a histogram of the data distribution. (*Source: United States Bureau of the Census,* Statistical Abstract of the United States,: 1980, 101st ed. *[Washington, D.C.: USPGO, 1980], p.467.)*

The arithmetic mean and the standard deviation are frequently used in analysis of geographic data. They have been introduced here as a way of describing data observations, but they can also be used in other areas of statistical analysis involving hypothesis testing. The arithmetic mean and the standard deviation are employed in a great many cartographic processing activities. (See Figure 5.6.)

Skewness

When the peak or mode of a distribution is displaced to either side of the mean in a distribution, it is referred to as *skewed.* If the bulk of the frequencies are found to be left of the mean (and a long tail of low frequencies to the right), it is positively skewed. If the long tail is to the left of the mean, and the greater portion of frequencies greater than the mean, it is negatively skewed. **Skewness** is the measure of the displacement, and is calculated this way:[7]

$$\text{skewness} = \frac{\Sigma(x - \overline{X})^3}{n\sigma^3}$$

Negative values indicate negative skewness and positive values positive skewness. The higher the value the greater the difference away from a normal distribution. Normal distributions have a skewness of 0.0.

Kurtosis

Kurtosis is a measure that describes the "peakiness" of a distribution. A flat distribution is one in which there is a nearly equal number of observations in each cell in the histogram. A distribution having a high "peakiness" is one in which the greater bulk of the observations are in one cell. Kurtosis is calculated using the formula:[8]

$$\text{kurtosis} = \frac{\Sigma(x - \overline{X})^4}{n\sigma^4}$$

A normal distribution has a kurtosis of 3.0. Values greater than 3.0 indicate "peakiness," and values below 3.0 a flatness in the histogram.

MEASURES OF AREAL CONCENTRATION AND ASSOCIATION

Since cartographers and geographers are concerned with the analysis of spatial patterns, they naturally examine data sets during preprocessing to discover any significant patterns of spatial variation. Some of the more useful ways of describing spatial patterns are discussed next.

THE COEFFICIENT OF AREAL CORRESPONDENCE

Researchers often need to assess the degree of correspondence between two areal distributions, usually to discover or demonstrate causal relationships. Figure 5.7, for example, shows two maps that present nominal data. One way to examine the areal correspondence—a measure of congruence—is to inspect them visually, but inconsistency will result because of the individual differences and biases of the observers. Thus, a *numerical* measure is preferable.

One such numerical measure is the **coefficient of areal correspondence,** computed as follows:[9]

$$C_a = \frac{\text{area covered jointly by both phenomena}}{\text{total area covered by the two phenomena}}$$

C_a varies from 1.0 (perfect correspondence) to 0 (no areal correspondence). Multiplication by 100 will yield a percentage. Table 5.9 contains the areas applicable to the example used in Figure 5.7. For this example, $C_a = 21$.

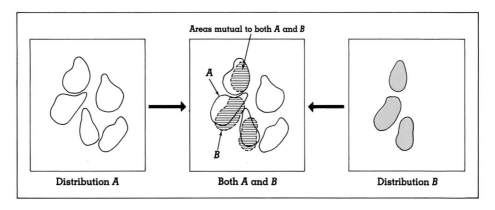

Figure 5.7   Coefficient of areal correspondence.
This index is a simple proportion measuring the amount of congruence between two areal distributions.

Table 5.9 Areas in Figure 5.7, the Example of the Coefficient of Areal Correspondence

Distribution A (sq km)
63
26
23
17
20
—
149

Distribution B (sq km) $Ca = \dfrac{42}{200}$
18                                  $= .21$
21
12
—
51

Total of areas mutual to A and B (sq km)
16
17
9
—
42

Areal means can reflect weighted values. A weighted average is computed in the same way as for grouped data:

$$\overline{X} = \frac{\Sigma fx}{\Sigma f} \qquad \overline{Y} = \frac{\Sigma fy}{\Sigma f}$$

The f is the weight at each x, y location. The denominator is the total of all the weights in the study areas.

This index is descriptive and does not necessarily suggest causal relationships. However, values approaching 1.0 may indicate that the areal variables are somehow related. Further inquiry would certainly be justified.

AREAL MEANS

Cartographers often deal with point patterns and have to describe them in some meaningful way. The *areal mean* is a useful way of illustrating the spatial balance point, in much the same way the mean describes balance along a number line. A plot of several areal means representing the same phenomenon over time can be useful in clarifying spatial change.

Areal Mean

In the simplest version of computing an **areal mean,** a Cartesian coordinate system is superimposed over the distribution of scattered points. (See Figure 5.8a.) The axes are scaled in convenient units with arithmetic intervals; the smaller the numbers, the easier the computations. Vertical and horizontal ordinates are then constructed from each point to each axis, and X- and Y-values are obtained for each point. After all X- and Y-values are noted, an arithmetic mean is computed for each. The pair of means is called the *areal mean.* It is plotted by using the same Cartesian coordinates.

Most geographical analyses involve point patterns (or centers of areas) that have weights attributed to them. Except at the simplest nominal scales, geographic point phenomena do not usually occur everywhere with equal value. An especially interesting study is to plot weighted means

over time to discover a spatially dynamic pattern. Weights are most often socioeconomic data such as income, production, sales, or employment data. Sometimes, data are available only for larger enumeration areas such as counties, census tracts, or blocks. In these cases, the data can be assumed to be aggregated at a central point within the area. (See Figure 5.8b.) The central point is then used as before in computing the weighted areal mean.

Advanced techniques, too lengthy to introduce here, are available to study other aspects of areal means—for example, centrographic analysis, including confidence plotting and hypothesis testing.

Standard Distance

Another useful statistic related to the areal mean is the **standard distance** (SD). For nonareal data, the standard deviation is the statistic ordinarily used to describe the dispersion of values around the mean. In the areal case, the equivalent is the standard distance, which is obtained by measuring the distance from each point to the areal mean, squaring each distance, summing these, dividing by N, and obtaining the square root.[10]

$$SD = \frac{\sqrt{\Sigma d^2}}{N}$$

where SD = standard distance, d is the distance from each point to the center, and N is the total number of points.

In practice, this involves considerable calculation—the distance from each point to center can be obtained using analytic geometry, which is time-consuming—and there is much potential for error. Other formulas have therefore been worked out, notably the following:

$$SD = \sqrt{\sigma y^2 + \sigma x^2}$$

where σy^2 is the variance of Y and σx^2 is the variance of X. The values of X and Y are obtained from a grid placed over the distribution of points (as in the areal mean).

Standard distances become useful when compared to others computed for identical data at different time periods.[11] Together with areal means, standard distances give very good quantitative assessment of point patterns. Standard distances may be plotted on the original point distributions using a circle, centered on the areal mean, with a radius equal to the standard distance. Used by themselves, however, these plotted circles have little utility.

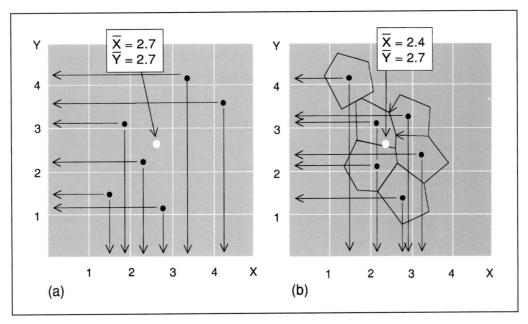

Figure 5.8   Plotting areal means.
Areal means for point distributions (a) or for centers of areas (b) are computed with equal ease. In each case, the points may be weighted by geographical values associated at each point.

A summary of the most important properties of areal means and standard distances includes the following:[12]

1. Both include all events (points) used in the analysis.
2. Because all observations are used, both are extremely sensitive to change in any one observation.
3. Because squares of distance are used in computing the standard distance, its value is strongly affected by points at extreme distances from the areal mean.

LOCATION QUOTIENT

A simple measure to illustrate geographical localization is the **location quotient.** The location quotient is a ratio of proportions and provides an index for determining geographical share. One normally assumes at the outset that the variable being measured is uniformly distributed. It is not difficult to calculate and can be used to determine if more sophisticated measures should be pursued. Moreover, quotients can be mapped easily and show at a glance the idea of geographical concentration.

The location-quotient concepts can be illustrated by example. A geographer wishes to determine if there is anything unusual about the distribution of food store sales in Maryland. Data are collected on food store sales, as a percent of all retail sales, for the state and for each county. (See Table 5.10).

First, the proportion of food sales to total retail sales in the state is determined (expressed as a percent):

$$\frac{5,718,377}{30,205,991} \times 100 = 18.9\%$$

Next, the food sales to total retail sales in each county is calculated. Each county's location quotient is then determined by a ratio between the county's proportion and that of the state. For example:

$$\textbf{Caroline} \;\; \frac{36.5}{18.9} = \textbf{1.93}$$

If the proportion of food sales to total sales is the same in a county as it is in the entire state, the location quotient would be 1.0, and the distribution of the proportion of food sales would be *uniform* throughout the state. In counties having a proportion less than the state's, a location quotient of less than 1.0 results, suggesting lower food sales than would be expected. A location quotient greater than 1.0 suggests that a county has a greater share of food sales than expected. The quotient cannot be less than 1.0. Of course, it is not likely for all counties to have a quotient of 1.0. What is unusual and bears checking is when all the counties having greater or lesser than 1.0 are grouped together geographically.

In the case of Maryland, food sales as a proportion of all retail sales is hardly uniform, and can be seen by mapping the quotient. (See Figure 5.9.) The counties of Calvert, Caroline, Charles, Kent, Queen Anne's, and especially Somerset, considerably exceed the quotient of 1.0 (all are in the top quartile). It may or may not be geographically significant that these counties (except Caroline and Charles) border the Chesapeake Bay. Somerset County deserves closer scrutiny at least. Food distributors and marketing teams may use this simple technique to guide them in any future or more sophisticated studies.

Table 5.10 Maryland: Food Store and Total Retail Sales by County and for the State, 1986

County	Total Retail Sales (× 1,000 Dollars)	Food Store Sales (× 1,000 Dollars)	Food Sales as Percent of Total Sales	Location Quotient
Allegany	481,668	95,513	19.8	1.05
Anne Arundel	2,534,556	427,237	16.8	.88
Baltimore	5,688,127	1,039,909	18.2	.96
Baltimore City	3,775,658	732,202	19.3	1.02
Calvert	107,384	32,754	30.5	1.61
Caroline	58,642	21,440	36.5	1.93
Carroll	498,542	120,438	24.1	1.28
Cecil	319,283	57,096	17.9	.95
Charles	318,420	85,140	26.7	1.41
Dorchester	182,872	33,311	18.2	.96
Frederick	771,251	169,330	21.9	1.16
Garrett	136,742	23,659	17.3	.92
Harford	905,157	186,551	20.6	1.09
Howard	752,198	141,776	18.8	.99
Kent	94,276	30,254	32.0	1.69
Montgomery	6,053,328	1,051,753	17.4	.92
Prince George's	5,253,177	1,024,023	19.5	1.03
Queen Anne's	109,947	33,444	30.4	1.61
St. Mary's	236,175	60,927	25.8	1.37
Somerset	61,924	27,387	44.2	2.34
Talbot	242,648	46,783	19.2	1.02
Washington	691,825	116,915	16.9	.89
Wicomico	581,718	88,810	15.3	.81
Worcester	350,473	71,725	20.4	1.08
State	30,205,991	5,718,377	18.9	1.00

Source of data: Rand McNally, Commercial Atlas and Marketing Guide, *119th ed. (Chicago: Rand McNally, 1988); location quotients calculated by author.*

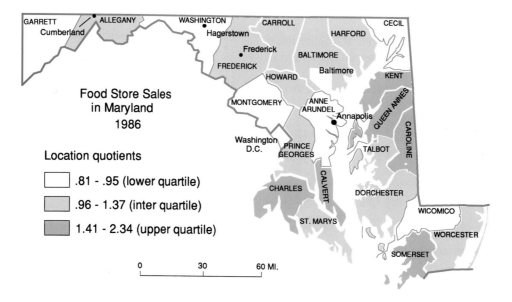

Figure 5.9 Location quotient comparing food sales to total retail sales in Maryland.
Countries having a location quotient greater than 1.0 are clustered in the east and south of the state, and four of those in the top quartile border the Chesapeake Bay.

Performing data calculations such as the location quotient is very nearly out of the realm of processing geographic data for mapping. In fact, it borders on data analysis. Nonetheless, if measures such as these are useful for the cartographer or the client, they should indeed be attempted.

MEASURING SPATIAL ASSOCIATION BY REGRESSION METHODS

An especially useful method for examining spatial association is **residuals from regression.** It is more complex than the foregoing methods but yields extremely valuable results.

The essential idea behind regression is to measure the degree of association between two magnitude (interval/ratio) variables. One variable normally is considered *independent* and the other *dependent,* because we are looking for causal relationships. A classic example is the relation of crop yield (dependent variable) to fertilizer application (independent variable). Regression methods in themselves do not establish causality and are useful only when applied to related variables.

Each pair of corresponding magnitudes for each variable is plotted on suitable graph paper to form a **scattergram.** (See Figure 5.10.) Normally, the independent variable is plotted on the horizontal axis and the dependent variable on the vertical axis. A **regression line** is a "best fit" line plotted on the graph, summarizing the mathematical relationship between the two variables. Numerous regression lines may be drawn; the simplest and most common is the straight, or linear, regression line.

Regression lines are plotted so that they minimize all deviations of the points from the line. (See Figure 5.10.) This is why they are termed *lines of best fit.* By plotting the line at different locations or at different slants, one can easily see how its position could affect the minimizing operation. The general **equation for a straight line** is

$$y = a + bx$$

For regression in the linear case, we use the following equations to determine values of a and b so that we might plot the *least-squares regression line:*

$$\Sigma y = a \cdot N + b \cdot \Sigma x$$
$$\Sigma xy = a \cdot \Sigma x + b \cdot \Sigma x^2$$

These equations can be solved simultaneously. The value for a may be found and substituted to find b. The value b is usually called the *regression coefficient*. With a and b known, various values of x and y may be selected and used in the general equation for the line in order to plot the regression line. One interesting feature of the least-squares regression line is that it passes through the means of both the X and Y variables. This can be used for a visual check of the plotted line.

In regression analysis, we are generally interested in how well the regression line summarizes the actual values

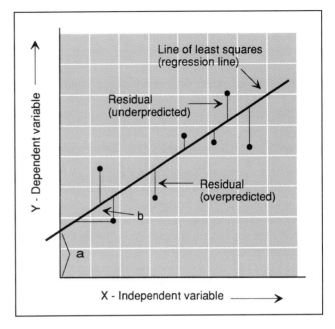

Figure 5.10   Components of the linear regression line. The a in the equation of a straight line ($y = ax + bx$) is the intercept of the line with the y-axis. The b refers to the slope of the line. Notice that both underpredicted and overpredicted residuals may be recorded. When a computed (expected) value of y is greater than the observed value, it is overpredicting.

of the dependent variable (Y). For each actual value of the independent variable (X), there are two corresponding values of y: y-observed and y-estimated from the regression line. The difference between these two is called the *residual* (or Z_i) value:

$$Z_i = (y\text{-observed}) - (y\text{-estimated})$$
$$Z_i = y - y'$$

The Standard Error of the Estimate
How well the regression line, or the computed values of a and b, fits the data can be estimated by calculating the **standard error of the estimate.** This is actually a standard deviation of the residual values discussed above. It is calculated as follows:

$$SEy' = \sqrt{\frac{\Sigma(y - y')^2}{N - 2}} \text{ or } \sqrt{\frac{\Sigma Z^2}{N - 2}}$$

The standard error expresses the *spread* of the plotted observed points around the regression line, as well as how accurate X is in predicting Y. It may be recalled that for normally distributed data, certain percentages of the observations fall between plus and minus 1, 2, or 3 standard deviations. Likewise, the observed values of Y fall between plus and minus one, two, or three standard errors from the regression line. About 68 percent of the observations fall between ±1 standard errors of the regression line, and about 95 percent between ±2 standard errors. Observations beyond ±2 are extremely rare—only 5 percent.

Correlation

Statistical **correlation** is a method of gauging the *strength* of the mathematical association between variables. The statistic is called r and varies between +1 and −1. Typically, plotted points on scattergrams of different r values will appear as in Figure 5.11. The formula for r is usually written:

$$r = \frac{\Sigma(y - Y)(x - X)}{\sqrt{\Sigma(y - Y)^2 \Sigma(x - X)^2}}$$

Examination of the scattergrams of Figure 5.11 illustrates that a perfect correlation can exist for either direct or inverse relationships. The correlation coefficient indicates how well X is performing as an independent variable relative to Y. The statistic r is dimensionless, not in the units of the original variables.

The value of r and its overall significance normally depend on the number of observations being compared. However, coefficients below about .25 are generally not indicative of any particular association between two variables. Between .25 and .50, there is a weak to moderate association, and r values between .50 and about .75 usually indicate a strong association. A coefficient of .75 or more suggests a very strong association between variables.

Mapping Residuals from Regression

Besides the numerical indices of association between two variables, correlation and regression analysis is important in *mapping residuals*. It is useful in identifying and locating those cases that are *not* well explained by the regression line (observed values beyond ±1 or ±2 standard errors of the estimate). The goal is to determine if their spatial patterns reveal other factors affecting their behavior in the data set.

Such analysis begins with calculation of the correlation coefficient, the regression line, and the standard error of the estimate. Mapping is then done if there is sufficient indication that it may yield further information. An example is provided in Tables 5.11 and 5.12, and in Figures 5.12, 5.13, and 5.14. A particular segment of the population (the independent variable, X) is thought to hold certain attitudes regarding health treatment for new-born infants: it is believed to increase infant mortality rates (the dependent variable, Y). Health authorities realize that these thoughts are not shared by all members of the population group but are not sure where the deviations occur most frequently.

A simple linear regression is performed, a correlation coefficient determined, and the standard error of the estimate calculated. A scattergram and regression line are plotted, with standard error bands added on each side of the regression line. (See Figure 5.13.) Another map (Figure 5.14) is developed to show three groups of residuals: group 1 of those map regions within ±1 standard error, group 2 of the regions between ±1 and ±2 standard errors, and group 3 of those regions greater than ±2 standard error. The health authorities wanted especially to locate the regions in group 3, because they would be most unique. Only region G fit into the latter group. This hypothetical study would conclude with a visit of the health authorities to region G for a field investigation, in the hope of developing better health-information methods.

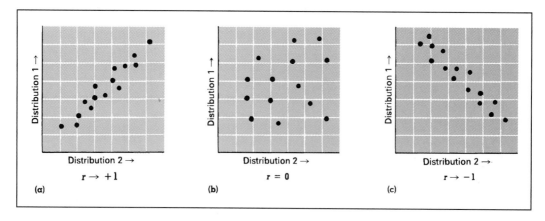

Figure 5.11 Scattergram patterns for different correlation (r) values.
In (a), the correlation approaches a strong positive relationship (that points would all fall on a line if r = 1.0). In (c), the relationship is strongly negative (inverse). No apparent linear correlation exists in (b), which yields an r value of 0.

Table 5.11 Residuals from Regression Table: Variables, Values, and Residuals

Map Region	Independent Variable (X)	Dependent Variable (Y)	Predicted Y(Ye)	Y Residuals (Yo-Ye)
A	5	5	21.25	16.25
B	10	22	25.30	3.30
C	35	55	45.55	9.45
D	30	40	41.55	1.50
E	45	35	53.65	18.65
F	50	50	57.70	7.70
G	20	100	33.40	66.60
H	50	10	57.70	47.70
I	50	95	57.70	37.30
J	15	10	29.35	19.35
K	15	30	29.35	.65
L	25	25	37.45	12.45
M	25	15	37.45	22.45
N	25	55	37.45	17.55
O	15	20	29.35	9.35
P	35	30	45.55	15.55
Q	45	65	53.65	11.35
R	55	50	61.75	11.75
S	40	50	49.60	.40
T	55	75	61.75	13.25
U	35	75	45.55	29.45

Table 5.12 Formula Values for Data in Table 5.11

X Variable	Y Variable
$N = 21$	$N = 21$
$\bar{X} = 32.38$	$\bar{Y} = 43.42$
$\Sigma x = 680$	$\Sigma y = 912$
$\Sigma x^2 = 26{,}850$	$\Sigma y^2 = 54{,}634$
$\sigma_x = 15.1$	$\sigma_y = 26.75$
$\Sigma xy = 33{,}445$	

$\Sigma y = a \cdot N + b \cdot \Sigma x$
$\Sigma xy = a \cdot \Sigma x + b \cdot \Sigma x^2$
$b = .810$
$a = 17.2$
$\Sigma(Y_o - Y_E) = 11{,}857.012$
r (correlation coefficient) $= .4593$
S_E (standard error of estimate) $= 24.98$

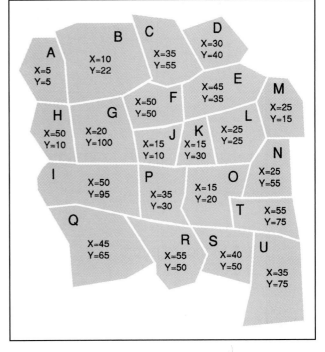

Figure 5.12 Data map used in mapping residuals from regression example.
It is extremely difficult to discern a relationship between X and Y by examination of a data map such as this, especially when the relationship is not very strong.

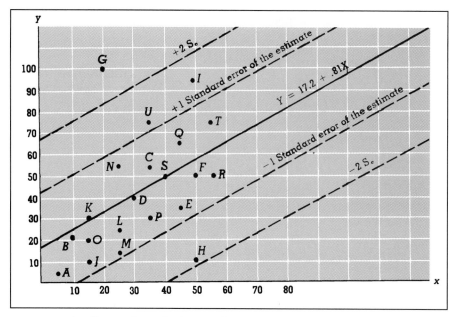

Figure 5.13 Scattergram, regression line, and standard error of the estimate sample plots for the sample data and problem from Table 5.12.

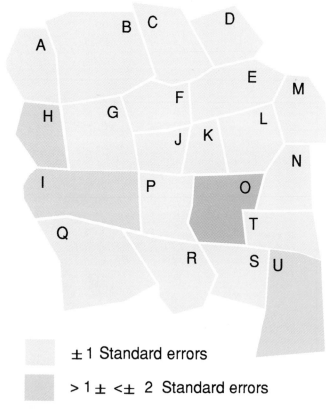

± 1 Standard errors

> 1 ± < ± 2 Standard errors

Figure 5.14 Map of residuals from regression.
The geographic areas having Y values considerably under- or overpredicted relative to the regression line are mapped with identifiable area symbols. These sections of the study area need to be investigated more closely. Further data may be needed to determine why the dependent variables in these sections behave as they do. Identifying deviate areas is a major application of residual mapping.

DATA CLASSIFICATION

Cartographers are called on to find ways to map large amounts of data. Recognizing the need for some sort of generalization and simplification, the cartographer is faced with a problem of data **classification.** Scientists classify objects or events for two main reasons:[13]

1. To reduce a large number of individuals (objects or events) to a smaller number of groups in order to facilitate description and illustration.
2. To define phenomena—classes about which general statements can be made.

All scientific disciplines, including geography, attempt classification to facilitate interpretation of the enormously complex world. Although classification may lead to loss of detail, it usually makes meaningful interpretation possible.
   Clarification is a primary goal of classification:

   Every classification or "Taxonomy" is a simple but implicit theory. If by trial and error we are able to produce classifications and maps of spatial phenomena which suggest or expose important relationships, then our choices are good choices. Our classifications have served their purposes.[14]

We can learn at least two important ideas from this statement. First, classification is for a specific purpose: to show something not presented before. It is not a means in itself, but a method to assist the researcher in uncovering spatial relationships. Second, there are no rigidly defined rules for classification in all circumstances. Any number of methods may be selected—even by trial and error.

Most cartographic data classification is of the agglomerative type. Data values are placed into groups based on their numerical proximity to other values along a number array. Classification is necessary because of the impracticability of symbolizing and mapping each individual value.[15] The goal is to do the grouping in a way that reveals the spatial patterns that serve the thematic purpose of the mapping activity.

CLASSIFICATION IS MORE

As an example of how classification leads to a loss of detail yet provides more interpretive power, consider the distribution of murder and nonnegligent manslaughter in the United States, shown in Figure 5.15. No classification of the data exists in Figure 5.15a. Those states above the average rate (8 per 100,000 people) have been rendered as a group in Figure 5.15b. They are set apart by an important criterion: all are above average. The geographical pattern shows that the heavily populated states east of the Mississippi River, the Old South states plus Texas and Oklahoma, and the western states of California, Nevada, Arizona, New Mexico, and Alaska are part of this class.

A different geographical picture emerges in Figure 5.15c, after classifying the two groups even further by computing an average of the above-average states and mapping only those states above this new average. The new group represents a class having the extreme values of the original data. Now the lower tier of the Old South states (plus Texas) appears as a region in the southeast. Four other states also remain in this group: Alaska, California, Nevada, and New Mexico.

Unclassed mapped data present a geographical pattern that is not particularly revealing and has little interpretive power. As the data are grouped into similar classes having identical numerical characteristics, there is an increase in their ability to convey information about which generalizations can be made.

Classification is introduced in this chapter because it is an integral part of data processing. A variety of methods of data classification are treated in Chapter 7, where choropleth mapping is discussed, but the methods can be applied to other mapping techniques as well.

(a)

(b)

(c)

Figure 5.15 Classifying data to reveal more about the distribution of nonnegligent manslaughter in the United States.
See the text for an explanation.

NOTES

1. Phillip Muehrcke, *Thematic Cartography* (Association of American Geographers, Resource Paper No. 19—Washington, D.C.: Association of American Geographers, 1972), p. 16.
2. Gouri K. Bhattacharyya and Richard Johnson, *Statistical Concepts and Methods* (New York: Wiley, 1977), p. 1; See also Gareth Shaw and Dennis Wheeler, *Statistical Techniques in Geographical Analysis* (New York: Halsted, 1994); and Robert Earickson and John Harlin, *Geographic Measurement and Quantitative Analysis* (New York: Macmillan, 1994).
3. Linton C. Freeman, *Elementary Applied Statistics for Students in Behavioral Science* (New York: Wiley, 1968). pp. 12–16.
4. Bhattacharyya and Johnson, *Statistical Concepts,* p. 30.
5. S. Gregory, *Statistical Methods and the Geographer* (London: Longman Group, 1978), p. 30.
6. Bhattacharyya and Johnson, *Statistical Concepts,* p. 36.
7. David Ebdon, *Statistics in Geography,* 2d ed. (Oxford, England: Basil Blackwell, 1985), pp. 28–31.

8. Ibid.
9. David Unwin, *Introductory Spatial Analysis* (London: Methuen, 1981), pp. 189–90.
10. Peter J. Taylor, *Quantitative Methods in Geography* (Boston: Houghton Mifflin, 1977), p. 27.
11. Ibid., pp. 27–30.
12. Ibid., pp. 29–30.
13. R. J. Johnston, *Classification in Geography* (London: Institute of British Geographers, 1976), p. 4.
14. Ronald Abler, John S. Adams, and Peter Gould, *Spatial Organization: A Geographer's View of the World* (Englewood Cliffs, NJ: Prentice Hall, 1971), p. 157.
15. This statement is generally accurate, although recent computer methods have allowed the mapping of unclassed data. This idea will be presented in more detail in Chapter 7 dealing with choropleth maps.

GLOSSARY

areal mean method of calculating the spatial balance of a set of data points, p. 100

arithmetic mean customary one-number descriptor of an interval/ratio data set; defined simply as, p. 95

$$\frac{\Sigma x}{N}$$

array in statistical terminology, an ordered arrangement of values; examples are ascending or descending arrays, p. 91

classification scientific reduction of a large number of individual observations, events, or numbers into smaller groups to facilitate explanation, pp. 106–107

coefficient of areal correspondence method of comparing areal spatial distributions; defined as a ratio, pp. 99–100

$$\frac{\text{area covered jointly by both phenomena}}{\text{total area covered by both phenomena}}$$

correlation method of showing the mathematical association between two or more variables, p. 104

data processing processing geographic data before mapping in order to reduce, enhance, retain key features, or show primary spatial patterns; preliminary activity of data symbolization, p. 90

equation for a straight line $y = a + bx$; important in regression and correlation analysis, p. 103

histogram a graphic way of presenting the frequency or relative frequency of occurrence of a variable, p. 91

interval/ratio data data resulting from measurement; most sophisticated data, p. 95

kurtosis numeric value indicating "peakiness" in a frequency distribution, p. 99

location quotient a measure of geographical concentration; illustrates deviation from assumed proportional share, p. 101–103

median that place in a ranked ordinal data set that neither exceeds nor is exceeded in rank by more than 50 percent of the observations; used to describe ordinal data, p. 94

mode the class in a frequency distribution containing the highest relative frequency; used to describe nominal data, p. 92–93

nominal scale data data that results from counting; simplest data level, p. 92

normal distribution frequency distribution represented by a bell-shaped curve; used as a basis for comparison in many statistical measures, p. 98–99

ordinal data data that provides a ranking; less than, more than decisions, p. 94

percentage a proportion multiplied by 100 for ease of numerical comparison, p. 91

percentile the place in a ranked data set that divides the number of observations into specified portion of all the observations, p. 94

proportion a special ratio that expresses the relationship between the amount in one class and the total in all classes; *percentage* is a proportion developed by multiplying the decimal fraction by 100, p. 90–91

quartile value one method of describing dispersion in an ordinal data set; defined as the interquartile range (i.e., the upper quartile minus the lower quartile) divided by 2, p. 94

ratio a fraction used to express the relationship between two variables, p. 90

regression line drawn on a graph to depict the relationship between two variables; linear regression is a common form, p. 103

relative frequency of the total frequency occurrence in a data set, the part belonging to a specific class, p. 91

residuals from regression differences between observed y (or x) values and those estimated by the regression line, p. 103

scattergram diagram containing a plot of data points, each of which has a value in two dimensions; a graphic way to illustrate mathematical correlation; also called a scatterplot, p. 103

skewness numeric value of deviation from the normal, or bell-shaped, frequency distribution, p. 99

standard deviation the square root of the variance; used to describe dispersion around the arithmetic mean in an ordinal/ratio data set; defined as, p. 97–99

$$\sqrt{\frac{\Sigma(x - \overline{X})^2}{N}}$$

standard distance   measure for depicting dispersion around an areal mean, p. 100–101

standard error of the estimate   the standard deviation of residual values; illustrates the dispersion of residuals (differences) of y (or x) around the regression line; used to measure the appropriateness of the regression line in describing the relationship between two variables, p. 103

statistics   concepts and methods used to collect and interpret data for drawing conclusions, p. 90

value   an individual numerical observation of a variable, p. 91

variable   raw data used in a statistical analysis, p. 91

variance   the statistic used most commonly to describe dispersion around the arithmetic mean in an ordinal/ratio data set; defined as, p. 97.

$$\frac{\Sigma(x - \overline{X})^2}{N}$$

variation ratio   used to define dispersion around the mode in a nominal data set; defined as, p. 94:

$$\frac{1 - f_{\mathrm{modal}}}{^-N}$$

READINGS FOR FURTHER UNDERSTANDING

Abler, Ronald, John S. Adams, and Peter Gould. *Spatial Organization: A Geographer's View of the World.* Englewood Cliffs, NJ: Prentice Hall, 1971.

Barker, Gerald M. *Elementary Statistics for Geographers.* New York: Guilford Press, 1988.

Bhattacharyya, Gouri K., and Richard Johnson. *Statistical Concepts and Methods.* New York: Wiley, 1977.

Blalock, H. M. *Social Statistics.* New York: McGraw-Hill, 1972.

Dixon, Wilfred J., and Frank J. Massey, Jr. *Introduction to Statistical Analysis.* New York: McGraw-Hill, 1957.

Ebdon, David. *Statistics in Geography.* 2d ed. Oxford: Basil Blackwell, 1985.

Freeman, Linton C. *Elementary Applied Statistics for Students in Behavioral Science.* New York: Wiley, 1968.

Gregory, S. *Statistical Methods and the Geographer.* London: Longman Group, 1978.

Hammond, Robert, and Patrick McCullough. *Quantitative Techniques in Geography.* Oxford: Clarendon Press, 1974.

Johnston, R. J. *Classification in Geography.* London: Institute of British Geographers, 1976.

King, Leslie J. *Statistical Analysis in Geography.* Englewood Cliffs, NJ: Prentice Hall, 1969.

Muehrcke, Phillip. *Thematic Cartography.* Association of American Geographers, Resource Paper No. 19. Washington, DC: Association of American Geographers, 1972.

Neft, David S. *Statistical Analysis for Areal Distributions.* Monograph Series Number Two. Philadelphia: Regional Science Research Institute, 1966.

Sneath, Peter H. A., and Robert R. Sokal. *Numerical Taxonomy.* San Francisco: Freeman, 1973.

Taylor, Peter J. *Quantitative Methods in Geography.* Boston: Houghton Mifflin, 1977.

Unwin, David. *Introductory Spatial Analysis.* London: Methuen, 1981.

CHAPTER

GEOGRAPHIC INFORMATION SYSTEMS*

CHAPTER PREVIEW

Geographic information systems, or GIS, is a rapidly growing field that incorporates knowledge from a number of disciplines. The technology is employed in natural resource management, demographic studies, health issues, emergency applications, and other areas requiring spatial problem solving or research. "Performing" GIS is a nonlinear process that involves acquiring, processing, analyzing, and displaying spatial data through the integration of computer software, hardware, and people.

Although much time and effort is expended in acquiring data and constructing the layers of a GIS database, the primary focus of geographic information systems is the analysis of spatial data and their associated attributes. The analysis may include data query, measurement, statistics, and operations on one or more of the layers of the GIS database, as well as exploration and analysis of graphic displays of spatial data. Graphic analysis enables a GIS analyst to visualize patterns and trends. In addition to computer display, hard copy output from GIS is often required. Output from GIS software is not always optimal, particularly because the software defaults often violate cartographic convention.

*Chapter authored by Elaine Hallisey Hendrix.
GIS Research Coordinator
Georgia State University

GEOGRAPHIC INFORMATION SYSTEMS DEFINED

Cartography and GIS have become increasingly intertwined in recent years. Today it is likely that a cartographer will be working with GIS, or at least alongside a GIS analyst in some capacity. Therefore it is important that a cartographer be introduced to the fundamental concepts of a GIS. This chapter provides this necessary foundation.

The definition of the term **geographic information system,** or **GIS,** has been the subject of debate for some time. GIS has a wide range of applications and therefore GIS operators use the technology differently depending on the requirements of their own particular area of study. Some would simply define geographic information systems as the use of a GIS software package. Others more broadly define GIS as the integration of specialized software, hardware, and people, designed to analyze geographically referenced, or spatial, data. Still others stress the importance of viewing GIS as a process.[1] The process in which spatial data are entered, stored, analyzed, and displayed is specific to the requirements of each project. The GIS process is, therefore, unique to each project. Considering these viewpoints, a geographic information system is defined as a computer-assisted process designed to acquire, store, analyze, and display spatial data and their attributes. The key component in this definition is the ability of a GIS to *analyze* spatial data. Analysis based on location is what distinguishes GIS from simple desktop mapping and database management.[2]

GIS has been described as an empowering technology, its development comparable to the introduction of the printing press, the telephone, or the first computer.[3] GIS is an important tool because it allows for the manipulation and display of geographic data in new ways[4] and improves the efficiency of more traditional spatial analysis. A rapidly growing industry, GIS incorporates knowledge from geography, geodesy, cartography, computer science, mathematics, statistics, and a number of other disciplines. With recent increases in the power of microcomputers and the increased availability of user-friendly GIS software, GIS has become accessible to a greater number of users. This chapter focuses primarily on microcomputer GIS.

APPLICATIONS OF GIS TECHNOLOGY

Any study that involves a spatial component may benefit from GIS, so there is a wide range of potential applications. GIS technology may be applied to natural resource management, land management, demographic studies, health issues, facilities management, street network applications, emergency applications, and many other fields. GIS is truly interdisciplinary in nature.

In natural resource management, for example, GIS has been used to aid in the development of a predictive model for the location of old-growth forests in northwest Arkansas.[5] Scientists determined that unaltered ancient forest remnants can still be found on sites with poor soils and steep, south-facing slopes. With GIS technology, they were able to query a spatial database containing information on soils, slope, and the direction of slope (aspect), to identify potential sites of old-growth forest. The researchers then field-checked the results to validate their predictive model. The technology has also been applied to coastal zone management.[6] Scientists used GIS to develop a coastal erosion model that is less location-specific than previous erosion models. The Landsat Pathfinder program uses GIS and remote sensing to study change in the world's tropical forest cover over space and time.[7] Mapping and quantifying deforestation will help scientists understand how human-induced changes affect the global environment. A study of the gradient of urbanization from New York City to northwestern Connecticut benefited from the application of GIS.[8] Researchers identified characteristics of urban and suburban environments based on the landscape structure of native forest vegetation.

GIS technology is also being used to study retail trade and market structures.[9] Applying GIS to the study of consumer demographics, retail-gravity models, and central-place tendencies allows the identification of market potentials and the development of new methods of predicting business survival. Different types of business can use this information to determine the likelihood of success in a particular area.

GIS technology is being applied to a variety of health issues, including the study of variation in disease frequency and health status, and measures of health care delivery and resource allocation.[10] Public health researchers, for example, have used GIS to map and study cancer mortality. Health care providers have employed GIS to locate new health care facilities based on demographics, need, competition, and other factors.

The city of Wilson, North Carolina, implemented a comprehensive GIS that is used by city employees involved in planning, public works, public utilities, the fire department, the police department, and other departments.[11] In addition to routing emergency squad cars, San Bernardino, California, police use GIS to analyze crime patterns and identify locations that require increased surveillance.[12]

Archaeologists also use GIS technology. The Royal Commission on the Ancient and Historical Monuments of Scotland has been using GIS for several years.[13] The commission created a database to enable the National Monuments Record of Scotland to be searched either spatially or as an index. With their GIS, it is possible to carry out a proximity search of all sites that fall within 100 meters of the coastline of Scotland in a few minutes as opposed to a manual search requiring many days. An evaluation of Chaco Anasazi roadways is presented at a World Wide Web **(WWW)** site generated at the University of California, Santa Barbara.[14] This research, also presented at a

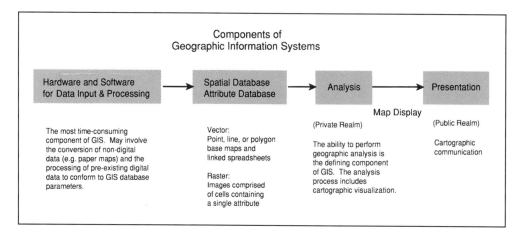

Figure 6.1   Components of a geographic information system.
Although the process of GIS generally flows in a linear fashion, there are many instances in which the process is not linear. After visualization of the data, for example, the analyst may determine that additional data input is necessary—some GIS databases are continually evolving. In some cases, such as emergency 911 systems, public presentation of data is not required.

poster session at the 1996 Society for American Archaeology in New Orleans, used GIS to explain that Chaco Anasazi road routes seem to conform with models focusing on religious and integrative road functions.

COMPONENTS OF A GIS

A typical geographic information system is comprised of spatial and attribute databases, hardware and software for data input, a geographical analysis system, and a map display system. The components of a GIS suggest the process-oriented nature of the discipline. (See Figure 6.1.) Data are input, manipulated, analyzed, and displayed in some fashion. It is important to remember that the process, examined in the following pages, is not always linear. Analysis, for example, may reveal that new data are required, necessitating new data input and manipulation.

DATA MODELS

A **spatial database** is a set of digital base maps that describes the "geography" of the earth's surface. In this context, the term geography refers to both physical and human-made features on the earth as well as imaginary boundaries, such as census tracts. The maps in the spatial database cover the same geographic extent and are "layered" on top of one another to determine relationships between the layers. (See Figure 6.2.) An individual layer in a spatial database is sometimes referred to as a **theme** or a **coverage**. An **attribute database** describes the characteristics or qualities of the features or images portrayed on the base maps. For example, a base map layer may show origin points of tornadoes in Georgia. (See Figure 6.3a.) The attributes associated with each tornado may include the year, month, day, and time of occurrence, the number of injuries and fatali-

ties, and other information for each tornado. In another example, a spatial database layer may consist of census tract boundaries in metropolitan Houston. (See Figure 6.3b.) The attribute database associated with the census tracts of the base map may include population per tract, average age per tract, median income per tract, or other information enumerated at the tract level.

The manner in which the spatial and attribute databases are associated depends upon the data model. A **data model** is an abstraction of the real world which incorporates only those properties thought to be relevant to an application.[15] In GIS, two major data models are used: the vector data model and the raster data model. (See Figure 6.4.)

The **vector data model** is based on points (or nodes), lines (or arcs), and polygons. Point features in a spatial database may represent convenience stores, line features may represent roads, and polygons may represent counties. Generally, point, line and polygon features are stored as separate layers. In addition, different *types* of point, line, and polygon features are usually stored as separate layers. For example, counties and census tracts are both polygon features, but they are stored as different GIS layers. In a vector-based GIS, the attributes of each feature in a layer are stored in a spreadsheet linked to the spatial database. As shown in Figure 6.4a, the spatial database has four polygon features with identifiers from 1 to 4 and a background with an identifier of 0. Linked to the spatial database is a spreadsheet containing values for attributes A, B, and C. These attributes could represent land-use codes, the number of mobile homes, or some other characteristic of each polygon.

Some vector files, such as the U.S. Bureau of the Census TIGER files (see Chapter 17), store **topology.** Topology is the relationships among points, lines, and polygons. Topological relationships include adjacencies, connectiv-

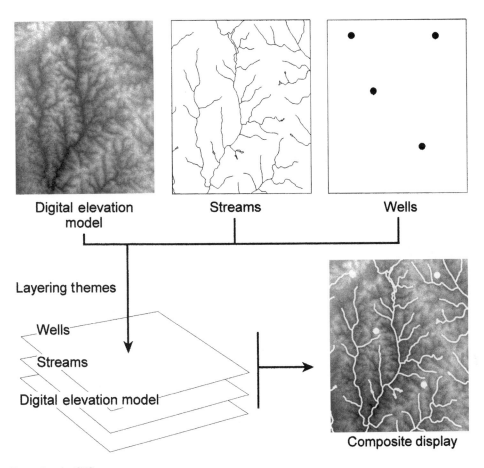

Figure 6.2   Layering in GIS.

ity, and containment. (See Figure 6.5.) Although some properties of a map change if the map is stretched and distorted (distances, angles, and relative proximities, for example) topological relationships remain constant. Because these topological relationships are stored, it makes data more effective for various kinds of spatial analysis. Stored information about adjacencies, for instance, enables a planner to query a spatial database about areas next to each other that may have different zoning codes. Connectivity allows transportation planners to model traffic flow through a street network. Containment data facilitate queries about small stands of Atlantic white cedars, for example, located within a larger population of some other single tree species.

The **raster data model** is based upon cells, or pixels, each cell containing a single attribute value, or *z*-value. GIS rasters are usually comprised of square cells, but may consist of any **tessellation** scheme. A tessellation is any regular shape, such as a square, rectangle, triangle, or hexagon, that covers a plane surface without any gaps. The two raster images in Figure 6.4b cover the same geographic area as the vector model shown in Figure 6.4a, but display two different attributes. Although Figure 6.4b depicts discrete data in raster format for comparison with the

vector data model, the cell-based nature of the raster model makes it particularly suitable for working with data that vary continuously over a surface, including temperature or elevation.

The type of application defines the data model employed. Raster-based GISs are often used for environmental applications involving continuous data, such as classifying land use-types from satellite imagery or calculating slope from elevation data. A special type of vector file, called a triangulated irregular network, or **TIN,** also models surfaces. (See Figure 6.6.) A TIN consists of irregularly shaped triangles. Each vertex for every triangle is encoded with an elevation value. In locations where the elevation is fairly constant, the triangles are large. In locations of rapid elevation change, the triangles are smaller. Vector GISs are more suited than raster GISs for tasks when precise location is important, such as tax mapping, route finding, or locating points. The vector model is also more suitable when consistent enumeration units, such as census tracts, are used because of the efficiency of storing the numerous attributes in a linked spreadsheet as opposed to much larger individual raster images.

Raster GISs, on the other hand, are more efficient than vector-based packages for certain overlay procedures,

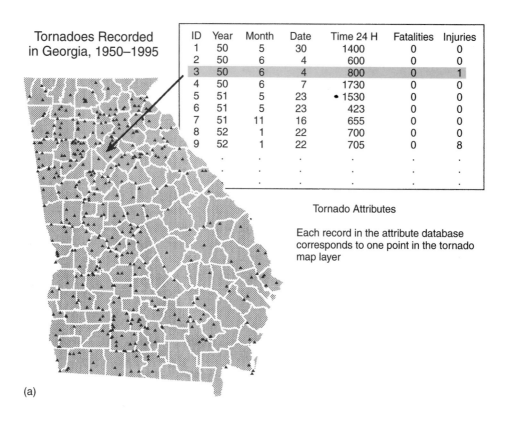

Tornadoes Recorded
in Georgia, 1950–1995

ID	Year	Month	Date	Time 24 H	Fatalities	Injuries
1	50	5	30	1400	0	0
2	50	6	4	600	0	0
3	50	6	4	800	0	1
4	50	6	7	1730	0	0
5	51	5	23	• 1530	0	0
6	51	5	23	423	0	0
7	51	11	16	655	0	0
8	52	1	22	700	0	0
9	52	1	22	705	0	8
.	.	.	.	.	.	.
.	.	.	.	.	.	.
.	.	.	.	.	.	.

Tornado Attributes

Each record in the attribute database
corresponds to one point in the tornado
map layer

(a)

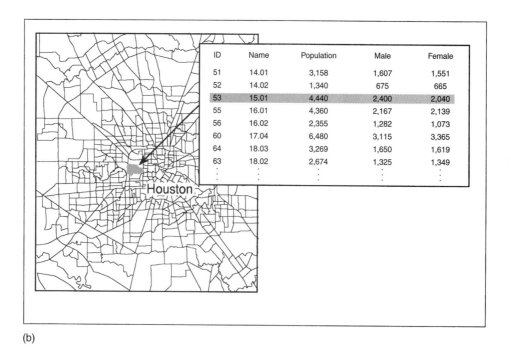

ID	Name	Population	Male	Female
51	14.01	3,158	1,607	1,551
52	14.02	1,340	675	665
53	15.01	4,440	2,400	2,040
55	16.01	4,360	2,167	2,139
56	16.02	2,355	1,282	1,073
60	17.04	6,480	3,115	3,365
64	18.03	3,269	1,650	1,619
63	18.02	2,674	1,325	1,349

Houston

(b)

Figure 6.3   Attributes records in GIS.
(a) Tornado attribute database. Tornadoes in Georgia and associated attributes for nine of the tornadoes. Each tornado is linked to one record. (b) Houston census tract attribute database. Houston census tracts and associated attributes for eight of the tracts. Each tract is linked to one record in the attribute database. The highlighted tract, for example, is linked to the highlighted record.

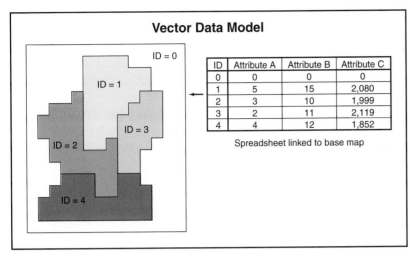

(a)

(b)

Figure 6.4 Data models in GIS
(a) The vector data model is based upon points, lines, and polygons. The example here displays four polygon features and a background polygon. Attributes associated with each polygon are displayed in a spreadsheet linked to the map. (b) The raster data model is based upon cells, each of which contains a single attribute value. The geographic area shown in the two examples, Display of Attribute A and Display of Attribute B, corresponds to that of the vector data model display.

which will be discussed in the section on analysis. When various types of spatial analysis are required, it is sometimes advantageous to perform some processing with a vector-based GIS and some with a raster-based GIS, translating data between the two GISs as necessary. Although individual GIS software packages are still primarily either vector- or raster-based, some vector packages now offer simple raster processing capabilities and some raster packages offer simple vector processing capabilities.

DATA ACQUISITION AND PROCESSING

Building a GIS database is the most time-consuming portion of any GIS project. Obtaining and processing data so that they are viable for geographical analysis usually requires considerable thought and effort. In most commercial GIS projects, up to 80 percent of the cost, in time and

labor, is related to building the database.[16] Data input is achieved in a variety of ways. In some cases the GIS operator creates his or her digital database from analog, or nondigital, sources. Map digitizing is the process of converting hard copy maps to digital format. Hard copy maps may be digitized using a digitizing board or a scanner. (See Chapter 17.) The use of either a digitizing board or scanner is dependent on the requirements of the project and the GIS data model used, vector or raster. Digitizing boards produce vector files, whereas scanners produce raster files. Global positioning system (GPS) receivers are often used to input location and elevation data. (See Chapter 2.) Some GIS software allows for either on-call or continuous logging of GPS data.

For the attribute database, tabular data may be entered into a spreadsheet or other type of file through keyboard entry. **Address matching,** or address **geocoding,** employs existing tabular and spatial data to produce new spatial data,

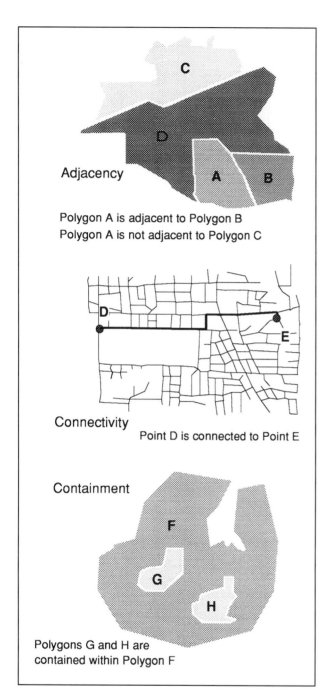

Adjacency

Polygon A is adjacent to Polygon B
Polygon A is not adjacent to Polygon C

Connectivity

Point D is connected to Point E

Containment

Polygons G and H are
contained within Polygon F

Figure 6.5 Topological Relationships: Adjacency,
Connectivity, and Containment in GIS

a point coverage. (See Figure 6.7) Address matching is the process of assigning a pair of geographic coordinates to a location based on an address in a table. Address matching requires a table of addresses to be geocoded, or the GIS operator must interactively key in the addresses to be geocoded. In addition, address matching requires a reference coverage for use in matching. A linear street-base coverage that includes street address ranges linked to each street segment may be used for reference in address matching. If a lower level of geographic precision is acceptable, a polygon layer,

such as a ZIP code coverage may be used for reference. A **point-of-event table** that contains *x* and *y* coordinates representing the locations of discrete "events," such as the origin of a lightning-caused fire or the location of a school, is importable into most GISs. The software reads the geographic coordinates in the table and creates a new point coverage.

During creation of a digital database, the GIS operator must be sure to adequately document both the process and the final product. The documentation, or **metadata,** is important because it provides a record of the original analog data sources and the process used to build the digital database for the originator and those who wish to use the data in the future. In addition to creating digital databases from analog sources, many GIS operators employ existing data sets. An enormous amount of digital data are available from sources such as private vendors, governments, and other organizations. These data are distributed in numerous formats and media, including diskettes, compact discs, and the Internet. (See Chapter 17.)

Spatial Data Formats

Most GIS software packages have proprietary data formats, but allow for the import and export of data in a number of tabular and spatial formats. Common tabular formats include simple text files or database files such as those in .dbf format. (See Chapter 17.) Common spatial data formats include AutoCAD .dxf, ARC/INFO .e00, DLG, DOQ, and a host of others. Although most GIS packages offer import and export capabilities, it is often confusing, difficult, and time-consuming to translate data from one format to another. In addition, valuable information can be lost in the translation. U.S. federal agencies now mandate the use of the Spatial Data Transfer Standard (**SDTS).**

The SDTS provides an intermediate exchange format requiring only one encoder and one decoder. (See Figure 6.8.) In other words, data in a proprietary format can be encoded, or exported, into SDTS intermediate format and recipients of the transferred data can decode, or import, the SDTS data into their own proprietary formats. Instead of having numerous import and export options, only SDTS intermediate format import and export options are required, thereby improving efficiency in database development. The U.S. Geological Survey, the U.S. Bureau of the Census, and the Army Corps of Engineers now produce and distribute spatial data in SDTS format. A goal of the U.S. federal government is to see commercial software companies develop SDTS translation software for their own proprietary formats.

In addition to being a prime source of digital data, the Internet is also useful for the exchange of information and ideas. Newsgroups and mailing lists, addressed in Appendix E, offer a forum for the discussion of a number of topics pertaining to cartography and GIS.

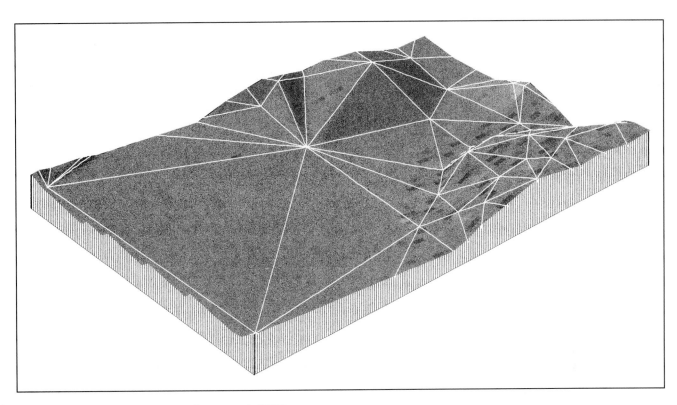

Figure 6.6 A triangulated irregular network (TIN).
Pictured here is a TIN layered on the surface it models. Note that the triangles are large in flat areas and small in locations of rapid elevation change.

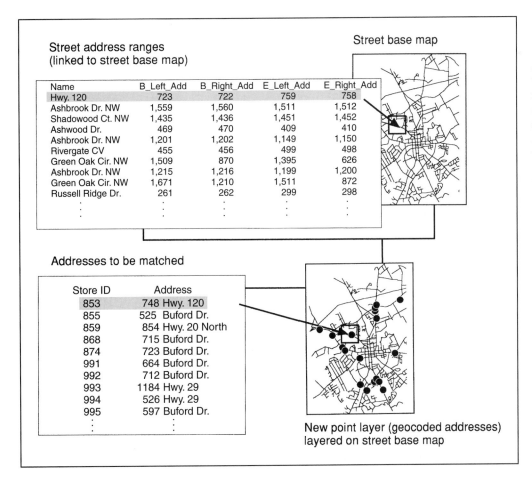

Street address ranges
(linked to street base map)

Name	B_Left_Add	B_Right_Add	E_Left_Add	E_Right_Add
Hwy. 120	723	722	759	758
Ashbrook Dr. NW	1,559	1,560	1,511	1,512
Shadowood Ct. NW	1,435	1,436	1,451	1,452
Ashwood Dr.	469	470	409	410
Ashbrook Dr. NW	1,201	1,202	1,149	1,150
Rivergate CV	455	456	499	498
Green Oak Cir. NW	1,509	870	1,395	626
Ashbrook Dr. NW	1,215	1,216	1,199	1,200
Green Oak Cir. NW	1,671	1,210	1,511	872
Russell Ridge Dr.	261	262	299	298

Street base map

Addresses to be matched

Store ID	Address
853	748 Hwy. 120
855	525 Buford Dr.
859	854 Hwy. 20 North
868	715 Buford Dr.
874	723 Buford Dr.
991	664 Buford Dr.
992	712 Buford Dr.
993	1184 Hwy. 29
994	526 Hwy. 29
995	597 Buford Dr.

New point layer (geocoded addresses)
layered on street base map

Figure 6.7 Address matching at the street level.
A point layer is created using a street layer that includes a linked spreadsheet of street address ranges, and a file containing addresses to be matched. In the highlighted example, store 853, with the address 748 Hwy 120, is located within the address range of 722 to 758 on Hwy 120.

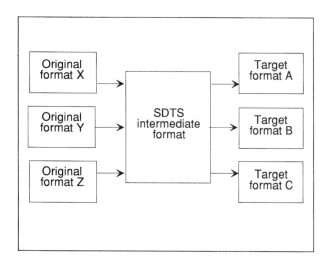

Figure 6.8   Spatial Transfer Standard (SDTS) .

Data Processing

After data are imported into a software package, they usually must be processed further. Attribute data must often be edited in some way. Records, for example, may need to be added to or deleted from a table, tables may need to be joined, or units of measure may need to be changed. (See Figure 6.9.)

Spatial data often require transformation of some kind. Existing spatial data are available in a wide variety of projections, grid referencing systems, and datums. (See Chapters 2 and 3.) Data may be obtained, for example, in a transverse Mercator projection with (UTM) coordinates based on the NAD83 datum or a Lambert conformal conic projection with state plane coordinates based on the NAD27 datum, or a host of other projection, grid, and datum parameters. **Geodetic coordinates,** or longitude and latitude spherical coordinates, are treated in many GISs as a special type of grid referencing system.[17] Geodetic coordinates (also called geographic coordinates in some GIS software packages) are spheroidal, but are treated as a plane coordinate system with the origin at the intersection of the prime meridian and the equator. (See Figure 6.10.) Longitude, x, and latitude, y, are both positive in the northeast quadrant and are both negative in the southwest quadrant. In the northwest quadrant, longitude is negative and latitude is positive and in the southeast quadrant, longitude is positive and latitude is negative. Although the projection is often specified in the GIS software as "none," the geodetic coordinates are displayed as in a **Plate Carree projection** where one degree of longitude is consistently equal to one degree of latitude. When displaying a map on a Plate Carree projection, distortion of shape, area, and scale increases with distance from the equator.

Layers in a database must be **georeferenced** to the same projection, grid, and datum. A GIS operator, for instance, has a digital boundary file of the counties of eastern Mississippi in a transverse Mercator projection using

a Mississippi, east state plane grid based on the NAD27 datum, expressed in feet. He or she obtains a boundary file of census tracts, covering the same geographic area as the counties, in geodetic decimal degrees based on the NAD83 datum. To use these two boundary files in a GIS database, he or she must transform the layers to be consistent in terms of projection (transverse Mercator or Plate Carree), grid (state plane or geodetic decimal degrees), and datum (NAD27 or NAD83). In another example, a GIS operator receives a digital orthophoto quarter quad (DOQQ) in the NAD83 datum and she or he wants to overlay a 1992 TIGER/line street base in the NAD27 datum. (See Figure 6.11.) The layers are both in geodetic decimal degrees, but do not align properly because of the difference in datums.

Other types of processing sometimes required in constructing a spatial database include **rubber sheeting** and **edge matching.** (See Figure 6.12 on page 121.) Often a data layer is obtained that carries no locational information. Rubber sheeting, technically referred to as **conflation,** is a georeferencing process that involves the use of ground control points to stretch a map so that it will register with other GIS layers. Ground control points are locations on the ground, such as street intersections, benchmarks, or building corners, with known x and y coordinate values. These control points are located on the map, fixed in place, and the remainder of the map is stretched or warped like a sheet of rubber, to register the map layer to the spatial database. Edge matching is the process of joining adjacent coverages. Lines, such as streets, or polygon boundaries may not join exactly and so must be adjusted to achieve as close to a seamless transition as possible.

Metadata

As stated previously, metadata is required in the construction of GIS databases, particularly those that include data from a variety of sources. In addition to details regarding projection, grid referencing system, and datum of a data set, the metadata should also contain information about resolution, scale, and data quality, and definitions of terms or codes used in the database.

Satellite imagery, often incorporated into a GIS database, is in raster format. A GIS operator must know the **cell resolution,** or cell size, to work with this remotely sensed imagery. SPOT multispectral satellite data, for example, have a spatial resolution of 20 meters (See Chapters 4 and 17.) In other words, each cell covers 20 meters by 20 meters on the ground. A USGS 7.5 minute digital elevation model (DEM) has a resolution of 30 meters: each cell, consisting of a single elevation value, represents 30 square meters on the ground. Cell resolution defines the level of analysis that can be performed and the display quality that will result. The larger the number of cells per area in a raster, the higher the resolution. With high resolution images, the analysis is more precise and the display is of better quality than those images with a lower resolution.

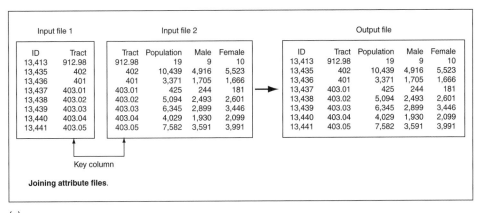

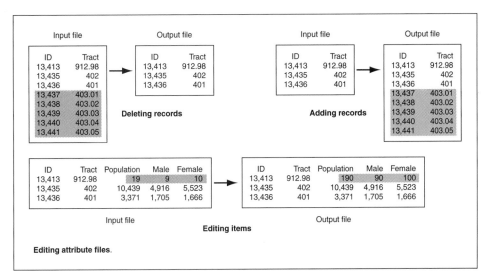

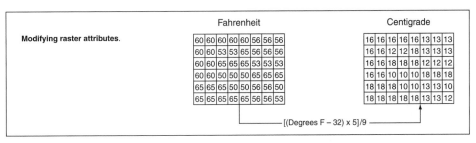

Figure 6.9 Joining, Editing, and Modifying GIS attribute files.
(a) Joining files. In this simple example of one-to-one join operation, Input file 1 is attribute data already linked to the spatial database. Input file 2 is attribute data to be merged, or joined, to the original database. The key column, or relate item, is a field found in both input files and is used to match the two files. The Output file contains all the fields from both input files. (b) Editing files. In these three examples, records are deleted and added, and individual items are edited. In addition to these types of editing, new files may need to be calculated and other types of processing may be required. (c) Modifying raster attributes. In this example, attribute data values are changed from one unit of measure to another.

Some GISs allow the operator to subdivide, or **disaggregate,** cells within an image. In an image that initially contained 500 cells, for example, each cell could be subdivided into four cells with the output image increasing by a factor of four to 2,000 cells. Generally in this type of disaggregation, the three new cells take on the attribute value of the original single cell so that an image with redundant values, requiring more computer storage space, results. Disaggregation of data is discouraged for analysis because it implies a level of precision that does not exist. Some

GISs allow the operator to drape imagery over an orthographic, or three dimensional, display. Disaggregation is used occasionally, however, to prepare data for some types of display. Satellite imagery with a resolution of 20 meters, for example, may be draped over a digital elevation model whose resolution has been disaggregated from an initial 30 meters to a 20-meter resolution. Scale must be considered in creation of a digital spatial database, because it affects the level of accuracy and precision in analysis and display.

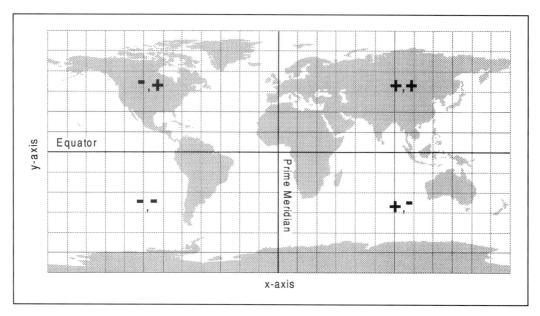

Figure 6.10 The Plate Caree projection in GIS.
Geodetic coordinates are usually displayed in a Plate Carree projection, as shown in this illustration. The method of expression is longitude, then latitude. The origin is located at the intersection of the prime meridian and the equator, so longitude is negative in the western hemisphere and latitude is negative in the southern hemisphere.

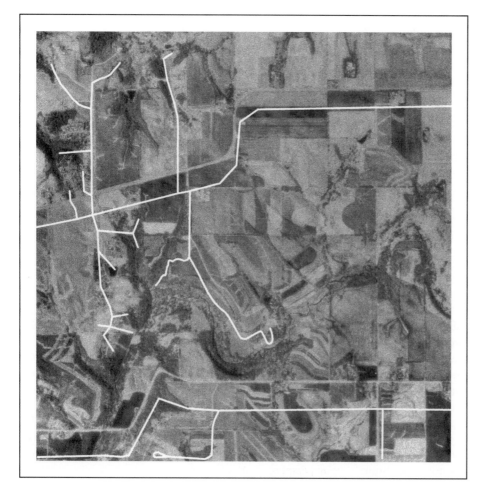

Figure 6.11 Inconsistent datums. In this example, a 1992 TIGER/Line street base in the NAD27 datum overlays a portion of a digital orthophoto quarter quad (DOQQ) in the NAD83 datum. The layers are obviously misaligned, particularly in the north, south direction.

Figure 6.12 Rubber sheeting and edge matching in GIS.

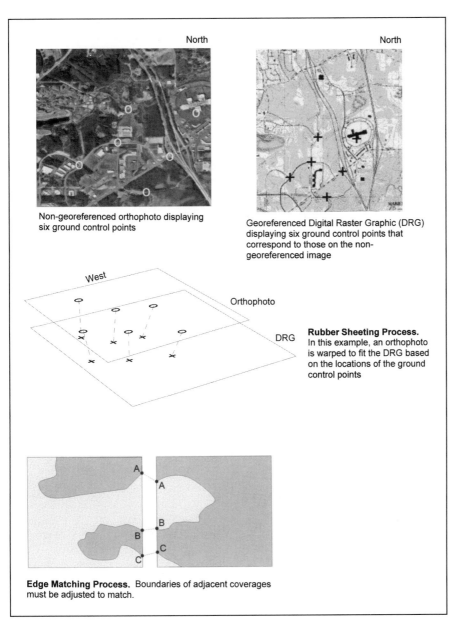

Non-georeferenced orthophoto displaying six ground control points

Georeferenced Digital Raster Graphic (DRG) displaying six ground control points that correspond to those on the non-georeferenced image

Rubber Sheeting Process. In this example, an orthophoto is warped to fit the DRG based on the locations of the ground control points

Edge Matching Process. Boundaries of adjacent coverages must be adjusted to match.

Metadata must include scale definitions because, in GIS processing, scale is often not apparent. As long as two layers are georeferenced correctly, scale differences in initial data sources are not always apparent when analyzing data. A GIS operator can zoom in, increasing the scale, to view a smaller area, and can then zoom back out, decreasing the scale, to view the entire map area. Although the scale changes easily, with no real thought necessary on the part of the operator, scale is still a key factor in data analysis and display. If studying a boundary dispute between two small counties with a shared boundary of 5 miles, for example, a theme originally digitized from a scale of 1:500,000 is inappropriate; this scale choice leads to a theme that is too generalized. On the other hand, if a display of the entire world is required, boundaries digitized at 1:500,000 are too detailed.

There is also the issue of layering themes derived from maps at different scales. A digital line graph (DLG) originally digitized at a scale of 1:100,000 may be layered on a county boundary digitized at a 1:250,000 scale. The concern when layering themes derived from maps at different scales is that the data from the smaller-scaled theme, the county boundary in this case, will be more generalized than the data from the larger-scaled theme. The accuracy and precision of the entire database are therefore limited to the scale of the smaller-scaled theme. *A good rule is to attempt to build a spatial database from data that are at fairly consistent scales.*

Metadata must also address data quality issues to allow a GIS operator to determine if the data are "good" enough. How were data gathered and processed? A GIS operator obtains a point layer, for instance, displaying the locations of manufacturers in North Carolina. He or she must know how the data were mapped. Did someone manually locate the manufacturers on a map from a list of addresses and, if so, which map was used? Was digital address matching utilized? If so, was the address matching at the street level, ZIP code level, or a higher level? At the street level, location is more precise than at the ZIP code level, but there is also a higher likelihood of errors in accuracy as opposed to data geocoded at the ZIP code level. Which street base or ZIP code map was used in the address matching process? The GIS operator may know of some existing error in a particular street or ZIP code base. Details regarding data quality are important in the decision to use a specific layer.

Terms and codes used in a data set must also be defined in the metadata. Each field in a linked attribute table must be identified. In some U.S. data sets, for example, enumeration units, such as counties or census tracts, are identified with Federal Information Processing Standards (**FIPS**) codes. The state of Georgia, for instance, has a FIPS code of 13 and Applying County within Georgia has a FIPS code of 001. The complete FIPS code for Applying County, Georgia, is 13001. Applying County is comprised of five census tracts numbered from 95.01 to 95.05. The complete identifier for tract 95.01 within Applying County, Georgia, is 130019501. The metadata must specify that the features in the data set are identified with FIPS codes and may include additional documentation describing how to decode a FIPS code.

The U.S. Federal Geographic Data Committee (**FGDC**), formed in 1990, has been responsible for the development of a content standard for digital spatial metadata.[18] The FGDC Content Standard for Digital Geospatial Metadata (i.e., the **FGDC Standard**), which has been in effect for all federal agencies since 1995, addresses documentation issues such as those discussed above. The purpose of the FGDC Standard is to provide consistency on the type of spatial metadata to capture. Metadata required include the following:

    Identification of the data set
    Information on data quality
    Spatial data organization (e.g., raster vs. vector)
    Spatial reference (e.g., projection, grid, datum)
    Entity (i.e., map features) and attribute definitions
    Distribution information
    Metadata reference (e.g.. date of metadata, contacts)

The FGDC Standard is only mandatory for U.S. federal agencies, but the FGDC hopes that the private sector adopts the FGDC Standard.

GEOGRAPHICAL ANALYSIS

After the database is constructed, geographical analysis begins. Although most desktop mapping packages offer a spatial and an attribute database, the ability to analyze data based on spatial characteristics is required for a *true* GIS. Geographical analysis is what separates GIS from desktop mapping. For instance, a GIS with the appropriate spatial

and attribute databases can find all areas of urban land use (one layer) associated with earthquake faults (a second layer). Another typical GIS analysis task is the ability to find features in one layer that are located within a specified distance of features in another layer. For example, if a map displays house locations (one layer), a GIS operator can determine the number of day care centers (a second layer) and attributes associated with the day care centers, within a 5-mile radius of each house.

Fundamental categories of analysis in GIS include data query; measurement; statistics; operations on single layers, including classification; and operations on multiple layers, including overlay procedures.

DATA QUERY

With a GIS, an analyst can determine what is at a particular location or, conversely, can query, or question, the GIS database regarding locations that satisfy a certain condition. (See Figure 6.13.)

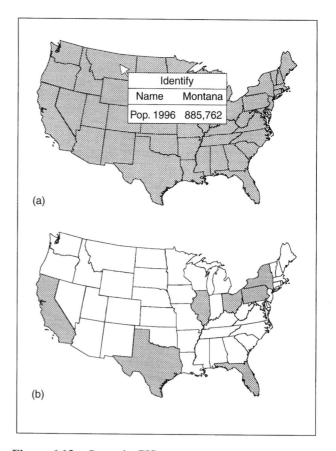

Figure 6.13   Query in GIS.
(a) Query by Location. The operator "points" to a feature to obtain attribute information for that feature (b) Query by Attribute. The operator queries the attribute database to determine which features satisfy a condition. In this example the query reads 'Which states have a 1996 population of greater than 10,000,000 persons?'

SAMPLE FGDC COMPLAINT METADATA

Metadata for Digital Orthophoto Quadrangles

These metadata describe the Digital Orthophoto Quadrangle holdings of the USGS for the conterminous United States. This is a data-set level implementation of the Content Standards for Digital Geospatial Metadata, and will be in place until file-specific information for individual digital orthophoto quadrangles is available.

Table of Contents

    Identification_Information
    Data-Quality_Information
    Spatial_Data_Organization_Information
    Spatial_Reference_Information
    Entity_and_Attribute_Information
    Distribution_Information
    Metadata_Reference_Information

Identification-Information

Citation:

    Citation_Information:
        Originator: U.S. Geological Survey
        Publication_Date:
        Title: Digital Orthophoto Quadrangles
        Geospatial_Data_Presentation_Form: remote-sensing image
        Publication_Information:
        Publication_Place: Reston, VA
        Publisher: U.S. Geological Survey

Description:

Abstract

    Orthophotos combine the image characteristics of a photograph with the geometric qualities of a map. The primary digital orthophotoquad (DOQ) is a 1-meter ground resolution, quarter-quadrangle (3.75-minutes of latitude by 3.75-minutes of longitude) image cast on the Universal Transverse Mercator Projection (UTM) on the North American Datum of 1983 (NAD83). The geographic extent of the DOQ is equivalent to a quarter-quad plus the overedge ranges a minimum of 50 meters to a maximum of 300 meters beyond the extremes of the primary and secondary corner points. The overedge is included to facilitate tonal matching for mosaicking and for the placement of the NAD83 and secondary datum corner ticks. The normal orientation of data is by lines (rows) and samples (columns). Each line contains a series of pixels ordered from west to east with the order of the lines from north to south. The standard, archived digital orthophoto is formatted as four ASCII header records, followed by a series of 8-bit binary image data records. The radiometric image brightness values are stored as 256 gray levels ranging from 0 to 255. The metadata provided in the digital orthophoto contain a wide range of descriptive information including format source information, production instrumentation and dates, and data to assist with displaying and georeferencing the image. The standard distribution format of DOQs will be JPEG compressed images on CD-ROM by counties or special regions. The reconstituted image from the CD-ROM will exhibit some radiometric differences when compared to its uncompressed original but will retain the geometry of the uncompressed DOQ. Uncompressed DOQs are distributed on tape.

Purpose:

    DOQs serve a variety of purposes, from interim maps to field references for earth science investigations and analysis. The DOQ is useful as a layer of a geographic information system and as a tool for revision of digital line graphs and topographic maps.

    Time-Period-of-Content:
        Time_Period_Information:
        Range_of_Dates/Times:
            Beginning_Date: 19940222
            Ending_Date: present
        Currentness_Reference: ground condition

    Notes: This is a quoted example of the first lines of metadata for a digital orthophoto quadrangle. This example complies with the FGDC Standard. A complete set of FGDC compliant metadata can be many pages long and includes details, for individual data sets, about the items listed in the Table of Contents.

Source: United States Geological Survey, 1988.

MEASUREMENT AND ROUTING

Measurement of the area and/or perimeter of map features is a common GIS task. A real estate analyst, for example, may be required to determine the area of a land parcel. Distance measurements are also a common GIS task. A furniture delivery scheduler may want to find the distance between two houses. With any GIS, a straight-line distance measurement is possible. Some vector GISs also allow the scheduler to use the connectivity of a network to identify the shortest route between the two houses and to measure the distance along that route. (See Figure 6.14.)

STATISTICAL ANALYSIS

Most GISs offer simple statistical analysis of attribute data, such as calculation of means and standard deviations. Some GIS software packages also offer advanced statistical analysis of spatial data, including operations such as multiple regression analysis between layers, analysis of spatial

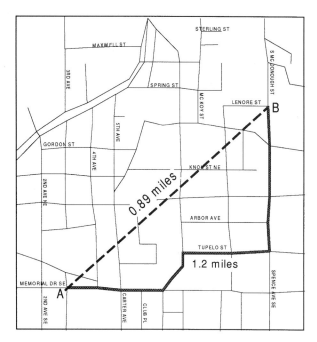

Figure 6.14   Distance Measurement and Routing. The straight line distance between intersections A and B, represented by the dashed line, is 0.89 miles. The shortest path along the road network, determined by the software and represented by the solid line, is 1.2 miles.

trends, and examination of spatial autocorrelation, which is a significant nonrandom arrangement of features.

OPERATIONS ON SINGLE LAYERS

Often in GIS analysis it is necessary to manipulate attribute data values of a single layer. If, for instance, a layer includes temperatures recorded in degrees Fahrenheit, the analyst may apply a conversion formula to temperature attribute to convert to degrees expressed in Centigrade. Classification of data in a layer is an important and frequently performed procedure in GIS analysis. Attribute data may be classified as in desktop mapping packages. Many GIS software packages offer quantitative attribute classification using a variety of classing schemes, including equal steps, quantiles, and user-defined schemes. (See Chapter 7.) GIS users may also classify qualitative, or categorical, data, such as land-use type or political party, into more generalized aggregations. For example, if working with land-use categories that include pasture, orchards, vineyards, deciduous forest, evergreen forest, and mixed forest, the user may reclassify the pasture, orchards, and vineyards to agriculture and the deciduous forest, evergreen forest, and mixed forest to forestland. Six categories are aggregated to two categories.

In addition to attribute classification, classification based on spatial characteristics is available with GIS technology. Some simple examples include calculating **slope,** defining **aspect,** and creating buffers. Stated mathemati-

cally, slope is the difference in elevation between two points, divided by the distance between the points. The greater the elevation range and the smaller the distance, the steeper the slope, which is expressed in degrees or percentages. Using DEMs with evenly spaced elevation values, raster GISs handle slope calculations fairly easily. Aspect is the direction of the slope. Aspect is usually expressed in degrees from north, with north set at 0 degrees, east at 90 degrees, south at 180 degrees, and west at 270 degrees. Buffering combines distance calculations and reclassification. A **buffer** is a zone surrounding a feature. Buffers can be employed in a number of cases. If, for example, a landfill cannot be sited within 200 meters of a stream, an analyst can use GIS buffering capabilities to measure the 200 meter distance from the stream and reclassify the study area to locations that are suitable or unsuitable for siting the landfill.

OPERATIONS ON MULTIPLE LAYERS

Operations on multiple layers, including **overlay** procedures, are also standard analysis tasks. It is occasionally necessary, for example, to "mask" locations outside a study area. A simple method when using rasters is to create a masking image that has z-values of 1 within the study area and z-values of 0 outside the study area. (See Figure 6.15.) The GIS operator then multiplies the original image and the mask image, resulting in an output image that contains the original z-values within the study area and z-values of 0 for background. The operator may retain background z-values of 0 or may choose to reclassify the background to some other value.

Overlay, a major element in GIS analysis, is the process of comparing the features and attributes of two or more layers. Overlay is analogous to placing "transparent" maps, encompassing the same geographical location, on top of one another to analyze the relationships among the various layers. Common vector overlay procedures, available in some desktop GISs, include point-in-polygon overlay, line-in-polygon overlay, and polygon overlay. (See Figure 6.16.) Point-in-polygon overlay entails the layering of a point coverage and a polygon coverage, to determine within which polygon each point is located. An analyst studying cases of residential lead poisoning might use point-in-polygon overlay to identify within which census tract each case falls to determine if there are any common demographic factors involved in the cases. Line-in-polygon overlay involves the layering of a line coverage with a polygon coverage and permits the analyst to identify the polygon or polygons within which each line is located. This type of analysis could be used to sum road miles by county.

Polygon overlay is the layering of two polygon coverages. Polygon features and attributes are merged, using either an intersection or union option, to create a new polygon coverage. With an **intersection operation,** only those areas that are common to both input coverages are in the

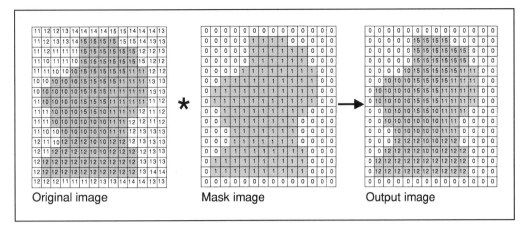

Figure 6.15   Masking

Overlay procedure used to mask locations outside of a study area. The original image is multiplied by the mask to produce an image where locations within the study area (shown on all the images in blue) retain their original cell values and locations outside the study area are given a cell value of 0.

output coverage. With a **union operation,** all areas of each input coverage are included in the output. If the appropriate layers are at hand, polygon overlay could be used to find out what percentage of land is available for development within each of the cities of a particular county.

Boolean overlay is another type of overlay operation. In raster GIS analysis, the cell values in a **boolean image** (also termed a binary or logical image) are either 1 or 0. Cells with a z-value of 1 represent areas where a condition is met and cells with a z-value of 0 represent areas where a condition is not met. The mask image described earlier is an example of a boolean image. Cells within the study area have a z-value of 1, whereas those with a value of 0 are outside the study area.

With boolean overlay, a GIS analyst can perform boolean, or logical, analysis. The most commonly used boolean operators are the intersection, or AND (not to be confused with addition), operation and the union, or OR operation. (See Figure 6.17.) As shown in Figure 6.17, the output of a boolean intersection operation is an image, covering the same geographic area as both input images, that contains cells with a z-value of 1 where the condition in corresponding cells in *both* input images is met, and cells with a z-value of 0 where the condition is not met in one, or the other, or both input images. The output of a boolean union operation is an image that contains cells with a z-value of 1 where the condition in corresponding cells is met in one, or the other, or both of the input images and cells with a z-value of 0 where the condition is not met in either of the input images.

Boolean analysis is often used in site suitability analysis. For example, an analyst needs to locate areas of coniferous forest growing on a particular soil type. (See Figure 6.18.) First he or she reclasses the vegetation layer: the cells representing coniferous forest are set to 1 and all other cells are set to 0. Then he or she reclasses

the soil layer, assigning a 1 to all cells representing the required soil type and a 0 to all other cells. Finally, he or she performs a boolean intersection on the two input layers to create an image with cells containing z-values of 1 in areas that satisfy both conditions.

An example of a typical problem, incorporating a number of the analysis techniques described above, is shown in the **cartographic model** in Figure 6.19. In performing spatial analysis, many GIS operators use cartographic modeling. A cartographic model is a graphic representation of the data files and analysis steps. A cartographic model aids the analyst in organizing and documenting the analysis process. Data files are shown in the rectangular boxes and analytical procedures are indicated near the arrows between the input and output data files.

The problem is as follows: engineers have proposed a route for a new highway. They have found, however, that they must locate the road at least 500 meters from stands of an endangered plant community. The engineers are aware that the plants are typically found on (1) south facing land, that is (2) moderately sloped and (3) has poor soil.

Using these three criteria, in addition to air photographs and remotely sensed imagery, the engineers devise an analytical process to predict the locations of these plants and to determine, through field checks of the identified areas, if the route requires modification. The analysis process requires three input layers covering the same geographic area: a digital elevation model (DEM), a soils layer, and the proposed route.

Using the DEM, slope is calculated in percent, and aspect is calculated in degrees. The slope output is reclassified, so that areas representing moderate slope (from 10 to 20 percent) are given a value of 1, and all other areas are given a value of 0. The output of this reclassification operation is a boolean layer. The aspect output is also reclassified so that areas representing south facing slopes (135 to

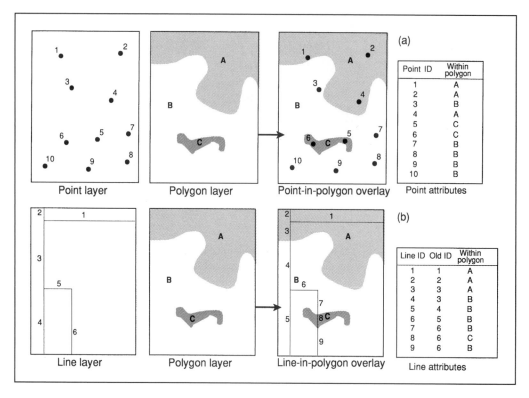

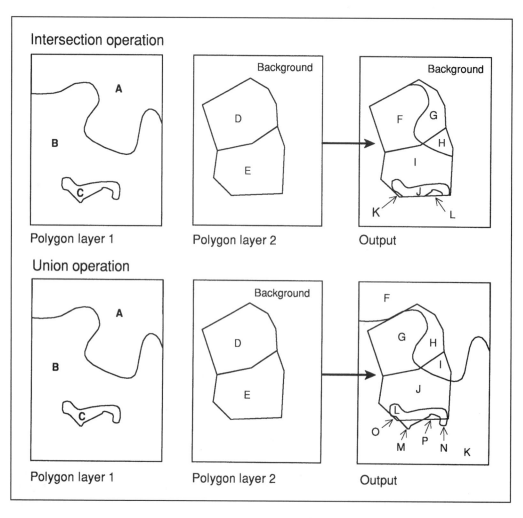

Figure 6.16 Overlay methods in GIS. (a) Point-in Polygon Overlay. Point and polygon coverages are layered to determine within which polygon each point is located. The point attribute database is generally modified to indicate the containing polygon for each point record. (b) Line-in-Polygon Overlay. Line and polygon coverages are layered to determine within which polygon, or polygons, each line is located. Lines that fall within more than one polygon are split to form new lines. The line attribute table is generally modified to indicate new line Ids, old line Ids, and the polygon within which each new line is located. (c) Polygon-on-Polygon overlay. With the intersection operation, only those areas that are common to both input polygon coverages are retained in the output coverage. In this case, all of the area of polygons D and E are output. Polygons A, B, and C subdivide D and E into seven new polygons and the background. With the union operation, all areas of each input image are retained. In this case, the overlay of polygons A, B, and C in Layer 1 and D and E in Layer 2, results in eleven new polygons in the output coverage. Attribute tables resulting from these operations include attributes from both input layers.

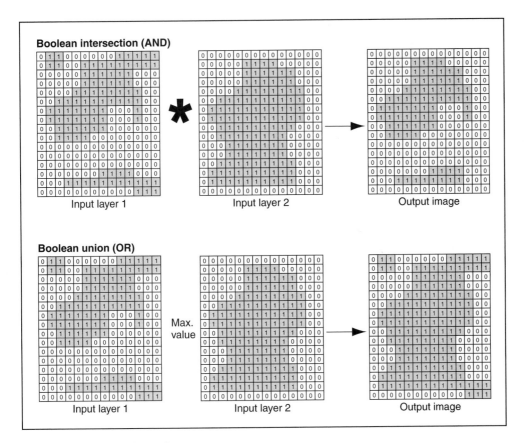

Figure 6.17   Boolean Operations.
Boolean images are comprised of cells containing z-values of 0 or 1. In boolean operations, a z-value of 1 means that user-specified conditions are met. The boolean intersection, accomplished by multiplying boolean images, results in an image that contains z-values of 1 in cells where conditions are met for both input images. The boolean union is accomplished by setting the output z-value of each cell to the maximum z-value of the corresponding cells of the input images. The boolean union results in an image where specified conditions are met in either one or both of the input images.

225 degrees) are assigned a value of 1 and all other areas are assigned a value of 0. The output of this reclassification operation is also a boolean layer. The soils layer is reclassified to create another boolean layer with poor soil types reclassified to 1 and other soil types reclassified to 0.

A boolean AND (intersection) overlay is performed on these three layers to output a layer displaying potential stands of the endangered plant communities. The road layer is used to create a 500-meter buffer surrounding the proposed route. The area within the buffer is assigned a value of 1 and the area outside the buffer is assigned a value of 0. To identify potential stands within the road buffer, another boolean AND overlay is performed. The final output meets the slope, aspect, and soil criteria, and displays the potential stands that are within 500 meters of the proposed highway. The engineers are now able to field-check the results and make modifications to the route if necessary.

MAP DISPLAY AND OUTPUT

As emphasized previously, the *analysis* of spatial data is the focus of geographic information systems. Analysis is the

key component in the process of GIS. During the analysis of spatial data, a number of temporary displays, not intended for presentation, are produced on the computer screen. Often these soft copy maps reveal patterns and relationships that are not visible in data tables and formulas (Note: The term soft copy refers to a computer display, whereas hard copy refers to printed or plotted output.) Alan MacEachren and Mark Monmonier state:

Human vision, instead of being considered a potential
source of bias, has come to be recognized as a powerful
tool for extracting patterns from chaos.[19]

They suggest that graphic analysis of data, which is increasing, has proven to be as effective as numerical analysis of data, which is decreasing.

Visual examination of these graphic displays may lead the GIS analyst to explore the data further and to pose new questions (a subject introduced in Chapter 1). The ability to explore data is facilitated by the development of powerful, user-friendly GIS software. The software available today enables an analyst to produce different types of maps, including both

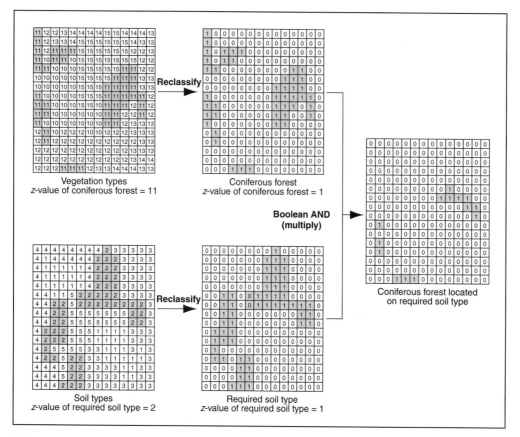

Figure 6.18   Boolean Analysis.

Vegetation and soil type images are reclassified to create boolean images, where z-values of 1 satisfy the required condition and z-values of 0 do not satisfy the required condition. The boolean images are multiplied to achieve a boolean AND, or intersection, where z-values of 1 represent locations that satisfy both conditions and z-values of 0 do not satisfy both conditions.

static and animated displays, much more quickly than manual mapping methods of the past. Analysts can view and study data in a variety of forms. When using visual thinking of this type, map display overlaps spatial analysis; there is no clear distinction between these two components of GIS processing.

Despite the increase in cartographic visualization, *many* users of GIS, particularly those outside research and academia, are required to produce static maps, both hard copy and soft copy, to communicate results to an audience of map readers. Unfortunately, GIS is often taught in disciplines that have traditionally had little connection to cartography. A large number of students of GIS are given no cartographic instruction in map communication. Oftentimes, map design is only an afterthought. With the proliferation of user-friendly GIS software, and the application of GIS technology to numerous disciplines, it is important to emphasize established cartographic conventions with regard to map design and communication.

Cartographic thinking and cartographic communication in GIS, briefly described above, are examined in the following sections. An example of private realm graphic analysis is explored and map communication issues for GIS analysts are addressed.

PRIVATE REALM GRAPHIC ANALYSIS

In this example of private realm graphic analysis, an analyst has obtained a point-of-event text file comprised of data regarding tornadoes in the state of Georgia from 1950 to 1995. (See Figure 6.20.) The text file includes one record for each tornado recorded during the time period. Each record, representing an individual tornado observed, contains a numerical identifier for the tornado, the year, month, date, and time of occurrence, the beginning latitude and longitude, the length of the path of the tornado in tenths of a mile, the number of fatalities and injuries, and the damage, in dollars, caused by the tornado. For those tornadoes with a path length of greater than 3 miles, an ending latitude and longitude are also recorded.

Although some potentially useful calculations, such as summing the number of tornado fatalities from 1950 through 1995, could be performed from the text data alone, the analyst decides to explore some of the spatial characteristics of the data. He or she maps the tornado data using the latitude and longitude fields. (See Figure 6.3a.) The analyst observes that most of the tornadoes are recorded in the northern and western part of the state.

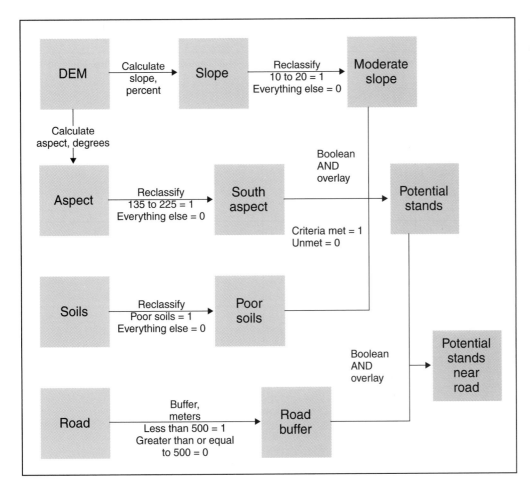

Figure 6.19
Cartographic Model.
This flowchart documents the steps required to determine potential locations of an endangered plant community within a specified distance of a proposed road.

Very few tornadoes are recorded in east, central Georgia and in several counties in the southeast. The observations lead to questions. Is this pattern related to the topography of the state, or is there some other cause?

The analyst maps tornado paths and path directions in search of additional patterns. (See Figure 6.21.) The majority of the tornadoes move from the southwest to the northeast. Why? Is the path of a tornado the same as the movement of weather fronts through the state? There are a few tornadoes with paths from west to east. Were there unusual weather patterns on those days or is there some data error?

Path and path direction are consistent throughout the state. The analyst sees nothing that explains the relatively low number of tornadoes in east, central, and southeast Georgia. He or she decides to map the data in a different manner. (See Figure 6.22a.) The map in Figure 6.22a displays the distribution of tornadoes as a density surface. The areas that are darker have had a higher density of recorded tornadoes than have the lighter areas.

The analyst, who is familiar with population distribution, immediately notices that the dark areas on the tornado density surface are in the same locations as Georgia cities and towns. He or she adds a point layer of major cities to the display and also creates a population density surface to

study the correlation further. (See Figure 6.22b.) High tornado density corresponds to high population density. The east, central parts of the state contain no cities with high population densities. The analyst examines a hard copy map of the state and finds that the small portion of the southeastern part of the state with low tornado density lies within the Okefenokee National Wildlife Refuge. In addition, smaller areas of low tornado density in central and northeastern Georgia, seem to fall within the boundaries of national forests. Low tornado density corresponds to low population density.

The distribution of population, upon first glance, *appears* to determine the distribution of tornados. Are tornadoes attracted to people? No. At this scale, for the state of Georgia, the distribution of tornadoes is probably random. *The method of recording tornadoes, direct human observation (more recently aided by radar), defines the pattern of tornado distribution.*

For this data set, if a tornado touches down in the forest and there is no one to see it, it does not exist. In other words, the tornado goes unrecorded. There is bias in the data-gathering technique. This conclusion leads to some considerations and questions regarding the use of the tornado data set and other point-of-event data. The tornado

Tornado attribute table

ID	Year	Month	Date	Time_24_h	BDDLAT	BDDLON	EDDLAT	EDDLON	Length_0	Fatalities	Injuries	Damage_COD
1	50	5	30	1,400	34.75	−84.92			5	0	0	3
2	50	6	4	600	33.68	−84.93			20	0	0	3
3	50	6	4	800	33.85	−84.25	33.85	−84.20	33	0	1	4
4	50	6	7	1,730	33.03	−84.15			20	0	0	3
5	51	5	23	1,530	31.30	−83.83	31.27	−83.68	92	0	0	4
6	51	5	23	423	31.83	−81.50			3	0	0	4
7	51	11	16	655	31.53	−84.17			5	0	0	3
8	52	1	22	700	31.27	−84.02	31.28	−84.00	19	0	0	5
9	52	1	22	705	31.28	−84.00	31.33	−83.90	68	0	8	5
.	.	.	.	.	.	.	.	.	.	.	.	.
.	.	.	.	.	.	.	.	.	.	.	.	.
.	.	.	.	.	.	.	.	.	.	.	.	.

Field descriptions

ID	Unique identifier
Year	Last two digits of year of occurrence
Month	Month of occurrence
Date	Date during month of occurrence
Time_24_h	On a 24-hour scale
BDDLAT	Beginning latitude in decimal degrees
BDDLON	Beginning longitude in decimal degrees
EDDLAT	Ending latitude in decimal degrees
EDDLON	Ending longitude in decimal degrees
Length_0	Path length of tornato in tenths of a mile
Fatalities	Number of fatalities
Injuries	Number of injuries
Damage_COD	1. <$50
	2. $50 to $500
	3. $500 to $5,000
	4. $5,000 to $50,000
	5. $50,000 to $500,000
	6. $500,000 to $5 million
	7. $5 million to $50 million
	8. $50 million to $500 million
	9. $500 million to $5 billion

Figure 6.20   Georgia tornado attribute table.

The first nine records of the Georgia tornado attribute table are displayed. A description of the attributes is required to understand the contents of each field. *(Source: The data shown here have been processed from tornado archive data available through the Storm Prediction Center of the National Oceanic and Atmospheric Administration at http://www.nssl.noaa.gov/~spc/arcjove/tornadoes/index.html. Some of the attributes in the original data set are not shown. In addition, longitude and latitude have been modified from sexegesimal degrees to decimal degrees.)*

data set is probably inappropriate for studying the true distribution of tornados in Georgia over time because there is a large probability that many tornadoes went unrecorded in areas with low population density. Are there analysts who are using these data inappropriately? Are there other point-of-event data sets that are similarly biased? The tornado data set could be suitable for studying tornado paths and path directions because path and direction are probably independent of population distribution. The data set is definitely useful for determining such information as number of tornado fatalities over time or for calculating the damage in dollars, per county, for a particular time span.

In this example, private realm graphic analysis enabled an analyst to determine that a data set is probably unsuitable for certain types of spatial analysis. Exploration of the data, and the search for patterns, also led to further questions regarding this type of point-of-event data.

PUBLIC REALM CARTOGRAPHIC COMMUNICATION

The primary focus of this text is public realm cartographic communication. Map design and cartographic convention are addressed throughout the book. For information regarding the projection, symbol type, mapping technique, text placement, or colors to employ to create an effective map, it is recommended that the reader consult the relevant chapter. This section focuses on *some* of the many design issues specific to desktop geographic information systems software.

When creating a map, if the user does not indicate a projection and grid system, a number of GIS software packages default to a Plate Carree projection. The Plate Carree projection, as previously stated, distorts the shape, area, and scale as the display moves away from the equator. There are countless examples of maps whose creators do not real-

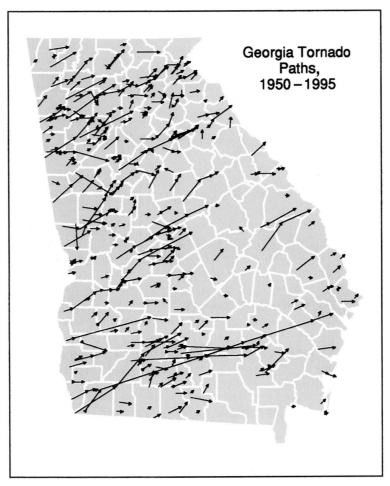

Figure 6.21   Georgia tornado paths, 1950-1995.
Note: Only those tornadoes with a path length of greater than three miles are
shown.

ize that they must indicate an appropriate projection. Dot density maps, in particular, require an equal area projection. Most software packages, however, will allow a user to output a dot density map without setting the projection.

At least one GIS software package automatically "chooses" a projection for the user, if one is not indicated, based upon the map scale or center. The software manual advertises that the user does not need to know anything about projections. This statement is highly debatable. It is necessary to understand the characteristics of projections to select the most appropriate projection for the thematic mapping technique used. The default projection is usually not the best.

The issue of scale is a concern in GIS output. As mentioned in the section on data acquisition, the scale of the original data source defines how the data may be analyzed. It also defines how the data may be output. If a map of lakes is digitized at a 1:250,000 scale, printed output from this digital map should not be at the 1:24,000 scale. The lakes would be too generalized.

The expression of scale is significant also. Graphic bar scales are best because if the map is resized, a common occurrence in GIS map production, the size of the bar scale changes along with the map. Verbal scales and representative fractions should be avoided. Most GIS software packages do, by default, provide graphic bar scales as a part of their output features. For very small-scale maps, such as a world map, scales of any kind should not be used because the scale changes over the map.

When creating choropleth maps, classing schemes should be examined carefully. There is usually a selection of different classing schemes, including the Jenks optimization method in some packages. Choropleth maps can usually be created quickly, so experimentation with different classing methods is fairly easy. The number of classes used is also a consideration. For choropleth mapping a small number of classes, five or six, is desirable, but often the default is nine or ten classes.

Another common problem in choropleth mapping in GIS is the default color schemes, or **palettes.** Choropleth mapping

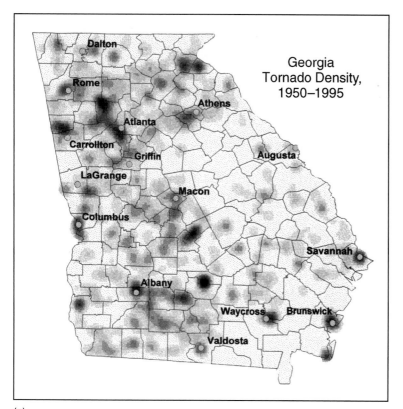

(a)

Figure 6.22   Visualization in GIS.
(a) Darker areas have recorded a higher number of tornadoes than have lighter areas. Note: These figures were created quickly, for "private thinking," to visualize a general trend. They use a Plate Carree projection which distorts the area of the state. For map communication, a more appropriate projection should be chosen. (b) The darker areas have a higher population density; the lighter areas have a lower population density.

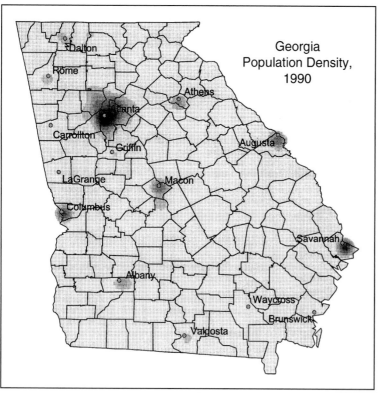

(b)

uses quantitative data, but often the default palette is qualitative in nature. (See Chapter 15.) For instance, the percentage of persons receiving Medicaid, by county for the United States, is mapped using the choropleth technique. The data are separated into five classes. Numerous palettes have been designed to be used with quantitative data. (See Chapter 15.) An example of a suitable palette is a gradation from light blue, representing the class with the lowest percentage of persons receiving Medicaid, to dark blue, representing the class with the highest percentage of persons receiving Medicaid. With some GISs, however, the default palette is comprised of different hues, such as red, green, blue, and yellow. The quantitative nature of the data is not communicated with this default palette and the changing class values are not reflected. The user must select an appropriate palette.

Conversely, there are instances when quantitative palettes are used as defaults for mapping qualitative data. Classed qualitative data, such as soil types or political parties, have no implied value. Qualitative data are therefore more effectively displayed with palettes comprised of different hues. The user must be aware of the type of data being used to select suitable palettes.

In addition to inappropriate palette defaults and choices, some users attempt to display too many classes of data when using qualitative data. A number of digital data sets, such as Land Use and Land Cover (LULC) data from the USGS, contain more than 35 categories of land use. The numerous classes may be used in analysis but, for communication purposes, these classes should be aggregated in some way so that the map reader can distinguish the different land use types.

It is crucial that the user understand how the different mapping techniques are achieved in a GIS. When using proportional symbols in mapping, for example, the symbol sizes should be calculated according to established cartographic convention as described in Chapter 9. The size of the symbol is directly related to the value it represents. Some desktop GISs, however, simply use arbitrary text point sizes.

Text placement is also an important issue in GIS output. Labeling points, lines, and polygons automatically is often imprecise or unacceptable cartographically. One problem is **overposting,** labeling for one feature placed on top of labeling for another feature. Another problem with automatic labeling of polygons is where to place the label. Because of these difficulties, many GIS users analyze their data within the GIS and then transfer maps and images to a more versatile graphics package for final cartographic enhancements. GIS output is improving, however. Most packages allow the user to move text that has been placed automatically. For points, some packages permit the user to set the automatic labeling to a specified location—above and to the right, below and to the right, and so forth. Others now allow labeling of text along the curve of a linear feature, such as a river or a road.

The important point to remember in creating cartographically sound output from a GIS is to be knowledgeable regarding design elements and cartographic convention. GIS software defaults are usually not the best choice when creating a map for communication purposes. It is the responsibility of the map creator to be aware of the limitations of the software concerning cartographic output.

THE PLACE OF GIS

At the beginning of this chapter GIS was defined as a computer-assisted process designed to acquire, store, analyze, and display spatial data and their attributes. Since the emergence of GIS, there has been debate about the "place" of GIS. Is cartography part of GIS or is GIS part of cartography?

Many in the GIS community, especially those from noncartographic backgrounds, consider cartography to focus on map compilation, design, and production, whereas GIS focuses on spatial analysis. From this perspective, cartography and GIS have a technical similarity, but a difference in objectives. The traditional concept of cartography, as primarily concerned with map design and production, separates GIS from cartography. From this perspective, cartography is applied only when mapped output is required. Cartography, according to this concept, is a small part of GIS.

Although hard copy maps have long been used to study spatial patterns, the rise of computer technology has enabled cartographers to visualize data more efficiently. Because of the increase in this type of data visualization, more recent models of cartography, including that of this text, incorporate cartographic analysis as part of the cartographic process. As emphasized previously, the analysis of spatial data is the primary focus of GIS. For this reason, many cartographers consider GIS a part of cartography that falls primarily within the private realm of cartographic thinking (see Chapter 1). GIS, from this viewpoint, is part of cartography.

Another recent topic of discussion among the GIS community is whether GIS is a tool or a science.[20] Wright, Goodchild, and Proctor summarized comments regarding the place of GIS from subscribers to GIS-L, an Internet mailing list for general GIS discussion. They determined that points of view range along a tool-science continuum: from GIS as a tool, to GIS as toolmaking, and finally to GIS as a science. Those who view GIS as a tool see it as the use of GIS software, hardware, and digital spatial data to advance some specific purpose. Toolmakers support the advancement of the capabilities and use of GIS. Those who claim GIS is a science contend that its ability to analyze data in new ways raises GIS to a science. Additionally, the fact that some in the GIS community are concerned with the analysis of issues raised by the use of GIS strengthens the argument for GIS as science. The acronym GIS, is now occasionally defined as "geographic information science." As the use of GIS increases, discussion about its place will continue.

NOTES

1. Geographer's Craft. Available: http://www.utexas.edu/depts/grg/gcraft/notes/intro/intro.html#Geog.

2. J. R. Eastman, *IDRISI for Windows User's Guide.* (Worcester, MA: Clark Labs for Cartographic Technology and Geographic Analysis, 1997), pp. 2–3.

3. Michael N. DeMers, *Fundamentals of Geographic Information Systems* (New York: Wiley, 1997), p. 7.

4. Ronald F. Abler, "Awards, Rewards, and Excellence: Keeping Geography Alive and Well," *The Professional Geographer* 40 (1988): 135–40.

5. David W. Stahle and Phillip L. Chaney, "A Predictive Model for the Location of Ancient Forests," *Natural Areas Journal* 14 (1994): 151–58.

6. Richard C. Daniels, "An Innovative Method of Model Integration to Forecast Spatial Patterns of Shoreline Change: A Case Study of Nags Head, North Carolina," *The Professional Geographer* 48 (May 1996): 195–209.

7. Walter Chomentowski, Bill Sales, and David Skole, "Landsat Pathfinder Project Advances Deforestation Mapping," *GIS World* (April 1994): 34–38.

8. Kimberly E. Medley, Mark J. McDonnell, and Steward T. A. Pickett, "Forest-Landscape Structure along an Urban-to-Rural Gradient," *The Professional Geographer* 47 (May 1995): 159–68.

9. Christopher C. Muller and Crist Inman, "The Geodemographics of Restaurant Development," *The Cornell H. R. A. Quarterly,* June 1994, pp. 88–95.

10. T. C. Ricketts et al., *Using Geographic Methods to Understand Health Issues* (AHCPR Publication No. 97–N013; NTIS No. PB97–137707—Rockville, MD: Agency for Health Care Policy and Research, March 1997).

11. Curtis Hinton, "North Carolina City Saves Time, Lives, and Money with Award-Winning GIS," *Geo Info Systems,* September 1997, pp. 35–37.

12. "Computer's Track the Criminal's Trail," *American Demographics,* January, 1994, pp. 13–14.

13. P. Dixon. 9/8/97 Posting at GISARCH-L July, 1996. Available: http://www.mailbase.ac.uk/lists/gisarch/1996-07/0000.html.

14. John Kantner. 9/9/97 An Evaluation of Chaco Anasazi Roadways. Available: http://www.sscf.ucs.edu/anth/project/lobo/SAA96. Available: kantner@sscf.ucsb.edu.

15. Donna J. Peuquet, "A Conceptual Framework and Comparison of Spatial Data Models," *Cartographica* 21 (1984): 66–113.

16. Michael N. DeMers, *Fundamentals of Geographic Information Systems* (New York: Wiley, 1997), p. 179.

17. Eastman, *IDRISI for Windows User's Guide,* pp. 4–35.

18. SDTS Task Force of United States Geological Survey, "Spatial Data Transfer Standard Senior Management Overview," Available: http://mcmcweb.er.usgs.gov/sdts/training.html.

19. Alan MacEachren and Mark Monmonier, Introduction, *Cartography and GIS* 19 (1992): 197–200.

20. Dawn J. Wright, Michael F. Goodchild, and James D. Proctor, "GIS: Tool or Science?" *Annals* (Association of American Geographers) 87 (June, 1997): 346–62.

GLOSSARY

address matching   the process of assigning a pair of geographic coordinates to a location based on an address in a table; requires a list of addresses to be geocoded as well as a reference coverage for use in matching, p. 115

aspect   the direction of the slope usually expressed in degrees from north with north set to zero, p. 124

attribute database   describes the characteristics or qualities of the features or images portrayed on the base maps; both an attribute and spatial database are necessary to form a GIS database, p. 112

boolean image   a raster comprised of cells containing the values 0 and 1, p. 125

boolean overlay   overlay of Boolean images, p. 125

buffer   a zone surrounding a feature, p. 124

cartographic model   a graphic representation of the data files and analysis steps; aids the analyst in organizing and documenting the analysis process, p. 125

cell resolution   in a raster, the size of a cell; the ground area a cell encompasses, p. 118

conflation   technical term for rubber sheeting, p. 118

coverage   a layer or theme in a GIS, p. 112

data model   an abstraction of the real world; in GIS, the two major data models are the raster model and the vector, p. 112

disaggregate   to subdivide data; generally discouraged because it implies a level of precision that may not exist, p. 119

edge matching   the process of joining adjacent coverages, p. 118

FGDC   the U.S. Federal Geographic Data Committee, formed in 1990; responsible for the development of a content standard for digital spatial metadata, p. 122

FGDC Standard   the FGDC Content Standard for Digital Geospatial Metadata, in effect for all U.S. federal agencies since 1995; provides consistency on the type of spatial metadata required, p. 122

FIPS   Federal Information Processing Standards; in addition to providing standard codes for U.S. geographic entities (e.g. states, counties, census tracts), FIPS addresses a wide range of computer system components, p. 122

geocode   to assign geographic coordinates to features in a layer, p. 115

geodetic coordinates   *x* and *y* coordinates, based on a sphere, expressed in degrees of longitude and latitude; projected as a Plate Carree projection, p. 118

geographic information system (GIS) a computer-assisted process designed to acquire, store, analyze, and display spatial data and their attributes, p. 111

georeference to relate the plane coordinates on a map to earth surface coordinates; layers in a spatial database must be georeferenced to the same projection, grid system, and datum, p. 118

intersection operation an overlay operation that outputs a theme consisting of only those features common to each of the input layers, p. 124

metadata data about data; data documentation, p. 116

overlay to analyze features and attributes among themes by layering the themes, p. 124

overposting labeling for one feature placed on top of labeling for another feature, p. 133

palette a color scheme in digital mapping, p. 133

Plate Carree projection a projection in which one degree of longitude is consistently equal to one degree of latitude; geodetic coordinates are often displayed in a Plate Carree projection, p. 118

point-of-event table table that contains x and y coordinates representing the locations of discrete events, p. 116

raster data model data model based upon cells, or pixels, each cell containing a single attribute value, or z-value; one of the two major data models used in GIS; (see **vector data model**), p. 112

rubber sheeting a georeferencing process that involves the use of ground control points to stretch a map so that it will register with other GIS layers; also known as conflation, p. 118

SDTS Spatial Data Transfer Standard; U.S. standard designed to improve the exchange of spatial data between different computer platforms, p. 116

slope the difference in elevation between two points, divided by the distance between the points, p. 124

spatial database a set of digital base maps that describes the "geography" of the earth's surface; both a spatial and an attribute database are required to form a GIS database, p. 112

tessellation any regular shape, such as a square, rectangle, triangle, or hexagon, that covers a plane surface without any gaps, p. 112

theme a layer or coverage in a GIS, p. 112

TIN triangulated irregular network; a special type of vector file, consisting of irregularly shaped triangles, that models surfaces, p. 112

topology in a vector-based GIS, the relationships between points, lines, and polygons; topological relationships include adjacencies, connectivity, and containment, p. 112

union operation an overlay operation that outputs a theme consisting of features found on at least one or both of the input layers, p. 125

vector data model data model based on points (or nodes), lines (or arcs), and polygons; one of the two major data models used in GIS; (see **raster data model**), p. 112

WWW World Wide Web; one facet of the Internet consisting of computers handling multimedia (plain text, graphics, video, and sound) documents, p. 111

z-value in a raster-based GIS, a cell value; as in Cartesian geometry, the horizontal axis of a raster grid is the x-axis, the vertical axis is the y-axis, and the third dimension, representing a surface, is the z-axis, thus the name z-value, p. 112

READINGS FOR FURTHER UNDERSTANDING

Abler, Ronald F. "Awards, Rewards, and Excellence: Keeping Geography Alive and Well." *The Professional Geographer* 40 (1988): 135–40.

Chomentowski, Walter, Bill Sales, and David Skole. "Landsat Pathfinder Project Advances Deforestation Mapping." *GIS World,* April, 1994, pp. 34–38.

"Computers Track the Criminal's Trail." *American Demographics,* January 1994, pp. 13–14.

Chrisman, Nicholas. *Exploring Geographic Information Systems.* New York: Wiley, 1997.

Daniels, Richard C. "An Innovative Method of Model Integration to Forecast Spatial Patterns of Shoreline Change: A Case Study of Nags Head, North Carolina." *The Professional Geographer* 48 (1996): 195–209.

Davis, Bruce. *GIS: A Visual Approach.* Santa Fe: OnWord Press, 1996.

DeMers, Michael N. *Fundamentals of Geographic Information Systems.* New York: Wiley, Inc., 1997.

Dixon, P. 9/8/97 Posting at GISARCH-L July, 1996. Available: http://www.mailbase.ac.uk/lists/gisarch/1996-07/0000.html.

Eastman, J. R. *IDRISI for Windows User's Guide.* Worcester, MA: Clark Labs for Cartographic Technology and Geographic Analysis, 1997.

Environmental Systems Research Institute, Inc. *Understanding GIS: The ARC/INFO Method.* Redlands, CA: ESRI, 1990–1995.

Geographer's Craft. Available: http://www.utexas.edu/depts/grg/gcraft/notes/intro/intro.html#Geog.

Goodchild, Michael F., and Karen K. Kemp, eds. *NCGIA Core Curriculum.* Santa Barbara: University of California, National Center for Geographic Information and Analysis, 1991.

Hearnshaw, Hilary M., and David J. Unwin, eds. *Visualization in Geographical Information Systems.* Chichester, England: Wiley, 1994.

Hinton, Curtis. "North Carolina City Saves Time, Lives, and Money with Award-Winning GIS." *Geo Info Systems,* September 1997: pp. 35–37.

Kantner, John. 9/9/97 An Evaluation of Chaco Anasazi Roadways. Available:

http://www.sscf.ucs.edu/anth/project/lobo/SAA96. Available: kantner@sscf.ucsb.edu.

MacEachren, Alan M., and Mark Monmonier. Introduction. *Cartography and GIS* 19 (1992): 197–200.

————, **and D. R. Fraser Taylor, eds.** *Visualization in Modern Cartography.* New York: Pergamon Press, 1994.

Medley, Kimberly E., Mark J. McDonnell, and Steward T. A. Pickett. "Forest-Landscape Structure along an Urban-to-Rural Gradient," *The Professional Geographer* 47 (1995): 159–68.

Muller, Christopher C., and Crist Inman. "The Geodemographics of Restaurant Development." *The Cornell H. R. A. Quarterly.* June, 1994, pp. 88–95.

Peuquet, Donna J. "A Conceptual Framework and Comparison of Spatial Data Models," *Cartographica* 21 (1984): 66–113.

————, **and Duane F. Marble, eds.** *Introductory Readings in Geographic Information Systems.* Bristol, PA: Taylor & Francis, 1990.

Ricketts, T. C., et al. *Using Geographic Methods to Understand Health Issues.* AHCPR Pub. No. 97-N013; NTIS, No. PB97-137707. Rockville, MD: Agency for Health Care Policy and Research, March 1997.

SDTS Task Force of United States Geological Survey. "Spatial Data Transfer Standard Senior Management Overview." Available: http://mcmcweb.er.usgs.gov/sdts/training.html.

Stahle, David W., and Phillip L. Chaney. "A Predictive Model for the Location of Ancient Forests." *Natural Areas Journal* 14 (1994): 151–58.

Star, Jeffrey, and John Estes. *Geographic Information Systems: An Introduction.* Englewood Cliffs, NJ: Prentice Hall, 1990.

Taylor, D. R. Fraser, ed. *Geographic Information Systems: The Microcomputer and Modern Cartography.* Oxford, England: Pergamon Press, 1991.

Tomlin, C. Dana. *Geographic Information Systems and Cartographic Modeling.* Englewood Cliffs, NJ: Prentice Hall, 1990.

Worboys, Michael F. *GIS: A Computing Perspective.* Bristol, England: Taylor & Francis, 1995.

Wright, Dawn J., Michael F. Goodchild, and James D. Proctor. "GIS: Tool or Science?" *Annals* (Association of American Geographers) 87 (1997): 346–362.

PART II

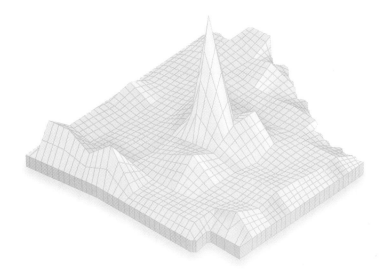

TECHNIQUES OF QUANTITATIVE THEMATIC MAPPING

The chapters in Part II cluster naturally to form the quantitative thematic mapping techniques portion of this book. Each chapter presents a different thematic technique, and each includes the rationale for its adoption. Part II begins with an examination of the choropleth map, perhaps one of the most widely used thematic map forms today. Enumeration data are symbolized by area patterns or colors in this form of map. Chapter 8 addresses the common dot map, with examples selected to emphasize the proper technique in its use. This kind of map is not encountered as often as in earlier times, except perhaps in agricultural atlases. Symbolizing quantities at points is presented in Chapter 9, which introduces the technique of mapping by proportional symbols. Much of the research literature of the past 30 years is devoted to this map technique; the issues are presented with current design standards.

Chapter 10 introduces volume, or isarithmic mapping. This technique is similar to the mapping of re-

lief by contours of elevation, and is a form of map familiar to many. The thematic cartographer will deal with the isoplethic variety where mapped values are assumed to occur at points, or with the isometric form where values do actually occur at points. Chapter 11 deals with value-by-area mapping, or mapping by transforming familiar geographical space into some other space, yielding a cartogram. This form of mapping is more abstract than real, yet can yield interesting results and is often used to attract the reader's attention. It is a useful pedagogical device in geography. The design of flow maps concludes the chapters in Part II. This two-century-old technique is often used for mapping the movement of things or ideas, connections and interactions, and spatial organization, and frequently finds its way into economic presentations. Flow maps are often referred to as "dynamic" maps. They present interesting design challenges for the cartographer.

CHAPTER

7

MAPPING ENUMERATION AND OTHER AREALLY AGGREGATED DATA: THE CHOROPLETH MAP

CHAPTER PREVIEW

Choropleth mapping is a common technique for representing enumeration data. The **choropleth map** *was introduced early in the nineteenth century. It was used by the Bureau of the Census in several statistical atlases in the last half of that century and has been a favorite of professional geographers and cartographers ever since. Its name is derived from the Greek words* choros *(place), and* pleth *(value). Choropleth mapping has also been called* area *or* shaded *mapping. For the most part, the rationale of the choropleth technique is easily understood by map readers. Major concerns of the cartographer are data classification, areal symbolization, legend design, and development of an appropriate base map. From a variety of classification methods, the designer is faced with selecting the one that best serves the purpose of the map and best depicts the spatial array of the data.*

Methods include constant and variable interval formats. One result of grouping data is the creation of different maps based on the classification scheme chosen by the designer. A new type of choropleth map is the unclassed variety, chosen when the map designer simply wants to give an unstructured picture of the spatial data. A form of map not used often today, the dasymetric map, is discussed because of its relation to choropleth mapping and to geographic information systems (GIS).

An entire chapter is devoted to this form of mapping because of its widespread use and appeal, not only for professionals, but for the general public as well. Because of this, the student must look at the choropleth map in considerable detail to learn its advantages and consider the standards for its use in a variety of mapping situations.

SELECTING THE CHOROPLETH TECHNIQUE

What guides cartographers in selecting one mapping technique over another when approaching a given design task? When is a given technique not appropriate? The following section describes under what conditions the choropleth map should be chosen.

MAPPING RATIONALE

The choropleth technique is defined by the International Cartographic Association as follows: "A method of cartographic representation which employs distinctive color or shading applied to areas other than those bounded by isolines. These are usually statistical or administrative areas."[1] Because this form of mapping is used to depict bounded areal classified or aggregated data (often defined by administrative areas), it is sometimes called *enumeration* mapping. (See Figure 7.1.)

Choropleth mapping may be thought of as a three-dimensional histogram or stepped statistical surface. (See Figure 7.2.) A choropleth map is simply a planimetric representation of this three-dimensional **data model.** In the model, the height of each prism is proportional to the value it represents. The planimetric way of looking at the model incorporates areal symbolization to depict the heights of the prisms. In black-and-white mapping, the higher prisms are

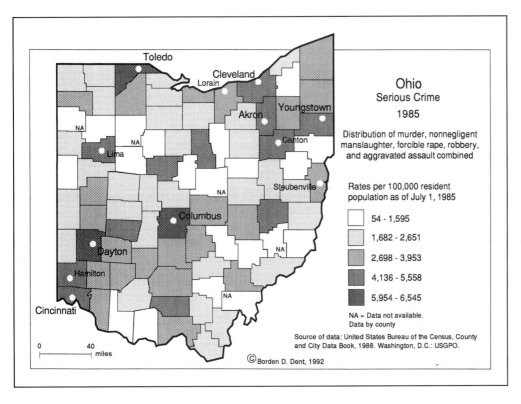

Ohio
Serious Crime
1985

Distribution of murder, nonnegligent manslaughter, forcible rape, robbery, and aggravated assault combined

Rates per 100,000 resident population as of July 1, 1985

54 - 1,595
1,682 - 2,651
2,698 - 3,953
4,136 - 5,558
5,954 - 6,545

NA = Data not available.
Data by county

Source of data: United States Bureau of the Census, County and City Data Book, 1988. Washington, D.C.: USGPO.

©Borden D. Dent, 1992

Figure 7.1 A typical thematic map.
Each enumeration unit, in this case a county, has an areal symbol applied to it, depending on the class in which its data value falls. Over the entire map, it is possible to determine spatial variation of the data. It is important to provide locational information on the base map, as is done here.

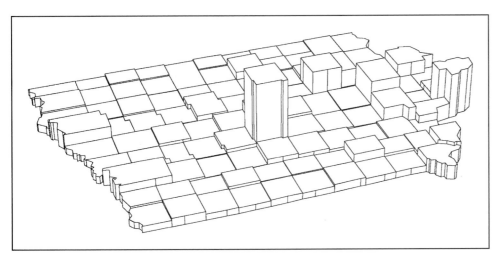

Figure 7.2 The data model concept in choropleth mapping.
In this conceptual model, each enumeration unit is a prism raised vertically in proportion to the value it represents.

normally represented by darker-area symbols; conversely, the lower prisms are represented by lighter-area symbols. It is helpful to think of this conceptual model when beginning a choropleth design task. When color choropleth maps are produced, considerable differences of opinion exist on how to apply hues as area symbols. This topic will be addressed later in the chapter.

Mapping techniques normally require that the cartographer collect data by statistical or administrative areas. An areal symbolization scheme is then devised for these values, and the symbols are applied to those areas on the map whose data fall into the symbol classes. (See Figure 7.3.) Techniques developed through computer mapping can alter this general procedure; the newer techniques will be introduced later in the chapter.

Map readers use choropleth maps in three ways: to ascertain an actual value associated with a geographic area, to obtain a sense of the overall geographical pattern of the mapped variable with attention to individual values, and to compare one choropleth map pattern to another. It has been argued convincingly that the reader who wants to find only individual values should simply consult a table of values, and not bother with the map.

Using two or more choropleth maps to compare geographical distributions (usually to look for positive correlations) is an acceptable application of choropleth mapping, but this subject is not treated in detail here. Our presentation concentrates on methods useful in the production of an individual choropleth map whose purpose is to portray a single geographical theme.

It must be pointed out that the choropleth map has come under close scrutiny by cartographers because for any given choropleth map solution there are really several others.

Which is best? There is never a conclusive test to assist the designer. Some cartographers suggest that we should never offer only one map for a particular data set, but several, so that the reader has the advantage of seeing other map solutions. An ethical alternative to providing many maps would be to add a statement to the map that tells the reader that this is but *one* of many mapping alternatives.

APPROPRIATENESS OF DATA

The choropleth technique should be selected only when the form of data is appropriate. Typically, and appropriately, choropleth maps are constructed when data occur or can be attributed to definite enumeration units, for example, statistical units used by the Bureau of the Census or administrative political subdivisions such as cities, counties, school districts, or others—that is, enumeration data. Geographic phenomena that are continuous in nature should not be mapped by the choropleth technique because their distributions are not controlled by political or administrative subdivisions. For example, to map average annual temperature by this method would not be appropriate, but to map the number of hospital beds per 1,000 people would be.

Enumeration data may be of two kinds: *totals* or *derived values* (rates or ratios). The number of people living in a census tract is an example of the former and average annual income is an example of the latter. Traditionally, it is not acceptable to map total values when using the choropleth technique. This convention is based on sound reasoning. In most choropleth mapping situations, the enumeration units are unequal in area. The varying size of areas and their mapped values will alter the impression of the distribution.

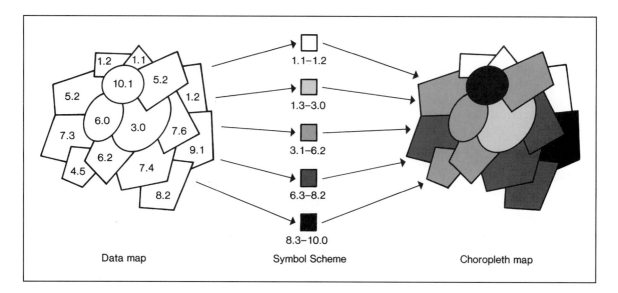

Figure 7.3 The choropleth technique.
Cartographic designers begin with a data model—a map of the enumeration units and all data values. Area symbol classes are selected to represent range-graded numerical classes. The final map is then developed by applying appropriate area symbols to the enumeration units, depending on each unit's data value.

(See Figure 7.4a.) In addition, uniform distributions may be masked when enumeration total values are used. (See Figure 7.4b.) It has therefore become customary to use *ratios involving area* or *ratios independent of area*. Thus data involving areas are standardized over the maps.

A familiar example of a ratio involving area is density, of which many kinds could be named. Population per square mile is often used, as is crop yield per acre. Ratios independent of area include per capita income, infant deaths per 100,000 live births, and so on. Proportions, including percentages, are also used. Much of the enumeration data available to cartographers is in the form of aggregated areal data, such as average value of farm products sold, median family income, average annual income, and average persons per household unit. These are treated in the same manner as ratios or proportions in choropleth mapping.

The most important assumption made in choropleth mapping is that the value in the enumeration unit is spread uniformly throughout the unit. (See Figure 7.5.) The top of each prism in the data model is horizontal and unchanging. Wherever one places a pencil point in the area, one finds the amount of the variable chosen to represent the entire area. Thus, the choropleth technique is insensitive to changes of the variable that may occur at scales larger than the chosen enumeration unit. If the variable is changing within the enumeration unit, the change cannot be detected on the choropleth map. (See the discussion about dasymetric mapping at the end of this chapter.)

In choropleth mapping, the boundaries of the chorograms have no numerical values associated with them. They function only to separate the enumeration areas and signify the geographic extent to which the enclosed area values apply. This is in contrast to the lines on the isopleth map which do have values, and which will be discussed in a later chapter.

WHEN TO USE THE CHOROPLETH MAP

The choropleth technique is appropriate whenever the cartographer wishes to portray a geographical theme whose data occur within well-defined enumeration units. If the data cannot be dealt with as ratios or proportions, they should not be portrayed by the choropleth technique. Also choropleth (or any other mapping technique) should not be used if the interest is to show actual, precise values within

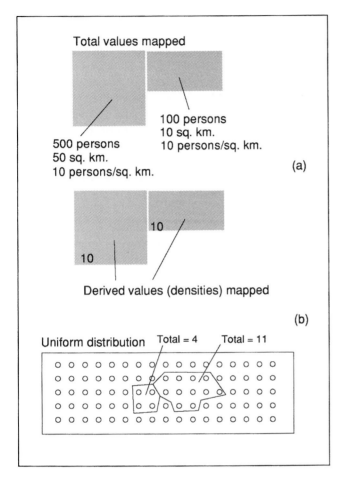

Figure 7.4   Total values should not be mapped by the choropleth method.

In (a), mapping data totals masks the even densities, because the areas are of unequal size. Uniform distributions, as in (b), also will be obscured when totals alone are mapped.

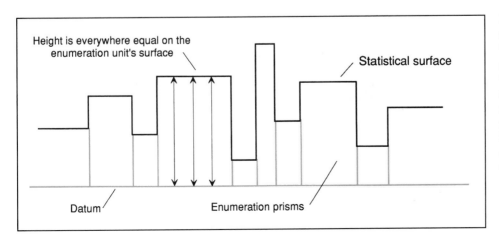

Figure 7.5   The stepped statistical surface.

The choropleth technique assumes a stepped statistical surface on which the value in each unit is constant. That is, the unit's surface is everywhere equidistant from the map datum plane. The view shown here is as if a vertical profile or a slice were taken out of the three-dimensional data model.

enumeration units. Choropleth mapping is simple and should be used only when its assumptions are acceptable to the cartographer and to the eventual reader. The method should not be made overly complex.[2]

PRELIMINARY CONSIDERATIONS IN CHOROPLETH MAPPING

Important considerations in the design of a choropleth map include thorough examination of the geographic phenomenon and its elements, map scale, number and kind of areal units, data processing, data classification, areal symbolization, and legend design.

Geographic Phenomena
All map design begins with careful analysis of just what it is that is being mapped. A careful designer assembles facts that will help in understanding the mapping activity. The novice may attempt to develop maps showing dairying, without knowing anything about cows! In a mapping problem to illustrate the geographical aspects of retailing, what measures should be used? Dollar sales, payrolls, and number of employees might be appropriate. What industrial or trade indices are commonly accepted and used by analysts? What surrogate measures might be used? What other geographical variables accompany the one being mapped? How does this phenomenon behave spatially or aspatially, with or without other geographic phenomena?

Cartographic designers must equip themselves with as much knowledge about the map subject as possible. In many cases, consultants are brought in during the early phases of design. Regardless of the individual steps taken, premapping research is necessary for effective map design, just as a product designer must measure the proportions of the human hand before designing an electric drill.

Map Scale
Map scale as it relates to choropleth design involves two considerations: necessity and available space.[3] Necessity dictates that the scale be sufficient to accommodate symbol recognition—the areal units must be large enough for the reader to see and differentiate areal patterns. In most cases, however, the cartographer is forced to operate in a map space smaller than the ideal. Available space sets constraints on the eventual size of the whole map, and consequently the size of each area symbol. The designer seeks the balance that best serves the purpose of the map.

Number and Kinds of Areal Units
For practically all choropleth mapping, the larger the number of areal units used for the entire study area, the more details of the geographical distribution the map can show. Scale is significant in this respect. Spatial detail is added as the number of enumeration units is increased; conversely, spatial coarseness increases as the number of units is decreased. (See Figure 7.6.)

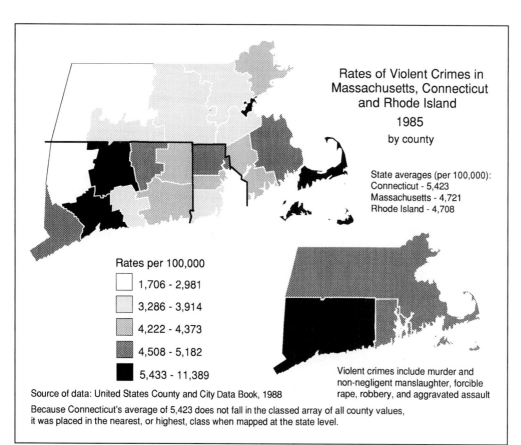

Figure 7.6 Choropleth map detail and the number of enumeration units. Greater distributional detail is possible with a larger number of enumeration units. In this case the distribution of crime in the three states is more clearly shown with the use of the smaller counties, than when only whole states are mapped.

Rates of Violent Crimes in Massachusetts, Connecticut and Rhode Island

1985

by county

State averages (per 100,000):
Connecticut - 5,423
Massachusetts - 4,721
Rhode Island - 4,708

Rates per 100,000

1,706 - 2,981

3,286 - 3,914

4,222 - 4,373

4,508 - 5,182

5,433 - 11,389

Violent crimes include murder and non-negligent manslaughter, forcible rape, robbery, and aggravated assault

Source of data: United States County and City Data Book, 1988

Because Connecticut's average of 5,423 does not fall in the classed array of all county values, it was placed in the nearest, or highest, class when mapped at the state level.

Symbolization also has a direct effect on determining the number of areal units. As the number of units increases, their sizes decrease, making it more difficult to differentiate symbols.

The choice of how many areal units to use depends also on such variables as time, cost, map purpose, map size and scale, and symbolization. Each design task will have its own set of constraints.

The kind of areal unit is usually dictated by map purpose, level of acceptable generalization, data availability, and the scale considerations mentioned above. For much choropleth mapping, the kind of areal unit is determined before actual mapping begins—and most often is dictated by availability of data. Cartographers seldom have the option of specifying areal units, which are usually those used by local, state, or federal census sources. This is not always undesirable, however; standardization leads to easier comprehension.

Data Processing

Ideally, the designer would like to live in a world in which data are presented in mappable form (though this would also eliminate the need for the designer). This is not the case. Choropleth mapping requires that data be in ratio or rate form, most often necessitating the processing of raw data according to the purpose of the map. Electronic calculators, microcomputers, or large computers can be used to ease the burden when dealing with large data sets. Spreadsheet computer software can be used effectively to process enumeration data and can be imported directly into many mapping programs.

Data processing may require consultation with experts familiar with the purpose of the map, such as the map client. Data processing is an integral part of the total design activity and deserves careful attention.

DATA CLASSIFICATION TECHNIQUES

Just a few years ago, before the computer plotting of thematic maps, the usual steps in the choropleth technique included combining mapped values into groups or classes, symbolizing each group with a unique areal symbol, and applying the appropriate areal symbol to each enumeration unit according to the class in which its value fell. The entire mapping process was then completed. This will henceforth be referred to as the **conventional choropleth technique.**

Values are grouped into classes to simplify mapped patterns for the reader. The number of classes became somewhat standardized when it was learned that map readers could not easily distinguish between more than 11 areal symbol gray tones.[4] In practice, no more than six classes are recommended,[5] and a minimum of four is also good practice. Actually, the reasons for classification have been better management of symbol selection and map readability, not for any inherent advantages in grouping data.[6]

With modern computer plotters and video displays, it is possible to symbolize each value by its own unique areal symbol (line, tone, or hue), thus producing an *unclassed* choropleth map.[7] Such maps have not become widely used because not all cartographers approve of the result of not classifying the data: an unstructured or ungeneralized view of the mapped phenomenon. Differences between the conventional choropleth technique and the unclassed variety will be addressed in the following paragraphs. Unclassed choropleth maps are referred to by some as *tonal maps.*

THE IMPORTANCE OF CLASSIFICATION

The classification operation performed for choropleth mapping behaves pretty much like a group of stacked sieves. (See Figure 7.7.) Each sieve acts as a class boundary, and only values of certain sizes are allowed to pass into one of several classes. As we look at classification techniques, it is wise to recall from the previous chapter that in the activity of classification we wish the results to be meaningful and revealing. The sieve analogy is presented simply to remind you that the class boundaries, or sieves, may be anywhere, but must be chosen wisely.

Assigning values to groups on the choropleth map is a form of data classification that leads to simplification and generalization.[8] Details may be lost, but as with all kinds of generalization, the results allow more information to be transmitted. Classification allows the designer to structure the message of the thematic communication. How well this is done depends largely on the designer's ability to understand the geographic phenomenon. It is not possible to structure (and hence generalize) messages with the unclassed method, except in the sense that the cartographer has chosen what to map.

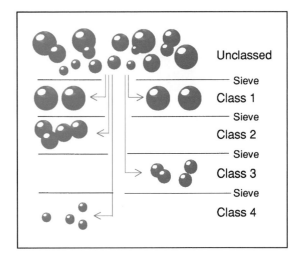

Figure 7.7   The sieve analogy in classification.
Each sieve functions as a screen allowing only balls of a certain size to drop through to the next level. Each sieve can be compared to taxonomic criteria established for the particular study. Spaces between the sieves become the classes, and the sieves are the class boundaries.

The purpose of the choropleth map dictates its form. If the map's main purpose is to simplify a complex geographical distribution for the purpose of imparting a particular message, theme, or concept, the conventional choropleth technique should be followed. On the other hand, if the goal is to provide an inventory for the reader in a form that the reader must simplify and generalize, then the unclassed form should be chosen.

METHODS OF DATA CLASSIFICATION

Before discussing individual methods of classifying data, it is necessary to examine what happens when values are grouped. In the three-dimensional data model, it is clear that grouping data values tends to smooth the model and reduce its irregularities. (See Figure 7.8.) This procedure, however, reduces the number of individual data values transmitted to the reader. The reader can determine only within a specified range of values what a particular enumeration unit's value is. Classification—determining class intervals and class boundaries—consists conceptually of passing planes through the imaginary model. The various methods of classification are simply ways of determining the vertical position of the planes.

There are numerous methods to assist the designer in selecting map classes. These may be grouped into four types: exogenous, arbitrary, idiographic, and serial.[9] **Exogenous data classification** includes schemes in which values not related to the way the data are arrayed are chosen to subdivide the values into groups. An income level selected to define poverty is one example. A certain disease-incidence rate may be critical for the epidemiologist, and this becomes a

class limit. **Arbitrary data classification** methods use regular, rounded numbers having no particular relevance to the distribution as class divisions, such as 10, 20, 30, 40, and so on. Usually this system is used for reasons of convenience. **Idiographic data classification** methods, long used by cartographers, are determined by particular events in the data set. The "natural breaks" method often used by cartographers is an example of the ideographic class. (See Table 7.1.) Quartiles are also frequently used. **Serial data classification** methods include standard deviational units, equal intervals, and arithmetic and geometric progressions. As the material in Table 7.1 indicates, the nature of the data set helps in selecting a method.

Another way to view the classing activity is to consider the map user. By doing so it is possible to form two groups of classing objectives—statistical and geographical.[10] In the former, the user views the map to gain some statistical attribute of the data array, such as medians, averages, standard deviations, or other statistical measures. Specific, limited numbers of significant planes are passed through the data model. In the second group of users, the map reader views the map for the purpose of seeing how well the map, and its classification, has replicated the original data array. Several classing techniques achieve this objective, such as natural breaks and arithmetic and geometric progressions. Here, reproduction of the data model is attempted with several well-selected planes. As with all forms of thematic mapping, map purpose must guide the way.

There is no one best way of devising class intervals and class boundaries for choropleth maps. Simplicity is a major goal, whether the system is devised on purely visual or more rigidly formed mathematical grounds.[11] It is also worth considering that the class interval system should

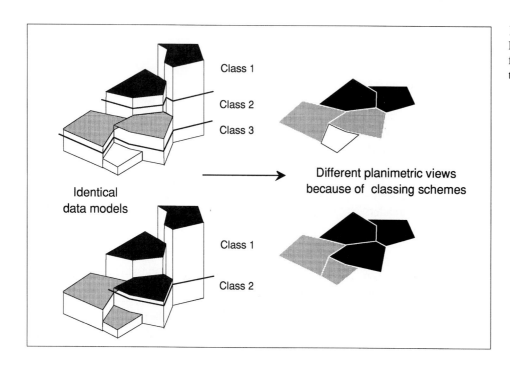

Figure 7.8 Planimetric maps. Different planimetric maps result from passing one or more planes through the data model.

Class 1
Class 2
Class 3

Identical data models

Different planimetric views because of classing schemes

Class 1
Class 2

Table 7.1   Summary of Major Class-Interval Methods

Method	Characteristics
Equal steps	Particularly useful when histogram of data array has a rectangular shape (rare in geographic phenomena) and when enumeration units are nearly equal in size. In such cases, produces an orderly map. Example of a constant interval.
Standard deviation	Should be used only when the data array approximates a normal distribution. The classes formed yield information about frequencies in each class. Particularly useful when purpose is to show deviation from array mean. Understood by many readers. Usually limited to six classes. Example of a constant interval.
Arithmetic progression	A variable, systematic, mathematical class-interval system. General form will vary depending on positive or negative form of the common difference. Used only when the shape of data array approximates the shape of an arithmetic progression.
Geometric progression	Same characteristics as arithmetic progressions. Very useful when frequency of data declines continuously with increasing magnitude—which commonly happens with geographic data.
Quantiles	Good method of assuring an equal number of observations in each class. Can be misleading if the enumeration units vary greatly in size, although this can be offset by areal weighting. Does not yield information about frequencies in each class.
Natural breaks	Good, graphic way of determining natural groups of similar values by searching for significant depressions in frequency distribution. Minor troughs can be misleading and may yield poorly defined class boundaries. Used in conjunction with other methods.
Optimal	An extension of the natural breaks method, but one that yields a quantitative index. The most appropriate optimal technique is one that forms classes that are internally homogeneous and at the same time retains heterogeneity among classes. As with the natural breaks method, class intervals are irregular and class boundaries may look unorganized. The optimal method is a numerical clustering activity.

include the full range of the data, have no overlapping classes, and reflect some logical division of the data array in order to portray the purpose of the map.

Developing class intervals and class boundaries for conventional choropleth maps remains largely an experimental activity. Each geographic phenomenon being mapped presents a unique set of circumstances and nuances for grouping. Mapping purposes vary, as does the intended readership. The cartographer must look at each new choropleth map from a fresh perspective, without feeling constrained to follow a given method of classification, *and it is wise to remember that there may be a fallacy in showing but one solution to a given choropleth map problem.*

CONSTANT INTERVALS

A common way of expressing data classes is the *equal-step* method, which is analogous to passing planes through the three-dimensional data model so that the vertical distances between them are equal. Devising class boundaries this way encloses equal amounts of the range of the mapped quantity within each class interval. Calculate equal-step class intervals as follows:

1. Calculate the range of the data (R):

$$R = H - L$$

where H is the highest reported value and L the lowest reported value.

2. Obtain the common difference (CD):

$$\text{CD} = \frac{\text{P}}{\textbf{number of classes}}$$

3. Obtain the class limits by calculating:

$$L + 1 \cdot \text{CD} = \textbf{first class limit}$$
$$L + 2 \cdot \text{CD} = \textbf{second class limit}$$
$$L + (n - 1) \cdot \text{CD} = \textbf{last class limit}$$

where n is the number of class limits, which will be 1 less than the number of classes.

Maps that are classed by this method generally have intuitive appeal. Their legends tend to appear orderly. If the enumeration units are nearly equal in size and the numerical distribution rectangular, such maps appear neat and organized. Unfortunately, most histograms are not rectangular.

Standard Deviations

If the data set displays a normal frequency distribution, class boundaries may be established by using its standard deviation value. Class intervals designated this way should be used only when it can be assumed that the reader understands them. Assessments are made about the number of values in each class. Class boundaries are compiled by computing the mean and standard deviation, then determining the boundaries by adding or subtracting the deviation from the mean. Usually no more than six classes are needed to account for most of the values in a normal distribution. Special symbolization problems arise with this method because class boundaries are arrayed around a central value instead of ascending from the lowest value, as is usually the case. (See Figure 7.9.) Nonetheless, this method yields constant class intervals because the standard deviation is unchanging.

VARIABLE INTERVALS

It is more common, and usually more desirable, for cartographers to develop variable intervals in the class system. Generally, these methods are better for depicting the actual distribution of the mapped quantity. Variable intervals are classified as either *systematic* or *irregular* forms.

Arithmetic and Geometric Intervals

These mathematically defined interval systems produce class boundaries and intervening distances that change systematically. They should be used only when a graphic plot of the mapped values tends to replicate mathematical progressions. (See Figure 7.10.) From this plot, it is possible to see if an orderly mathematical function exists by comparing the shape of the curve to that of a typical arithmetic or geometric progression.

An *arithmetic progression* (identical in idea with the equal-step method described above) is defined as

$$a, a + d, a + 2d, a + 3d, \ldots a + (n - 1)d$$

where

- a = first term
- d = common difference
- n = number of terms
- l = last term = $a + (n - 1)d$

(Note: The word *term* above and on the next page refers to a given numerical value of the set of values in the entire progression.)

A *geometric progression* is defined as

$$a, ar, ar^2, ar^3, \ldots ar^{n-1}$$

where

- a = first term
- r = common ratio
- n = number of terms
- l = last term = ar^{n-1}

Different expressions of d and r will yield a variety of progression curves. Some worked problems appear in Appendix D.

Quantiles

Quantile class-interval systems produce irregular variable intervals, such as quartiles (25 percent), quintiles (20 percent), deciles (10 percent), or any similar value. Developing class boundaries of quantiles assures an equal number of values in each class and minimizes the importance of the class boundaries. If the enumeration areas vary consider-

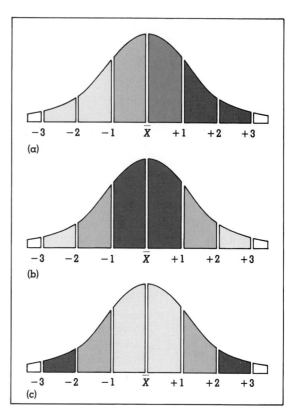

Figure 7.9 Alternative ways of symbolizing classes by standard deviations.

In (a), the classes range in visual importance from -3 to +3 in a continuum. In (b), greater importance is given those values farthest from the mean. In (c), greater importance is assigned to those values farthest from the mean. The purpose of the map will dictate the choice of a symbolization method. Because of the bidirectional nature of the standard deviation, however, there appears to be little intuitive appeal for method (a).

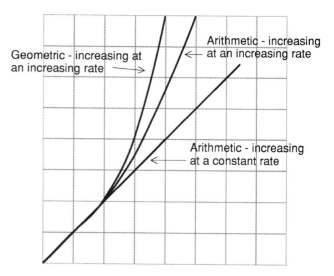

Figure 7.10 Common mathematical progressions.

A graphic plot (or array) of the data values on arithmetic paper is compared visually to common mathematical progression plots. This can assist in selecting an interval plan.

ably in area, this method can be misleading, although the values can be weighted by area.

Class boundaries are produced in this way:

1. Array all values in ascending order.
2. Solve for *K,* the number of values in each class:

$$K = \frac{\textbf{number of enumeration areas}}{\textbf{number of classes}}$$

3. Beginning with the lowest value, *K* values are included in the first class, *K* values in the next class, and so on. The class boundary is normally the mean value between the two adjacent values separating adjoining classes.

Groups Based on Similarities— Natural Breaks

Most classification tasks faced by cartographers involve *numerical classification,* in which numbers along an imaginary number line are grouped according to some criteria of similarity. The cartographer usually attempts to form number groups so that the numerical differences within groups are less than the differences between groups. Regardless of the method, most cartographers agree that *any* method that strives to class the data by these criteria is superior to those that do not.

Visual Inspection   One method calls for the construction of a **graphic array.** (See Figure 7.11.) A graphic array, a kind of rank-order graph, is begun by using arithmetic grid paper and graduating the horizontal axis so that it accommodates the number of observations in the variable array. The vertical axis is graduated to accommodate the range of data values. For this example, we are using the values displayed in Table 7.2. Producing the graphic array requires that the values in the data set are arranged (sorted) in ascending order.

Construction of the plot begins by placing a dot on the graph corresponding to each value's position in the array on the horizontal axis, and correctly placed according to its position on the vertical axis. After all dots are located on the graph, they may or may not be joined by a line. This completes the construction of the graphic array.

Data classing is accomplished by visual inspection of the dots or line. Class limits are selected where the slope of the line changes dramatically. These points tend to define the boundaries between clusters of similar values in the array, and help achieve the goal in numerical classing as defined on the previous page. These points can also be seen from a tabular list, but the graphic plot greatly assists in the endeavor. As before, any natural selection may be influenced by the number of classes imposed on the project. Choices must be made, however, within these constraints because an unlimited number of classes is possible.

Developing class boundaries by construction and visual inspection of a graphic array is troublesome for very large data sets. Just where it becomes too much trouble depends

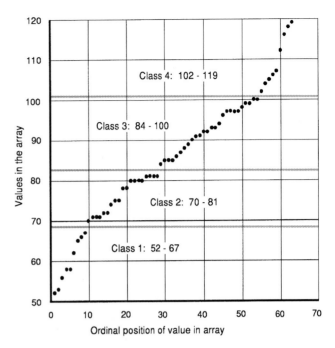

Figure 7.11   Graphic array.
Values on the data set are plotted in an array and class boundaries are identified at places where slopes change remarkably.

Table 7.2	Data Table for Graphic Array		
52	75	86	97
53	75	86	98
56	78	88	99
58	78	89	99
58	80	90	100
62	80	91	100
65	80	91	102
66	80	92	104
67	81	92	105
70	81	93	106
71	81	93	107
71	81	94	112
71	84	96	113
72	85	97	116
72	85	97	118
74	85	97	119

on the project, and on whether there is a computer close by to assist the cartographer. Computer solutions eliminate the tedium of numerical processing, but do not make choices.

Jenks Optimization   A method of classification that has become widely accepted was introduced by George F. Jenks.[12] This procedure, now frequently called the **optimization method,** incorporates the logic most consistent

Table 7.3   An Example of the Optimization Method of Developing Choropleth Class Limits

Array Values	Deviations from Array Mean $(x-\bar{X})$	Squared Deviations from Array Mean $(x-\bar{X})^2$	Solution 1 (Special Case with a Class for Each Value)	$(x-\bar{X}c)^2$	$(x-\bar{Z}c)^2$		Solution 2 $(x-\bar{Z}c)$	$(x-\bar{Z}c)^2$		Solution 3 $(x-\bar{Z}c)$	$(x-\bar{Z}c)^2$
2	−4.54	20.61	$Z_1=2$	0	0		−1.0	1.0	$Z_1=2.5$	.5	.25
3	−3.54	12.53	$Z_2=3$	0	0 → $Z_1=3.0$		0	0		5	.25
4	−2.54	6.45	$Z_3=4$	0	0		1.0	1.0		−1.0	1.0
5	−1.54	2.37	$Z_4=5$	0	0		.5	.25 $Z_2=5.5$	$Z_2=5.0$	0.0	0
6	−.54	.29	$Z_5=6$	0	0 → $Z_2=5.5$		.5	.25		1.0	1.0
7	.46	.21	$Z_6=7$	0	0		0	0		−.33	.11
7	.46	.21	$Z_7=7$	0	0 → $Z_3=7.0$		0	0 $Z_3=7.33$		.33	.11
8	1.46	2.13	$Z_8=8$	0	0		1.5	2.25		.67	.45
9	2.46	6.05	$Z_9=9$	0	0		−.5	.25		−1.0	1.0
10	3.46	11.97	$Z_{10}=10$	0	0 → $Z_4=9.5$		.5	.25 $Z_4=10.0$		0.0	0
11	4.46	19.89	$Z_{11}=11$	0	0		1.5	2.25		1.0	1.0

$\bar{X}=6.54$

$\text{SDCM} = \Sigma\Sigma(x-\bar{Z}_c)^2 = 0$ $\text{SDCM} = \Sigma\Sigma(x-\bar{Z}_c)^2 = 7.5$ $\text{SDCM} = \Sigma\Sigma(x-\bar{Z}_c)^2 = 5.17$

$\text{SDAM} = \Sigma(x-\bar{X})^2 = 82.72$ $\text{GVF} = \dfrac{\text{SDAM}-\text{SDCM}}{\text{SDAM}} = 1.0$ $\text{GVF} = \dfrac{\text{SDAM}-\text{SDCMS}}{\text{SDAM}} = .909$ $\text{GVF} = \dfrac{\text{SDAM}-\text{SDCM}}{\text{SDAM}} = .937$

---

with the purpose of data classification: forming groups that are internally homogeneous while assuring heterogeneity among classes. Computers are normally necessary to perform the calculations, except in the rare case of only a few values—where an electronic calculator is still indispensable. Whenever possible, this method must be explored.

The measure produced by this optimization technique is called the **goodness of variance fit (GVF)**.[13] The procedure is a minimizing one in which the smallest sum of squared deviations from class means is sought. The steps in computing the GVF are as follows (see Table 7.3):

1. Compute the mean ($\bar{X}$) of the entire data set and calculate the sum of the squared deviations of each observation (x_i) in the total array from this array mean:

$$\Sigma\,(x_i-\bar{X})^2$$

This will be called SDAM (squared deviations, array mean).

2. Develop class boundaries for the first iteration. Compute the class means ($\bar{Z}_c s$). Calculate the deviations of each x from its class $(x_i-\bar{Z}_c)$ mean, square these, and calculate the grand sum

$$\Sigma\Sigma\,(x_i-\bar{Z}c)^2$$

This will be called SDCM (squared deviations, class means).

3. Compute the goodness of variance fit (GVF):

$$\text{GVF} = \frac{\text{SDAM} - \text{SDCM}}{\text{SDAM}}$$

The computed difference between SDAM and SDCM is the sum of squared deviations between classes.

4. Note the value of GVF for iteration 1. The goal through the various iterations is to *maximize* the value of GVF.

5. Repeat the above procedures until the GVF cannot be maximized further.

A choropleth map with a class for each value is considered to be one containing no errors. In this case, a class mean would be identical to the one value in the class, so the squared deviation would be 0. Across all classes, then, SDCM = 0.0, and GVF = 1.0, the maximum value of GVF. Of course, we do not have classes for each value in conventional choropleth mapping. Real solutions for GVF will be less than 1.0, and "best" solutions will only tend toward 1.0.

It should be apparent that applying an optimization technique requiring experimentation with class intervals is a timely process. However, a "one-pass" (no iteration) solution is available in a computer program available from Michigan State University. (See box, "Jenks Classing Program.")

In the previous chapter, the measures of kurtosis and skewness were defined as numerical ways of describing the structural characteristics of data distributions. In one

JENKS CLASSING PROGRAM

A computational, interactive microcomputer program is available that includes a "one-pass" solution to the Jenks optimization method of class interval determination. Written by James B. Moore, Richard E. Groop, and George F. Jenks, version 3 (copyright 1983–1988) is available from the Department of Geography, Michigan State University.

This program allows the user to input data from the keyboard or a disc file. Data can be edited and presented on screen or printer, and may be saved on disc for later use. Six data classing options appear on the menu: Jenks method, quartiles, even steps, standard deviations, nested means, and user-defined. Any number of classes may be specified, but may not exceed the number of observations. Output includes a graphic array plot, with class boundaries appearing as boxes in the array, if desired. For the method specified, a table showing class number, number of observations in each class, limits of each class, mean of each class, and class variance is produced.

Total variance (the sum of class variances) and low and high values of the data array are presented.

Classes produced by the Jenks method option are optimal—that is, for the number of classes specified, no better class interval solution is possible that yields a smaller total variance. Minimum within-class error and maximum distance between classes result. The algorithms used in this program are derived from W. D. Fisher and John A. Hartigan. GVF values are not produced by the program.

There are many computational advantages to this program. Results are produced quickly, and comparison between classing methods is easily done by inspecting the total variances. This program is a welcome companion for the quantitative thematic cartographer.

See Richard E. Groop, *Jenks: An Optimal Data Classification Program for Choropleth Mapping*. Technical Report 3 (East Lansing: Department of Geography, Michigan State University, March 1980.

study conducted to determine the relative effectiveness of the Jenks optimization method, when compared to four other classing methods (quartile, equal-interval, standard deviation, and natural-breaks), the following results were achieved:[14]

1. For the quartile method, GVF scores decreased with increasing skewness, and decreased similarly with increasing kurtosis.
2. For the equal-interval method, GVF scores showed little correspondence with skewness, that is, GVF scores remained little changed with increasing skewness. GVF scores did not change greatly with increasing kurtosis, but did show to be somewhat better with flat distributions.
3. For the standard deviation method, GVF scores generally declined with increasing skewness, but not always. The relationship between kurtosis and GVF was similar.
4. For the natural-breaks method, GVF varied greatly with skewness and kurtosis. "In general, distributions with high coefficients of skewness and kurtosis class the data more accurately than those which are close to normal."
5. For the Jenks optimization method, high GVF scores result for distributions ranging from low to high skewness, and from low to high kurtosis. "This procedure produces consistently high GVF indices irrespective of the degree of skewness and kurtosis because the method selects classes on the basis of the GVF criteria."
6. None of the methods proved wholly reliable because the GVF did vary, but the Jenks optimization method proved best overall.

The implications for the designer are apparent. The structural characteristics of the data distribution to be classed must be considered before classing is begun. Skewness and kurtosis should be calculated, and GVF scores computed. *Here is the point: It appears abundantly clear that any numerical clustering procedure is better used than none at all, at least when the geographical objective of classing is of primary concern.*

Iteration and optimization methods are cumbersome if not impossible without the employment of computers. For small mapping tasks, they can be approached with only an electronic calculator, although this soon becomes troublesome. The most reasonable solution is to reduce beforehand the number of possible iterations by visual inspection of the histogram of the array of values, so that only a limited number will need to be processed.

DIFFERENT MAPS FROM THE SAME DATA

One consequence of having a variety of methods of classing is that different maps can result from the same data. (See Figure 7.12.) An alteration in either the number of classes or the class limits can cause such change. The designer has the task of mapping the data in the manner that best represents the distribution of the geographic phenomenon. This requires thorough knowledge of the mapped phenomenon and awareness that the map reader's perception of choropleth map patterns is the result of several conditions:[15]

1. The map reader's experience with cartographic materials.

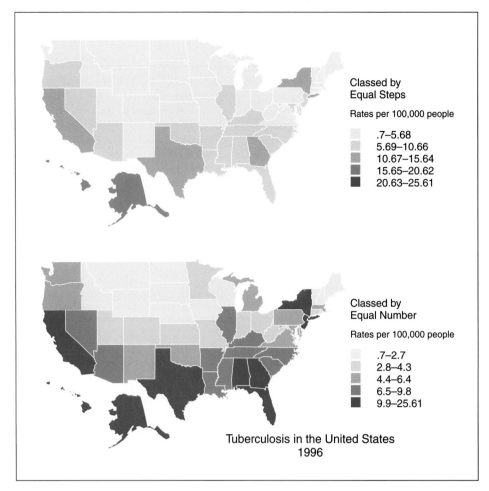

Figure 7.12   Different choropleth maps from the same data.
Different maps result because of the way their data have been classed. In this case, tuberculosis distribution in the United States takes on different patterns as a result of classing methods.

2. The cartographic method used to portray the distribution.
3. The complexity of the map pattern.

   Pattern complexity has been studied by cartographic researchers, and most agree that complexity should be reduced. Pattern complexity is defined in these terms:

1. Number of regions (contiguous enumeration units within the same class)

2. Fragmentation index
3. Aggregation index
4. Contrast index
5. Size disparity index

Space does not permit a detailed explanation of these indices. However, at least one researcher has discovered that map readers tend to judge choropleth maps having different classing schemes as similar in all pattern-complexity indices. This has led him to conclude that only *one* index,

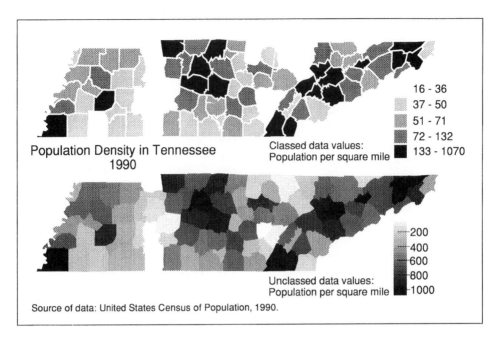

Population Density in Tennessee
1990

Classed data values:
Population per square mile

16 - 36
37 - 50
51 - 71
72 - 132
133 - 1070

Unclassed data values:
Population per square mile

200
400
600
800
1000

Source of data: United States Census of Population, 1990.

Figure 7.13   Classed and unclassed choropleth maps.
A five class map as in the top map, shows a different data distribution than the unclassed version shown in the bottom map. Note that the legends are treated differently. The influence of enumeration boundaries in the unclassed map is reduced with smaller scales.

perhaps the number of regions, can be used to compare pattern complexity among different maps.[16]

Cartographic designers need to examine closely the visual pattern that results from a particular classing method. Although no exact indices cover all cases, it is best to look for a pattern that results in a *simple* picture. The purpose of the map, the background and experience of the reader, and the mapping method must all be considered.

UNCLASSED CHOROPLETH MAPS

The **unclassed choropleth map,** made possible with the advanced technologies of computer plotting, CRT display, and facsimile, can remove the classification burden from the designer.[17] (See Figure 7.13.) In such maps, however, the cartographer loses the ability to direct the message of the communication.[18] Map generalization, simplification, and interpretation are left to the reader. There is increasing evidence that readers can make reasonably good judgments of the mapped values by inspecting the legends that accompany the map bodies and, specifically, that the unclassed method can convey values more accurately to most map readers than classed maps having fewer than six classes.[19]

The authors of one research effort tested the effectiveness of classed and unclassed choropleth maps, with both crossed-line (cross-hatched) and solid-fill (tinted) area symbols.[20] They wanted to determine the relative effectiveness of each with respect to value discrimination (the ability to discern differences between patterns), value estimation, and ability to regionalize (from the patterns on the map). The results are summarized here:

1. In value estimation, classed choropleth maps were significantly better than unclassed, regardless of the symbol type used. The authors concluded, "classed maps have a distinct and statistically significant advantage over unclassed maps in value-estimation tasks."
2. In region-estimation tasks, the results were far less clear. "Overall, there is no significant difference between classed and unclassed choropleth maps, whether crossed-line or solid-fill symbology is used."
3. In value-discrimination tasks, crossed-line symbols are more accurate than solid-fill designs (both in and out of map contexts). "In value-estimation . . . crossed-line shading was significantly better within a map context on both classed and unclassed maps."

These authors also tested maps to determine if increasing the number of data units leads to different responses to both classed and unclassed maps. The answers to this question were not so unequivocal. In conclusion, the authors state:

If one intuitively prefers generalizing data into classes for choropleth maps and accepts the Jenks Optimal system as resolving the long-standing problem of what is the "correct" classification method, this study provides evidence to support continued classification in choropleth maps. If, however, one is less sanguine over the classification problem, the levels of accuracy attained with unclassed maps, and the failure of classed maps to perform significantly better overall in the region recognition task, provide evidence for the effectiveness of unclassed choropleth maps. Crossed-line symbology has been shown to perform better than the solid-filled method. What is now required is further empirical work to validate and expand present findings. In particular, extension of experimentation to include dot-matrix symbolization, color sequences, and testing of subjects with computer-generated map images on monitors deserve attention.[21]

After considerable deliberation, I have concluded that "unclassed" choropleth maps are not choropleth maps at all, but some other, not as yet clearly defined, kind of quantitative map. They do nothing by intent to preserve the integrity of the enumeration unit, and most often eliminate the discreteness of the units altogether—the quality that serves to define the choropleth map. Perhaps they should be called "continuous-tone" quantitative maps, a term mentioned in the literature and briefly by the authors quoted on page 150.[22]

Cartographic designers must look to the purpose of their design activity and make individual judgments regarding the use of unclassed maps. They may be used in conjunction with ordinary classed maps to provide the reader with an unstructured inventory of the mapped area. Other possible uses (other than as a main map) may be discovered.

LEGEND DESIGN, AREAL SYMBOLIZATION, AND BASE-MAP DESIGN

Class-interval specification is not the only activity that occupies the designer's time while preparing a choropleth map. Other design considerations include legend design, symbol selection and scaling, base-map preparation, and attention to reproduction details.

SOURCES OF MAP-READING ERROR AND THE NEED FOR ACCURATE DESIGN RESPONSE

Reading and interpreting a single choropleth map are at best difficult assignments. The design must be planned carefully to reduce the reading task and eliminate as many sources of error as possible. One of the chief sources of error results from **range grading** of the map data—that is, creating classes. Effective classification and corresponding legend design will help eliminate this source of method-produced error. Another source of considerable error lies in the behavioral characteristics of readers in dealing with graded-tone series. Fortunately, studies have been done to help the designer minimize perceptual error. Yet another area that creates problems is the choice of areal symbols; some symbols are easier to read and less distracting to the eye. Finally, the provision of adequate base materials for the reader is an area that has not received much attention from cartographic researchers. Unfortunately, most choropleth maps do not have enough base information.

RANGE GRADING, LEGEND TREATMENT, AND EFFECTIVE DESIGN

A significant amount of error in reading choropleth maps arises from range grading the data array. In Figure 7.14a, the reader can be certain only that the value in the middle chorogram is somewhere between 6 and 16. Because this classification scheme does not have vacant classes, further error is introduced. For example, the reader might guess that a value of 15 occurs in class 2, but it does not because the value 15 is not in the data array.

Better alternatives exist for this condition. (See Figure 7.14b.) It would be more desirable to report the class breaks in the legend so that only the values actually existing in each class form the legend scores. The reader's estimate of the actual value is then narrowed, resulting in greater accuracy. There is no reason for widening the ranges of each class to serve continuity when map-reading error will result.

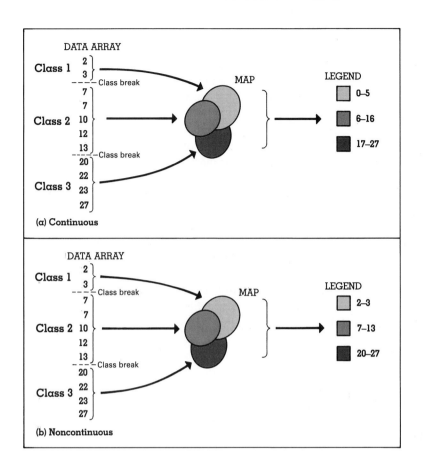

(a) Continuous

(b) Noncontinuous

Figure 7.14   Continuous and noncontinuous legend and classing designs.

Map interpretation errors may result from the continuous design (a), because the reader is led to believe that data values may exist in a class. Less confusion results when the legend is scaled in a noncontinuous fashion (b), because only existing data values are used in forming the legend classes.

Another way to improve legend design is always to include a frequency histogram of the data array. Class symbolization and class boundaries may be included as part of the histogram. (See Figure 7.15.) This gives the reader additional information about the aspatial qualities of the data array, and provides background on how the class intervals and boundaries have been selected. In light of the communication goals of thematic mapping, it is important to provide this element for the reader. The inclusion of class means, or other measures of central tendency for each class, is likewise desirable. These may be labeled alongside class boundaries. (See Figure 7.15.)

SYMBOL SELECTION FOR CHOROPLETH MAPS

Selecting symbols for choropleth maps places the cartographer in a philosophical thicket. There are two opposing points of view on the issue of whether it is desirable for the map reader to make easy distinctions among the areal symbols, thus maintaining the integrity of each chorogram on the map. If so, the designer chooses symbols (1) *whose patterns* are different (dots, lines, hachures); (2) *whose texture* varies (small spacing between elements to large spacing of the elements in the same pattern); or (3) *whose orientation* changes (vertical, horizontal, or oblique to the map border); and (4) that at the same time display a light-to-dark gradation series in keeping with the overall technique of the choropleth map.

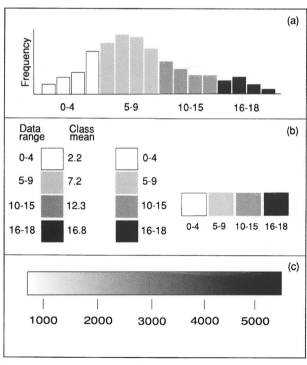

Figure 7.15   Various legend designs for choropleth maps. Class symbols applied to a data histogram, as in (a), are desirable when possible. Different forms of conventional legends are shown in (b), with one case showing class means. A legend for an unclassed map is offered in (c).

The other side of the argument is that the differentiability of areal symbols is not as important as the spatial variation of tone *over the whole* map. This position concentrates on showing the overall geographical distribution, whereas the former is concerned chiefly with providing an areal table. A smooth gradation of tone is easily accomplished today by the use of computer software. (See Figure 7.16.)

Dot patterns cause little eye irritation during reading because they do not cause the eye to wander. Line patterns, on the other hand, should be avoided if at all possible. Black lines are harsh, cause severe irritation, and can mislead the eye to trace paths along their axes. However, screening black line patterns (causing them to look gray) is particularly useful, especially when considerable other base material is provided on the map.

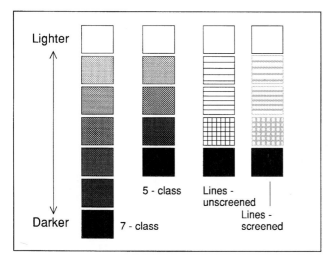

Figure 7.16 Area symbols for choropleth maps.
In most cases, we strive to make the pattern's values range from light to dark. For the line patterns visual irritation can be reduced by screening the lines, as has been done here for the color sequence.

The technique of conventional choropleth mapping requires the cartographer to symbolize each enumeration unit according to where its value falls in the range graded series of values. Because of the nature of scaling and legend design, such area symbols take on a stepped appearance. Area symbols are usually chosen from screen choices provided by cartographic or graphic software, or from film negatives during photo preparation of the maps. The perception of these shading tones is intended to reflect the numerical values they represent.

Empirical studies have demonstrated that perception of tones in graded series is not linear.[23] In a series of printed tones, most map readers do not perceive the tones as varying from light to dark in perfect correspondence with the physical stimulus (the percentage of area inked). Moreover, the perception of gray tones in a graded series is affected by whether the individual elements of the texture of the area symbol are perceptible. In this text, we discuss only those patterns whose textures are so fine that the individual dots are not perceptible, especially because these patterns are superior. Perceptible patterns are usually those that have fewer than 75 rows of dots per inch.

Perception of Areal Symbols— Indistinguishable Pattern

The perception of value of tonal areas having no recognizable pattern has been studied by cartographers and psychologists alike for several decades.[24] The results of several studies suggest that the equal-value gray scale of A. E. O. Munsell is most accurate for perception of tonal areas that have no distinguishable pattern. (See Figure 7.17.) The physical stimulus is the percentage of reflectance (the ratio of reflected light to incident light falling on the surface) and the response is perceived value. The relationship between perceived value and the reflectance can be approximated by this formula:[25]

$$V = 2.467R^{-33} - 1.636$$

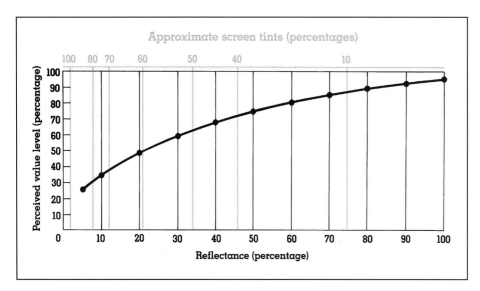

Figure 7.17 The equal-value gray scale of A. E. O. Munsell. The graph illustrates the relationship between percentages of reflectance of a tonal area and perceived value. Applied to the graph are approximate screen-tint values associated with various reflectances. *(Source of screen tints: Carleton W. Cox, "The Effects of Background on the Equal Value Gray Scale,"* Cartographica 17 *[1980]: 53–71.)*

where *V* is perceived value and *R* is reflectance. The value produced can be transformed into a more useful value-reflectance scale by multiplying the answer by 10.

Differences in gray tones at the darker end of the scale are more noticeable than equal differences at the lighter end. Therefore, the actual differences in tone must be made greater at the lighter end if appropriate value perception is to be obtained. A graph of the relationship can be used to determine appropriate amounts; little or no computation is needed.

The following steps can be followed where the areal symbols are black and light gray at the extremes:

1. Determine the mean value of each class.
2. Scale the mean values by proportionality to a new perceived-value scale, from 0 to 100 percent.
3. For each value to be represented, determine the reflectance by computation or from a graph of the value-reflectance relationship. (See Figure 7.17.)
4. For each class, choose a screen tint whose reflectance approximates that obtained in step 3.

When map designers create choropleth maps on which map readers are faced with the task of distinguishing gray tones of finely textured screens between classes, they can be guided by the results of a very useful study suggesting the following relationship:[26]

$$P = .398 \times W^{1.2}$$

where *P* is a succession of evenly spaced numbers from 0 to 100, and *W* is the gray tones in percent area white (which then must be subtracted from 100 to yield gray tones in percent area inked). Table 7.4 presents the calculated percent area inked for six equal-contrast gray scales. For example, if your are doing a five class choropleth map, you would select percentages of grey in the computer software for each class symbol that most closely approximates the numbers in the table.

Texture, the number of rows of repeating dots per inch, of area patterns also influences the perception of gray tones. For example, in one study more than half the subjects thought that a 133-line, 60 percent screen was darker than a 65-line 70 percent screen.[27] It appears clear that texture must be carefully selected when choosing area patterns, and that equal-value curves be applied only in cases where texture differences are limited.

PROVIDING ADEQUATE BASE INFORMATION

One of the principal failures of choropleth map designers has been a reluctance to place adequate base information on the map. This no doubt results from the difficulty of making the base-map elements compatible with often-used dark or black areal symbols at the high end of the symbol spectrum. It becomes virtually impossible to communicate the base information in an adequate way. The designer is thus faced with a choice of alternatives: providing minimal base material or adding base information and changing the symbolization scheme to accommodate it.

As with most other decisions in thematic cartography, the map purpose and the experience and needs of the reader will dictate the choice. A great many choropleth maps are made for the general public—people who want to be able to tie the choropleth map pattern to other features that can provide locational cues. Unless the reader is especially familiar with the mapped area, much of the information-carrying potential of the choropleth map is lost.

Scale, coupled with map purpose, can act as a constraint in the choice of providing base information. A choropleth map of the United States that uses counties as enumeration areas has the capacity to show regional variations of the mapped phenomenon, if this is the main purpose. On the other hand, a marketing analyst working in an urban area might want political boundaries, interstate highways, major thoroughfares, or the competition plotted. The designer must not make arbitrary decisions; the needs of users and map clients must be taken into consideration. Designers faced with providing adequate base information need to choose areal symbols carefully and avoid the black and dark hues. In black-and-white mapping, line patterns can be screened to make the line-pattern effect less objectionable and to gray the patterns to accommodate overprinting. In color mapping,

Table 7.4 Percent Area Inked of Gray Tones for Six Equal-Contrast Gray Scales

Five-Tone	Six-Tone	Seven-Tone
0	0	0
21.0	15.8	12.4
44.5	33.5	26.3
70.7	53.3	41.8
100	75.3	59.1
	100	78.4
		100

Eight-Tone	Nine-Tone	Ten-Tone
0	0	0
10.0	8.2	6.8
21.2	17.4	14.5
33.6	27.6	23.1
47.6	39.1	32.6
63.2	51.9	43.3
80.6	66.2	55.2
100	82.2	68.5
	100	83.4
		100

Source: Glen R. Williamson, "The Equal Contrast Gray Scale," The American Cartographer 9 *(1982): 131–39, table 3.*

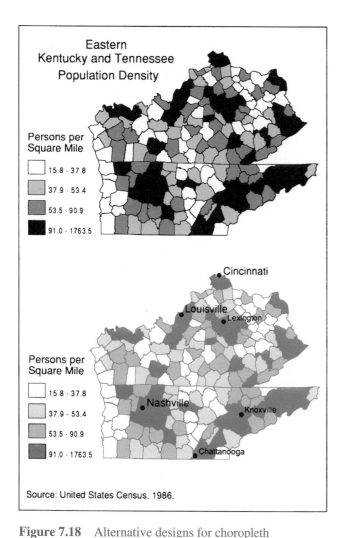

Figure 7.18   Alternative designs for choropleth symbolization and base-map material.

It is difficult to provide base material when dark hues (or black) are used in the symbol array (a). A solution is provided in (b) by printing the areal symbols in color so that base material is more distinguishable. The alternative in (b) should be used only when it is compatible with the purpose of the map.

more care is needed in providing a color spectrum (probably less saturated hues) that will accommodate overprinting. (See Figure 7.18.) Surprisingly, this is not all that difficult with a little experimentation. You are encouraged to look at Figure 7.1 again for an example. One cartographer has addressed the importance of providing base information this way:

> *The simplistic notion that base data are merely elements of graphic design, subject to the whims of individual map makers, has reinforced the notion that these data are separate, unrelated map components. The cartographic solution to the problem of base-data selection must be to evaluate the relative appropriateness of all data to map purpose; the same criterion should be used in selection of both base and thematic data—their power to contribute to map purpose.*[28]

DASYMETRIC MAPPING

A form of mapping somewhat linked to choropleth and enumeration mapping is called **dasymetric** mapping. The dasymetric map has been used for some time, as early as the 1820s, though not often.[29] This form of mapping has been mostly identified with population mapping (actually population density), having been used little for mapping other geographical distributions since its inception. Discussion of this method by cartographers and population geographers probably reached an apex in the 1920s and 1930s. Dasymetric mapping is defined as follows:

> *Simply defined, the dasymetric method portrays a statistical surface as a series of zones of uniform statistical value separated by escarpments of rapid change in values (i.e., zones of high gradient).*[30]

So what the cartographer does is to define areas on the map having similar numerical values, separated by "edges" having steep numerical gradients. The method requires intensive labor in production and many resources and appropriately selected related maps and data. But what really sets the method apart, says Professor George F. McCleary, is the *intention and execution* of the compilation process applied to the mapping activity by the cartographer.[31]

Several different types or forms of the dasymetric method have been identified, but the one this author considers most appropriate today is the one requiring *internal adjustment*:

> *In developing a dasymetric surface by the methods of internal differentiation, the cartographer brings to bear all of the elements which have an influence on the pattern of the distribution of the dasymetric zones and escarpments of the variable which he is mapping. In this way, he effectively simulates the location of the boundaries of the uniform zones as well as the values of these segments of surface.*[32]

Although dasymetric mapping has not been practiced often in thematic mapping in the last thirty years, the method has application in the world of Geographic Information Systems today because its principles simulate what the cartographer does when layers, as filters, are applied in GIS mapping. More work needs to be done with this form of mapping and GIS applications.

NOTES

1. E. Meynen, ed., *Multilingual Dictionary of Technical Terms in Cartography* (International Cartographic Association, Commission II—Wiesbaden, West Germany: Franz Steiner Verlag, 1973), p. 123.

2. The discussion in this chapter does not include bivariate choropleth mapping. In this form of mapping, two data sets are portrayed. Each enumeration unit is symbolized according to which bivariate class (such as a combined class based on income and education levels) it falls into. For more information on these types of maps, see Vincent P. Barabba, "The Graphic Presentation of Statistics," in *The Census Bureau: A Numerator and Denominator for Measuring Change,* Technical Paper 37, U.S. Bureau of the Census (Washington, DC: USGPO, 1975), pp. 89–103; Lawrence W. Carstensen, Jr., "Bivariate Choropleth Mapping: The Effects of Axis Scaling," *American Cartographer* 13 (1986): 27–42; and Stephen Lavin and J. Clark Archer, "Computer-Produced Unclassed Bivariate Choropleth Maps," *American Cartographer* 11 (1984): 49–57.

3. Jean-Claude Muller, *Mathematical and Statistical Comparisons in Choropleth Mapping* (unpublished Ph.D. dissertation, Department of Geography, University of Kansas, Lawrence, 1974), p. 21.

4. George F. Jenks and Duane S. Knos, "The Use of Shading Patterns in Graded Series," *Annals* (Association of American Geographers) 51 (1961): 316–34.

5. Mark S. Monmonier, *Maps, Distortion, and Meaning,* Resource Paper No. 75–4. (Washington, DC: Association of American Geographers, 1977), p. 27. An interesting review of the many proposals for how many classes should be used on choropleth maps is contained in Patricia Gilmartin and Elizabeth Shelton, "Choropleth Maps on High Resolution CRTs: The Effects of Number of Classes and Hue on Communication," *Cartographica* 26 (1989): 40–52.

6. Kurt E. Brassel and Jack J. Utano, "Design Strategies for Continuous-Tone Area Mapping," *American Cartographer* 6 (1979): 39–50.

7. Waldo R. Tobler, "Choropleth Maps without Class Intervals," *Geographical Analysis* 5 (1973): 262–65.

8. George F. Jenks and Michael R. C. Coulson, "Class Intervals for Statistical Maps," *International Yearbook of Cartography* 3 (1963): 119–34.

9. Ian S. Evans, "The Selection of Class Intervals," *Transactions* (Institute of British Geographers) (1977): 98–124.

10. Michael R. C. Coulson, "In the Matter of Class Intervals for Choropleth Maps: With Particular Reference to the Work of George Jenks," *Cartographica* 24 (1987): 16–39.

11. Mark S. Monmonier, "Contiguity—Biased Class—Interval Selection: A Method for Simplifying Patterns on Statistical Maps," *Geographical Review* 62 (1972): 203–28. The issue of class-interval selection on statistical maps is an old one. As early as 1869 a person by the name of Ficker, at the Seventh International Statistical Congress held in The Hague, argued that some system of natural breaks be used to group data. He took this position against a popular one held at the time that equal steps be used. For an interesting account of this history, see H. G. Funkhouser, "Historical Development of the Graphical Representation of Statistical Data," *OSIRIS* 3 (1937): 270–404.

12. The Jenks optimization method of classing has its roots in a technique published in George F. Jenks and Fred C. Caspall, "Error on Choropleth Maps: Definition, Measurement, Reduction," *Annals* (Association of American Geographers) 61 (1971): 217–44. A publication appeared in George F. Jenks, *Optimal Data Classification for Choropleth Maps* (Occasional Paper No. 2. Department of Geography, University of Kansas, May 1977), in which Jenks identified the Fisher algorithm as the best solution for a grouping technique [Walter D. Fisher, "On Grouping for Maximum Homogeneity," *American Statistical Association Journal* 53 (1958): 789–98], and operational programs of the method appeared in John A. Hartigan, *Clustering Algorithms* (New York: Wiley, 1975). The GVF first appeared in Arthur Robinson, Randall Sale, and Joel Morrison, *Elements of Cartography,* 4th ed. (New York: Wiley, 1978), pp. 178–79, and was attributed to personal communication with Jenks. Jenks himself never published any papers on the GVF or any more on the actual operational techniques used by Fisher. Richard E. Groop, while at the University of Kansas in the late 1970s, worked on the algorithm with Jenks and converted the program from mainframe Fortran to Tectronics Basic, and later published the report, Richard E. Groop, *Jenks: An Optimal Data Classifications Program for Choropleth Mapping,* Technical Report 3 East Lansing: Michigan State University, Department of Geography, March 1980). The procedures of the algorithm were not published in the report. The latest version of the Jenks optimization method was developed by James B. Moore, Richard E. Groop, and George F. Jenks for microcomputer applications and is available from Michigan State University.

13. Jenks and Coulson, "Class Intervals," pp. 16–39.

14. Richard M. Smith, "Comparing Traditional Methods for Selecting Class Intervals on Choropleth Maps," *The Professional Geographer* 38 (1986): 62–67.

15. Mark S. Monmonier, "Measures of Pattern Complexity for Choropleth Maps," *American Cartographer* 1 (1974): 159–69.

16. Kang-tsung Chang, "Visual Aspects of Class Intervals in Choropleth Mapping," *Cartographic Journal* 15 (1978): 42–48.

17. A recent version of unclassed map is the facsimile-produced map made with modern weather-satellite imagery-processing equipment. A report of this method can be found in Jean-Claude Muller and John L. Honsaker, "Choropleth Map Production by Facsimile," *Cartographic Journal* 15 (1978): 14–19.

18. This point of view has been stressed by Michael W. Dobson, "Choropleth Maps without Class Intervals? A Comment," *Geographical Analysis* 5 (1973): 358–60. Others have voiced nearly identical positions.

19. Michael P. Peterson, "An Evaluation of Unclassed Cross-Line Choropleth Mapping," *American Cartographer* 6 (1979): 21–37.

20. Katherine Mak and Michael R. C. Coulson, "Map-User Response to Computer-Generated Choropleth Maps: Comparative Experiments in Classification and Symbolization," *American Cartographer* 18 (1991): 109–24.

21. Ibid.

22. Ibid. An interesting discussion on designing legends for bivariate (two-data distribution), unclassed choropleth maps appears in Helen Ruth Aspaas and Stephen J. Lavin, "Legend Design for Unclassed, Bivariate, Choropleth Maps," *American Cartographer* 16 (1989): 257–68. These maps are not discussed here. The discussion of unclassed choropleth maps often focuses on only the technology of producing as many areal symbols as there are data values. However, the problem is broader than this. The map reader's ability to *discriminate* among so many areal symbols is also a critical function in reading these maps and is a question that is not altogether well researched. This point is addressed in Robert G. Cromley, "Classed versus Unclassed Choropleth Maps: A Question of How Many Classes," *Cartographica* 32 (1995): 15–27. Several professional printers that I have dealt with over the years are reluctant to use dot screens in adjacent areas that are closer than 20 percent, and I find that 15 percent often tests map readers' abilities. With white and solid as the end classes, that will limit the number of classes to eight.

23. Jenks and Knos, "The Use of Shading Patterns," pp. 316–34.

24. A. Jon Kimerling, "A Cartographic Study of Equal Value Gray Scales for Use with Screened Gray Areas," *American Cartographer* 2 (1975): 119–27; Carleton W. Cox, "The Effects of Background on the Equal Value Gray Scale," *Cartographica* 17 (1980): 53–71.

25. Cox, "The Effects of Background on the Equal Value Gray Scale," pp. 53–71.

26. Glen R. Williamson, "The Equal Contrast Gray Scale," *The American Cartographer* 9 (1982): 131–39.

27. Richard M. Smith, "Influence of Texture on Perception of Gray Tone Map Symbols," *The American Cartographer* 14 (1987): 43–47.

28. Dennis E. Fitzsimons, "Base Data on Thematic Maps," *The American Cartographer* 12 (1985): 57–61.

29. George F. McCleary, Jr., *The Dasymetric Method in Thematic Cartography* (unpublished Ph.D. dissertation, Department of Geography, University of Wisconsin–Madison, 1969), p. 45.

30. Ibid., p. 212.

31. Ibid., p. 19.

32. Ibid., p. 201.

GLOSSARY

arbitrary data classification for convenience, simple rounded numbers are chosen as class boundaries, p. 144

choropleth map a form of statistical mapping used to portray discrete data by enumeration units; area symbols are applied to enumeration units according to the values in each unit and the symbols chosen to represent them, p. 139

conventional choropleth technique form of choropleth map in which the data array is classed into groups and the groups are symbolized on the map with areal symbols, p. 143

data model conceptual three-dimensional model in which prisms are erected vertically from each chorogram in direct proportion to its mapped value, p. 139–140

dasymetric map a form of quantitative map similar to the choropleth map except that zones of uniform values are used, separated by escarpments of rapid value change, pp. 156–157

exogenous data classification approach to data classification in which certain critical values not associated with the data set are selected as class boundaries, p. 144

goodness of variance fit (GVF) index devised by Jenks; an example of the optimization method in data classification, pp. 148–150

graphic array ordered graphic plot of data values to assist in class boundary determination, pp. 147–148

idiographic data classification particular events in the data array are selected as class boundaries; these events may be troughs (depressions) in the data histogram, p. 144

optimization method classing method in which within-class error is minimized and between-class differences are maximized, pp. 148–150

pattern complexity measure of complexity of repeating elements in a visual image; used in studying choropleth maps, p. 150

range-grading result of the data classification activity often used in choropleth map symbolization, p. 153

serial data classification mathematical functions best fitted to the data array are chosen to determine class boundaries, p. 144

unclassed choropleth map the data array is not classed into groups; the data are symbolized by continuously changing proportional area symbols, pp. 150–152

READINGS FOR FURTHER UNDERSTANDING

Aspaas, Helen Ruth, and Stephen J. Lavin. "Legend Design for Unclassed, Bivariate, Choropleth Maps." *American Cartographer* 16 (1989): 257–68.

Barabba, Vincent P. "The Graphic Presentation of Statistics." In *The Census Bureau: A Numerator and Denominator for Measuring Change.* U.S. Bureau of the Census, Technical Paper No. 37. Washington, DC: USGPO, 1975.

Brassel, Kurt E., and Jack J. Utane. "Design Strategies for Continuous-Tone Areal Mapping." *American Cartographer* 6 (1979): 39–50.

Castner, Henry W., and Arthur H. Robinson. *Dot Area Symbols in Cartography: The Influence of Pattern on Their Perception.* American Congress on Surveying and Mapping, Cartography Division, Technical Monograph No. CA–4. Washington, DC: American Congress on Surveying and Mapping, 1969.

Chang, Kang-tsung. "Visual Aspects of Class Intervals in Choropleth Mapping." *Cartographic Journal* 16 (1978): 42–48.

Coulson, Michael R. C. "In the Matter of Class Intervals for Choropleth Maps: With Particular Reference to the Works of George Jenks." *Cartographica* 24 (1987): 16–39.

Cox, Carlton W. "The Effect of Background on the Equal Value Gray Scale." *Cartographica* 17 (1980): 53–71.

Cromley, Robert G. "Classed versus Unclassed Choropleth Maps. A Question of How Many Classes." *Cartographica* 32 (1995): 15–28.

Dobson, Michael W. "Choropleth Maps without Class Intervals? A Comment." *Geographical Analysis* 5 (1973): 358–60.

Evans, Ian S. "The Selection of Class Intervals." *Transactions* (Institute of British Geographers, 1977): 98–124.

Fisher, Walter D. "On Grouping for Maximum Homogeneity." *American Statistical Association Journal* 53 (1958): 789–98.

Groop, Richard E. *Jenks: An Optimal Data Classification Program for Choropleth Mapping.* Department of Geography, Technical Report 3. East Lansing: Michigan State University, 1980.

Hartigan, John A. *Clustering Algorithms.* New York: Wiley, 1975.

Jenks, George F. "Generalization in Statistical Mapping." *Annals* (Association of American Geographers) 53 (1963): 15–26.

————, "Contemporary Statistical Maps—Evidence of Spatial and Graphic Ignorance." *American Cartographer* 3 (1976): 11–19.

————, *Optimal Data Classification for Choropleth Maps.* Department of Geography, Occasional Paper No. 2. Lawrence: University of Kansas, 1977.

————, and Fred G. Caspall. "Error on Choropleth Maps: Definition, Measurement, Reduction." *Annals* (Association of American Geographers) 61 (1971): 217–44.

————, and Michael R. C. Coulson. "Class Intervals for Statistical Maps." *International Yearbook of Cartography* 3 (1963): 119–34.

————, and Duane S. Knos. "The Use of Shading Patterns in Graded Series." *Annals* (Association of American Geographers) 51 (1961): 316–34.

Kimerling, A. Jon. "A Cartographic Study of Equal Value Gray Scales for Use with Screened Gray Areas." *American Cartographer* 2 (1975): 119–27.

————, "The Comparison of Equal-Value Gray Scales." *American Cartographer* 12 (1985): 132–42.

Lavin, Stephen, and J. Clark Archer. "Computer-Produced Unclassed Bivariate Choropleth Maps." *American Cartographer* 11 (1984): 49–57.

Lloyd, Robert, and Theodore Steinke. "Visual and Statistical Comparison of Choropleth Maps." *Annals* (Association of American Geographers) 67 (1977): 429–36.

MacEachren, Alan M. *Some Truths with Maps: A Primer on Symbolization and Design.* Washington, DC: Association of American Geographers, 1994.

Mackay, R. R. "An Analysis of Isopleth and Choropleth Class Intervals." *Economic Geography* 31 (1955): 71–81.

McCleary, George F., Jr. *The Dasymetric Method in Thematic Cartography.* Unpublished Ph.D. dissertation. Madison: University of Wisconsin, Department of Geography, 1969.

Meynen, E., ed. *Multilingual Dictionary of Technical Terms in Cartography.* International Cartographic Association, Commission II. Wiesbaden, West Germany: Franz Steiner Verlag, 1973.

Monmonier, Mark S. "Contiguity-Biased Class-Interval Selection: A Method for Simplifying Patterns on Statistical Maps." *Geographical Review* 62 (1972): 203–28.

————, "Measures of Pattern Complexity for Choropleth Maps." *American Cartographer* 1 (1974): 159–69.

————, *Maps, Distortion, and Meaning.* Resource Paper No. 75–4. Washington, DC: Association of American Geographers, 1977.

————, "Flat Laxity, Optimization, and Rounding in the Selection of Class Intervals." *Cartographica* 19 (1982): 16–26.

————, *How to Lie with Maps.* Chicago: University of Chicago Press, 1991.

————, *Mapping It Out: Expository Cartography for the Humanities and Social Sciences.* Chicago: University of Chicago Press, 1993.

Muller, Jean-Claude. "Mathematical and Statistical Comparisons in Choropleth Mapping." Unpublished Ph.D. dissertation. Lawrence: University of Kansas, Department of Geography, 1974.

————, and John L. Honsaker. "Choropleth Map Production by Facsimile." *Cartographic Journal* 15 (1978): 14–19.

Olson, Judy M. "Autocorrelation and Visual Map Complexity." *Annals* (Association of American Geographers) 65 (1975): 189–204.

Peterson, Michael P. "An Evaluation of Unclassed Cross-Line Choropleth Mapping." *American Cartographer* 6 (1979): 21–37.

Robinson, Arthur H., Randall Sale, and Joel Morrison. *Elements of Cartography.* 4th ed. New York: Wiley, 1978.

Scripter, Morton W. "Nested Means Map Classes for Statistical Maps." *Annals* (Association of American Geographers) 60 (1970): 385–93.

————, "Influence of Texture on Perception of Gray Tone Map Symbols." *American Cartographer* 14 (1987): 43–47.

Smith, Richard M. "Comparing Traditional Methods for Selecting Class Intervals on Choropleth Maps." *Professional Geographer* 38 (1986): 62–67.

Tobler, Waldo R. "Choropleth Maps without Class Intervals." *Geographical Analysis* 5 (1973): 262–65.

Tufte, Edward R. *The Visual Display of Quantitative Information.* Cheshire, CT: Graphics Press, 1983.

Williams, Robert L. "The Misuse of Area in Mapping Census-Type Numbers." *Historical Methods Newsletter* 9 (1976): 213–16.

Williamson, Glen R. "The Equal Contrast Gray Scale." *American Cartographer* 9 (1982): 131–39.

CHAPTER

8

MAPPING POINT PHENOMENA: THE COMMON DOT MAP

CHAPTER PREVIEW

Mapping discrete geographic phenomena can be accomplished by dot mapping, also called areal-frequency mapping. Its main purpose is to communicate variation in spatial density. In use for over 100 years, the method is popular because of its simple mapping rationale. Although the simplest form is one-to-one mapping, where one dot represents one item, it is more common for one symbol to represent multiple items (one-to-many mapping). Significant design decisions involve the placement of dots and the selection of dot value and size. Legend design is also important, *especially because empirical research indicates that map readers do not perceive dot numerousness or density in a linear relationship. Representative densities must therefore be shown in the legend. It is now possible to generate dot maps by computer—a process that is very cost-effective, at least for agencies producing many dot maps. The Bureau of the Census has produced computer-generated dot maps for some of its agricultural map products. Dot mapping is now possible using software written for microcomputers, although the methods used are far from adequate.*

From the previous chapter's focus on choropleth mapping, a form of area mapping, we now proceed to treat point mapping, notably dot-distribution maps. These are also quantitative maps; information pertaining to density is gained by visual inspection of the spatially arrayed symbols to arrive at relative magnitudes. Precise numerical determination of density is subordinate in this form of mapping. Although any point symbol can be used, it has become customary to use small dots—thus the name **dot mapping.**

MAPPING TECHNIQUE

Dot-distribution maps were introduced as early as 1863,[1] so it is not surprising that their use and acceptance is widespread. In the simplest case, the technique involves the selection of an appropriate point symbol to represent each discrete element of a geographically distributed phenomenon. The symbol form does not change, but its number changes from place to place in proportion to the number of objects being represented. Precise location of each symbol is critical to the method. As in other forms of thematic mapping, a variety of types have evolved over the years.

A CLASSIFICATION OF DOT MAPS

Professor Richard E. Dahlberg found that dot maps can be classified according to the way the dots are used:[2]

1. *Dot equals one.* In this case, there is a one-to-one correspondence between object and symbol. There may or may not be precise control over dot location. (See Figure 8.1a.)
2. *Dot equals more than one.* In the many varieties of this group, one symbol refers to more than one object. (See Figure 8.1b.)
   a. Dots are located so that the original data could be recovered by counting the dots.
   b. Dots are positioned at the centers of gravity of the objects they represent.
   c. Dots are used as shading symbols, either in a uniform pattern resembling a dot screen or across a statistical unit to represent smooth transition.
   d. Dots represent a part of stratified data, such as rural population, with urban populations represented by proportional point symbolization.
   e. Dots that represent a large part of the total data are rather large; dots may be arranged for counting rather than for the purpose of illustrating precise location.

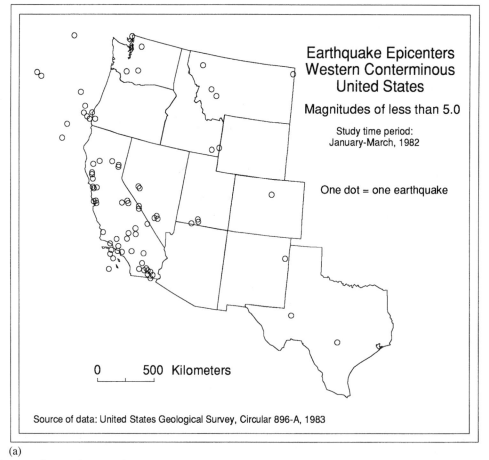

(a)

Figure 8.1a Dot equals one dot mapping.
In this form of dot mapping, there is one symbol for each occurence.

f. Two distributions are shown on the same map, each represented by dot symbols.

g. Dots represent a distribution mapped along with another feature, to show functional relationships.

ADVANTAGES AND DISADVANTAGES OF DOT MAPPING

The number of dot maps appearing in professional journals and other publications seems to have declined over the last three decades, at least those done by hand, probably because of the high cost of production. Dot maps are very time-consuming to construct and require considerable research. Although computer production is now possible, it has not yet led to a proliferation of new dot maps. This may occur when the advantages and disadvantages of the method are better understood.

The advantages of dot mapping include:

1. The rationale of mapping is easily understood by the map reader.

2. It is an effective way of illustrating spatial density, especially suitable for showing geographic phenomena that are discontinuous (consisting of discrete elements).

3. Original data may be recovered from the map if the map has been designed for that purpose.

4. More than one data set may be illustrated on the same map. This is recommended only if there is a distributional or functional relationship between the sets.

5. Final execution by computer means is quick and rather straightforward and presents few production or reproduction problems.

Possible disadvantages of dot mapping would include:

1. Map interpretation is not one-to-one; perception of relative density is not linear.

2. The method is time-consuming if done by hand and therefore costly. Considerable research into factors controlling dot distribution is necessary, and provision must be made for the acquisition of ancillary resource materials.

3. When the map has been designed for maximum portrayal of relative spatial density, it is practically impossible for the reader to recover original data values.

Although dot mapping has seen a decline, perhaps because of these several disadvantages, many cartographers still rank the method quite high. For example, these words by Michael R. C. Coulson make the case:

It is not the dot size, or value, or even placement, that give the real power to the dot map—assuming some reasonable

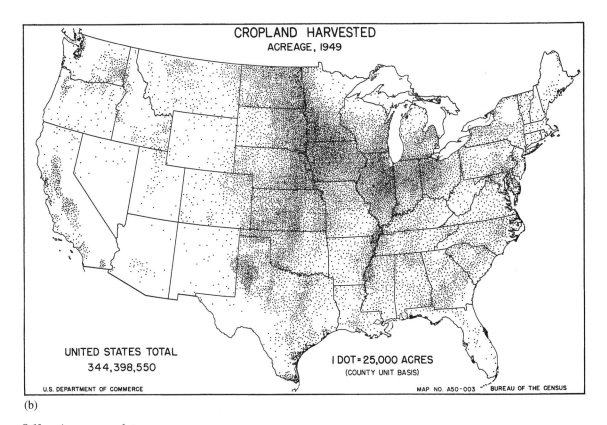

(b)

Figure 8.1b   A common dot map.
The elements of the common dot map include point symbols, political boundaries, and a legend that includes dot value and, if possible, representative densities taken from the map. The reader gets the idea that the amount of the item varies from place to place. This particular map, drawn by hand without the aid of a computer, does not have the newer, recommended legend showing representative densities. *(Source: United States Bureau of the Census,* Portfolio of U.S. Census Maps, 1950 *[Washington, D.C.: USGPO, 1953], p. 16.)*

decisions have been made. Rather, the power of the dot map is in the overall pattern of the distribution that is revealed.[3]

In summary, map designers must weigh these advantages and disadvantages carefully before deciding to apply dot mapping to a given project. As in all design choices, the map purpose, readership, and costs are central concerns in making the decision.

THE MAPPING ACTIVITY

The ideal dot map would be at a scale that allowed for one-to-one mapping, that is, one symbol for each mappable element. This requires an amount of informational and geographical detail not normally at the disposal of the cartographic designer. It is possible when individualized field studies have been carried out, but most map designers are not so fortunate. Usually they must construct the dot map with data gathered in enumeration form, which requires a different approach.

An excellent recent example of a one-to-one dot map, however, is the map AIDS in Los Angeles County produced at the Center for Geographic Studies, California State University at Northridge (Publications in Medical Geography No. 3, 1989). Students are urged to review this highly detailed one-to-one dot map. It was designed under the supervision of Professor William Bowen.

In most dot-mapping circumstances, a many-to-one situation exists: each dot represents more than one mapped element. In this case, each dot can be thought of as a **spatial proxy** because it represents at a point some quantity that actually occupies geographical space.[4] (See Figure 8.2.) These geographical spaces are called **dot polygons** or territorial domains. They do not normally appear on the final map but are assumed to exist around each dot, making up the total enumeration area.

It is essential in dot mapping that associated resource materials showing functionally related variables be gathered together prior to mapping, so that greater precision in dot placement can be achieved. Each mapping task has its own peculiar set of related materials. Examples are physical relief diagrams; soil maps; maps showing lakes, ponds, rivers, and streams; land-use maps; national forest or park maps; and military reservation maps. The related distribution maps tend to operate as filters and serve to control the placement of the dots. (See Figure 8.3.)

For example, referring to county-level road maps at a scale of about 1 mile to the inch, where individual houses can be shown, is useful in guiding the placement of symbols on a population dot map. Other helpful resource maps in population dot mapping include forest and parkland maps, military reservation maps, and physical relief maps. Maps that show the location of towns are frequently useful, as are resource maps that show where population is large and where it is limited.

After all relevant information is at hand, the designer can begin constructing the map. This involves decisions regarding scale, dot value, and dot size. These are closely interrelated in the construction of the map; a decision regarding one affects the others. In manual production of dot maps, work generally begins in pencil on worksheets. In this phase, the preliminary decisions are tested, largely by trial and error.

Size of Enumeration Unit

As previously mentioned, dot mapping usually involves the presentation of geographical quantities that have been collected by statistical area, usually political or administrative units. Under most conditions, the smaller the statistical unit in relation to the overall size of the map, the greater will be the accuracy of the final dot distribution.[5] (See Figure 8.4.) Smaller statistical units mean a smaller territorial domain

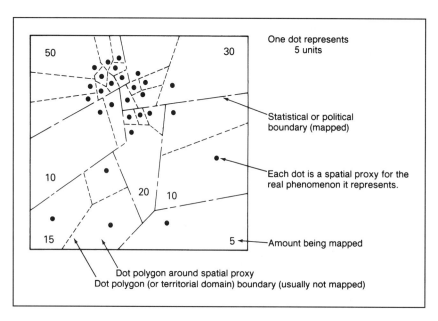

Figure 8.2 The conceptual elements of a dot map.

Each dot functions as a spatial proxy for the item it represents. Around each spatial proxy is its polygon or territorial domain. Political or statistical units are normally included to provide the reader with spatial cues, but the boundaries of territorial domains are not. These domains help the mapmaker in conceptualizing and producing the map.

for each dot, reducing the chance of locational error. Large-scale maps (small earth areas) require quite small statistical units. In fact, it is often found that common statistical units used for enumeration data (block, tract, or county-sized units) are too large for dot mapping at large scales. Professional cartographers normally use intermediate-to-small scales for dot mapping.

Cartographers know that scale is relative in all mapping activities. This is especially true for dot mapping. As we will point out below, most dot mapping today is done by computer means, and dots are placed within enumeration units *randomly,* thus accurate locational control is lost, *except that provided by small enumeration units.* Ideally, of course, an enumeration unit for each dot would be desirable, but not practical. To control the location of dots as accurately as possible in a computer-drawn dot map, strive for the smallest enumeration unit that is practical.

Dot Value and Size

Closely related to map scale is the determination of **dot value**—the numerical value represented by each dot. (See Figure 8.5.) Of course, one-to-one mapping involves no choice, because each symbol represents only one item. In many-to-one dot mapping, extreme care must be exercised in selecting dot value and size, because they affect the map reader's impression of accuracy. Small dots with small values cause a dot map to appear more accurate than it may actually be. The reverse is also true: large dots with large values appear crude and give the impression of unprofessional compilation.

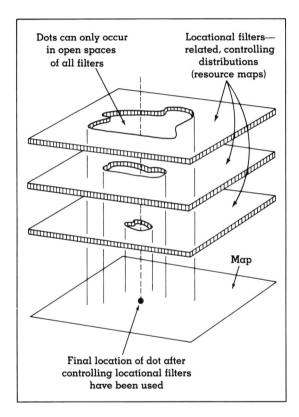

Figure 8.3   Dot placement.
Locational filters (related distributions) determine the location of dots. Each dot's territorial domain can exist only in that part of the map which has not been influenced by other distributions.

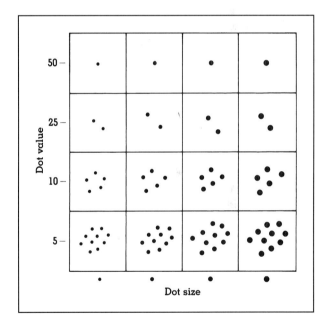

Figure 8.5   The effect of changing dot value and size.
The visual effects of changing dot value and size can be dramatic. The extremes of emptiness and crowdedness are easily seen.

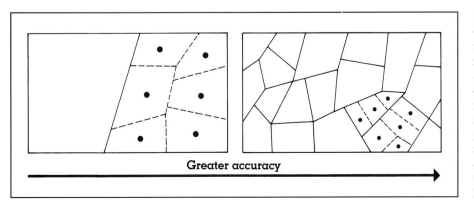

Figure 8.4   Accuracy.
Greater accuracy can be achieved by using statistical units that are smaller in relation to the entire mapped area. Statistical units provide locational control to a great degree. The more numerous (hence smaller) these units are, relative to the mapped area, the greater will be the degree of locational accuracy produced. The territorial domains of the dots decrease in size with smaller enumeration units; greater locational accuracy is thus possible.

It is difficult to list definite rules for selection of dot values and size, but here are some *general* guidelines:

1. Choose a dot value that results in two or three dots being placed in the statistical area that has the least mapped quantity.
2. Choose a dot value and size such that the dots just begin to **coalesce** in the statistical area that has the highest density of the mapped value. (See Figure 8.6.)
3. It is preferable to select a dot value that is easily understood. For example, 5, 500 and 1,000 are better than 8, 49, or 941.
4. Select dot value and size to harmonize with the map scale so that the total impression of the map is neither too accurate nor too general. This will require experimentation.

Dot scaling begins with an examination of the range of the data. The smallest value should be divided first by 2, then by 3. For example, if the lowest value in the array is 500 units, this divided by 2 will yield a dot value of 250; if divided by 3, the dot value will be 166. Any dot value greater than 500 would eliminate a dot in this lowest enumeration area and should not be used.

Suppose a value of 250 were selected for this example. This preliminary dot value should be divided into the number

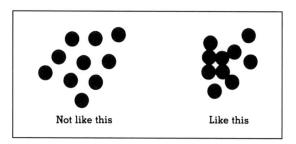

Figure 8.6   Dot coalescence.
Dots should begin to coalesce in the densest part of the map.

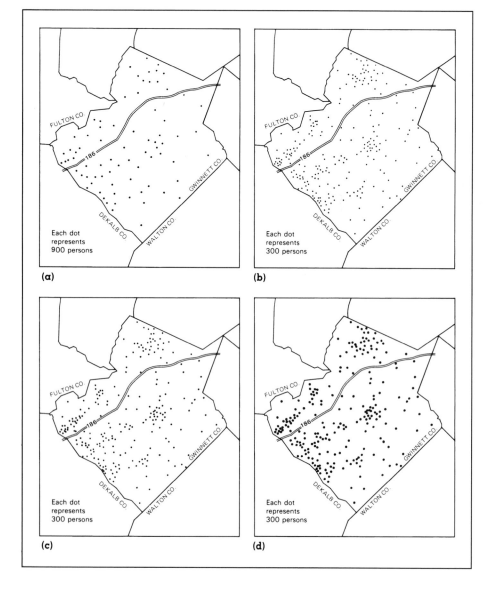

Figure 8.7   Dot value and size. Changes on dot value and size will produce very different maps. The dot value in (a) is too large, leading to a map that appears too empty. In (b), the dot size is too small, again leading to a map that looks too empty. Dot size in (c) is still too small, though larger than in (b). Dot value in (d) remains at 300 persons, but dot size is now larger, giving a better overall impression for the map.

of units present in the enumeration area that contains the highest value. The result is the total number of dots that must be placed in the highest enumeration area; it may be a large number. At this time, the designer begins to consider dot size. What size dots, when the correct number of them are placed in the highest enumeration area, will result in coalescence there?

During these trial-and-error tests, the designer should develop some working sketches of representative enumeration areas, at final map scale, and have several dot sizes at hand with which to experiment. The enumeration area containing the largest number of dots is likely to be the one with the highest density. It is useful to know beforehand which enumeration units contain the highest and lowest densities of the mapped quantity. Experimental designs can then be tested on these extreme cases.

In summary, the designer first adopts a unit value for each dot, divides this value into the data total in each enumeration area to obtain the number of dots to be placed there, selects a dot size, and proceeds to place the dots. Tabular listing helps organize this material.

Several different dot maps can result from identical data because the selection of dot value and size is subjective. (See Figure 8.7.) There is no quantitative index to tell the designer which is best. Knowledge of the distribution plus an intuitive sense of what looks right are the standards against which any dot map must be judged.

Counting Up Method

Traditional counting methods in dot mapping include first the assignment of a dot value (after the question of coalescence is dealt with), as just described, then this value is divided into each enumeration unit to determine the number of dots for that unit. Unfortunately, this often leads to error, as the *remainders* are never mapped. For example, if there are 4,500 persons in an enumeration unit, and the dot value is 4,000, then that unit would get one dot, and the remainder of 500 persons would never get mapped (the error). This might be thought of as "counting down."

Another method is one that "counts up." In this method, developed by the United States Census Bureau for their nighttime view of the United States 1990 population distribution map, a smaller enumeration unit is used, and the population of each is added until the dot value is reached, and then the area gets a dot. The "remainder" is then added to other units until another dot value is reached.[6] (See Figure 8.8.) In this way, remainders are accounted for.

Dot Placement

In most circumstances, the general rule is to locate the dots as close to the real distribution as possible, using the **center of gravity principle.** (See Figure 8.9.) Because each dot is a spatial proxy, it must be located so as to best represent all of its referents. This requires considerable knowledge of the real distribution, and relevant resource materials are of invaluable assistance in this respect.

Geographic phenomena rarely occur uniformly over space, so a geometric arrangement of the dots within an enumeration area is not recommended. (See Figure 8.10a.) Only under unusual circumstances, or when the real distribution dictates it, should such a pattern be used. It is also good practice not to allow the political boundaries of the enumeration areas to exercise too much control over the location of the dots. (See Figure 8.10b.)

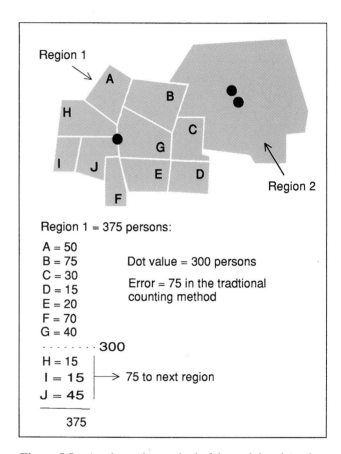

Region 1 = 375 persons:

A = 50
B = 75   Dot value = 300 persons
C = 30
D = 15   Error = 75 in the tradtional
E = 20   counting method
F = 70
G = 40
- - - - - - 300
H = 15
I = 15   } → 75 to next region
J = 45

375

Figure 8.8   An alternative method of determining dot value.

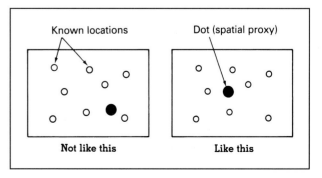

Figure 8.9   Proper placement of dots.
Dots should be placed at the approximate center of gravity of the elements they represent.

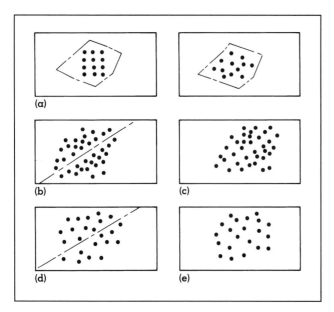

Figure 8.10   Acceptable and unacceptable methods of dot placement.

In (a), the irregular pattern on the right is superior to the regular one on the left. Enumeration unit boundaries should not exercise too much control over the location of the dots; when the boundary in (b) is removed, its effect is still felt (c). The dots in (d) have been located in such a way that, if the boundary is removed, its control will not be noticed (e).

Legend Design

Every dot-distribution map should have a legend containing two components: a statement indicating the unit value of the dot and a set of at least three squares (or other areas) that illustrate three different densities taken from the map. (See Figure 8.11.) It is good practice for the legend to indicate that a dot represents a value but does not equal it. The three squares may be arranged in any fashion that is unambiguous yet harmonious with the remainder of the map, but the legend must be designed for clarity.

DOT MAP PRODUCTION

Today, dot maps are usually produced by computer means, and the laborious task of placing each individual dot on the map sheet is all but a thing of the past. Moreover, the cartographers need not worry about the perfection of each dot, as they once did when manual drafting means were used. If the cartographer is faced with the manual production of a dot map, there are suitable art materials at hand to ease the production burden.

VISUAL IMPRESSIONS OF DOT MAPS

I suggested earlier that reader perception of dot densities is not a one-to-one or linear relationship. This section treats this important concept in more detail.

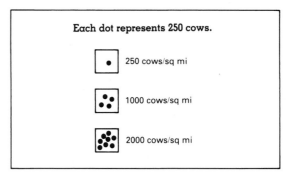

Figure 8.11   Designing dot legends.

A typical dot legend includes a notation of dot value and should include representative densities across the range of data values. The size of the density boxes and the densities represented are determined in accordance with the map scale.

QUESTIONS OF NUMEROUSNESS AND DENSITY

The reader of a dot map is asked to judge relative densities and to make certain assessments about the spatial distributions of these densities. In most dot mapping we do not expect the reader to recover the original data—an impossibility if there is coalescence. The dot map is used to present data at the ordinal level in the sense that all the reader must do is to judge that there is more of the item in one place and less of it in another. **Apparent density** is the subjective reaction of the map reader to the physical stimulus of the actual density of dots per unit area.[7]

Numerousness is the subjective reaction to the physical number of objects in the visual field, without actually counting the objects during perception. Dot maps are usually selected as a way of illustrating the spatial distribution of discrete objects; the reader is not normally expected to judge the actual number. Cartographers select this mapping technique to convey the spatial variableness of density, and nothing more.

Cartographic research has demonstrated conclusively that most map readers underestimate the number of dots on a dot map. By extension, apparent densities are likewise guessed on the low side.[8] The evidence suggests that, when a person perceives an area as having 10 more dots than another with 1 dot, the actual ratio is probably closer to 15 or 16 to 1.

Whatever dot value and dot size are used it should be borne in mind that an area full of dots on the map will carry a psychological impression that on the ground the same area is full of whatever is being mapped, and in the same way, too sparse a sprinkling of dots conveys an unmistakable impression of emptiness.

Source: G. C. Dickinson, *Statistical Mapping and the Presentation of Statistics* (London: Edward Arnold, 1973), p. 49.

This holds positive lessons for the cartographic designer. The following design ideas are suggested by this research.[9]

1. Include a legend with examples of low, middle, and high densities—visual anchors. Provide at least a high-end anchor, preferably full-range anchors, to counteract the map reader's tendency to underestimate relative densities.
2. When recovery of original data is important, do not use coalescing dots; noncoalescing dots improve the estimation of dot number.

RESCALING DOT MAPS

There has been some investigation by cartographers into ways of **rescaling** dot maps to allow more accurate reading. New dot values are assigned to compensate for the reader's tendency to underestimate. One method calls for each dot to represent a range of values, not simply a unit value.[10] The method is very time-consuming; it appears to be particularly applicable to computer-generated maps.

COMPUTER-GENERATED DOT MAPS

Producing dot maps by automated methods has attracted the attention of cartographers during the last three decades. One program, developed by Wendell K. Beckwith under the supervision of Professor Joel Morrison, was used for the production of many dot maps for the *Census of Agriculture, Graphic Summary, of 1969.*[11] The program utilized county data, areas of each county, and centroids (geographic coordinates) for each county in the United States. For each map, the program computed the number of dots for each county (after a dot value was selected), and the dots in each county were positioned for greater accuracy by using the 16 land-use categories as mapped in the *National Atlas of the United States of America.* The land-use classification information was converted into nearly half a million cells to provide

controls over the placement of dots. The current version of Beckwith's program retains most of the conceptual elements of the original, but has added algorithms that take into account the impact of political boundaries on the location of the dots. The Bureau of the Census completed the publication of the *Graphic Summary* to accompany the 1978, 1982, 1987, and 1992 censuses of agriculture, each including approximately 100 computer-generated dot maps. (See Figure 8.12.) These maps are completely machine-produced, even their lettering and other peripheral information. The version of the program used in 1969 did not produce these elements automatically.[12] The newest publication, called the *Agricultural Atlas of the United States,* for 1992, carries forth this tradition of computer-generated dot maps.[13]

Dot maps produced by software programs written for microcomputers are now possible, but they do not have the design flexibility as do those designed for large mainframe or minicomputers. (See Figures 8.13 and 8.14.) For example, programs written for smaller computers do not accommodate locational filters.

MICROCOMPUTER DOT MAPPING

Choropleth and proportional symbol mapping programs for microcomputers are more common than those that produce dot maps. However, the productions of such maps are possible, although they come with considerable constraint. Several mapping and GIS programs, for example Maptitude (a product of Caliper Corporation), ArcVIEW (a product of ESRI), and Map Viewer (a trademark of Golden Software), to name a few, are in common use today. That they include the dot-mapping option at all is a wonder, and certainly the companies deserve credit for doing so.

Accurate dot mapping requires considerable knowledge about locational controls over dot placement. This requirement unfortunately cannot be met with low-cost microcomputer mapping software. The solution is usually to provide for

DOT DISTRIBUTION MAPS

A dot map uses a dot to represent the number of a phenomenon found within the boundaries of a geographic area. In addition, the cartographer usually attempts to show the pattern of distribution within the area by placing the dots where the phenomenon is most likely to occur. Two types of dot maps appear in this report: the first one shows location of a phenomenon and the second shows increase/decrease data as it relates to the last agricultural census. Placement of dots used each of the following system tools: a land-use filter, weighting factors, and scan lines. The source of the land-use filter (a vector-based digital file containing the 16 land-use categories) was provided by the USGS. The land-use file was updated from the TIGER file to reflect the extent of the 1990 Urban-

ized Areas. Weighting factors were assigned to each land-use based on the likelihood of each agricultural activity occurring within a specific land-use. The land-use filter was converted into scan lines. The mapping software matched the data to the appropriate county and placed dots within the scan lines of the land-use filter with the highest weighting factors for that phenomenon. The dot distribution maps are monochrome (black and white) and the increase/decrease dot maps are black and red, respectively.

Source: United States Bureau of the Census, *Census of Agriculture.* Vol. 2, Subject Series, pt. 1., *Agricultural Atlas of the United States* (Washington, DC: USGPO, 1995).

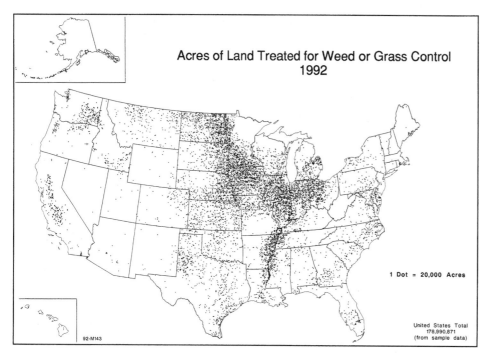

Acres of Land Treated for Weed or Grass Control
1992

1 Dot = 20,000 Acres

United States Total
178,990,871
(from sample data)

92-M143

Figure 8.12   Dot map produced by computer.
The computer output is a plot tape that drives a COM (computer output on microfilm) unit, which plots the map directly on publication-size 310-mm microfilm. *(Source: United States Bureau of the Census, 1982 Census of Agriculture, vol 5, Special Reports, pt. 1, Graphic Summary [Washington, DC: USPGO, 1995], p. 93.)*

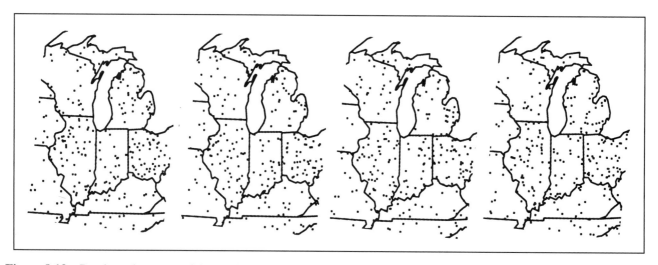

Figure 8.13   Random placement of dot symbols leads to different maps.
The random method of dot placement, used by many computer mapping programs, is "quick and dirty" but leads to dot maps with many errors.

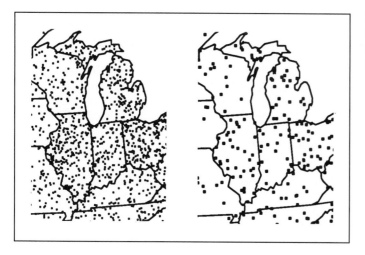

Figure 8.14   Different dot maps result from different dot sizes and values.
Fortunately, most computer mapping programs offer the user flexibility in specifying dot sizes and values.

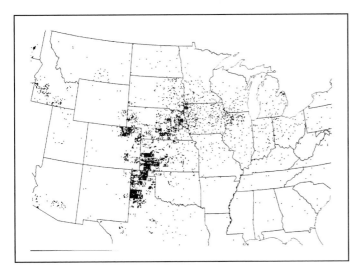

Figure 8.15   Undesireable effect of boundaries on dot placement.

the random placement of dots within an enumeration district, which is the way these programs go about their business of dot mapping. As a consequence, without warning, for example, you may find dots representing people clustered in the Mojave Desert or the Carson Sink! Figure 8.13, for example, shows four different dot maps all produced from the same data, but by the random placement of dots. Also, the effects of boundaries on the location of dots may be a consequence. (See Figure 8.15.)

As mentioned earlier, cartographers are advised to use as small an enumeration area as possible to control the location of the randomly positioned dots.

Much to the credit of the software producers, however, they do allow the designer to have control over two important dot-mapping variables: dot value and dot size. Figure 8.14 presents two maps produced with the same data but with different dot sizes and values. Thus, the cartographer has some control over the final appearance of the map. Until more sophistication is possible with such programs, the designer is cautioned to provide a statement on the map that the dots have been located randomly.

NOTES

1. R. P. Hargreaves, "The First Use of the Dot Technique in Cartography," *Professional Geographer* 13 (1961): 37–39.
2. Richard E. Dahlberg, "Towards the Improvement of the Dot Map," *International Yearbook of Cartography* 7 (1967): 157–67.
3. Michael R. C. Coulson, "In Praise of Dot Maps," *International Yearbook of Cartography* 30 (1990): 61–91.
4. Ibid.
5. T. W. Birch, *Maps: Topographical and Statistical* (Oxford: Oxford University Press, 1964), p. 154.
6. Elaine Hallisey Hendrix, *Electronic Dot Map Preparation: An Examination of Design and Method,* (unpublished masters practicum research paper,

Department of Geography, Georgia State University, Atlanta, 1995).
7. Judy M. Olson, "Rescaling Dot Maps for Pattern Enhancement," *International Yearbook of Cartography* 17 (1977): 125–37.
8. Robert W. Provin, "The Perception of Numerousness on Dot Maps," *American Cartographer* 4 (1977): 111–25.
9. Ibid.
10. Olson, "Rescaling Dot Maps," pp. 125–37.
11. United States Bureau of the Census, *Census of Agriculture,* 1969, vol. 5, Special Reports, pt. 15, *Graphic Summary* (Washington, DC: USGPO, 1973), p. 14.
12. The information on the dot mapping program at the Bureau of the Census was generously shared with the author by Frederick R. Broome, Chief, Computer Mapping Staff, Geography Division, United States Bureau of the Census.
13. United States Bureau of the Census, *Census of Agriculture, 1992,* vol. 2, Subject Series, pt. 1, *Agricultural Atlas of the United States* (Washington, DC: USGPO, 1995), p. xii.

GLOSSARY

apparent density   the subjective response to the physical stimulus of the actual density of the objects in a visual field; density is underestimated in reading dot maps, p. 168

center of gravity principle   placement of the dot in its territorial domain in the way that best represents all the individual elements in the domain, p. 167

coalescence   merging of the dots in dense areas; the visual integrity of individual dots is lost, p. 166

dot mapping   method of producing a map whose purpose is to communicate the spatial variability of density of discrete geographic data; also called areal frequency mapping, p. 162

dot polygon the area represented by each dot on a dot map; also referred to as the dot's territorial domain, p. 164

dot value the numerical value represented by each dot in dot mapping; sometimes referred to as unit value, p. 165

numerousness the subjective response to the physical stimulus of the actual number of objects in a visual field; not the result of the conscious activity of counting; number is underestimated in reading dot maps, p. 168

rescaling method used to enhance dot maps by assigning new dot values to compensate for the reader's underestimation of density, p. 169

spatial proxy the dot's function of representing geographical space, p. 164

READINGS FOR FURTHER UNDERSTANDING

Birch, T. W. *Maps: Topographical and Statistical.* Oxford: Oxford University Press, 1964.

Coulson, Michael R. C. "In Praise of Dot Maps." *International Yearbook of Cartography* 30 (1990), 61–91.

Dahlberg, Richard E. "Towards the Improvement of the Dot Map." *International Yearbook of Cartography* 7 (1967): 157–67.

Dickinson, G. C. *Statistical Mapping and the Presentation of Statistics.* London, England: Edward Arnold, 1973.

Geer, Sten de. "A Map of the Distribution of Population in Sweden: Method of Preparation and General Results." *Geographical Review* 12 (1922): 72–83.

Hargreaves, R. P. "The First Use of the Dot Technique in Cartography." *Professional Geographer* 13 (1961): 37–39.

Hendrix, Elaine Hallisey. "Electronic Dot Map Preparation: An Examination of Design and Method." Unpublished masters practicum research paper. Atlanta: Georgia State University, Department of Geography, 1995.

MacKay, R. Ross. "Dotting the Dot Map: An Analysis of Dot Size, Number, and Visual Tone Density." *Surveying and Mapping* 9 (1949): 3–10.

Olson, Judy M. "Rescaling Dot Maps for Pattern Enhancement." *International Yearbook of Cartography* 17 (1977): 125–37.

Provin, Robert W. "The Perception of Numerousness on Dot Maps." *American Cartographer* 4 (1977): 111–25.

Smith, Guy-Harold. "A Population Map of Ohio for 1920." *Geographical Review* 18 (1928): 422–27.

United States Bureau of the Census. *Census of Agriculture, 1969.* Vol. 5, Special Reports, pt. 15, *Graphic Summary.* Washington, DC: USGPO, 1973.

———, *Census of Agriculture, 1992.* Vol. 2, Subject Series, pt. 1, *Agricultural Atlas of the United States.* Washington, DC: USGPO, 1995.

CHAPTER

9

FROM POINT TO POINT:
THE PROPORTIONAL SYMBOL MAP

CHAPTER PREVIEW

Proportional or graduated quantitative point symbols have long been used in thematic mapping. The circle is one of the most popular forms, probably because of its compact size and ease of construction; others include the square and the triangle. Cartographic and psychological research into how map readers estimate the size of symbols indicates that the physical elements of area and volume are underestimated during perception. This has led to apparent-magnitude scaling methods. Recent research has concluded that underestimation can be controlled by careful legend design, even though absolute scaling is employed, so

that apparent-magnitude scaling methods may no longer need to be used. A method that eliminates the need for any true direct scaling is to present symbols in range-graded series, using point symbols of differential size, as in ordinal scaling. Regardless of the method of scaling, graduated point symbols must be rendered graphically so that they appear as dominant, unambiguous figures in perception. Proportional symbol maps can also be generated by computers, but the designer must still understand the basic concepts of this kind of mapping.

This chapter deals with one of the most flexible and popular thematic map techniques ever devised: *proportional point symbol mapping,* also called *graduated* or *variable* point symbol mapping. The conceptual basis for this technique is easily understood by most map readers. Its popularity among cartographers is likely to continue for some time but, like other processes in cartographic design, this method must be clearly understood if effective designs are to be created. Although much has been learned about this form of mapping in the last 30 years (notably in the area of symbol scaling), there is no single generally accepted design approach.

CONCEPTUAL BASIS FOR PROPORTIONAL POINT SYSTEM MAPPING

Proportional point symbol mapping is based on a fundamentally simple idea. The cartographer selects a symbol form (circle, square, or triangle) and varies its size from place to place, in proportion to the quantities it represents. Map readers can form a picture of the quantitative distribution by examining the pattern of differently sized symbols. (See Figure 9.1.) This simple approach can be spoiled by overloading the symbol form with too much information or by selecting scaling methods inappropriate to perceptual principles. The basic concepts of this form of quantitative thematic mapping are nonetheless easily grasped by most map readers. This is no doubt why the method has reached its present level of popularity.

WHEN TO SELECT THIS METHOD

There are two commonly accepted instances when graduated point mapping is selected by the cartographer: when data occur at points and when they are aggregated at points within areas. (See Figure 9.2.) Of course, data occurring at points is to be interpreted relative to map scale. Practically any magnitude can be symbolized this way, including totals, ratios, and proportions. Density, however, is normally symbolized by the choropleth technique. Whenever the goal of the map is to show relative magnitudes of phenomena at specific locations, the graduated symbol form of mapping is appropriate.

The kinds of data that can be mapped are varied: total population, percentage of population; black population; value added by manufacturing, retail sales, employment, any agricultural commodity, tonnage shipped at ports, and so on. Historically, proportional symbol maps have been used extensively to portray economic data, but magnitudes from the physical and cultural worlds also may be illustrated.

A BRIEF HISTORY OF PROPORTIONAL POINT SYMBOLS

The earliest history of proportional point symbols on maps is really a history of the use of the **graduated circle;** other symbol forms were not used until later. In 1801, William Playfair used graduated circles to depict statistical, noncartographic,

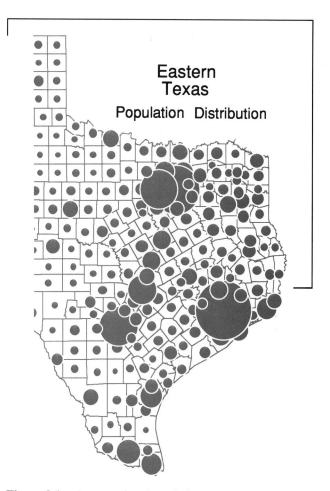

Figure 9.1 A proportional symbol map.
In this case proportional circles have been used at the centers of countries to represent map quantities (population).

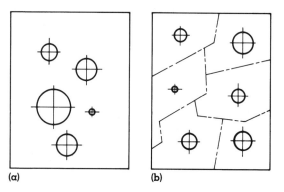

Figure 9.2 Two kinds of location and the use of proportional symbols.
In (a), circles are placed at the exact points where quantities being mapped are located (cities, towns, stores, etc.). Area data assumed to be aggregated at points, as in (b), may also be represented by proportional point symbols. In each case, the symbols are used to show different quantities at distinct locations.

I n the case of graduated circle maps. . . . Some might argue that the immediate visual impression is of vital importance; others might suggest that any serious map user will be inclined to attempt some form of quantitative comparison between symbols. Although such stances may be considered contradictory, in essence they both possess a quantitative basis. However, they tend to differ on three counts: firstly, the former takes a more holistic viewpoint, while the latter is more concerned with specific detail; secondly, the former relies less heavily than the latter upon the existence of a suitably informative legend; thirdly, the former aims at the nominal and ordinal levels of scaling, while the latter ranges between the ordinal and ratio levels.

Source: T. L. C. Griffin, "Group and Individual Variations in Judgment and Their Relevance to the Scaling of Graduated Circles," *Cartographica* 22 (1985): 21–37.

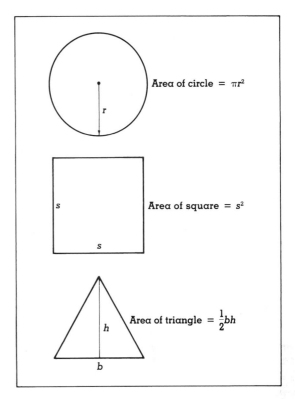

Figure 9.3   Three two-dimensional symbols (circle, square, and triangle) and their area formulas.

data.[1] It is most interesting that Playfair used circle *areas* rather than scaling them to diameters or circumferences. We persist in using this scaling method today. It was apparently not until 1837 that circles were employed on maps. Henry Drury Harness used them to map city populations.[2] In 1851, they were used by August Peterman in the mapping of city populations in the British Isles.[3] Professor Arthur Robinson states that they were used by Minard, in France, by the year 1859.[4] Minard's circles portrayed port tonnages.

From these beginnings, the use of proportional symbols on thematic maps gradually increased. Graduated symbols were adopted slowly in the United States. It was not until the early part of the twentieth century that professional geographers and cartographers began using them to any degree. They appeared first in professional journals. The graduated circle continues to be the form of proportional symbol most often used, although other choices are possible.

Many cartographers use bar graphs, line graphs, or other linear graphs as proportional symbols. The history of this form of point symbolization is unclear. I discourage this practice because the spatial distribution of phenomena becomes muddled. It is virtually impossible to see relative magnitudes at points when, for example, a bar may stretch nearly halfway across the map. Proportional point symbol mapping requires compact geometrical symbols at points.

A VARIETY OF CHOICES

Circles, squares, and triangles are the most common forms of proportional symbols, with the circle being the dominant form. In each case, *area* is the geometric characteristic that is customarily scaled to geographical magnitudes. (See Figure 9.3.) Circles have no doubt become more common because of several advantages they have over the other symbols:

1. Their geometric form is compact.
2. Historically, circle scaling was less difficult than for other geometrical forms (with the exception of the

square), because the square root of the radius can be used. Today, computers have eliminated any need for manual calculation of symbol sizes and areas.
3. Circles are more *visually stable* than other symbol forms and thus cause little eye wandering.

Squares are used increasingly because computer software can easily generate them. Before the widespread use of desktop mapping, squares were difficult to construct because they have straight sides, and they must be oriented to the map border, to other symbols, or to other structural elements of the map. This required greater production time and cost and provided opportunities for production error. The square is also considered less visually stable than the circle. It is not without advantages, however—these will be discussed later in the context of scaling. Triangles have disadvantages similar to those of squares; they are used even less.

It is not uncommon for two proportional forms to illustrate two distributions on the same map. This is discouraged because it introduces too much complexity to the map and can undermine the communicative effort. A better solution would be to develop two separate maps.

THREE-DIMENSIONAL SYMBOLS

Cartographers and geographers have experimented with point symbols of three-dimensional appearance, including spheres, cubes, and other geometrical volumes. (See Figure 9.4.) The use of **three-dimensional point symbols** can

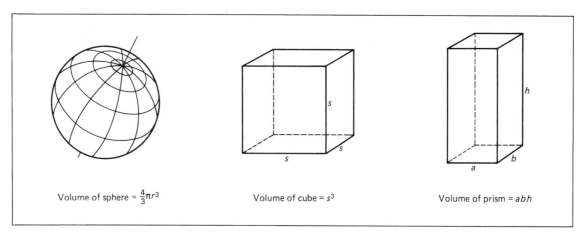

Figure 9.4   Three three-dimensional symbols (sphere, cube, and rectangular prism) and their volume formulas.

result in very pleasing, plastic, and visually attractive maps. Their chief difficulty lies in the inability of most map readers to gauge their scaled values correctly. Potential solutions to these shortcomings will be discussed more fully on the following pages.

In addition to the visual effects, another important advantage of three-dimensional symbols lies in their scaling. With proportional circles, areas are scaled to the square root of the data (Table 9.1), but three-dimensional symbols, including the sphere, ordinarily have their volumes scaled to the cube root of the data. The net result is that, in three-dimensional mapping, the necessary range of symbol sizes is reduced. Greater data ranges can thus be handled on the map when spherical symbols are used. Less crowding of symbols also results, because the areas covered by symbols are reduced.

The difficulty is with reader perceptual scaling. Most map readers cannot correctly judge visually the relative sizes of quantitative three-dimensional symbols. We respond visually to the areas covered on the map by three-dimensional-looking symbols and do not "see" relative volumes. Cartographers have not explored the use of three-dimensional-*looking* symbols, especially spheres, scaled to the areas they occupy on the map sheet. Likewise, as we will discuss below, the best alternative may be to employ three-dimensional-looking symbols, but *range-graded,* rather than having each data value symboled by its own symbol.

PROPORTIONAL SYMBOL SCALING

We may now more formally introduce several ideas concerning how map readers perceive quantitative symbols. For several decades, psychologists, cartographers, and graphical method statisticians have researched the mechanisms whereby map readers perceptually scale quantitative symbols. The researchers study response patterns to stimuli—the symbols and their characteristics such as length, area, and volume. The research attempts to document carefully how stimulus and response interact, usually by way of mathematical description. These studies are usually referred to as *psychophysical investigations.*

Table 9.1   Differences in Mapped Areas of Circles and Spheres

Symbol	Formula	Quantity Represented		Radii		Ratio of Large Radius to Small Radius	Area on Map Covered by Small Symbol (sq units)	Area on Map Covered by Large Symbol (sq units)	Ratio of Area of Large Symbol to Area of Small Symbol
		Small	Large	Small Symbol	Large Symbol				
Circle	πR^2	100	1,000	.1	.3162	3.162	.0628	.6278	9.99
Sphere	$\frac{4}{3}\pi R^3$	100	1,000	.1	.2154	2.154	.0628	.2913	4.6

PSYCHOPHYSICAL EXAMINATION OF QUANTITATIVE THEMATIC MAP SYMBOLS

The general psychophysical relationship between visual stimulus and response can be described by this formula:[5]

$$R = K \cdot S^n$$

where S is stimulus, R is response, n is an exponent describing the mathematical function between S and R, and K is a constant of proportionality defined for a particular investigation. This relationship is sometimes referred to as the **psychophysical power law** (because $R = S$ raised to a power, n).

The results of the research efforts suggest overall that most people do not (or cannot) respond to the geometric properties of quantitative symbols in a linear fashion. (See Figure 9.5.) A linear response pattern means that if 10 units of stimulus are viewed, 10 units of response will be measured, and this one-to-one relationship will hold throughout the range of all stimuli. It has been found, overall, that length is correctly perceived, but area and volume are not. The geometrical property of area is increasingly underestimated as higher magnitudes of stimulus are judged. Volume is underestimated to a greater degree than area.

These findings have had considerable impact on map designers. The initial results, especially in the case of proportional circles, caused quick adoption of methods to adjust symbol sizes to compensate for the underestimation. This led to perceptually based **apparent-magnitude scaling.** (See Figure 9.6.) The pioneering work of Professor James Flannery must be cited in this context; his contributions to scaling adjustments became the standard for nearly 30 years.[6]

Psychophysical studies have had a different impact on designers. They suggest that apparent-magnitude scaling might be abandoned in favor of **absolute scaling,** but with more careful attention given to legend design.[7] Absolute

scaling is simply the direct proportional scaling of magnitudes to the symbol's area. For example, if we have two values, 6,400 and 1,600, that are to be symbolized and scaled by proportional circles, we must first set up a proportion, remembering that the area of a circle is πR^2 (R = radius of the circle), thus:

$$\frac{\pi R_1^{\,2}}{\pi R_2^{\,2}} = \frac{\textbf{data value 1}}{\textbf{data value 2}}$$

This reduces to

$$\frac{R_1^{\,2}}{R_2^{\,2}} = \frac{\textbf{data value 1}}{\textbf{data value 2}}$$

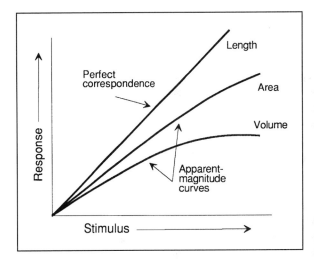

Figure 9.5   Apparent-magnitude curves for length, area, and volume.
In the visual psychophysical world, the area and volume of geometric figures are underestimated. Length is usually correctly perceived. Underestimation means that for every unit of physical stimulus perceived, less than unity is reported in response.

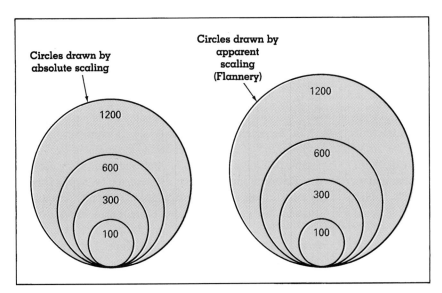

Figure 9.6   Absolute and apparent scaling of proportional circles.
Because the areas of circles are not perceived linearly, Flannery has introduced a correction factor, causing circles to appear larger as their values increase. This produces a range of circle sizes scaled by apparent magnitude.

and then, taking the square root of both sides of the equation, we get:

$$\frac{R_1}{R_2} = \frac{\sqrt{\text{data value 1}}}{\sqrt{\text{data value 2}}}$$

For this example, we next have

$$\frac{R_1}{R_2} = \frac{\sqrt{6,400}}{\sqrt{1,600}}$$

and then,

$$\frac{R_1}{R_2} = \frac{80}{40} = \frac{2}{1}$$

or the radius of value 1 is twice that of value 2.

In practice a radius is selected to represent the smallest data value that yields a circle that looks good on the map; then a radius is calculated for the largest data value to determine if the circle size is appropriate and, if so, all intermediate circle sizes for the remaining values are calculated. If not, the process is begun again and a final set of circle sizes is computed and drawn.

For example, suppose you had a data range of values with 10 being the smallest, and 150 the largest, and decided to render the smallest circle with a radius of .1 inch:

$$\frac{\pi R_1{}^2}{\pi R_2{}^2} = \frac{10}{150}$$

$$\frac{R_1{}^2}{R_2{}^2} = \frac{10}{150}$$

$$\frac{(.1)^2}{(R_2)^2} = \frac{10}{150}$$

$$\frac{.1}{R_2} = \frac{10}{150}$$

$$\frac{.1}{R_2} = \frac{3.16}{12.24}$$

$$3.16\, R_2 = 1.224$$

$$R_2 = .387$$

If you like the computed size of the largest circle size, then you will compute the sizes of all intermediate circles. If not, you must redo from the beginning until you arrive at a circle-size set that looks good on your map.

Not too many years ago, textbooks on cartography included graphic devices to compute circle sizes. These were easily constructed by first constructing a horizontal scale graduated linearly to include the data values in the array. Three or four circles were scaled and placed on the horizontal bar at their respective locations at perpendiculars drawn from the horizontal bar. The tops of these circles were joined by a curved line. Intermediate circle radii could then be determined by erecting perpendicular lines at values on the linear scale, and radii ascertained by noting the length of the perpendiculars to the curved line—not absolutely precise, but adequate. Today it is probably just as easy to work up circle sizes by spreadsheet programs.

The Proportional Circle

Scaling for circles has received more attention from psychologists and cartographers than scaling for any other form of symbol, because of the circle's popularity as a proportional symbol. When Flannery made clear that circle sizes need to be adjusted to compensate for underestimation, a standard was adopted based on his findings.

The psychophysical relationship, expressed as $R = K \cdot S^n$, is usually transformed into logarithmic form:

$$\log R = \log a + b \cdot \log S$$

This is an equation for a straight line where a is the constant K (and the intercept on the R-axis), b is equivalent to the slope of the line (and the n of the original power equation), and S and R are stimulus and response, as before. By regression methods, a line will then be fitted to the experimental data, and a value for n is produced.

In the original equation $R = S^n$ (omitting K for simplicity), if a perfect relationship exists between R and S, then the exponent n would be 1.0. Flannery's experimental data yielded a value of $n = .8747$. For example, if a circle of area value 2 is judged experimentally, it would be seen as having an area of

$$2^{.8747} = 1.833$$

Flannery arrived at his *adjustment* factor by transposing the equations. In logarithmic terms, the square root of number N is $(\log n)(.5)$. To compensate for underestimation, the symbols are enlarged somewhat by using .5716 instead of .5. This value is tied directly to the value of .8747 for n.

The procedure to scale circles this way includes the following steps (the method used by Flannery):

1. Obtain logarithms of data values.
2. Multiply these log values by .5716 (or rounded to .57).
3. Determine the antilogarithms of each.
4. Scale all values to the new set of antilogarithms in ordinary fashion.

With the availability of handheld calculators and computer spreadsheets that easily raise any number to any power, it is unnecessary to use logarithms in these calculations. See Table 9.2 for an example of circles scaled by absolute methods and those by apparent-scaling methods. In this table, the reason for the calculated difference between the two final circle-size outcomes is that for linear scaling, the exponent used is .5, but for the apparent-magnitude scaling, it is .5716.

In the psychophysical power law, n expresses how the response can be described relative to the stimulus. We might assume that n is a constant for circle judgment tests, but it is not. Research indicates that n fluctuates because of these factors: testing method, instructions, standard (or anchor) stimulus used, and the range of stimuli used in the experiments.[8] These n values range from .58 to 1.20. It is now evident that no one correction factor developed for an apparent-magnitude scale is the answer. This is especially true in light of parallel findings which show that nearly correct responses can be achieved with carefully designed legends.

Table 9.2 Worked Example of Linear and Apparent-Magnitude Scaling of Circle Sizes

Data Value (n)	Linear Scaling, Exponent = .500	Apparent-Magnitude Scaling, Exponent = .5716
20	4.47	5.54
30	5.47	6.99
40	6.62	8.24
50	7.06	9.30
.		
.		
.		
200	14.14	20.66

Linear Problem	Apparent-Scaling Problem

$$\frac{\pi R_1^2}{\pi R_2^2} = \frac{20}{200}$$

$$\frac{R_1}{R_2} = \frac{\sqrt{20}}{\sqrt{200}}$$

$$\frac{(.1)}{R_2} = \frac{4.47}{14.141}$$

$$4.47(R_2) = (.1)(14.141)$$

$$4.47(R_2) = 1.414$$

$$R_2 = .316$$

$$D_2 = .63$$

$$\frac{\pi R_1^2}{\pi R_2^2} = \frac{20}{200}$$

$$\frac{R_1}{R_2} = \frac{\sqrt{20}}{\sqrt{200}}$$

$$\frac{(.1)}{R_2} = \frac{5.514}{20.488}$$

$$5.54(R_2) = (.1)(20.488)$$

$$5.54(R_2) = 2.048$$

$$R_2 = .369$$

$$D_2 = .738$$

Available studies also indicate that apparent-magnitude scaling is not that much more efficient than absolute scaling if the range of required circle-size judgments is 10 or less.[9] One final caution, too, is worthy of note. Some map readers simply cannot make proportional symbol judgments, even if apparent scaling is used. Map designers will need to weigh all these factors when making decisions regarding this form of quantitative map.

It must be pointed out, finally, as already shown in Figure 9.6, that selecting the apparent-magnitude scaling method requires much greater map space, and this is often a critical factor in design, as space is usually limited.

Perception of Circles among Circles Most cartographic research on the judgment of circle size has measured rather abstractly how people perceive circle size in nonmap settings. Moreover, and more troubling, the influences of neighboring circles on size judgment have been largely overlooked. However, at least one study has investigated this phenomenon and has produced these interesting results.[10]

1. When a circle is seen among circles that are smaller, it is seen on the average about 13 percent larger than it would ordinarily be seen if judged among circles that are larger. (See Figure 9.7a.)
2. Circles surrounded by circles of the same size were judged both high and low relative to the surrounding circles.
3. The effect of surrounding circles can be reduced if internal borders are used on the map. (See Figure 9.7b.)

Thus circle judgment is affected by the size of neighboring circles on the map. At present no generally applicable solution is available to control these undesirable effects. If repeated design alternatives show that this problem cannot be

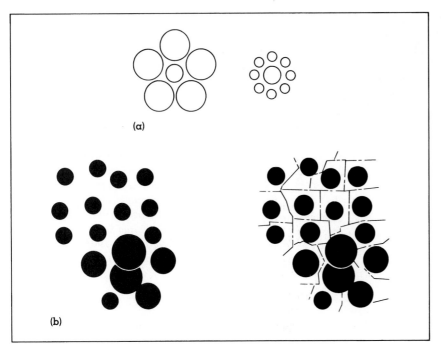

(a)

(b)

Figure 9.7 The effect of neighboring symbols on the judgment of circle size. In (a), the inner circles in the right and left drawing are of the same size. Because of the contrast effects produced by neighbors, the inner circle on the right appears larger than its counterpart on the left. In (b), the provision of internal boundaries reduces the contrast effect of neighboring symbols. Unfortunately, cartographic designers have little control over variables of this kind.

solved, perhaps a different form of map should be constructed. Designers need, at least, to know that this problem can arise so that troublesome areas on the map can be treated with greater care.

One very important aspect of proportional symbol maps is their ability to show spatial numerousness. (See Figure 9.8.) This feature is extremely effective when symbol sizes are carefully selected for the chosen base map. Although overlapping symbols make it difficult to estimate individual symbol sizes, a sense of visual cohesiveness results, and this may lead to memorability, certainly a goal in cartographic communication.[11] This dilemma, to show overall pattern versus specific symbol quantity, usually can be dealt with by careful examination of map purpose. If not, one solution is to provide an areal table alongside the map and, in the long run, develop special scaling methods that take into account symbol coverage.[12]

The Square Symbol

Graduated squares are used in quantitative point symbol mapping in much the same fashion as circles. The areas of the squares are scaled to data amounts yielding simple proportional relationships between the square roots of data and the sides of the squares:

$$\frac{S_1^{\,2}}{S_2^{\,2}} = \frac{\text{data value 1}}{\text{data value 2}}$$
$$\frac{S_1}{S_2} = \frac{\text{data value 1}}{\text{data value 2}}$$

Squares have never received quite the attention by researchers that circles have, probably because they are not as

commonly used, and prior to computer production, they were more difficult to produce. Visually, they introduce a rectangularity to the map's design which is often difficult to coordinate with other map elements. (See Figure 9.9.) Nonetheless, they can show geographical distribution—clustering, dispersion, or regularity—as well as circles.

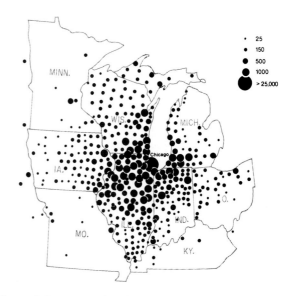

Figure 9.8 Proportional symbol maps show spatial numerousness.
In this illustration the spatial variableness of newspaper circulation *(Chicago Tribune)* is clearly shown by the proportional symbol-mapping method. *(Source: Reprinted courtesy of Duxbury Press.)*

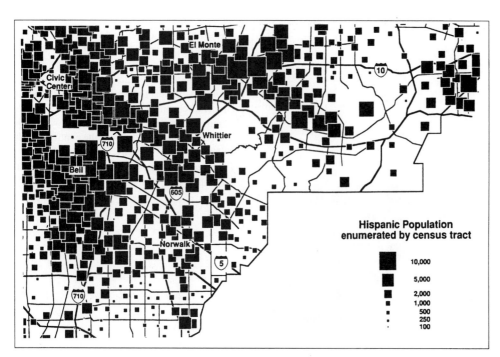

Figure 9.9 Proportional symbol map using squares. Some cartographic designers dislike using squares because they introduce "retangularity" on the map. However, squares are perceived more accurately than circles. In this particular map the squares clearly show clustering. *(Detail of a map,* Hispanic Population in Metropolitan Los Angeles County, California, *published by the Center for Geographical Studies, Department of Geography, California State University, Northridge, 1991. Used with permission.)*

One important consideration for the selection of squares is that their proportional areas are nearly perfectly perceived. In one study, the exponent value n in the psychophysical power law was experimentally determined to be .93.[13] Other psychological and cartographic studies suggest that the areas of squares are more accurately judged than circle areas. This does not mean that circles should be abandoned in favor of squares. With appropriately designed legends, both symbols can communicate well. The selection should be made in concert with other design elements and mapping goals.

RANGE GRADING: A PROBABLE SOLUTION

Apparent-magnitude scaling methods (the introduction of corrective factors into absolute scales) are too coarse for universal adoption. They are based on the "average" map reader who does not exist outside mathematical tables. Moreover, the landmark corrective factor of Flannery, although quite important in the early assessment of symbol scaling, is not applicable in all cases because it was developed from a single psychophysical study with one experimentally derived exponent. Subsequent investigations have demonstrated that the exponent itself varies depending on test conditions.

Legend designs have also been examined closely. This avenue of research is based on the theory of **adaptation level.** This theory holds that perceived judgments are based on an adaptation level (a neutral reference point) and that the perceptual anchors from which judgments are made affect this point.[14] Two **anchor effects** have been shown to exist: *contrast* and *assimilation.* The contrast effect is displacement of perceptual judgment away from the anchor; by assimilation, perceptual judgments are displaced toward the anchor. The size of the anchor has an effect on how perceptual judgments of the symbols on the map are made. In the case of proportional circles and squares, judgments are displaced toward the anchor, whether it is a large or a small anchor. This is evidence that assimilation is taking place.

One study has produced important findings regarding anchor effects in proportional circles and squares. There are important design implications.[15]

1. When graduated circles are scaled according to the apparent value (after Flannery), there is greater underestimation of symbol sizes when a small legend value is used, less overestimation when a large symbol is used, only slight overestimation of circle sizes when *three* legend circles are used, and fairly accurate estimates when middle-size legend circles are used. The variation in responses is less when three legend symbols are used.

2. For squares, small- and middle-size legend squares lead to underestimations. (The squares are scaled to the square root of the data, with no corrective factor.) A large legend square yields overestimations, and three

legend squares (small, middle, and large) lead to better estimates overall—principally at the higher end of the scale. Cox, author of the study, suggests that perhaps only a middle-size and a large square symbol are needed.

3. *Apparent scaling does not compensate for underestimation.* The use of several differently sized legend symbols, scaled conventionally (absolute scaling), may be sufficient for accurate judgment of circle and square size.

It has become abundantly clear that certain proportional point symbol formats are not good design solutions. Thus,

1. Never provide apparent scaling for circles unless several circle sizes are used throughout the data range in the legend.

2. Do not provide absolute scaling for circles when only one legend (anchor) circle is used.

3. Do not scale any symbols to their volumes (cube roots of the data).

Traditional, proportional point symbol mapping called for each value in the data to be represented by an absolutely scaled symbol (**square root scaling** for two-dimensional symbols and **cube root scaling** for three-dimensional symbols). As experimental studies showed that perception of these symbols was not linear, apparent scaling was introduced. Subsequent research has shown that apparent scaling may not be the total answer either. It appears that range grading proportional symbols may be a better choice. Certainly, classifying data to show spatial distribution is an appropriate design approach.[16]

Range grading is achieved in a manner similar to the development of classes for choropleth maps. The data array is divided into groups; each group is represented by a proportional symbol that is clearly distinguishable from other symbols in the series. In this scaling method, symbol-size discrimination is the design goal, rather than magnitude estimation. Range grading of proportional symbols is not a new alternative to symbol scaling; it was first suggested by Professor Meihoefer in the late 1960s.[17] His studies led him to devise a set of 10 circle sizes that were easily and consistently discriminated by his subjects and were applicable to most small-scale thematic maps. (See Table 9.3 and Figure 9.10.) Research since Meihoefer supports the use of range grading, especially the work done by Heino.[18]

An alternative array of circles useful for a range-graded series is presented in Figure 9.11. I have determined after repeated use of the Meihoefer series that the circles with radii of 9.77 and 11.28 mm, and to some degree the one with a radius of 13.82 mm, are not that easily distinguished on actual small-scale maps. The circles presented in Figure 9.11 have not been tested rigorously in an experimental setting, but have proved useful in practice. Others might wish to explore using them. As with the Meihoefer series, any five adjacent circles may be selected, or for that matter, any three or four, and although not recommended, any adjacent six may be selected. Range grading of proportional point symbols raises

Table 9.3   Acceptable Range-Graded Circle Sizes for Small-Scale Thematic Maps

Radius of Circle (mm)	Area of Circle (sq mm)	Difference (mm)
1.27	5.0	
1.99	12.5	7.5
2.82	25.0	12.5
3.99	50.0	25.0
5.64	100.0	50.0
7.98	200.0	100.0
9.77	300.0	100.0
11.28	400.0	100.0
13.82	600.0	200.0
15.96	800.0	200.0

Source: Hans Joachim Meihoefer, "The Utility of the Circle as an Effective Cartographic Symbol," Canadian Cartographer 6 *(1969): 105–17.*

the issue of class interval determination, just as with choropleth class intervals. The most applicable method appears to be the optimization approach. Just as in determining choropleth class intervals, the goal is to devise classes that have the greatest degree of internal homogeneity and a high degree of heterogeneity among classes.[19] Although Meihoefer used 10 circles in his research, I recommend using perhaps only four or five classes. The legend may appear in a variety of ways, but it must be clear that the circles are range graded and do not represent individual values. (See Figure 9.12.)

GRAPHIC DESIGN CONSIDERATIONS FOR PROPORTIONAL POINT SYMBOL MAPS

Several issues relating to the design of proportional point symbols have been presented, focusing on the conceptual basis of their use and on scaling methods. Other design concerns deal more with their graphic treatment, discussed in this concluding section.

THE NATURE OF THE DATA

Cartographers must deal with various attributes of the statistical data that is set before them and select a mapping method that best represents the particular aspect of the data. The data attributes have been described as density, clustering or dispersion, extent, orientation, and shape (of the statistical distribution). Proportional point symbol mapping should not be selected when data variation is small or when the nature of the classing method renders a map of homogenous appearance. (See Figure 9.13.) The

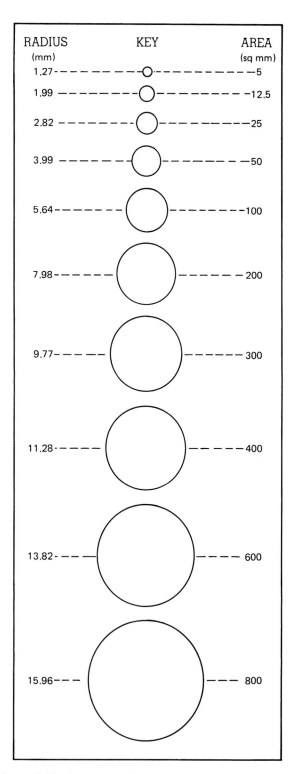

Figure 9.10   Representative circle sizes useful in a range-graded series for small-scale thematic maps.
Each circle is easily differentiated from its neighbor—the fundamental criterion in a range-graded series. No more than five classes are advisable for this form of scaling. When using this chart, any adjacent set of circles may be selected from the array.

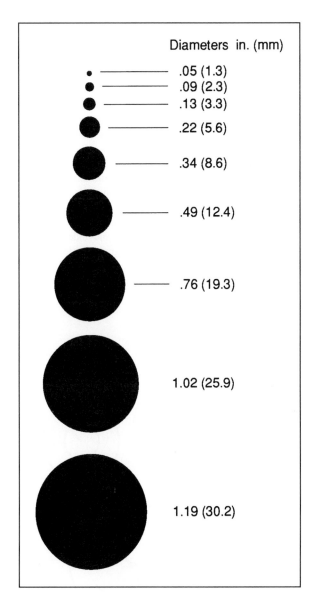

Diameters in. (mm)

.05 (1.3)
.09 (2.3)
.13 (3.3)
.22 (5.6)
.34 (8.6)
.49 (12.4)
.76 (19.3)
1.02 (25.9)
1.19 (30.2)

Figure 9.11   Empirically derived circle sizes useful in a range-graded series on a small-scale thematic map. Although useful in practice, these circles have not been experimentally tested. Any set containing three to six adjacent circles may be used.

result will be a monotonous map that looks dull and undiscriminating (although it may be technically accurate). The intuitive appeal of the design can be the criterion of a map's appropriateness in such cases.

OVERLOADED PROPORTIONAL POINT SYMBOLS

Maps and their quantitative symbols are unique mechanisms for the communication of spatial concepts. All too often, however, cartographic designers fall into the trap of pushing the system beyond its capacities. Because it is possible to cram a lot of information into one symbol, the de-

signer often tries for too much of a good thing. **Symbol overload** results when it is no longer easy for the reader to make judgments about the quantitative nature or spatial pattern of the distribution. Unfortunately, overload is reached at a point much earlier than most designers realize.

Probably more than any other symbol, the proportional circle has been misused by unthinking designers. Because it can be sized, segmented, colored, sectored, or whatever, why not do all of these on one map? (See Figure 9.14.) Unfortunately, the map reader can no longer see the distribution or its intent clearly and easily. The cartographer should not have troubled himself or herself or the reader. In at least two maps in *The National Atlas of the United States of America,* a proportional circle carries three different data sets.[20] These maps lose their thematic goals and exhibit complex graphic displays that almost totally destroy their communicative efforts.

A good design approach is to limit the number of variables symbolized by proportional point symbols to one, possibly two, but never three or more. When two or more variables are being represented, keep only one variable at the interval/ratio level and the other at the nominal or ordinal level. Do not use two at the interval/ratio level, because this leads quickly to excessive complexity and symbol overload.

GRAPHIC TREATMENT OF PROPORTIONAL SYMBOLS

Consideration already given to the scaling of circles and squares applies equally to black or gray-tone circles (and probably squares). Gray-tone treatments do not require special considerations. Cartographers find several advantages in using gray proportional symbols in black-and-white mapping.[21]

1. The amount of information can be expanded by showing two distributions, one with black symbols and one with gray.
2. The use of gray-tone symbols allows the cartographer greater freedom for expression in design.
3. Gray-tone symbols allow for better ordering of information on the map. Symbols can be rendered in tones of gray in proportion to the values they represent.
4. Some symbols can be de-emphasized (gray) relative to others (black).

According to one cartographic investigation of the effect of color on the perception of graduated circles, it is apparent that different hues do not affect the estimation of circle-size differences.[22]

Open and "cut-out" circle treatments have been carefully examined by cartographic researchers. (See Figure 9.15.) Groop and Cole, for example, made these discoveries:[23]

1. Transparent circles (rendered so that edges of both are visible even after overlap) are more accurately perceived than cut-out circles and are seen similarly to circles having no overlap.

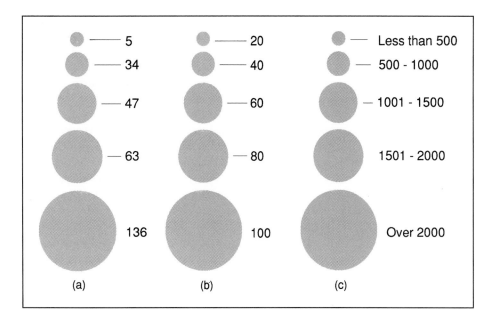

Figure 9.12 Different ways to illustrate range-graded series by legend circles.

In (a), a representative number is presented, perhaps an average of each homogeneous class. No rational relationship between numbers is apparent. In (b), an attempt to provide preferred numbers and even intervals is shown. Finally, in (c), data class boundaries of a continuous distribution are shown adjacent to each class circle. No one method is preferred. The goal is to make the legend readable and understandable and the range-graded series apparent. It may be necessary to explain the classing scheme in a legend note.

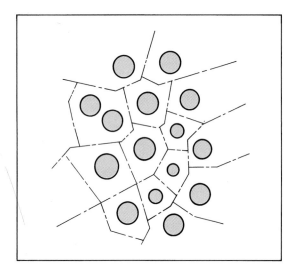

Figure 9.13 Inappropriate use of proportional symbols
Maps in which unvarying geographic data are symbolized by proportional symbols are montonous and reveal little. Designers should not use this mapping metnod with distributions of this kind.

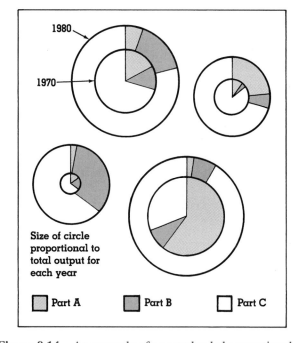

Figure 9.14 An example of an overloaded proportional point symbol, using a segmented circle (showing proportional parts) for two different years.
The map reader is expected to make an easy and quick assessment of the change in the geographical distribution of each part, at several locations. The map designer should seek simple solutions rather than use symbols so unwisely.

2. With cut-out circles (in which the overlap obscures part of one circle), there is a strong relationship between the amount of overlap and estimation error.

Because of the complex nature of many proportional symbol maps, the effects of overlapping, tightly clustered circles on readers' estimations of circle sizes are not known with certainty. In some cases, the reader is asked to estimate the magnitudes of individual symbols, in others, range-graded series. The designer must be aware that the reader's judgment is influenced by the graphic treatment of the symbols.

Another interesting research project suggests that map readers have preferences regarding the way graduated cir-

cles are rendered on the map.[24] For the map readers tested in the study, maps having circles with little contrast to their surroundings, or that are open with line work showing through, proved least popular. (See Table 9.4.) Black-filled circles were judged best, and gray-filled symbols also did well. The author of the test remarks that the map readers seemed to be more influenced by the features of trans-

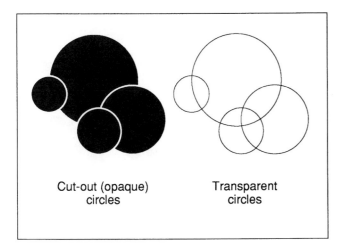

Figure 9.15   Cut-out and transparent circles.
Readers tend to make errors of perceptual judgment with cut-out symbols, in proportion to the amount of the circle obscured by the overlap. Transparent circles, on the other hand, are perceived about as well as circles that do not overlap. The designer must weigh these results against other design elements and the purpose of the map. Cut-out circles add a three-dimensional (plastic) quality to the map, whereas a map containing many transparent circles looks flat and uninteresting.

Table 9.4   Preferences for Proportional Circles in Different Map Environments

Judged	Map Environment (Percent)					
	F	**C**	**B**	**E**	**D**	**A**
Best (rank 1 of 6 ranks)	30	26	26	16	1	1
Worst (rank 6 of 6 ranks)	4	5	0	0	39	51

A = Transparent circles with linework showing through
B = Gray-filled circles with linework showing through
C = Black-filled circles with linework (in white) showing through
D = Opaque white-filled circles
E = Opaque gray-filled circles
F = Opaque black-filled circles

parency and opacity of the circles and less by whether the circles were black or gray.

The same author also studied the effects of different circle environments on symbol-size judgments. He concluded by stating, "Certainly, the second experiment showed that if experimental conditions are held constant then group differences in the judgment of white or black circles are minimal."[25] He believed that this would also hold true for gray circles. The results of the study tend to support the viewpoint later in this book (Chapter 13) that readers respond favorably to symbols with high contrast that appear as visual figures in perception.

It is important that proportional point symbols be made to appear as strong figures in perception. The graduated point symbol should be clear and unambiguous in meaning, its edges sharp and not easily confused with background material, and its surface character easily differentiated from other surfaces and textures on the map. Its scaling should be precise, not confusing for the reader. The symbol should be made to stand out from its surroundings, both graphically and intellectually. If these simple guidelines are followed, the final map will perform well in the communication of spatial concepts.

COMPUTER MAPS

It is possible to generate computer maps using proportional point symbols. (See Figure 9.16.) Design considerations remain the same, however, with respect to computer maps. Questions of scaling (absolute vs. apparent), range grading,

and graphic appearance of the circles must be confronted by the designer; the development of computer programs to produce these maps has not reduced the burden. In fact, the ease with which these maps may be produced on color CRT devices, for example, should spur further research into the psychophysical nature of quantitative point symbols and color. Absolute scaling is much less time-consuming in computer-generated maps. This may make this option more appealing but also cause a dilemma, given the present trend to present circles in range-graded fashion. Designers sensitive to the needs of clients and map readers alike must make the difficult design decisions, whether or not computers are employed.

NOTES

1. James J. Flannery, *The Graduated Circle: A Description, Analysis, and Evaluation of a Quantitative Map Symbol* (unpublished Ph.D. dissertation, Department of Geography, University of Wisconsin, 1956), p. 7.
2. Arthur H. Robinson, "The 1837 Maps of Henry Drury Harness," *Geographical Journal* 121 (1955): 440–50.
3. Flannery, *The Graduated Circle,* pp. 8–9.
4. Arthur H. Robinson, "The Thematic Maps of Charles Joseph Minard," *Imago Mundi* 21 (1967): 95–108; and Arthur H. Robinson, *Early Thematic Mapping in the History of Cartography* (Chicago: University of Chicago Press, 1982), pp. 207–8.
5. Kang-tsung Chang, "Circle Size Judgment and Map Design," *American Cartographer* 7 (1980): 155–62.

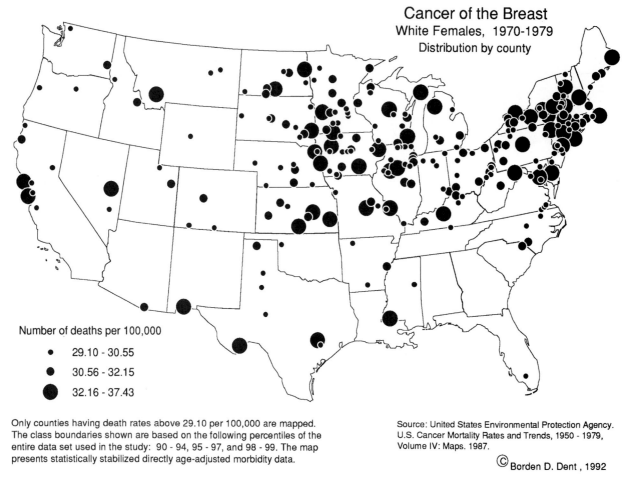

Cancer of the Breast
White Females, 1970-1979
Distribution by county

Number of deaths per 100,000

- 29.10 - 30.55
- 30.56 - 32.15
- 32.16 - 37.43

Only counties having death rates above 29.10 per 100,000 are mapped.
The class boundaries shown are based on the following percentiles of the
entire data set used in the study: 90 - 94, 95 - 97, and 98 - 99. The map
presents statistically stabilized directly age-adjusted morbidity data.

Source: United States Environmental Protection Agency.
U.S. Cancer Mortality Rates and Trends, 1950 - 1979,
Volume IV: Maps. 1987.

©Borden D. Dent , 1992

Figure 9.16   An example of a computer-generated proportional symbol map.
This map was produced by ***Designer*** Software (a trademark of Micrografx, Inc.), and printed on a *QMS* laser-jet 300-dpi printer.

6. Flannery, *The Graduated Circle.* Professor Robert L. Williams of Yale University was also working on scaling methods as Flannery was doing his research. Flannery refers to his work in the dissertation.

7. Chang, "Circle Size Judgment," pp. 155–62.

8. Among the studies that have direct bearing on this subject are Kang-tsung Chang, "Visual Estimation of Graduated Circles," *Canadian Cartographer* 14 (1977): 130–38; Carleton W. Cox, "Anchor Effects and the Estimation of Graduated Circles and Squares," *American Cartographer* 3 (1976): 65–74; and James J. Flannery, "The Effectiveness of Some Common Graduated Point Symbols in the Presentation of Quantitative Data," *Canadian Cartographer* 8 (1971): 96–109.

9. T. L. C. Griffin, "Groups and Individual Variations in Judgment and Their Relevance to the Scaling of Graduated Circles," *Cartographica* 22 (1985): 21–37.

10. Patricia P. Gilmartin, "Influences of Map Context on Circle Perception," *Annals* (Association of American Geographers) 71 (1981): 253–58.

11. Michael P. Peterson, "Evaluating a Map's Image," *American Cartographer* 12 (1985): 41–55; see also Terry A. Slocum, "A Cluster Analysis Model for Predicting Visual Clusters," *The Cartographic Journal* 21 (1984): 103–11.

12. Peterson, "Evaluating a Map's Image," pp. 41–55.

13. Paul V. Crawford, "The Perception of Graduated Squares as Cartographic Symbols," *Cartographic Journal* 10 (1973): 85–88.

14. Cox, "Anchor Effects," pp. 65–74.

15. Ibid.

16. George F. Jenks, "Contemporary Statistical Maps— Evidence of Spatial and Graphic Ignorance," *American Cartographer* 3 (1976): 11–19.

17. Hans Joachim Meihoefer, "The Utility of the Circle as an Effective Cartographic Symbol," *Canadian Cartographer* 6 (1969): 105–17.

18. Aarre Heino, "The Presentation of Data with Graduated Symbols," *Cartographica* 32 (1995): 43–50.

19. Michael W. Dobson, "Refining Legend Values for Proportional Circle Maps," *Canadian Cartographer* 11 (1974): 45–53.

20. United States Geological Survey, *The National Atlas of the United States of America* (Washington, DC: USGPO, 1970), pp. 223, 240.

21. Paul V. Crawford, "Perception of Gray-Tone Symbols," *Annals* (Association of American Geographers) 61 (1971): 721–35.

22. Richard E. Lindenberg, *The Effect of Color on Quantitative Map Symbol Estimation* (unpublished Ph.D. dissertation, Department of Geography, University of Kansas, Lawrence, 1986), p. 121.

23. Richard E. Groop and Daniel Cole, "Overlapping Graduated Circles: Magnitude Estimation and Method of Portrayal," *Canadian Cartographer* 15 (1978): 114–22.

24. T. L. C. Griffin, "The Importance of Visual Contrast for Graduated Circles," *Cartography* 19 (1990): 21–30.

25. Ibid.

GLOSSARY

absolute scaling   directly proportional scaling of area or volume of symbols to data values, pp. 177–178

adaptation level   in psychological theory, a neutral reference point on which perceived judgments are based; affected by the perceptual anchors from which judgments are made, p. 181

anchor effects   the size of the visual anchor affects the estimate of magnitude of an unknown symbol; the contrast effect causes estimation to be away from the anchor, and the assimilation effect causes judgments to be toward the anchor, p. 181

apparent-magnitude scaling   scaling of proportional symbols that incorporates correction factors to compensate for the normal underestimation of a symbol's area or volume, p. 177

cube root scaling   when three-dimensional symbols are used, they are scaled to the cube root of the data values, p. 181

graduated circle   a popular form of proportional point symbol; compact and easy to construct, p. 174

proportional point symbol mapping   type of quantitative thematic map in which point data are represented by a symbol whose size varies with the data values; areal data assumed to be aggregated at points may also be represented by proportional point symbols, p. 174

psychophysical power law   the mathematical expression that describes the relationship between stimulus and response; general form is

$$R = K \cdot S^n$$

where R is response, S is stimulus, n is exponent describing the relationship between R and S, and K is a constant, p. 177

range grading   a symbol represents a range of data values; differently sized symbols are chosen for differentiability for each of several ranges in the series of values being mapped, pp. 181–182

square root scaling   when two-dimensional symbols are used, they are scaled to the square root of the data values, p. 181

symbol overload   too much information in a symbol, so that readers have difficulty making assessments about the quantitative nature of the data, p. 183

three-dimensional point symbol   any point symbol made to appear three-dimensional, e.g., a sphere or cube, pp. 175–176

READINGS FOR FURTHER UNDERSTANDING

Alexander, John W. *Economic Geography.* Englewood Cliffs, NJ: Prentice-Hall, 1963. Students are encouraged to examine this book for its many good examples of proportional symbol maps, especially those on pp. 300, 303, 335, and 405.

Carstensen, Lawrence W. "A Comparison of Single Mathematical Approaches to the Placement of Spot Symbols." *Cartographica* 24 (1987): 46–63.

Chang, Kang-tsung. "Visual Estimation of Graduated Circles." *Canadian Cartographer* 14 (1977): 130–38.

———. "Circle Size Judgment and Map Design." *American Cartographer* 7 (1980): 155–62.

Clarke, John I. "Statistical Map Reading." *Geography* 44 (1959): 96–104.

Cox, Carleton W. "Anchor Effects and the Estimation of Graduated Circles and Squares." *American Cartographer* 3 (1976): 65–74.

Crawford, Paul V. "Perception of Graduated Squares as Cartographic Symbols." *Cartographic Journal* 10 (1973): 85–88.

Dobson, Michael W. "Refining Legend Values for Proportional Circle Maps." Canadian Cartographer 11 (1974): 45–53.

———. "Benchmarking the Perceptual Mechanism for Map-Reading Tasks." *Cartographica* 17 (1980): 88–100.

Dodd, Donald B. *Historical Atlas of Alabama.* Tuscaloosa: University of Alabama Press, 1974.

Ekman, G., and K. Junge. "Psychophysical Relations in the Perception of Length, Area, and Volume." *Scandinavian Journal of Psychology* 2 (1961): 1–10.

———**, R. Lindman, and W. William-Olson.** "A Psychophysical Study of Cartographic Symbols." *Perceptual and Motor Skills* 13 (1961): 335–68.

Flannery, James J. *The Graduated Circle: A Description, Analysis, and Evaluation of a Quantitative Map Symbol.* Unpublished Ph.D. dissertation, Madison: University of Wisconsin, Department of Geography, 1956.

———. "The Effectiveness of Some Common Graduated Point Symbols in the Presentation of Quantitative Data." *Canadian Cartographer* 8 (1971): 96–109.

Gilmartin, Patricia P. "Influences of Map Context on Circle Perception." *Annals* (Association of American Geographers) 71 (1981): 253–58.

Griffin, T. L. C. "Group and Individual Variation in Judgment and Their Relevance to the Scaling of Graduated Circles." *Cartographica* 22 (1985): 21–37.

———. "The Importance of Visual Contrast for Graduated Circles." *Cartography* 19 (1990): 21–30.

Groop, Richard E., and Daniel Cole. "Overlapping Graduated Circles: Magnitude Estimation and Method of Portrayal." *Canadian Cartographer* 15 (1978): 114–22.

Guelke, Leonard. "Perception, Meaning and Cartographic Design." *Canadian Cartographer* 16 (1979): 61–69.

Heino, Aaare, "The Presentation of Data with Graduated Symbols." *Cartographica* 32 (1995): 43–50.

Jenks, George F. "The Evaluation and Prediction of Visual Clustering in Maps Symbolized with Proportional Circles." In *Display and Analysis of Spatial Data,* eds. J. C. Davis and M. J. McCullogh. New York: Wiley, 1975, pp. 311–27.

———. "Contemporary Statistical Maps—Evidence of Spatial and Graphic Ignorance." *American Cartographer* 3 (1976): 11–19.

Keates, J. S. *Understanding Maps.* New York: Halsted Press, 1982.

Lindenberg, Richard E. *The Effect of Color on Quantitative Map Symbol Estimation.* Unpublished Ph.D. dissertation. Lawrence: University of Kansas, Department of Geography, 1986.

Meihoefer, Hans Joachim. "The Utility of the Circle as an Effective Cartographic Symbol." *Canadian Cartographer* 6 (1969): 105–17.

———. "The Visual Perception of the Circle in Thematic Maps—Experimental Results." *Canadian Cartographer* 10 (1973): 68–84.

Peterson, Michael P. "Evaluating a Map's Image." *American Cartographer* 12 (1985): 41–55.

Robinson, Arthur H. "The 1837 Maps of Henry Drury Harness." *Geographical Journal* 121 (1955): 440–50.

———. "The Thematic Maps of Charles Joseph Minard." *Imago Mundi* 21 (1967): 95–108.

———. *Early Thematic Mapping in the History of Cartography.* Chicago: University of Chicago Press, 1982.

———, **Randall Sale, and Joel Morrison.** *Elements of Cartography.* 4th ed. New York: Wiley, 1978.

Slocum, Terry A. "A Cluster Analysis Model for Predicting Visual Clusters." *Cartographic Journal* 21 (1984): 103–11.

Tufte, Edward R. The Visual Display of Quantitative Information. Cheshire, CT: Graphics Press, 1983.

United States Geological Survey. *The National Atlas of the United States of America.* Washington, DC: USGPO, 1970.

CHAPTER
10
MAPPING GEOGRAPHICAL VOLUMES: THE ISARITHMIC MAP

CHAPTER REVIEW

Isarithmic mapping involves mapping a real or conceptual three-dimensional geographical volume with quantitative line symbols. This form of quantitative thematic map dates back to the mid-sixteenth century, when isobaths were first charted. Today, two forms are recognized: isometric and isoplethic. Each involves the planimetric mapping of the traces of the intersections of horizontal planes with the three-dimensional surface. The isarithm is placed by "threading" it through a series of control points at which magnitudes are assumed to exist. Control points are locations where measurements are taken or places are chosen to represent unit areas. All isarithmic maps contain error, as do other quantitative thematic maps, but the designer can learn to recognize potential sources of error and reduce their effects on the overall map. In the total map design, isarithmic lines should be placed at the top of the visual/intellectual hierarchy and made to appear as figures in perception. Legends should be clear and unambiguous and should specify isoline units. There are no particular manual production problems; machine-produced isometric and isoplethic maps have become quite common.

Six distinct kinds of quantitative thematic maps are treated in this book: choropleth maps, dot maps, proportional symbol maps, isarithmic maps, value-by-area cartograms, and flow maps. Of these, the isarithmic map may be the most difficult conceptually. Isarithmic mapping requires fresh thinking. In the past, this form of mapping has enjoyed greater popularity than it does now, but it is still used widely in both professional and popular publications. Isarithmic mapping is an important part of the cartographer's repertoire and should therefore be mastered.

THE NATURE OF ISARITHMIC MAPPING

The basic concepts, diversity, and history of isarithmic mapping are first introduced so that the full range of the activity and its product, the isarithmic map, will be better understood.

FUNDAMENTAL CONCEPTS

An **isarithmic map** is a planimetric graphic representation of a three-dimensional volume. (See Figure 10.1.) The graphic image or map that results from **isoline mapping** is a system of *quantitative line symbols* that attempt to portray the undulating surface of the three-dimensional volume. Regardless of the items being mapped or any complexities associated with the map content, a third dimension must exist or be assumed to exist if a mapping technique is to be called *isarithmic.*[1]

The model may be a scaled-down version of a real three-dimensional volume, or it may be an abstract mental construct representing some varying geographical distribution. On the one hand, for example, the lithosphere is a real geographical volume whose top forms what is called the *topographic surface.* It is possible to make a scaled-down model of the volume, or a portion of it. An attempt to map the surface, using quantitative line symbols, results in an isarithmic map. On the other hand, the mind can form a construct of a geographic quantity that has volume—say, population density—which is high in some areas and low in others. This abstract mental construct also can be mapped isarithmically. Either real or abstract models can provide the content for isarithmic maps, but the third dimension must be present or assumed.

Another requirement of the isarithmic technique is that the volume's surface be continuous in nature, rather than discrete or stepped.[2] Geographic phenomena such as the locations of factories are discrete—values do not occur between points. Temperature, however, exists everywhere, both at and between observation points. Some geographic phenomena, such as population density, can be assumed to exist everywhere and thus can be mapped isarithmically. (See Figure 10.2.)

It may be useful to compare the common frequency histogram and the frequency polygon. Chapter 5 presented the histogram, which graphically portrays the occurrence of statistical data using bars with horizontal tops. An extension of the histogram is the **frequency polygon,** which is used to represent statistical data that are continuous in nature. It is developed by joining the centers of each bar of the histogram with straight lines, forming a polygon. The model represented by the isarithmic map, whether of a real volume (lithosphere), or abstract (such as population density), can be likened to a three-dimensional frequency polygon.[3] (See Figure 10.3.)

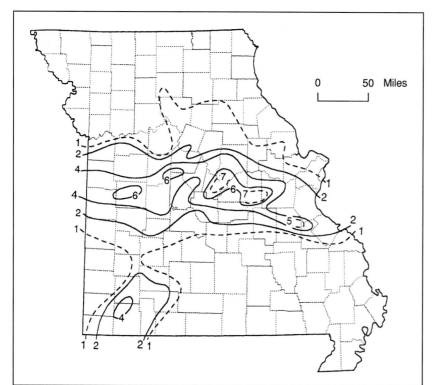

0   50   Miles

Figure 10.1   Isarithmic map.
Twenty-four hour rainfall, in inches, in central Missouri, ending July 7, 1993. This is an example of an isarithmic map. The isarithmic map is a planimetric representation of a three-dimensional volume. The isolines are traces of horizontal planes passing through the top surface of the volume a certain distance above the map's base. In this case the dashed isolines represent lines not part of the regular interval selected for the map.
(Source: United States Department of Commerce, National Weather Service, National Disaster Survey Report. The Great Flood of 1993 Washington, DC: USGPO, 1994, p. 3–23)

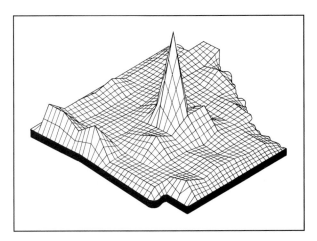

Figure 10.2 Arkansas population-density model.
Population density may be thought of as a continuous
phenomenon, so the surface of the three-dimensional model of it
can be mapped arithmically.

ISARITHMIC FORMS AND TERMINOLOGY

Isarithmic maps can have two distinct forms: the **isometric map** and the **isoplethic map.** The construction of each is achieved in similar fashion, but the nature of the data from which they are generated is quite distinct. Isometric maps are generated from data that occur or can be considered to occur at points; isoplethic maps result from mapping data that occur over geographic areas.

A distinct organization of data types and isarithmic forms has developed.[4] (See Figure 10.4.) Data that can occur at points, to be mapped isometrically, can be divided into *actual* and *derived* values. Actual data values include such things as temperature, precipitation, and elevation. Such values are generally obtained by recording instruments or by other means of point sampling in the field. Derived values are subdivided into two groups. One group includes such statistical measures as means and measures of dispersion; the other deals with such magnitudes as proportions and ratios.

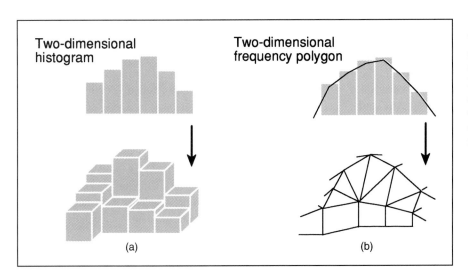

Figure 10.3 Surfaces.
The normal two-dimensional frequency
diagram or histogram in (a) produces a
three-dimensional stepped surface.
Isarithmic mapping can be more
appropriately compared to a three-
dimensional continuous surface that is
developed from a two-dimensional
frequency polygon, as in (b).

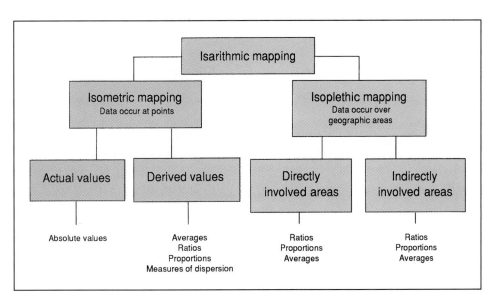

Figure 10.4 Data types
and isarithmic form.
Notice that absolute values are
never mapped isoplethically.

Isopleth maps are generated from data that occur over geographic areas called **unit areas.** The values that can be represented include ratios that directly or indirectly involve area—for example, population density or crop yield per acre. (See Table 10.1.) Professional cartographers do not illustrate absolute values isoplethically. If data are being used that represent areas, they must first be converted into ratios or proportions that involve the areal magnitude. For example, if population totals by census tract are being mapped, these should be changed into density values before mapping.

Isopleth mapping is more difficult conceptually than isometric mapping, especially for the beginner. It is not easy to conceive of an areal magnitude existing along a line. How can people per square mile exist in such a fashion? It is perhaps best not to think of these values in this way, but rather to envisage the line as a *surface element* below which is found the volume of population density. The following discussion of construction methods may also assist in understanding the conceptual basis behind this form of mapping.

THE BASIS OF ISARITHMIC CONSTRUCTION

The mechanical methods of isarithmic construction are straightforward but require a grasp of the basis behind the technique. Isarithmic mapping depicts the surface of a volume by quantitative line symbols. In both isometric and isoplethic mapping, construction of the lines begins by imagining a series of pins erected at the **data points** so that they extend vertically in proportion to the magnitude they represent at each point. (See Figure 10.5.) This is the vertical scale. The *tops* of all the pins will form the surface of the new volume. The line symbols are the planimetric traces of the intersections of hypothetical planes

with this three-dimensional undulating surface. The lines represent the surface, not the volume.

The vertical positions of the hypothetical planes are selected relative to a **datum** adopted for the particular map. The range of data values represented at the points dictates the value of the datum (usually zero). Each plane has an assumed value associated with it depending on its placement relative to the vertical scale (the same scale used when erecting the pins proportional to the magnitudes they represented). The magnitude or value of the isarithmic lines represents their vertical distance from the datum. Because the planes are constructed parallel to the datum, each isarithm will maintain an *unchanging magnitude* or vertical distance from the datum. Actually, these isarithms are traces of locations on the surface that are equally distant from the datum.

The total effect of these traces or isarithms on the surface is to show the varying amounts of the volume beneath the surface. The interpretation of the **pattern of isarithms** is the critical element in reading these maps. Pattern elements are *magnitude, spacing,* and *orientation.*

A BRIEF HISTORY OF ISARITHMIC MAPPING

Isarithmic mapping in each of its two forms has had a fairly long history, at least in relation to thematic mapping in general. **Isobaths** (isometric lines showing depth of the ocean floor) were first used as far back as 1584.[5] This early use was no doubt the result of the urgent need for reliable map information for commercial and military navigation.

In 1777, the **isohypse** line was proposed by Meusnier as a way of depicting surface features. An actual map using the isohypse was made by du Carla-Dupain-Triel in 1782. The isohypse is an isometric line; isopleths were not used until somewhat later. **Isogones,** lines showing equal magnetic

Table 10.1 Sample Ratios Useful in Isopleth Mapping		
Type	**Dividend and Divisor**	**Examples of Ratios**
Percentages, proportions	Same units, or with dividend a portion of the divisor	$\dfrac{20 \text{ acres wheat}}{200 \text{ acres cultivated land}} =$ 10% of cultivated land in wheat
Density	Different units, with the divisor the total area of the statistical division	$\dfrac{3,500 \text{ people}}{100 \text{ square miles}} =$ population density of 35 people per square mile
General ratios	Different units	$\dfrac{3,500 \text{ people}}{70 \text{ square miles of cultivated land}} =$ population density of 50 people per square mile of cultivated land

Source: Adapted from J. Ross MacKay, "Some Problems and Techniques in Isopleth Mapping," Economic Geography *27 (1951): 1–9.*

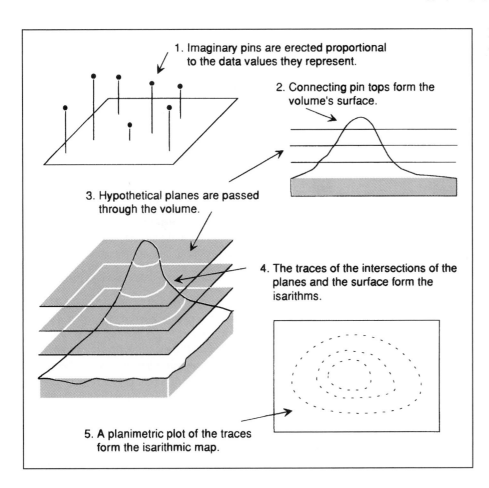

1. Imaginary pins are erected proportional to the data values they represent.

2. Connecting pin tops form the volume's surface.

3. Hypothetical planes are passed through the volume.

4. The traces of the intersections of the planes and the surface form the isarithms.

5. A planimetric plot of the traces form the isarithmic map.

Figure 10.5   The conceptual development of an isarithmic map.

declination, were first used around 1630 by Borri, an Italian Jesuit.[6] They were used on a thematic map by Edmond Halley in 1701. The renowned naturalist-scientist-geographer Alexander von Humboldt mapped equal temperatures using isometric lines called **isotherms.**

Professor Arthur Robinson, in tracing the genealogy of the isopleth, places its inception with Léon Lalanne, a Frenchman, in 1845.[7] Although based on the use of isometric lines, such as isobaths, isohypses, and isotherms, this marked a clear departure because point data representing area were part of the method. Lalanne's description of the method is still accurate today.

From these beginnings, the isometric and isoplethic techniques gained wide usage. Isopleth mapping came later and is clearly based on isometric examples. Isopleth mapping was adopted by American professional geographers late in the nineteenth century and earned a place of prominence during the early decades of this century.[8] Its use peaked just before World War II. During these early decades, the isopleth technique was most often used in agricultural mapping, showing intensity of crop production and rural population density.[9] The number of differently named isolines has reached sizable proportions. (See Table 10.2.)

Isoplethic mapping has declined among professional geographers during the last few decades. Not only has it been used less and less, but scholarly research concerning the isopleth has also declined. Several reasons for this decline have been given, the most notable of which is uncertainty of its scientific nature in the context of today's greater precision in mapping.[10] However, any thematic map is a form of generalization, so the isoplethic method is surely as conceptually correct as others.

The use of isometric maps has not declined appreciably and the conceptual basis on which they are formed has not received as much criticism as in the case of isopleths. The isometric map (and to a lesser degree the isoplethic map) was one of the first kinds to be produced by automated methods.

WHEN TO SELECT THE ISARITHMIC METHOD

Isarithmic mapping should be selected only if the advantages of its use contribute to achieving the goals of the mapping task. Certain additional requirements must be met before adopting this method:

1. The mapped data must be in the form of a geographical volume, or must be assumed to be voluminous, and must have a surface that bounds the volume.

2. It must be feasible to consider the mapped phenomena continuous in nature; discrete phenomena cannot be mapped isarithmically.

Table 10.2   List of Isoline Names

Isobath	Depth below a datum (e.g., mean sea level)
Isogonic line	Magnetic declination
Isocline	Magnetic dip (inclination) or angle of slope
Isohypse (contour)	Elevation above a datum (e.g., mean sea level)
Isodynamic line	Intensity of the magnetic field
Isotherm	Temperature (usually average)
Isobar	Atmospheric pressure (usually average)
Isohyet	Precipitation
Isobront	Occurrence of thunderstorms
Isanther	Time of flowering of plants
Isoceph	Cranial indices
Isochalaz	Frequency of hail storms
Isogene	Density of a genus
Isospecie	Density of a species
Isodyn	Economic attraction
Isohydrodynam	Potential water power
Isostalak	Intensity of plankton precipitation
Isovapor	Vapor content in the air
Isodynam	Traffic tension
Isophot	Intensity of light on a surface
Isoneph	Degree of cloudiness
Isochrone	Travel time from a given point
Isophene	Date of beginning of a plant species entering a certain phenological phase
Isopectic	Time of ice formation
Isotac	Time of thawing
Isobase	Vertical earth movement
Isohemeric line	Minimum time of (freight) transportation
Isohel	Average duration of sunshine in a specified time
Isodopane	Cost of travel time

Source: These names were selected from a larger list found in Norman J. W. Thrower, Maps and Man *(New York: Prentice Hall, 1972), Appendix B.*

3. The cartographer must fully understand the distribution being mapped. A casual acquaintance with the phenomenon will not suffice to develop a sensitive and accurate isarithmic solution to the mapped data. Extensive background research is generally required.

Various advantages to the isarithmic technique must be weighed in the selection process:

1. Isarithmic mapping shows the *total* form of a spatially varying phenomenon.
2. The method is commensurable (although less so at small scales) and graphic at the same time.

3. It is flexible and can easily be adapted to a variety of levels of generalization or degrees of precision.
4. The technique is easily rendered by using automated methods.

The cartographic designer chooses the isarithmic technique on the basis of these advantages, as weighed against those of other methods. Of course, such matters as data availability, base-map availability, and scale influence the ultimate selection. Beyond these considerations, other constraints are imposed by the method. Erwin Raisz, a noted cartographer and strong influence in the discipline during his life, once said, "Making true and expressive isopleth maps is something of an art and requires the best geographical knowledge of the region."[11] Another observer has expressed the same concern:

The drawing of an isopleth map is not a simple and straightforward procedure, although it is often treated as such. It is frequently subject to greater individual variations in judgment and skill than is usual with isometric maps. This arises from several different factors inherent in each type of map. In drawing an isometric map, the cartographer does not choose the positions of the numbers which he is plotting and he is often guided in his placement of lines by a knowledge of the distribution which he is mapping. Thus the draftsman of a contour map is aided by his "topographic sense." In isopleth mapping, the cartographer deals with ratios for areas and not points, and he cannot always rely upon any "sense" to assist him in the placement of his lines.[12]

ISARITHMIC PRACTICES

Like other forms of quantitative thematic mapping, the isarithmic variety contains elements and design strategies that are unique to the method. These are discussed in this section, along with sources of error in isarithmic mapping.

ELEMENTS OF ISARITHMIC MAPPING

The cartographer must master all elements of the isarithmic process, because they directly influence the quality of the finished map. The elements take on different significance in each mapping activity.

Placing the Isarithms—The General Case

In the construction of an isarithmic map, it is not usually necessary to erect pins or draw horizontal planes. Building a pin model is sometimes helpful in understanding a difficult, complex surface. The pin-and-plane model is, however, important conceptually. It should be thoroughly understood before beginning the planimetric placement of the isarithms with respect to the array of data points.

Methods of projective geometry form the basis of isarithm placement. (See Figure 10.6a.) The method assumes

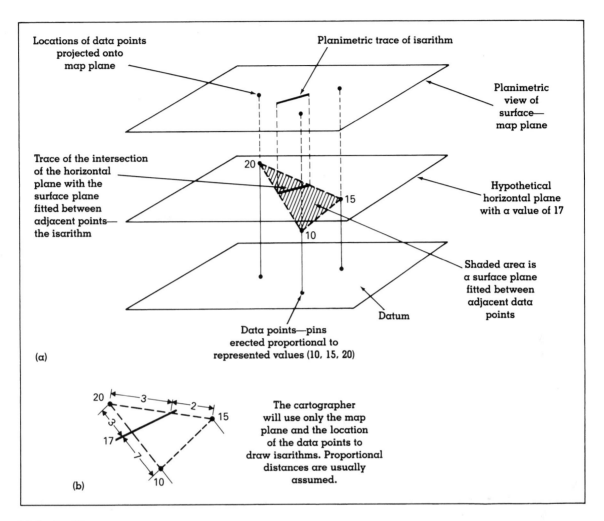

Figure 10.6   Isarithms.
Methods of projective geometry provide the basis for the planimetric location of isarithms.

that some generalized surface plane is interposed between adjacent data points. Where a hypothetical horizontal plane intersects this new surface plane, a trace is formed. The orthogonal projection of this trace is made to the map plane, intersecting the surface proportionally between the data points. In actuality the cartographer deals only with the map plane, the array of data points, and their values. (See Figure 10.6b.) The trace of the isarithm of a certain magnitude is ordinarily placed by assuming linear distances between the data points.

First, all values of the horizontal planes are chosen. The cartographer then completes the rough map by "threading" the isolines through the array of data points. In the next step, these sharp, straight-line segments are *smoothed*—generalized—to give the appearance of a continuously varying, undulating surface. (See Figure 10.7.)

This procedure describes how virtually all isarithmic maps are made. The most notable exception is placement of the isoline between the data points—a matter of *interpolation,* addressed more fully later in the chapter. Although the above procedure is generally followed, certain choices among the elements will affect the accuracy and appearance of the final map.

Locating Data Points

Data points, also called **control points,** have their locations specified by grid notation. Exact position can be determined either by an *x–y* rectangular coordinate reference or by geographic coordinates. Specification of location differs between isometric mapping and isoplethic mapping.

In isometric mapping, the positions of control points can usually be specified exactly, because the positions of recording instruments are well known. Problems usually relate to the uneven spacing of these points, because the lack of uniform spacing leads to varying precision in interpolation. In cases where the cartographer does have control over the pattern of spacing, the selection of a point pattern that yields a regular *triangular* net is most desirable.

Isopleth data point location is a more difficult procedure. A data point is usually selected for each enumeration district of the study and a value assigned which represents an average magnitude for each area. The magnitude is some ratio or proportion that directly or indirectly involves area. The location of the data point within the area is a matter of some concern, because it will affect the accuracy and appearance of the whole map.

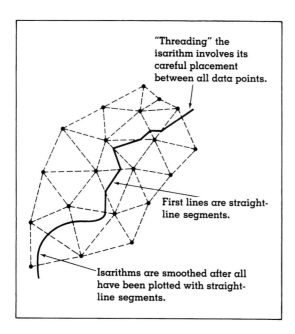

Figure 10.7   Plotting isarithms.
Isarithms are first drawn as straight-line segments and then smoothed before final rendering.

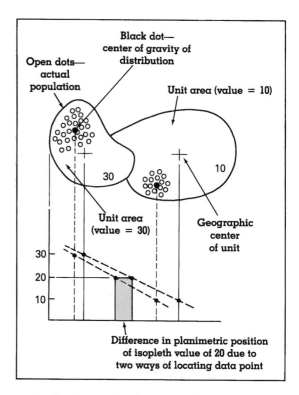

Figure 10.8   Data point placement affects the final map. Planimetric displacement may result from alternate ways of locating data points on isopleth maps. Over an entire map, the discrepancy can be considerable.

The actual location of the distribution affects the accurate position of the data point. Because the data point is selected to represent the distribution in the entire unit area, its location should reflect the spatial attributes of the distribution. If the unit area is regularly shaped and the distribution within it is known to be uniformly and evenly spaced, the geographical center of the unit area may be chosen for the location of the data point. The exact measurement of the center is not required; a visual approximation will suffice.

If the distribution is spatially skewed and clustered within the unit area, the location of the data point should reflect this pattern. The control point should be placed in the center of gravity of the actual distribution. When the geographic phenomenon is spatially skewed, the two distinct data point location schemes can bring about considerable difference in the placement of an isoline. (See Figure 10.8.) When applied over an entire map, the discrepancy can be quite extensive; completely different maps will result. Most cartographic scholars agree that a data point's location should be adjusted in the direction of the distribution. This requires a thorough understanding of the distribution being mapped.

Unit Areas in Isopleth Mapping

Unit areas in many isopleth mapping projects are civil enumeration districts or political area units: states, counties, minor civil divisions, and census tracts. In any given study, the cartographer must deal with several questions regarding the size, number (for a specified scale, number is a function of size), and shape of the unit areas.

Size and number are inseparable, and there is a functional relationship between them and overall map precision. Whenever a large number of unit areas comprise a map at small scale, the need for precision in locating isopleths drops off. Conversely, when a small number of units is used at a large mapping scale, the need for precision increases. Unfortunately, there are no quantitative guidelines for how many isopleths are required for a given level of precision at various map scales. Intuition, understanding of the distribution, and the purpose of the map determine the decision.

When unit areas are of the enumeration variety, it is good practice to use only those that are similarly shaped and sized. (See Figure 10.9.) Irregularity of size in particular can lead to uneven precision over the map and should be avoided. Considerations of control point location make it inadvisable to use unit areas of highly differing shapes. Especially for highly clustered distributions, the uneven spacing of control points can lead to a map containing varying levels of precision.

In rare cases, the cartographer can specify the shape and number of the unit areas prior to mapping. This generally occurs only in experimental designs in research. The best overall solution in such cases is to adopt a hexagonal pattern, for two reasons. First, this geometrical shape fills space. Second, connecting the centers of hexagons produces a triangular net, which is desirable in interpolation for isopleth location. (See Figure 10.9d.)

Interpolation Methods

Probably the key activity in isarithmic mapping is **interpolation** for isoline placement—a procedure for the careful positioning of the isolines in relation to the values of the data points. A number of options exist; with the introduction of automated mapping, these options are proliferating

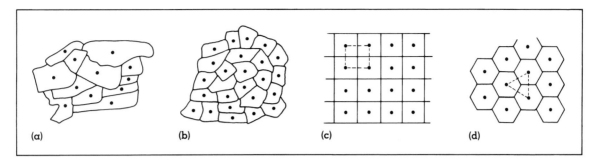

Figure 10.9   Different unit-area patterns.
Irregularly sized and shaped enumeration areas used as unit areas, as in (a), are less desirable than uniformly sized and shaped ones, as in (b). When unit-area shapes can be designated by the cartographer, square area, as in (c), should be avoided, since the pattern of control points leads to interpolation problems. A better solution is the hexagon-shaped unit area in (d), because the resultant data point pattern is triangular.

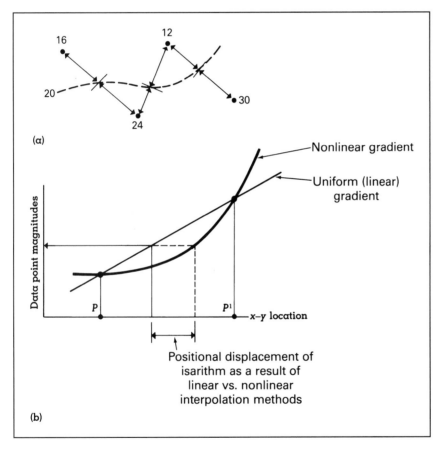

Figure 10.10   The interpolation model and the planimetric location of the isarithm.
Linear interpolation is standard in isarithm placement, as in (a). Planimetric displacement can result from the gradient selected for interpolation between adjacent data points, as in (b).

rapidly. For the sake of organization, this important subject is divided into manual and automated methods.

Manual   Through projective geometry, the trace of the intersection of a hypothetical plane with a surface segment results in a line that falls proportionally between the data points. This is true, but the case described in Figure 10.6 was a linear fit. In *linear interpolation* between data points on the map plane, the cartographer positions an isarithm with a certain value between adjacent control points at proportionate distances from each. (See Figure 10.10a.) The linear fit assumes that the gradient of change in magnitude between adjacent data points is even and regular. Equal change in the unit over equal horizontal distances produces a uniform gradient.

Adoption of this method assumes that the distribution being mapped changes in linear fashion, with a uniform gradient between data points. This may not be the case, however; many geographic phenomena do not behave this way. An example is population density in urban areas, which displays a more complicated pattern of change over distance.[13] Adopting a nonuniform gradient in preference to a linear one can result in considerable local displacement of the isoline. (See Figure 10.10b.) However, because of the underlying complexity of interpolating isoline position using a nonuniform gradient between all data points, and because the exact pattern of change from place to place is often unknown prior to mapping, most manual methods of interpolation assume uniform or linear change between data points.

This practice can be defended. Because linear interpolation is used throughout the whole map, the total form of the mapped distribution is not badly represented. Because most gradients of change are made up of both convex and concave portions along the slope, the linear or uniform gradient is a kind of compromise. Any error that results from its use is likely to appear randomly throughout the map.[14] Error resulting from the linear approach can be decreased by intensifying the density of data points, thus reducing their spacing and making the linear fit more precise between adjacent data points.

Another interpolation problem deals with choice of alternatives. This often results where the data points are arranged in rectangular fashion and two opposite data values are higher and two values are lower than the isoline being interpolated.[15] (See Figure 10.11a.) The only reasonable solution is to average each pair, then average these as a new value for the center of the data points. Triangular patterns (often difficult to find when enumeration unit areas are used) never result in this troublesome choice of alternatives and should be selected whenever possible.

Automated   The introduction of computer mapping and digital plotting has made rapid and iterative calculations possible, and isarithmic mapping has become much easier.

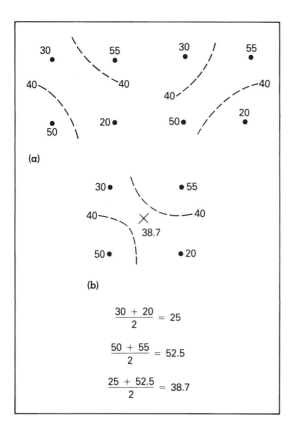

(a)

(b)

$$\frac{30 + 20}{2} = 25$$

$$\frac{50 + 55}{2} = 52.5$$

$$\frac{25 + 52.5}{2} = 38.7$$

Figure 10.11   Choice of alternatives in interpolation. When two opposite data points have higher values than two other opposite points, arranged in a square pattern, two interpolations are possible, as in (a). Averaging opposite pairs to determine an intermediate value, as in (b), is one solution to this dilemma.

The conceptual basis remains the same, but automation makes it possible to vary interpolation schemes in accordance with the mapping task at hand.

Computer-generated isarithmic mapping usually proceeds in two steps. **Primary interpolation** involves the calculation of data point values (z-values) at all locations on a fine-mesh grid matrix for the mapped area. This interpolation stage may include any one of a number of mathematical models: inverse-distance weighted, planar, quadratic, or cubic.[16] The next or **secondary interpolation** stage positions the isarithms with respect to the grid matrix of interpolated points. This stage locates the lines by linear interpolation and is generally accurate to within the tolerances of the plotter's capabilities.[17]

In typical manual methods of interpolation, the cartographer considers only the values of adjacent data points to determine the position of an isoline or a point to be interpolated. Computer interpolation during the primary stage is likely to embrace several points at a time. The inverse-distance weighting model is popular, having been used for some time, and now used in the popular computer-mapping program called SURFER.[18] The basis of this method is that points closer to the one being interpolated will have a greater effect on its value than points farther away. Any number of adjacent points may be used. A general formulation is[19]

$$Z_p = \frac{\sum\limits_{i=1}^{n} \dfrac{Z_i}{d_i}}{\sum\limits_{i=1}^{n} \dfrac{1}{d_i}}$$

where Z is the height (value) of the ith point, d_i is the distance from the point to the point being interpolated (Z_p), and n is the number of points used in the interpolation.

Computer-Generated Isarithmic Maps

Microcomputer-driven isoline map production is making rapid advances and offers to lessen the burden of manual production of these map forms. It is possible with only modest cost to add contouring programs to the workstation, and this increases considerably the options for the designer. It is the flexibility contained in such programs that expands the view for cartographers and is providing them with an incentive to do more of this kind of geographic mapping.

The SURFER program (a trademark of Golden Software, Inc., of Golden, Colorado) is discussed here. This program produces isarithmic maps, three-dimensional fish-net-type plots, and shaded relief maps from irregularly spaced data points (each containing x-, y-, and z-values). In the latest version of SURFER, the number of data points (triplets) that may be processed is limited only by the available memory in the computer. In the GRID portion of SURFER, a grid is superimposed over the array of data points and the z-values at the intersections of the grid are interpolated. The user can specify any number of interpolations for the gridding method, including the Kriging (regional) solution, which takes more time. In the inverse-distance option, any weighting power (the default is 2) greater than 1 and equal to or less than 10 may be chosen. Search-area parameters may also be set.

TOPO is the part of SURFER that produces contour maps and these lines may be either isoplethic or isometric, depending on the data used. The method of producing the isolines, of course, applies to either case. (See Figure 10.12.) TOPO is interactive and the user may select such features as contour interval, contour limits, frequency of labeled lines, smoothed or unsmoothed lines, dashed lines, hachure marks, line label sizes, ticks and labels at study area boundaries, and other variables. Original data points may or may not be posted. Map size can be varied for screen viewing or for computer peripheral devices, and a title may be added.

Gridded point values produced in GRID may also be used as input to SURFER which develops three-dimensional fish-net plots of the gridded surface. (See Figure 10.13.) The viewing azimuth (with respect to the original grid) as well as the viewing angle (angular height above the horizon) may be selected by the user. Stacked contours or fish-net views are possible. Some of the menu options are perspective or orthographic projection, contour interval, vertical scaling and horizontal sizing, posted data points, top or bottom views of the surface (or both), and others. A title and directional legend may be added. A geographic boundary, or other line phenomena (such as streams and roads) may be added with a special file.

Plotted outputs from SURFER-TOPO are useful for creating isarithmic maps either as final or draft-quality versions. The cartographic designer, for example, might wish to use the TOPO plot simply as a fair drawing and enhance it for final reproduction art. In this way, the laborious task of manual isoline interpolation is given over to the computer, which is efficient. Or, as is more likely the case, the map file can be imported into another graphics program and enhanced there. The three-dimensional plots from SURF may be used to amplify isoline maps (e.g., to assist the map reader in understanding the form of the isarithmic surface), and they are especially useful in teaching the principles of contour map reading.[20]

SURFER operates in an IBM environment that requires only a minimum 320K of RAM, a DOS of 2.0 or higher, and at least one disc drive (Version 6 is now written for Windows 95 and NT (a trademark of Microsoft Corporation) operating systems. Output is to a monitor, plotter, or printer. Drivers are supplied with the program. Laser printers are supported if they contain an HPGL (Hewlett-Packard graphics language) interface. Data input is by keyboard (the data editor that is supplied is easy to use), ASCII file, SLYK (.SLK), Excel (.XLS), or from LOTUS (.WKS) files (a trademark of the Lotus Development Corporation). USGS DEM files can be input into SURFER as well, making the program extremely versatile. The sample maps used here (Figures 10.12 and 10.13) are produced by a 300-dpi QMS-810 laserjet printer (QMS is a trademark of Quality MicroSystems, Inc.). Maps may be produced in panel sections that, when attached together, can yield a map up to 32 in. (80 cm) square.

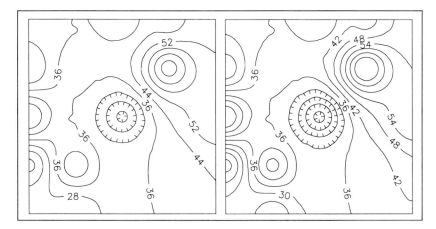

Figure 10.12   Computer-generated isoline maps are possible in desktop mapping. These two maps were produced using the SURFER program (a trademark of Golden Software, Inc., of Golden, Colorado).

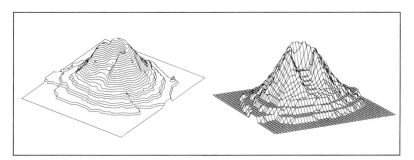

Figure 10.13   Computer-generated three-dimensional surfaces. These particular views are of the summit of Mt. St. Helens, after the eruption in 1989. These two drawings were produced using the SURFER program (a trademark of Golden Software, Inc., of Golden, Colorado).

Although cartographers now have rich resources such as SURFER to assist them in thematic map production, principles must be understood. Computer programs such as SURFER will not make decisions regarding the appropriateness of the method for the data at hand or help the cartographer as well they should.

The Selection of Isarithmic Intervals

It is impossible to map all isarithms because of space and time constraints. The cartographer is faced with the task of choosing the more appropriate ones and a suitable number of them. An **isarithmic interval** must be selected for the map.

The only reasonable and logical approach is to select a uniform interval. Isoline values of 2, 4, 6, 8, and 10 are appropriate, but 2, 5, 9, 14, and 20 are not. To put this another way, the hypothetical horizontal planes that pass through the three-dimensional surface should be *uniformly spaced vertically*. Only through this approach can the isolines show the total or integrated form of the three-dimensional surface. (See Figure 10.14.) This results in a particularly knotty problem at times, because most geographical distributions are skewed. In mapping population density, for example, extremely high values at urban centers lead to very crowded isopleths around cities. At best, the designer can choose map scale and isopleth interval carefully to avoid crowding; the result will be a compromise.

Another aspect of interval selection is the determination of the lowest isoline value. Because of the uniform interval, the lowest isoline greatly influences where subsequent isolines fall in relation to the whole map. Will the chosen interval show the distribution best? No universal rules apply here. As in so many other areas of thematic cartography, experimentation, geographical knowledge, study, and experience are necessary.

A graphic procedure can assist in interval selection. The **graphic array** method, mentioned in Chapter 7, plots the array of data point values in ascending order, along a horizontal axis to discover sharp changes in values in the data set. Isoline values are most revealing if they are selected to coincide with abrupt changes in slope on the graphic array. All places should be considered, and an interval chosen that best accommodates the places of rapid change and provides enough detail to show the form of the surface. An interval that provides too many isolines generally results in a map that looks "too accurate." A wide interval appears too generalized and reveals little. This is especially true in cases where a large number of isolines have been interpolated from a small number of control points.

In most cases, a wise choice of interval requires detailed knowledge of the mapped distribution. The cartographic designer should therefore do ancillary studies, plot rough dot maps of the distribution, or perform other

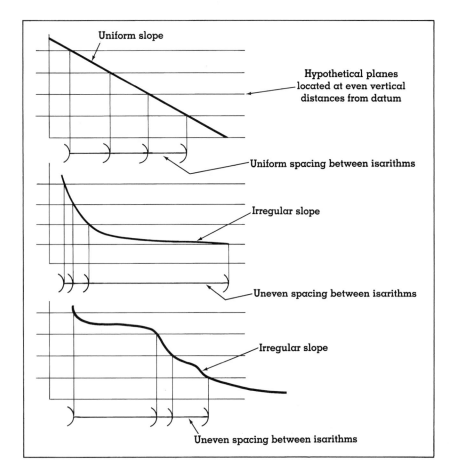

Figure 10.14 Isarithm spacing. The spacing of planimetrically viewed isarithms determines the gradient of change of the surface. Isarithms spaced close together indicate rapid change (steep slope), and those spaced farther apart suggest a relatively gradual slope. Experience in reading isarithmic maps is required.

research to learn the attributes of the distribution so that the isarithmic interval produces the best possible map. The final map represents only one of many possible solutions; it should therefore be based on all available information.

SOURCES OF ERROR IN ISARITHMIC MAPPING

Every thematic map contains some sort of error. Error can come from a variety of sources, four of which have been documented: CRM, method-produced, production, and map reading and analysis.[21] **CRM errors** result from collection, recording, and the manipulation of machines or techniques, and range from faulty recording instruments to the use of inappropriate statistical measures. **Method-produced errors** result from the cartographic technique. In isarithmic mapping, anything from the assumption to the placement of the isolines can lead to errors. **Production errors** are caused by the person or machine rendering the map: imprecise manual drafting or other graphic mistakes, incorrect registration, printing shortcomings, and the like. **Map reading and analysis errors** are generally caused by the interpretive limitations of the map reader. These range from psychophysical aspects to cognitive considerations.

Isarithmic mapping is subject to all these sources of error. The types and sources of error differ somewhat between isometric and isoplethic mapping, although some error sources are mutual to both kinds of mapping. Of course, the designer wishes to reduce error as much as possible, but it is not feasible to eliminate all map error.

The chief sources of CRM errors for isoplethic mapping relate to the quality of the data, which may be affected by observational bias or sampling errors. Observational error can be caused by either humans or machines. Faulty instruments or poor instrument-reading practices can lead to erroneous z-values. Bias or persistent errors may be the result of recording instruments that produce consistently high or low values. Sampling error is of considerable concern. Because the observations (z-values) used represent only one set from a much larger universe, the manner of selection (sampling) can become a source of error. In isoplethic mapping, the arrangement of the unit areas is usually not within the control of the cartographer; they are predetermined enumeration districts. The designer should therefore alert the map reader to the pattern of control points used. If the pattern of unit areas can be governed by the cartographer, one that yields a triangular pattern of control points is desirable because this will result in fewer interpolation errors.[22]

Method-produced errors for isarithmic maps, those caused by technique, differ between isoplethic maps and isometric maps. (See Table 10.3.) Sources of error can be found in the quantity of the data, the form of the data, and the interpolation model. In general terms, the quantity of data for both isometric and isoplethic maps is determined simply by the number of control points. Most cartographers agree that the greater the number, the less the potential for error. No definite number can be recommended for every mapping task, but 25 appears to be too few, and for most mappings a number between 49 and 100 is appropriate.[23]

No one mathematical interpolation model can be specified for all mapping cases. Certainly the model, whether linear or some other type, will affect the number of errors. It appears, at least for isoplethic mapping, that the linear interpolation model yields only *random* errors and therefore can be used with a certain degree of assurance.[24]

For isometric mapping, fewer errors result from an areal sampling method that distributes control points in a stratified, systematic, unaligned, scattered fashion, which is more uniform in spacing than random. In many instances, the distribution of data points is not controlled by the cartographer. The x and y positions are often determined by previously established networks of recording instruments.

The size and shape of the unit area and the method of assigning control points all affect the degree of accuracy of isopleth maps. "Map accuracy" may be a misleading term with such maps; there is no real measurable surface with which to compare the representation. Only general comments can be made about possible errors caused by the selection of the elements. If cartographers have control over their selection (and they most often do not), the preferable shape is hexagonal, for two reasons. First, a system of adjacent hexagons can completely cover a geographic area. Second, the centers of the hexagons, when connected by straight lines, form a triangular pattern, which results in fewer alternative choices during interpolation.

Table 10.3   Sources of Method-Produced Error in Isarithmic Mapping

	Map Type	
Error Source	**Isometric**	**Isoplethic**
Quantity of data	Number of data points	Number of data points
Form of data	Locational scatter of points	Size of unit area
		Shape of unit area
		Method of point assignment
Interpolation model	Mathematical formulation	Mathematical formulation

Source: Adapted from Joel L. Morrison, Method-Produced Error in Isarithmic Mapping. *Technical Monograph No. CA-5 (Washington, DC: American Congress on Surveying and Mapping, 1971), pp. 11–13.*

Most cartographers agree that the distribution should control the placement of control points. Less error is introduced if this plan is followed. The size of unit areas can affect accuracy and thereby reduce error, but specification of size is interconnected with map scale and the level of generalization planned for the map. At a given level, the greater the size of the unit area, the more potential for error.

The relationships of size, shape, and number of control points are critical to isopleth mapping and to potential error. At best, however, our knowledge is incomplete:

> At this stage of our knowledge, it is safe to conclude that the quality of isopleth mapping is greatly affected by the complexity of the original distribution, the variation in the pattern of unit areas (sizes and shapes), and the number of control points used. It is still not possible, however, to define precisely the manner in which these variables affect the fidelity of a map. For example, one cannot prepare a comprehensive summary which includes a complete classification of surfaces, unit area characteristics, and hexagonal patterns on the one hand and the probable measures of the quality of the isopleth maps on the other. The major problem seems to be that the essential spatial characteristics of quantitative distributions and unit area patterns have not yet been numerically defined, and further understanding is needed concerning the effects of sample point location and generalization of areal data on the precision of the isopleth map.[25]

Production errors in isarithmic mapping are caused by either manual or computer means during the preparation of the final map. Errors caused by registration problems, misaligned media, or imprecise drafting techniques can all lead to inaccurate final renderings. In computer mapping, the establishment of specifications that exceed the machine's capabilities cause final production errors.

An isarithmic map that is perfect in every technical sense—having as little error as possible—is still subject to error caused by the inability of the reader to interpret it accurately. Of those discussed, isoplethic maps are perhaps the most difficult to understand, because the logic behind the method is the most abstract. Map reading and analysis error can be reduced by training and experience.

PREPARING THE FINISHED ISARITHMIC MAP

Several design considerations related to the appearance of the completed isarithmic map must be learned. In addition, several production methods and media are especially useful in isarithmic map production.

DESIGN ELEMENTS

At least three design elements are usual in isarithmic maps: lines, labeling, and legend designs.

Making Isolines Appear as Figures

It is good design to make all isolines on the map appear dominant, as figures in perception. Thus they should be placed highest in the visual hierarchy. The easiest way to accomplish this is to render the lines as solid color, and screen much or all of the remaining map information. (See Figure 10.15a.) In this way, the total form of the distribution

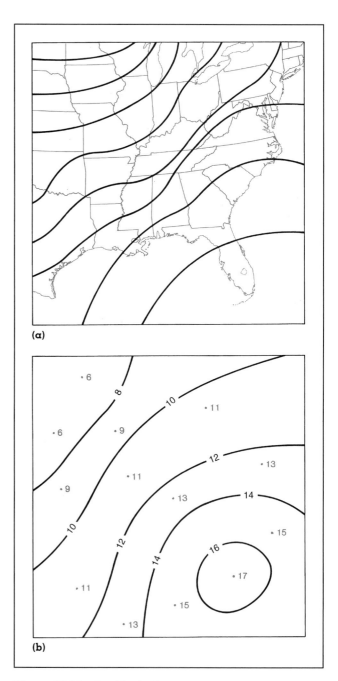

(a)

(b)

Figure 10.15 Isarithmic lines.
Isarithmic lines should be treated to make them appear dominant in map reading. In (a), the base material has been screened to reduce its contrast, thereby visually strengthening the isarithms printed in solid color. Data point locations and values may also be screened, as in (b), so that they will not interfere with the isarithms' function of showing total form.

becomes immediately recognizable, and the less important but necessary base-map material does not interfere. In some circumstances, it may be necessary to provide control point locations, and even their values, on the map. The points and values may then be screened to reduce their contrast and thereby lessen their figural strength. (See Figure 10.15b.)

If screening is not a design option, the importance of all isolines may be heightened by making the isolines noticeably heavier than other line work on the map. (See Figure 10.16.) As in other forms of thematic mapping, contrast should be developed to emphasize the visual and intellectual hierarchy.

Isoline Labels
Isolines should be labeled periodically to make reading easy. Do not overlabel; this reduces the statistical surface shown

by the lines. (See Figure 10.17.) Labels should be easy to read, but not so large as to dominate the map. Beginning students often make an easily avoidable mistake: placing labels upside down. Labels at the ends of isolines are appropriate, but each should be placed consistent with the flow of the line to which it is associated. This plan is violated in some places on the map in Figure 10.16. Can you see where? The recommended design strategy is provided in Figure 10.17.

Legend Design
Legends should be designed for clarity. The legend material most often contains only verbal statements. These must describe at least (1) the units of the isolines (e.g., frost-free days, days of sunshine, average annual precipitation in inches, persons per square mile) and (2) the isoline interval. In those instances when shading is applied to areas between isolines, the tints, the isoline interval, and the representative values can be combined in one legend. (See Figure 10.18.) The practice of applying shading between isolines is not recommended because it gives the impression of a stepped surface, which is very misleading.

Frequent questions are raised about shading between isolines, most notably regarding *hypsometrically tinted* elevation maps, where different colors are placed between the contours of elevation. By all accounts the method will probably continue. This, however, does not make the technique logically correct. By placing the *uniform areal colors* between the isolines the reader can get the impression of a stepped surface, which it is not. This author contends that it is improper to place uniform patterns or colors between isolines on planimetric isarithmic maps.

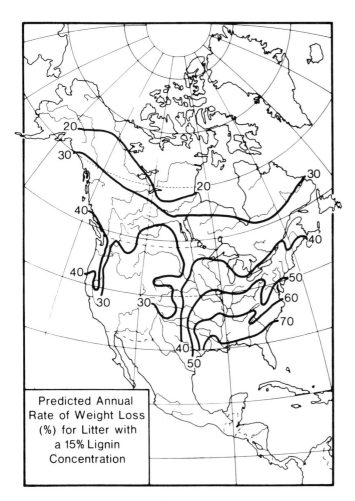

Figure 10.16   Isolines as figures.
Contrast of line weight is used to make isolines appear as figures on the map. On this map the isarithmic interval is 10 percent (rate of predicted annual weight loss). In most cases the cartographer has properly labeled the ends of the isolines, although in three instances this is not the case. See the text for further explanation of labeling practices. (*Reprinted by permission of the* Annuals [*Association of American Geographers*] *74* [1984]*: 557, fig.3, V. Meentemeyer.*)

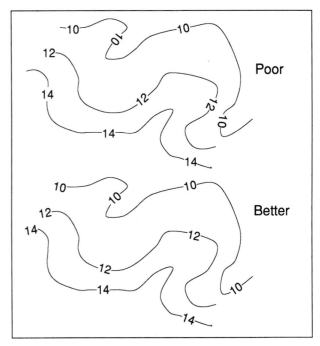

Figure 10.17   Design alternatives for labeling isarithms.

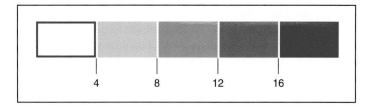

Figure 10.18 Legend design for isarithmic maps. If areas between isolines are covered with a pattern or tint, these should be incorporated into the legend; the isarithmic interval also can be shown.

PRODUCTION METHODS

Production of isoline maps does not involve particularly difficult or unique circumstances. For those not fortunate enough to have computer-production facilities, manual rendering methods include inking, scribing, and the use of flexible border tapes. When inking, it is advisable to employ either rigid plastic (French) curves or flexible curves to reduce undesirable irregularities in the line. An alternative to inking in positive art preparation is the use of crepe border tape, which is available in a variety of widths, easy to apply, and effective. Care must be taken, however, that fuzzy edges are eliminated when using these tapes. Flexible-head engravers are used in scribing and generally produce high-quality lines.

Machine production of isolines is usually accomplished with pen plotters, which apply ink to paper, or with automated scribe plotters. The latter can use regular scribe blades to etch the scribe film, and new versions of this process employ laser beams to etch the scribe film. Many software programs allow for the output to be sent to laser printers which produce very good quality prints on paper.

NOTES

1. Mei-Ling Hsu and Arthur H. Robinson, *The Fidelity of Isopleth Maps* (Minneapolis: University of Minnesota Press, 1970), p. 4.
2. Ibid.; and George F. Jenks, "Generalization in Statistical Mapping," *Annals* (Association of American Geographers) 53 (1963): 15–26. It may be pointed out that there are other ways of mapping continuous data although not in the line category. See for example, Richard E. Groop and Paul Smith, "A Dot Matrix Method of Portraying Continuous Statistical Surfaces," *American Cartographer* 9 (1982): 123–30; and Stephen Lavin, "Mapping Continuous Geographical Distributions Using Dot-Density Shading," *American Cartographer* 13 (1986): 140–50.
3. Calvin F. Schmid and Earle H. MacCannell, "Basic Problems, Techniques, and Theory of Isopleth Mapping," *Journal of the American Statistical Association* 50 (1955): 220–39.
4. Arthur Robinson, Randall Sale, and Joel Morrison, *Elements of Cartography,* 4th ed. (New York: Wiley, 1978), pp. 224–25.
5. The principal source for this discussion of the history of the isarithmic method is Arthur H. Robinson, "The Genealogy of the Isopleth," *Cartographic Journal* 8 (1971): 49–53; see also Arthur H. Robinson, *Early Thematic Mapping in the History of Cartography* (Chicago: University of Chicago Press, 1982).
6. Ibid.
7. Ibid.
8. Philip W. Porter, "Putting the Isopleth in Its Place," *Proceedings of the Minnesota Academy of Science* 25–26 (1957–58): 372–84.
9. Wellington D. Jones, "Ratios and Isopleth Maps in Regional Investigation of Agricultural Land Occupance," *Annals* (Association of American Geographers) 20 (1930): 117–95.
10. Arthur H. Robinson, "The Cartographic Representation of the Statistical Surface," *International Yearbook of Cartography* 1 (1961): 53–61; see also Mark P. Kumler and Richard E. Group, "Continuous-Tone Mapping of Smooth Surfaces," *Cartography and Geographic Information Systems* 17 (1990): 279–89.
11. Erwin Raisz, *Principles of Cartography* (New York: McGraw-Hill, 1962), p. 201.
12. J. Ross MacKay, "Some Problems and Techniques in Isopleth Mapping," *Economic Geography* 27 (1951): 1–9.
13. Truman A. Hartshorn, *Interpreting the City: An Urban Geography* (New York: Wiley, 1980), pp. 216–17.
14. Hsu and Robinson, *The Fidelity of Isopleth Maps,* p. 11.
15. J. Ross MacKay, "The Alternative Choice in Isopleth Interpolation," *The Professional Geographer* 5 (1953): 2–4.
16. Joel L. Morrison, "Observed Statistical Trends in Various Interpolation Algorithms Useful for First Stage Interpolation," *Canadian Cartographer* 11 (1974): 142–59.
17. Joel L. Morrison, *Method-Produced Error in Isarithmic Mapping* (Technical Monograph No. CA-5—Washington, DC: American Congress on Surveying and Mapping, 1971), pp. 15–16.
18. David Unwin, *Introductory Spatial Analysis* (New York: Methuen, 1981), p. 172.
19. Ibid.
20. William R. Buckler, "Computer Generated 3-D Maps: Models for Learning Contour Map Reading," *Journal of Geography* 87 (1988): 49–58.
21. Morrison, *Method-Produced Error,* pp. 3–4.
22. Hsu and Robinson, *The Fidelity of Isopleth Maps,* p. 8.

23. Morrison, *Method-Produced Error,* p. 61.
24. Hsu and Robinson, *The Fidelity of Isopleth Maps,* p. 11.
25. Ibid., p. 73.

GLOSSARY

control points   data points, p. 195

CRM errors   result from collection, recording, or the manipulation of machines and techniques, p. 201

data points   points at which magnitudes occur, or are assumed to occur, and from which isarithmic maps are constructed; also called *control points;* each point's magnitude is referred to as its *z*-value, p. 192

datum   the bottom-most horizontal base adopted for an isarithmic map; usually has a value of zero, p. 192

frequency polygon   a graph showing the frequency distribution of continuous data; a three-dimensional frequency polygon can be compared with geographical volume, p. 190

graphic array   a graph containing the plot of an array of data point values in ascending order; useful in selecting isoline intervals, p. 200

interpolation   procedure for the careful positioning of isolines between adjacent data points, p. 196

isarithmic interval   the vertical distance between the hypothetical horizontal planes passing through the three-dimensional geographic model; selection of the interval determines the degree of generalization and detail of the map, p. 200

isarithmic map   a planimetric graphic representation of a three-dimensional volume, p. 190

isobaths   isometric lines showing the depth of the ocean, p. 192

isogones   isometric lines showing equal magnetic declination, pp. 192–193

isohypse   isometric line showing the elevation of land surfaces above sea level, p. 192

isoline mapping   isarithmic mapping, p. 190

isometric map   one form of isarithmic map; made from data that occur at points, p. 191

isoplethic map   one form of isarithmic map; constructed from geographic data that occur over area, p. 191

isotherms   isometric lines showing equal temperature, p. 193

map reading and analysis errors   caused by the inability of the reader to interpret map accurately, p. 201

method-produced errors   result from the cartographic technique, p. 201

pattern of isarithms   arrangement of isarithms, especially their magnitude, spacing, and orientation, to reveal the surface configuration of the geographical volume, p. 192

primary interpolation   first step in machine interpolation of data points; involves the mathematical interpolation of all points on a fine-mesh grid adopted for the map, p. 198

production errors   result from either manual or machine rendering of the final map, p. 201

secondary interpolation   second step in machine interpolation of data points; involves the location of isolines with respect to values at points previously determined through primary interpolation; usually a linear model, p. 198

unit areas   geographic areas containing the statistical data from which isoplethic maps are generated, p. 192

READINGS FOR FURTHER UNDERSTANDING

Barnes, James A. "Central Areas and Control Points in Isopleth Mapping." *American Cartographer* 5 (1978): 65–69.

Blumenstock, David I. "The Reliability Factor in the Drawing of Isarithms." *Annals* (Association of American Geographers) 43 (1953): 289–304.

Buckler, William R. "Computer Generated 3-D Maps. Models for Learning Contour Map Reading." *Journal of Geography* 87 (1988): 49–58.

Groop, Richard E., and Paul Smith. "A Dot Matrix Method of Portraying Continuous Statistical Surfaces." *American Cartographer* 9 (1982): 123–30.

Hartshorn, Truman A. *Interpreting the City: An Urban Geography.* 2d ed. New York: Wiley, 1992.

Hsu, Mei-Ling. "The Isopleth Surface in Relation to the System of Data Derivation." *International Yearbook of Cartography* 8 (1968): 75–87.

————, **and Arthur H. Robinson.** *The Fidelity of Isopleth Maps.* Minneapolis: University of Minnesota Press, 1970.

Jenks, George F. "Generalization in Statistical Mapping." *Annals* (Association of American Geographers) 53 (1963): 15–26.

Jones, Wellington D. "Ratios and Isopleth Maps in Regional Investigation of Agricultural Land Occupance." *Annals* (Association of American Geographers) 20 (1930): 177–95.

Kraak, Menno-Jan, and Ferjan Ormeling. *Cartography: Visualization of Spatial Data.* Essex, England: Addison-Wesley Longman, 1996.

Lavin, Stephen. "Mapping Continuous Geographical Distributions Using Dot-Density Shading." *American Cartographer* 13 (1986): 140–50.

MacKay, J. Ross. "Some Problems and Techniques in Isopleth Mapping." *Economic Geography* 27 (1951): 1–9.

————. "The Alternate Choice in Isopleth Interpolation." *Professional Geographer* 5 (1953): 2–4.

————. "Isopleth Class Intervals: A Consideration in Their Selection." *Canadian Cartographer* 7 (1963): 42–45.

Monmonier, Mark S. *Computer-Assisted Cartography: Principles and Prospects.* Englewood Cliffs, NJ: Prentice Hall, 1982.

Morrison, Joel L. *Method-Produced Error in Isarithmic Mapping.* Technical Monograph No. CA-5. Washington, DC: American Congress on Surveying and Mapping, 1971.

———. "Observed Statistical Trends in Various Interpolation Algorithms Useful for First Stage Interpolation." *Canadian Cartographer* 11 (1974): 142–59.

Porter, Phillip W. "Putting the Isopleth in Its Place." *Proceedings of the Minnesota Academy of Science* 25–26 (1957–58): 372–84.

Raisz, Erwin. *Principles of Cartography.* New York: McGraw-Hill, 1962.

Robinson, Arthur H. "The Cartographic Representation of the Statistical Surface." *International Yearbook of Cartography* 1 (1961): 53–61.

———. "The Genealogy of the Isopleth." *Cartographic Journal* 8 (1971): 49–53.

———. *Early Thematic Mapping in the History of Cartography.* Chicago: University of Chicago Press, 1982.

———, **Randall Sale, and Joel Morrison.** *Elements of Cartography.* 4th ed. New York: Wiley, 1978.

Schmid, Calvin F., and Earle H. MacCannell. "Basic Problems, Techniques, and Theory of Isopleth Mapping." *Journal of the American Statistical Association* 50 (1955): 220–39.

Thrower, Norman J. W. *Maps and Man.* Englewood Cliffs, NJ: Prentice Hall, 1972.

Unwin, David. *Introductory Spatial Analysis.* New York: Methuen, 1981.

CHAPTER

11

THE CARTOGRAM: VALUE-BY-AREA MAPPING

CHAPTER PREVIEW

Erwin Raisz called cartograms "diagrammatic maps." Today they may be called cartograms, value-by-area maps, anamorphated images, or simply spatial transformations. Whatever name one uses, cartograms are unique representations of geographical space. Examined more closely, the value-by-area mapping technique encodes the mapped data in a simple and efficient manner with no data generalization or loss of detail. Two forms, contiguous and noncontiguous, have become popular. Mapping requirements include *the preservation of shape, orientation, contiguity, and data that have suitable variation. Successful communication depends on how well the map reader recognizes the shapes of the internal enumeration units, the accuracy of estimating these areas, and effective legend design. Complex forms include the two-variable map. Cartogram construction may be by manual or computer means. In either method, a careful examination of the logic behind the use of the cartogram must first be undertaken.*

We are accustomed to looking at maps on which the political or enumeration units (e.g., states, counties, or census tracts) have been drawn proportional to their geographic size. Thus, for example, Texas appears larger than Rhode Island, Colorado larger than Massachusetts, and so on. The areas on the map are proportional to the geographic areas of the political units. (Only on non-equal-area projections are these relationships violated.) It is quite possible, however, to prepare maps on which the areas of the political units have been drawn so that they are proportional to some space other than the geographical. For example, the areas on the map that represent states can be constructed proportional to their population, aggregate income, or retail sales volume, rather than their geographic size. Maps on which these different presentations appear have been called *cartograms, value-by-area maps, anamorphated images,*[1] and *spatial transformations.*

This chapter introduces this unique form of map. In these abstractions from geographic reality, ordinary geographic area, orientation, and contiguity relationships are lost. The reader is forced to look at a twisted and distorted image that only vaguely resembles the geographic map. Yet cartograms are being used more and more by professional geographers to uncover underlying mathematical relations, general models, and other revealing structures.[2] Cartographers likewise use them for communication of these ideas. Their eventual success as a communication device rests on the ability of the map reader to restructure them back into a recognizable form. Regardless of these complexities, cartograms are popular. Their appeal no doubt results from their attention-getting attributes.

THE VALUE-BY-AREA CARTOGRAM DEFINED

All **value-by-area maps,** or **cartograms,** are drawn so that the areas of the internal enumeration units are proportional to the data they represent. (See Figures 11.1 and 11.2.) This method of encoding geographic data is unique in thematic mapping. In other thematic forms, data are mapped by selecting a symbol (area shading or proportional symbol, for example) and placing it in or on enumeration units. In the area cartogram, the actual enumeration unit and its size carry the information.

Value-by-area cartograms can be used to map a variety of data. Raw or derived data, at ratio or interval scales, census data, or specially gathered data can be mapped in a cartogram. Because of the method of encoding, there is no data generalization. No data are lost through classification and consequent simplification. In terms of data encoding, the value-by-area cartogram is perhaps one of the purest forms of quantitative map, because no categorization is necessary during its preparation. Unfortunately, data retrieval is fraught with complexity, and readers may experience confusion because the base map has been highly generalized.

BRIEF HISTORY OF THE METHOD

As with so many other techniques in thematic mapping, it is difficult to pinpoint the beginning of the use of value-by-area maps. An early version was apparently used by Levasseur in his textbooks in both 1868 and 1875. To quote Funkhouser:

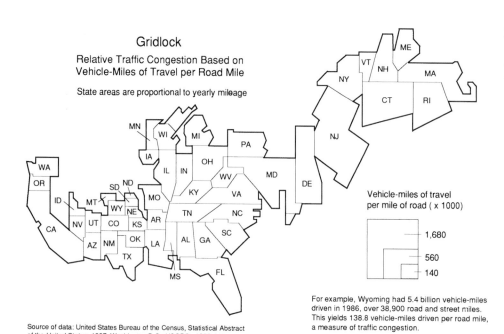

Gridlock

Relative Traffic Congestion Based on
Vehicle-Miles of Travel per Road Mile

State areas are proportional to yearly mileage

Vehicle-miles of travel
per mile of road (x 1000)

1,680

560

140

For example, Wyoming had 5.4 billion vehicle-miles
driven in 1986, over 38,900 road and street miles.
This yields 138.8 vehicle-miles driven per road mile,
a measure of traffic congestion.

Source of data: United States Bureau of the Census, Quatistical Abstract
of the United States, 1987. Washington D.C.: USGPO. 1986 data

Figure 11.1 Typical value-by-area cartogram.
(Cartogram designed by Bernard J. vanHamond. Used by permission.)

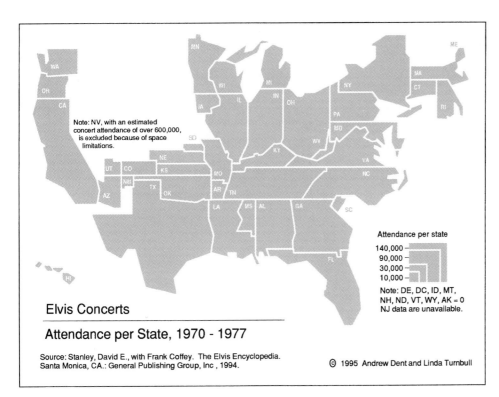

Figure 11.2 Elvis concerts attendance per state, 1970–77. A contiguous value-by-area cartogram showing unique data. This map reveals that unique and rarely mapped data can be the subject of cartogram mapping and can attract unusual attention. *(Map compiled by Andrew Dent and Linda Turnbull, Georgia State University. Used by permission.)*

These include colored bar graphs showing the number of inhabitants per square kilometer of the countries of Europe, the school population per hundred inhabitants, the number of kilometers of railroad per hundred square kilometer of territory, etc.; squares proportional to the extent of surfaces, population, budget, commerce, merchant marine of the countries of Europe, the squares being grouped about each other in such a manner as to correspond to their geographical position. (Author's emphasis)[3]

Although not called a value-by-area cartogram by Levasseur, the appearance of the actual graph seems to support the idea that it was indeed such a cartogram. Others have traced the idea of the cartogram to both France and Germany in the late nineteenth and early twentieth centuries respectively.[4] Erwin Raisz was certainly among the first American cartographers to employ the idea; he wrote on the subject 50 years ago.[5] Cartogram construction techniques were treated by Raisz through several editions of his textbook on cartography.[6] In 1963, Waldo Tobler discussed their theoretical underpinnings, most notably their projection system, and concluded that they are maps based on unknown projections.[7] Cartograms have been used in texts and in the classroom to illustrate geographical concepts; their role in communication situations has been investigated.[8]

Since their introduction, cartograms have been used in atlases and general reference books to illustrate geographical facts and concepts,[9] but no book has been devoted entirely to these interesting maps.

This chapter treats area cartograms only. Linear transformations (such as in Figure 1.8 in Chapter 1) are also possible, but are not discussed here.

TWO BASIC FORMS EMERGE

Two basic forms of the value-by-area cartogram have emerged: contiguous and noncontiguous. (See Figure 11.3.) Each has its own set of advantages and disadvantages, which the designer must weigh in the context of the map's purpose.

Contiguous Cartograms

In **contiguous cartograms,** the internal enumeration units are adjacent to each other. Although no definitive research exists to support this position, it appears likely that the contiguous form best suggests a true (i.e., conventional) map. With contiguity preserved, the reader can more easily make the inference to continuous geographical space, even though the relationships on the map may be erroneous. Making the cartogram contiguous, however, can make the map more complex to produce and interpret, for both manual and computer solutions.

Several advantages may be listed for the contiguous form:

1. Boundary and orientation relationships can be maintained, strengthening the link between the cartogram and true geographical space.
2. The reader need not mentally supply missing areas to complete the total form or outline of the map.
3. The shape of the total study area is more easily preserved.

The disadvantages of the contiguous form include:

1. Distortion of boundary and orientation relationships can be so great that the link with true geographical space becomes remote and may confuse the reader.

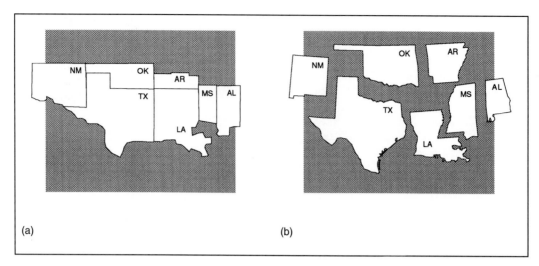

Figure 11.3   Contiguous and noncontiguous cartograms.
Contiguous cartograms like (a) are compact, and boundary relations are attempted. In noncontiguous cartograms, such as (b), enumeration units are separated and positioned to maintain relatively accurate geographic location.

2. The shapes of the internal enumeration units may be so distorted as to make recognition almost impossible.

Noncontiguous Cartograms

The **noncontiguous cartogram** does not preserve boundary relations among the internal enumeration units. The enumeration units are placed in more or less correct locations relative to their neighbors, with gaps between them. Such cartograms cannot convey continuous geographical space and thus require the reader to infer the contiguity feature.

There are nonetheless certain advantages in using noncontiguous cartograms:

1. They are easy to scale and construct.
2. The true geographical shapes of the enumeration units can be preserved.
3. Areas lacking mapped quantities (gaps) can be used to compare with the mapped units, for quick visual assessment of the total distribution.[10]

The disadvantages of the noncontiguous cartogram include:

1. They do not convey the continuous nature of geographical space.
2. They do not possess an overall compact form, and it is difficult to maintain the shape of the entire study area.

MAPPING REQUIREMENTS

Communication with cartograms is difficult at best, because it requires the reader be familiar with the geographic relations of the mapped space: the total form of the study area as well as the shapes of the internal enumeration units. This task may not be too difficult for students in the United States when the mapped area is their homeland and the internal units are states, but how many students in this coun-

try are familiar with the shapes of the Mexican states or those of the African nations? Likewise, are European students that knowledgeable about the shapes of the Canadian provinces or the states of the United States? On the other hand, by the very fact that they are unfamiliar with the mapped areas, map readers may pay more attention to the map than they otherwise would.

The situation can even be complex when mapping close to home. How many Tennessee residents know or could recognize the shapes of the counties in Tennessee? Georgia has 159 counties, Texas more than 200. Fortunately, most professional cartographers realize the futility of mapping little-known places with cartograms.

Cartograms can present a unique view of geographical space. Raisz stated many years ago that cartograms "may serve to right common misconceptions held by even well-informed people."[11] Harris and McDowell have suggested that the value-by-area map is a good way to teach about geographical distributions.[12] Tentative evidence indicates that map readers can obtain information from value-by-area maps as effectively as from more conventional forms. For this to happen, however, certain qualities of the true geographic base map must be preserved during transformation. The first of these is the **shape quality.** Preservation of the general shape of the enumeration units is so crucial to communication that the cartogram form should not be used unless some approximation of true shape can be achieved.

Conventional thematic maps are developed by placing graphic symbols on a geographic base map. Regardless of the form of the thematic presentation, the symbols are tied to the geographical unit with which the data are associated. Thus, for example, graduated symbols are placed at the centers of the states. Value-by-area maps, however, are unique in that the thematic symbolization also forms the base map. In a way, the enumeration units are their own graduated symbols, in addition to carrying the information of the conventional base map. On an original geographic base map of, for example, the United States, each state contains four kinds of information—size, shape, orientation, and contiguity. (See Figure 11.4.) In value-by-area mapping, only size is transformed; the other elements are preserved as nearly as possible. Contiguity is somewhat special and may not be as important as the others in map reading.

Individual unit shapes on the cartogram must be similar to their geographical shapes. It is through shape that the reader identifies areas on the cartogram. Shape is a bridge that allows the reader to perceive the transformation of the original. (See Figure 11.5.) If the reader cannot recognize shape, confusion results and comprehension is difficult, if

not lost altogether. The designer's problem is deciding how far it is possible to go along a continuum between shape preservation and shape transformation before the enumeration unit becomes unrecognizable to the majority of readers.

Geographical **orientation** is another important element in value-by-area mapping. Orientation is the internal arrangement of the enumeration units within the transformed space. Because the reader must be familiar with the geographic map of the study area to interpret a cartogram properly, the cartographer must strive to maintain recognizable orientation. When distortion of internal order occurs, communication surely suffers. How frustrating it would be to see Michigan below Texas!

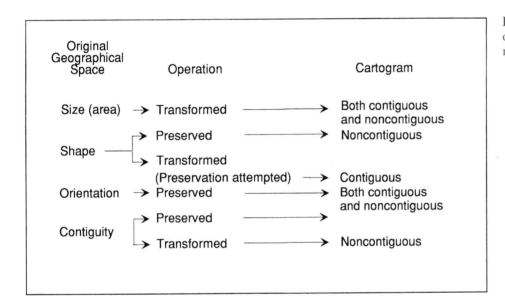

Figure 11.4   Ideal cartographic operations in value-by-area mapping.

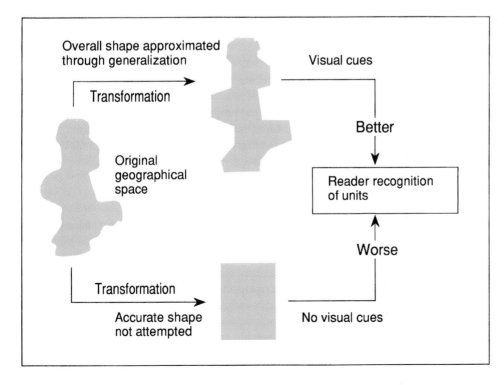

Figure 11.5   The importance of shape in cartogram design. Shape preservation provides necessary visual cues for efficient reader recognition of original spatial units.

Contiguity as an element in cartogram development relates, of course, only to the contiguous form. When producing this kind, it is desirable to maintain as closely as possible the original boundary arrangement from true geographical space. Of the elements mentioned thus far—shape, order, and contiguity—it appears that contiguity is the least important in terms of communication. It is likely that map readers do not use understanding of geographic boundary arrangements in reading cartograms. How many of us, for example, know how much of Arkansas is adjacent to Texas? On noncontiguous varieties, of course, contiguity per se cannot be preserved. It is possible, however, to maintain loose contiguity by proper positioning of the units, although gaps remain between the units.

Of the qualities mentioned (shape, order, and contiguity), shape is by far the most important. Use the value-by-area cartogram technique only where the reader is familiar with the shapes of the internal enumeration units. Do not overestimate the ability of the reader in this regard. Well-designed legends can be helpful, as discussed later in this chapter.

Data Limitations

Although value-by-area maps present numerous possibilities for the communication of thematic data, they are not without their limitations. Within the three principal ways of symbolizing data for thematic maps—point, line, and area—cartograms fall most comfortably into the category of area. Area is the element that must vary within the cartogram, so there are obvious limits outside of which one should not attempt this kind of representation. The limits are dictated by the data and their variability. It would be fruitless to map data that are exactly proportional to the areas of the enumeration units of the geographic base. (See Figure 11.6.) The cartogram would then replicate the original. At the other extreme, there could be a single enumeration unit having the same area as the entire "transformed" space, in which case no internal variation would be shown. No cartogram (or any other map) would be needed. Within these general limits, there exists a range of possibilities.

The chief goal of the cartogram is to illustrate a thematic distribution in dramatic fashion, which requires that the data be compatible with the map's overall purpose. The data set should be compared to the enumeration units on the geographical base. If reversal is evident (large states having small numerical value or vice versa), the cartogram is likely to be worthy of execution. Two measures, the linear regression and rank-order correlation indices, provide a degree of quantitative support. Unfortunately, these methods fall short in that they provide only overall indices of association; they do not indicate variation or agreement between data pairs within the total set. A statistical regression analysis may prove useful, but arbitrary limits must still be selected. More informal ways of determining appropriateness are easily workable.

Whatever procedure is chosen to determine the appropriateness of a data set for cartogram construction, such a determination should always be made before such a map is

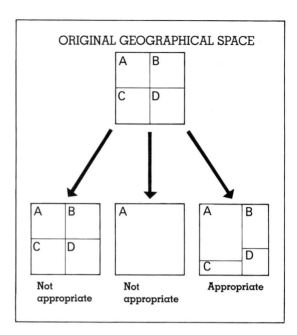

Figure 11.6   Data limitations and value-by-area mapping. If the original data lead to spatial transformation that is unchanged from the original (as on the left), the value-by-area technique is inappropriate. Also inappropriate would be those cases resulting in only one enumeration unit remaining after transformation (as in the center). Most suitable would be those instances when original data are transformed into new spatial arrangements dramatically different from the original (as on the right).

begun. Those not familiar with such maps often launch into a construction, only to find the results rather disappointing. If the map does not illustrate the distribution in a visually dramatic way, it is best abandoned.

COMMUNICATING WITH CARTOGRAMS

Success in transmitting information by the value-by-area technique is not guaranteed. There are at least three problem areas: shape recognition, estimation of area magnitude, and the stored images of the map reader. The designer should be familiar with the influences of each on the communication task.

RECOGNIZING SHAPES

It is by the shape of objects around us that we recognize them. We often identify three-dimensional objects by their silhouettes, and we can label objects drawn on a piece of paper by the shapes of their outlines. This holds true for recognition of outlines on maps. For example, South America can be seen as distinct from the other continents. The shape qualities of objects that make them more recognizable are simplicity, angularity, and regularity.[13] Simple geometric forms such as squares, circles, and triangles are easily identified. Shapes to which we can attach meaning are also easy to identify.

In the production of value-by-area maps, the cartographer ordinarily attempts to preserve the shapes of the enumeration units. How this is done is crucial to the effectiveness of the map. Many of the elements that identify the shape of the original should be carried over to the new generalized shape on the cartogram. The places along an outline where direction changes rapidly appear to be those that carry the most information about the form's shape.[14] Therefore, such points on the outline should be preserved in making the new map. These points can be joined by straight lines without doing harm to the generalization or to the reader's ability to recognize the shape. (See Figure 11.7.)

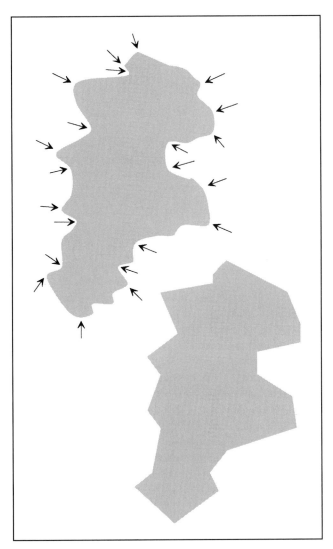

Figure 11.7   Straight-line generalization of the original shape.

Important shape cues are concentrated at points of major change in direction along the outline, as indicated here in the upper drawing. These points should be retained in transformation as a guide in the development of a reasonable straight-line generalization to approximate the original shape, as done here in the lower drawing.

ESTIMATING AREAS

Because each enumeration unit in a cartogram is scaled directly to the data it represents, no loss of information has occurred through classification or simplification. If any error results, it is to be found somewhere else in the communication process—most likely in the reader's inability to judge area accurately. The psychophysical estimation of area magnitudes is influenced by the shapes of the representative areas used in the map legend.

Research suggests that for effective communication of area magnitudes, the shapes of the enumeration units should be irregular polygons (not amorphous shapes) and that at least one square legend symbol should be used at the lower end of the data range.[15] It is best to provide three squares in the legend, one at the low end, one at the middle, and one at the high end of the data range. Of course, the overall communication effort may fail because the distortions from true shapes brought about by the method can interfere with the flow of information.

A COMMUNICATION MODEL

It has been stressed thus far that communicating geographic information with cartograms is difficult unless certain rules are followed. First, shape-recognition clues along the outline of enumeration units must be maintained. Second, if the cartographer cannot assume that the reader knows the true geographical relationships of the mapped area, a geographic inset map must be included. Third, the cartographer should provide a well-designed legend that includes a representative area at the low end of the value range.

These three design elements are placed in a generalized communication model of a value-by-area cartogram in Figure 11.8.[16] In this view, design strategies should accommodate the map-reading abilities of the reader. In Step 1, all the graphic components are organized into a meaningful hierarchical organization so that the map's purpose is clear.

Accurate shapes of the enumeration units are provided in Step 2 by retaining those outline clues that carry the most information—the places where the outline changes direction rapidly.

In the United States, people are exposed from early childhood to maps of the country through classroom wall maps, road maps, television, and advertising. Recently, satellite photographs have added to the already clear images of the country's shape in the minds of the population. How well these images are formed varies from individual to individual. Some people have well-formed images not only of the shape of the United States, but also of the individual states; others have difficulty choosing the correct outline from several possible ones. Successful cartogram communication may well rest on the accuracy of the reader's image of geographical space. Without a correct image, the reader cannot make the necessary match between cartogram space and geographical space. Confusion results if this connection is not made quickly.

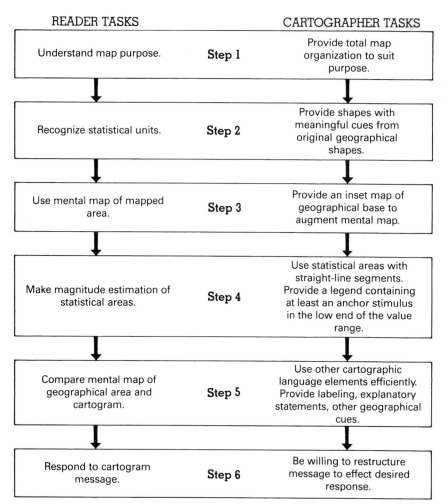

Figure 11.8 Cartographer and reader tasks in a generalized value-by-area cartogram communication model. Many of the steps are likely to occur simultaneously, not sequentially—especially Steps 2 through 5. *(Source: Borden D. Dent, "Communication Aspects of Value-by-Area Cartograms,"* American Cartographer 2 [1975]:*154–68.)*

In Step 3, the readers search through the represented geographic areas in an attempt to match what they see with their stored images.[17] Because the reader's stored images may be inaccurate, the designer should include a geographic map of the cartogram area in an inset map.

The map reader in Step 4 estimates the magnitudes of the enumeration units by comparing them with those presented in the legend. Effective legend design makes this task easier. Anchor stimuli in the legend should be squares, including at least one at the low end of the value range.

In Step 5, written elements, such as labels and explanatory notes, are included to assist the map reader in identifying parts of the map that may be unfamiliar at first. Finally, the designer should be willing to restructure the message to make the communication process better (Step 6). Inasmuch as the cartographer may not know what the reader thinks, because the cartagrapher and reader are usually separated in time and space, the first five tasks become even more important.

Advantages and Disadvantages

Unfortunately, cartograms have not been studied in enough detail to reveal exactly what impresses map readers about them or exactly how they are read. Preference-testing research has discovered that cartograms do communicate spatial information, are innovative and interesting, display re-markable style, and present a generalized picture of reality. Value-by-area maps are often stimulating, provoke considerable thought, and show geographical distributions in a way that stresses important aspects. On the other hand, they are viewed as difficult to read, incomplete, unusual, and different from reader's preconceptions of geographical space. Probably the most serious drawback is that no established methodology leads to consistent results. No two people devise identical cartograms of the same area. (This may be considered a strength rather than a drawback.) For the untrained map reader, the new configurations can cause visual confusion, detracting from the purpose of the map rather than adding to it.

The advantages of this thematic mapping technique are:[18]

1. To shock the reader with unexpected spatial peculiarities.
2. To develop clarity in a map that might otherwise be cluttered with unnecessary detail.
3. To show distributions that would, if mapped by conventional means, be obscured by wide variations in the sizes of the enumeration areas.

Disadvantages include:

1. Some map readers may feel repugnance at the "inaccurate" base map that results from the study.

2. Map readers may be confused by the logic of the method unless its properties are clearly identified.

3. Specific locations may be difficult to identify because of shape distortion of the enumeration areas.

TWO-VARIABLE CARTOGRAMS

The discussion thus far has concerned only the use of a single data set (variable), but it is possible to illustrate two or more data sets on a single cartogram. For example, on a cartogram of the United States in which the states are represented proportional to their populations, the cartographer can render individual states by gray tones, as on a choropleth map. The state areas may be represented as belonging to classes in another distribution. (See Figures 11.9 and 11.10.) This appears to be a very compatible representation of two distributions, as both relate to area. A choropleth map presupposes an even distribution throughout each enumeration unit, as does a cartogram. This form of **two-variable value-by-area cartogram** has been used successfully in mapping the spatial variation of socioeconomic data in Australian cities.[19]

Other second variables can be accommodated on cartograms by graduated point-symbol schemes. The second distribution can be represented by placing a graduated symbol within each enumeration unit of the cartogram. The reader must make the visual-intellectual comparison between the size of the enumeration unit and the size of the scaled symbol. This may be difficult for some readers at first. Although little research has been done on either method, it

would seem likely that they should be used only where there is a high degree of mathematical association between the two data sets. They certainly deserve further inquiry.

Another use related to two-variable mapping is to show how much of a total area is occupied by internal geographic divisions. (See Figure 11.11.) In this instance, the reader is asked to compare area proportions, and shape preservation is not often of central concern. The sizes of the internal areas are drawn proportional to the data being mapped.

CARTOGRAM CONSTRUCTION

There are two ways of producing value-by-area cartograms: manually and by computer technology. At present, more maps are probably generated by manual methods.

MANUAL METHODS

Manual techniques for the construction of value-by-area maps are quite simple. Suppose a cartographer wishes to construct a cartogram of total United States population. First, the total population is recorded for each state. The cartographer must then decide what the total area for the transformation is to be, and what proportion of the total population is represented by each state. Then the area for each state is computed on the basis of its share. (See Table 11.1.) Drafting can then begin. The cartographer must draft each state, preserving the shapes of the states while making their areas conform to the values computed. Of course, exact shapes are not preserved in contiguous cartograms.

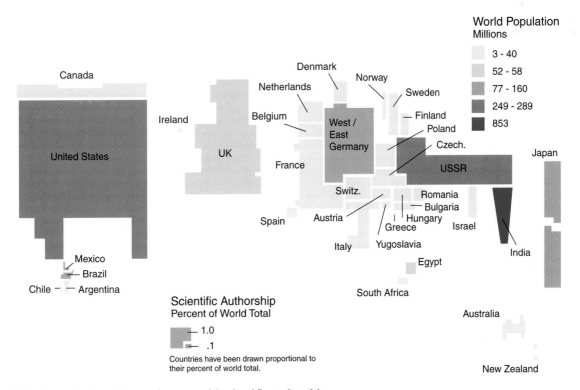

Figure 11.9   Contribution of countries to world scientific authorship.
(Source: Anthony R. deSouza, "Scientific Authorship and Technological Potential" (editorial), Journal of Geography *[July/August 1985]: 138. Reprinted by permission of the National Council for Geographic Education. Population layer added later and not part of the original map.*

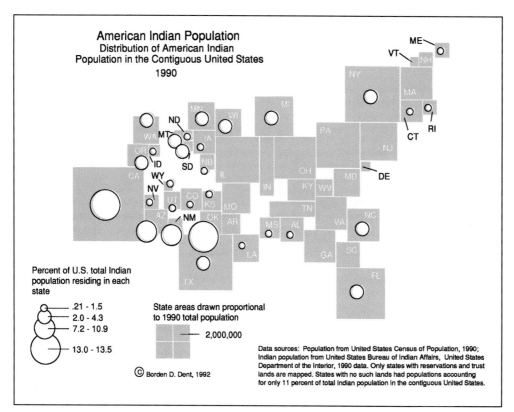

American Indian Population
Distribution of American Indian
Population in the Contiguous United States
1990

Percent of U.S. total Indian
population residing in each
state

.21 - 1.5
2.0 - 4.3
7.2 - 10.9
13.0 - 13.5

State areas drawn proportional
to 1990 total population

2,000,000

Data sources: Population from United States Census of Population, 1990;
Indian population from United States Bureau of Indian Affairs, United States
Department of the Interior, 1990 data. Only states with reservations and trust
lands are mapped. States with no such lands had populations accounting
for only 11 percent of total Indian population in the contiguous United States.

© Borden D. Dent, 1992

Figure 11.10 Value-by-area cartogram with superimposed distribution. Placing a second variable over a population cartogram may reveal interesting new patterns, or patterns not evident if mapped on geographical space. Experimentation is the key idea. Here it is clear that American Indians are concentrated in these states having relatively small total population (except California).

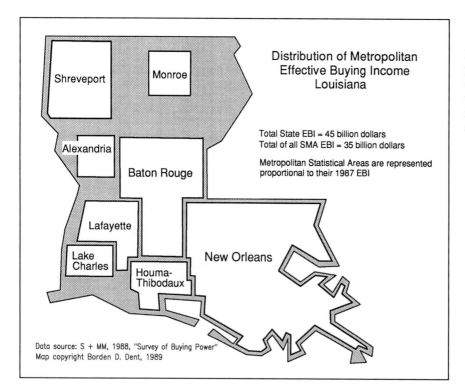

**Distribution of Metropolitan
Effective Buying Income
Louisiana**

Total State EBI = 45 billion dollars
Total of all SMA EBI = 35 billion dollars

Metropolitan Statistical Areas are represented
proportional to their 1987 EBI

Shreveport

Monroe

Alexandria

Baton Rouge

Lafayette

Lake
Charles

Houma-
Thibodaux

New Orleans

Data source: S + MM, 1988, "Survey of Buying Power"
Map copyright Borden D. Dent, 1989

Figure 11.11 Cartogram to show geographical proportion. In this presentation SMAs are drawn proportional to their buying power and are shown relative to the total buying power of the state. Shapes of the SMAs are not as important in this form of cartogram, although relative location is.

To facilitate the drafting of the states, it is convenient to begin by computing what some small areal division represents in terms of population. For example, the population of every .01 square inch can be calculated by dividing the total population into the total number of square inches determined for the cartogram (this unit size is selected simply because of the convenience of obtaining this grid paper). By dividing the population determined for each .01 square inch unit into the state's total population, the number of these .01 **counting units** can be ascertained. The cartographer need only

Table 11.1   Data Sheet for a Population Cartogram of the United States

State	1980 Population	Number of Counting Units	State	1980 Population	Number of Counting Units
Alabama	3,890,006	60	Montana	786,690	14
Alaska	400,481	6	Nebraska	1,570,006	25
Arizona	2,717,866	42	Nevada	799,184	14
Arkansas	2,285,513	35	New Hampshire	920,610	14
California	23,668,562	350	New Jersey	7,364,158	116
Colorado	2,888,834	46	New Mexico	1,299,968	21
Connecticut	3,107,576	49	New York	17,557,288	277
Delaware	595,225	9	North Carolina	5,874,429	91
Florida	9,739,992	154	North Dakota	652,695	11
Georgia	5,464,265	88	Ohio	10,797,419	168
Hawaii	965,935	15	Oklahoma	3,025,266	49
Idaho	943,935	15	Oregon	2,632,663	42
Illinois	11,418,461	175	Pennsylvania	11,866,728	186
Indiana	5,490,179	84	Rhode Island	947,154	14
Iowa	2,913,387	46	South Carolina	3,119,208	49
Kansas	2,363,208	35	South Dakota	690,178	11
Kentucky	3,661,433	57	Tennessee	4,590,750	74
Louisiana	4,203,972	67	Texas	14,228,383	242
Maine	1,124,660	18	Utah	1,461,037	25
Maryland	4,216,446	67	Vermont	511,456	11
Massachusetts	5,737,037	91	Virginia	5,346,279	84
Michigan	9,258,344	147	Washington	4,130,163	67
Minnesota	4,077,148	63	West Virginia	1,949,644	32
Mississippi	2,520,638	39	Wisconsin	4,705,335	74
Missouri	4,917,444	77	Wyoming	470,816	7

Total population (excluding District of Columbia and Puerto Rico) = 222,670,654. Total map area adopted in cartogram = 35 sq in. Counting unit size adopted for project = .01 sq in. Total number of counting units = 3,500. For each state, a ratio of the state's population to the national population was determined. The ratio was applied to the 3,500 total counting units to compute the number of units assigned to the state. For computation in this table, population figures were rounded to the nearest thousand.

arrange these small counting units until the shape of the state is approximated. (See Figure 11.12.) After the shape is achieved, the cartographer may wish to check the accuracy of the state's area by a quick planimeter measurement. Digital readout planimeters are available for such uses.

Each state's shape is adjusted and fitted to adjacent states until the cartogram is completed. The shape of the entire study area must be roughly preserved throughout. This is not difficult but is time-consuming and often frustrating. It is wise to construct the larger enumeration units first, then the smaller ones. If odd shapes result, the noncontiguous cartogram may be selected.

A question is often raised about how to treat enumeration areas with zero value. It is this author's opinion that having two or three areas with zero value should not prevent the map from being made. Those areas having zero values should be omitted from the cartogram, *but their names should be listed in a note at the bottom of the map as*

having zero values so that they could not be mapped. This informs the map reader that they were not *forgotten*. In a sense, this "other" space of the cartogram areas with zero value has simply collapsed. Perhaps there are other solutions, but this author knows of none.

Constructing a noncontiguous cartogram involves a slightly different procedure after computations are made. A conventional (generalized, if desired) base map is drawn. By using an optical reducer-enlarger, the states can be reproduced at their proportionate sizes relative to one that has the same size as on the true base map.[20] After the individual state areas are determined and rough shapes are formed, the cartographer positions the state outlines on a draft map to form the shape of the total study area. Relative geographical position of each state is sought. The newly sized states may be positioned in accordance with the centers of the states on a conventional map. Of course, the advantage of the noncontiguous form is the preservation of individual state shapes.

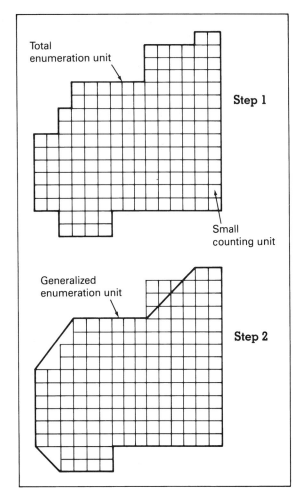

Figure 11.12 Constructing the cartogram. Small counting units are used to "build" the size and shape of the enumeration units (e.g., countries, states, counties) in Step 1. Step 2 involves smoothing to the approximate final shape.

COMPUTER SOLUTIONS

Computer programs are available for the generation of contiguous spatial transformations, notably one by Tobler,[21] and another by two Russian cartographers, Gusein-Zade and Tikunov.[22] The chief drawback of these programs is their inability to preserve shapes accurately, because the goal is to achieve contiguity and equal densities throughout. They also reduce flexibility in design. For the noncontiguous type, the size of polygons can be scaled in a variety of ways, including the use of optical reducer-enlarger projectors and photocopiers.

Cartograms, and especially computer solutions, can take many forms. Perplexed by what he thought to be inadequate mapping of the British census, social geographer and cartographer Daniel Dorling has experimented with a variety of forms to represent census statistics. He has said, for example, "The information in the census concerns not land but people and households. In visualizing these, a primary aim can be that each person and each household is given equal

> T heir main disadvantage [of cartograms] is that they are unfamiliar, but we do not learn from familiarity.
> Source: Daniel Dorling, "Map Design for Census Mapping," *The Cartographic Journal* 30 (1993): 167–183.

representation in the image."[23] His solution, which was facilitated by computer, was to draw a circle in each ward in Britain so that each circle was proportional to the population that it *represented.* Each circle was placed as nearly as possible to its original geographical neighbor as possible. This solution is quite unique and the final image provides a startling view of the population.

At least one author suggests that computer solutions may not be desirable because "the novelty of an automated approach may lead to intemperate haste in its utilization, whereby both the merits and weaknesses of topological transformation may be submerged in the deluge of products."[24] As in other computer applications in cartography, the machine can greatly reduce time and drudgery of production, but it must not replace or interfere with the designer's choices.

NOTES

1. V. S. Tikunov, "Anamorphated Cartographic Images: Historical Outline and Construction Techniques," *Cartography* 17 (1988): 1–8.
2. Peter Haggett, *The Geographer's Art* (Oxford, England: Blackwell, 1990), pp. 55–56.
3. H. Gray Funkhouser, "Historical Development of the Geographical Representation of Statistical Data," *Osiris* 3 (1937): 269–403: quotation from p. 355.
4. John M. Hunter and Jonathan C. Young, "A Technique for the Construction of Quantitative Cartograms by Physical Accretion Models," *Professional Geographer* 20 (1968): 402–6.
5. Erwin Raisz, "The Rectangular Statistical Cartogram," *Geographical Review* 24 (1934): 292–96.
6. Erwin Raisz, *General Cartography* 2nd ed. (New York: McGraw-Hill, 1948), pp. 257–58; and Erwin Raisz, *Principles of Cartography* (New York: McGraw-Hill, 1962), pp. 215–21.
7. Waldo R. Tobler, "Geographic Area Map Projections," *Geographical Review* 53 (1963): 59–78; see also Waldo R. Tobler, *Map Transformations of Geographic Space* (unpublished Ph.D. dissertation, Department of Geography, University of Washington, Seattle, 1961), p. 146.
8. Borden D. Dent, "Communication Aspects of Value-by-Area Cartograms," *American Cartographer* 2 (1975): 154–68.
9. There are numerous examples of such atlases. The following are particularly interesting: Tony Loftas, ed., *Atlas of the Earth* (London England,: Mitchell Beazley,

1972); Rezine Van Chi-Bonnardel, *The Atlas of Africa* (New York: Free Press, 1973); and Michael Kidron and Ronald Segal, *The State of the World Atlas* (New York: Simon and Schuster, 1981); cartograms have also been used to explore ways of presenting census data, as found in Danial Dorling, "Map Design for Census Mapping, *The Cartographic Journal* 30 (1993):167–83.

10. Judy M. Olson, "Noncontiguous Area Cartograms," *Professional Geographer* 28 (1976): 371–80.
11. Raisz, "The Rectangular Statistical Cartogram," pp. 292–96.
12. Chauncey Harris and George B. McDowell, "Distorted Maps. A Teaching Device," *Journal of Geography* 54 (1955): 286–89.
13. Borden D. Dent, "A Note on the Importance of Shape in Cartogram Communication," *Journal of Geography* 71 (1972): 393–401.
14. Ibid.
15. Borden D. Dent, "Communication Aspects," pp. 154–68.
16. Ibid.
17. Searching stored map images was addressed in an early paper: Borden D. Dent, "Postulates on the Nature of Map Reading" (paper presented at the annual meeting of the Georgia Academy of Science, 1976).
18. T. L. C. Griffin, "Cartographic Transformation of the Thematic Map Base," *Cartography* 11 (1980): 163–74.
19. Ibid.
20. Olson, "Noncontiguous Area Cartograms," pp. 371–80.
21. Waldo R. Tobler, "A Continuous Transformation Useful for Districting," *Annals* (New York Academy of Sciences) 219 (1973): 215–20.
22. Sabir M. Gusein-Zade and Vladimir S. Tikunov, "A New Technique for Constructing Continuous Cartograms," *Cartography and Geographic Information Systems* 20 (1993):167–73.
23. Dorling, "Map Design for Census Mapping," pp. 167–83; see also Daniel Dorling, "Visualizing Changing Social Structure from a Census," *Environment and Planning* 27 (1995):353–78; and Daniel Dorling, "Cartograms for Visualizing Human Geography," in eds. Hilary M. Hearnshaw and David Unwin, *Visualization in Geographical Information Systems,* (New York: Wiley, 1994), pp. 85–102.
24. Griffin, "Cartographic Transformation," pp. 163–74.

GLOSSARY

cartogram   name applied to a variety of representations; used synonymously with value-by-area map or spatial transformation, p. 208

contiguous cartogram   a value-by-area map in which the internal divisions are drawn so that they join with their neighbors, p. 209

counting unit   small spatial unit used in the manual preparation of value-by-area maps, p. 216–217

noncontiguous cartogram   a value-by-area map in which the internal divisions are drawn so that their boundaries do not join their neighbors; internal units appear to float in mapped space, p. 210

orientation   the internal arrangement of the enumeration unit within the total transformed region; cartogram communication relies heavily on the map reader's knowledge of the geography of the study area, p. 211

shape quality   a bridge allowing the reader to perceive the new value-by-area transformation of the original geographic base map; shape recognition is critical—without it, confusion results and communication fails, p. 210

two-variable value-by-area cartogram   a value-by-area map on which a second, related variable is mapped using area shading (chorograms) or graduated symbols, p. 215

value-by-area map   name applied to the form of map in which the areas of the internal enumeration units are scaled to the data they represent, p. 208

READINGS FOR FURTHER UNDERSTANDING

Burrill, Meredith. "Quickie Cartograms." *Professional Geographer* 7 (1955): 6–7.
Cole, John P., and Cuchlaine A. M. King. *Quantitative Geography.* London: Wiley, 1968.
Cuff, David J., John W. Pauling, and Edward T. Blair. "Nested Value-by-Area Cartograms by Symbolizing Land Use and Other Proportions." *Cartographica* 21 (1984): 1–8.
Dent, Borden D. "A Note on the Importance of Shape in Cartogram Communication." *Journal of Geography* 71 (1972): 393–401.
———. "Communication Aspects of Value-by-Area Cartograms." *American Cartographer* 2 (1975): 154–68.
Eastman, J. R., W. Nelson, and G. Shields. "Production Considerations in Isodensity Mapping." *Cartographica* 18 (1981): 24–30.
Getis, Arthur. "The Determination of the Location of Retail Activities with the Use of a Map Transformation." *Economic Geography* 39 (1963): 1–22.
Griffin, T. L. C. "Cartographic Transformation of the Thematic Map Base." *Cartography* 11 (1980): 163–74.
———. "Recognition of Areal Units on Topological Cartograms." *American Cartographer* 10 (1983): 17–28.
Haro, A. S. "Area Cartogram of the SMSA Population of the United States." *Annals* (Association of American Geographers) 58 (1968): 452–60.

Harris, Chauncey. "The Market as a Factor in the Localization of Industry in the United States." *Annals* (Association of American Geographers) 44 (1954): 315–48.

————, **and George B. McDowell.** "Distorted Maps, a Teaching Device." *Journal of Geography* 54 (1955): 286–89.

Hunter, John M., and Melinda S. Meade. "Population Models in the High School." *Journal of Geography* 70 (1971): 95–104.

————, **and Johnathan C. Young.** "A Technique for the Construction of Quantitative Cartograms by Physical Accretion Models." *Professional Geographer* 20 (1968): 402–6.

Kelly, J. "Constructing an Area-Value Cartogram for New Zealand's Population." *New Zealand Cartographic Journal* 17 (1987): 3–10.

Kidron, Michael, and Ronald Segal. *The State of the World Atlas.* New York: Simon and Schuster, 1981.

Loftas, Tony, ed. *Atlas of the Earth.* London, England: Mitchell Beazley, 1972.

Monmonier, Mark S. *Maps, Distortion, and Meaning.* Association of American Geographers, Resource Paper No. 75-4. Washington, DC: Association of American Geographers, 1977.

————. "Nonlinear Reprojection to Reduce the Congestion of Symbols on Thematic Maps." *Canadian Cartographer* 14 (1977): 35–47.

Olson, Judy M. "Noncontiguous Area Cartograms." *Professional Geographer* 28 (1976): 371–80.

Raisz, Erwin. "The Rectangular Statistical Cartogram." *Geographical Review* 24 (1934): 292–96.

————. *General Cartography.* New York: McGraw-Hill, 1948.

————. *Principles of Cartography.* New York: McGraw-Hill, 1962.

Rowley, Gwyn. "Landslide by Cartogram." *Geographical Magazine* 45 (1973): 344.

————. "The World: Upside Down, Inside Out." *The Economist,* December 22, 1984, pp. 19–24.

Tobler, Waldo R. *Map Transformations of Geographic Space.* Unpublished Ph.D. dissertation. Seattle: University of Washington, Department of Geography, 1961.

————. "A Continuous Transformation Useful for Districting." *Annals* (New York Academy of Sciences) 219 (1973): 215–20.

————. "Geographical Area and Map Projections." *Geographical Review* 53 (1963): 59–78.

Tufte, Edward R. *The Visual Display of Quantitative Information.* Cheshire, CT: Graphics Press, 1983.

Van Chi-Bonnardel, Rezine. *The Atlas of Africa.* New York: Free Press, 1973.

CHAPTER

12

DYNAMIC REPRESENTATION: THE DESIGN OF FLOW MAPS

CHAPTER PREVIEW

Maps that show linear movement between places are called flow maps and, because of this quality, are sometimes referred to as dynamic maps. Symbols on quantitative flow maps are lines, usually with arrows to show direction, that vary in width. Some flow lines are uniform in thickness and these are often referred to as "desire lines." Flow mapping began with maps done by Henry Drury Harness in 1837. Little flow mapping was done by government agencies in the United States in the nineteenth century, but with the impetus supplied by the study of economic geography, flow mapping became common in geography text- *books in the early decades of the twentieth century. Flow maps used in textbooks are divided into three classes: radial, network, and distributive. Particular attention must be paid to total map organization and figure-ground principles when designing flow maps because of their complex graphic structures. The projection for the flow map, line scaling and symbolization, and legend design are the chief design elements in flow mapping. Unique solutions, including computer graphics, should be explored in the design stages of producing a flow map.*

THE PURPOSE OF FLOW MAPPING

Maps showing linear movement between places are commonly called **flow maps,** or sometimes **dynamic maps.** Flow line symbolization is used when the cartographer wants to show what kind of (qualitative) or how much (quantitative) movement there is between two places. For the quantitative variety, the widths of the flow lines, or bands, connecting the places are drawn proportional to the quantity of movement represented. Any time the cartographer wishes to show movement between places, and has the data to support this theme, a flow map is appropriate. Its purpose may be to show movement of actual items, or movement of ideas.

In most instances, except for the *desire line* case, the cartographer attempts to show the actual route taken by the movement, although this may be difficult because of map scale and the level of generalization selected for the map.

Actually, because of the nature of the symbolization process, and the inherent complexity of these maps, all flow line maps become highly generalized. In the last edition of his book, Raisz places them into a class of maps called cartograms.[1] However, not to be confused with value-by-area cartograms, flow maps would be more correctly called linear cartograms.

Flow line symbols may be used on maps to show nonquantitative movement, as mentioned earlier. The lines are unscaled and generally have arrowheads to indicate direction of movement. Lines are usually of uniform thickness. Maps showing shipping routes, airline service, ocean currents, migration flows, and other similar presentations are examples. Early nonquantitative flow maps date back to the late eighteenth century. (See Figure 12.1.) It is possible to map different types of flow data on one map by using different line characters.

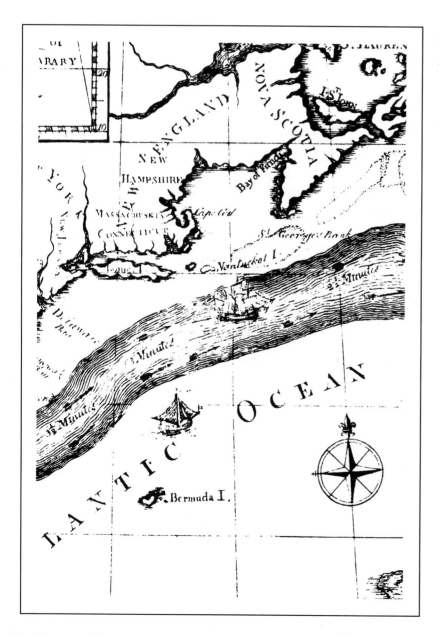

Figure 12.1 Chart of the Gulf Stream, commissioned by Benjamin Franklin, 1786. An early nonquantitative flow map.

QUANTITATIVE FLOW MAPS

Lines on quantitative flow maps are scaled such that their widths are proportional to the amounts they represent. (See Figure 12.2.) Weight, volume, value (dollar), and amount or frequencies are often the units used on quantitative flow maps. Data may be in nominal, ordinal, or interval levels. Absolute or derived values may be used. Direction may or may not be shown on the flow lines. International commodity flows are frequently mapped by quantitatively scaled lines, such as overseas movements of grains, ores, and produce. In many instances, the cartographer does not place amounts on the maps (as in a legend), but relies on the map reader to judge relative amounts visually. (See Figure 12.3.) However, direction of movement is frequently important, so arrowheads are likely part of the symbolization.

Traffic Flow Maps

Mapping traffic flows is uniquely suited to flow line symbolization and is the oldest form of this kind of map. Varying line widths are used to symbolize the number of vehicles passing over portions of rail, water, road, air, or even bird flyways. (See Figure 12.4.) The number of vehicles passing a certain point over the last 24 hours is a typical example. A state traffic map, for another example, may show flow bands proportional to the annual average 24-hour traffic volumes. It is common to see **traffic flow maps** without directional symbols. They are often used to show the organizational and hierarchical nature of urban systems.[2]

Desire Line Maps

Desire line maps represent a special case of quantitative flow map and are unique in that they do not attempt to portray the actual routes followed or the type of transportation used.[3] Typically, they illustrate social or economic interaction by use of straight lines connecting points of origin and destination. (See Figure 12.5.) Often desire lines represent the movement of one person between points (and a number of pairs are drawn on the same map), but they can be drawn between enumeration units and in that case symbolize aggregated movement data. In the former, each line has the same width, and in the latter, line widths are scaled to magnitudes as in typical flow mapping.

The application of desire line maps is best found in those instances where nodal geographical patterns are to be focused on, or possibly where urban hierarchies are being stressed.[4] They have been used frequently in this latter instance to show shopping or commuting structures.

Demarcation between flow maps and desire line maps is often not a clear division. Because of limitations of scale and the inherent generalization that takes place in symbolization, not to say anything about the lack of data,

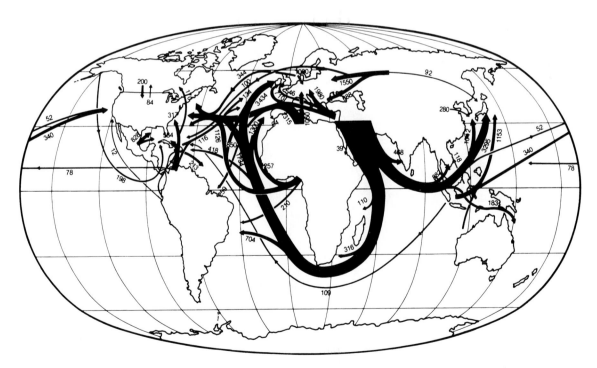

Arrows Indicate Origin and Destination
But Not Necessarily Specific Routes

Figure 12.2 International crude oil flow, 1980 (thousand barrels per day).
No legend is used on this map, but the lines are labeled to indicate the magnitude of flow. (*Source: U.S. Department of Energy, Energy Information Administration, 1981 International Energy Annual.*)

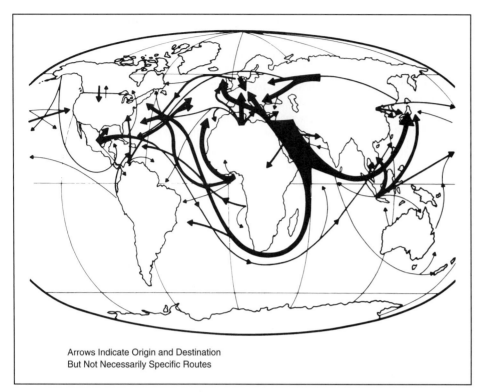

Figure 12.3 International crude oil flow, 1985.
No legend appears on the map and the magnitude of flow lines is not labeled. The reader is to gain an overall impression of the pattern of distribution. (*Source: U.S. Department of Energy, Energy Information Administration, 1986 International Energy Annual.*)

Arrows Indicate Origin and Destination
But Not Necessarily Specific Routes

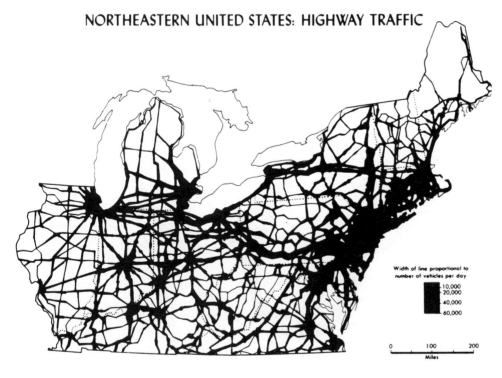

NORTHEASTERN UNITED STATES: HIGHWAY TRAFFIC

Width of line proportional to
number of vehicles per day

10,000
20,000
40,000
60,000

0   100   200
Miles

Figure 12.4 A typical traffic flow map.
In many instances such as this, direction of flow is not mapped. Width of lines shows number of vehicles passing in both directions for a specified time period. (*Reprinted by permission of Prentice Hall, Inc.*)

it is often impossible to show exact routes. Generalized, often diagrammatic, routes that may look like desire lines are selected for flow line maps, although the cartographer sets out to do something else. As with other forms of mapping, the purpose of the map will set the stage and ultimately dictate choices.

HISTORICAL HIGHLIGHTS OF THE METHOD

Our discussion here is limited by space and coverage, but a few remarks about the history of flow mapping are necessary to place this form of mapping in the correct historical

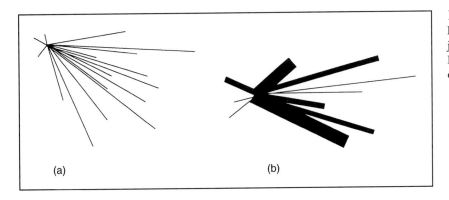

(a)

(b)

Figure 12.5 Desire line maps.
In (a), the lines are of uniform thickness (often joining origin and destination of the traveler). In (b), the desire lines vary in thickness based on aggregate interaction.

perspective. Quantitative flow mapping began in the "golden age" of statistical cartography in western Europe, in the two decades preceding the middle of the nineteenth century. Never prevalent in agency mapping in the United States, this form of map was utilized heavily in economic geography textbooks throughout the first half of the twentieth century.

EARLY FLOW MAPS

Early statistical cartography had its start in the late 1700s, although a few such maps appeared before that time. One notable entry before the late 1700s was the famous isogonic maps made by Halley in England in 1701.[5] Another was the first contour map showing depths in the English Channel made by Phillippe Buache in France in 1752.[6] By the late 1700s and early 1800s, however, a variety of statistical maps had begun to emerge. By 1835, western Europe had entered the "golden age" of geographic (statistical) cartography, a period lasting until roughly 1855.[7] It should be pointed out that little statistical mapping took place in the United States before 1850.[8]

The earliest quantitative flow maps apparently were done by Henry Drury Harness when he prepared the atlas to accompany the second report to the Railway Commissioners of Ireland in 1837.[9] On one map the relative number of passengers in different directions by regular conveyance was shown, and on the other, the relative quantities of traffic in different directions. In both instances the widths of the flow lines or bands were drawn proportional to the mapped quantities. Actually, the data were derived values. On the traffic conveyance map, the width of the lines is proportional to the average number of passengers weekly. These maps did not show exact routes, but showed straight lines of varying thicknesses connecting points (cities and towns). The maps of Harness remained unknown for nearly a century.[10]

Within 10 years of the production of the flow maps of Harness, Belpaire of Belgium and, more notably, Minard of France began publishing flow maps of essentially similar designs to those of Harness.[11] Charles Joseph Minard was more productive than Belpaire, and Minard's interests were primarily in the areas of economic geography. Minard apparently had no contact with geographers or cartographers, although he was instrumental in popularizing the flow line technique among statisticians. By his own account, he was interested in showing quickly, by visual impression, numerical accuracy,[12] so much so that he often overgeneralized other portions of his maps so that the flow lines themselves would command attention. This design aim is still relevant today. Although Minard did not invent the flow map, one contemporary authority has said that he did bring "that class of cartography [flow maps] to a level of sophistication that has probably not been surpassed."[13]

Minard produced some 51 maps, most of which were flow maps.[14] The mapped subjects of the flow maps were varied, and included such topics as people, coal, cereal, mines, livestock, and others. He mapped the distribution or flow of these commodities not only in France, but worldwide as well. His style, especially on the maps showing movement of travelers on principal railmaps in Europe (1862), is the same as is used today. Perhaps the most unique and provocative of his illustrations is the flow chart showing the demise of Napoleon's army in Russia; as suggested by Tufte, "It may well be the best statistical graph ever drawn."[15]

FLOW MAPS IN ECONOMIC GEOGRAPHY

From the time of Harness, Belpaire, and Minard quantitative flow maps have been used to map patterns of distribution of economic commodities, people (passengers), and any number of measures of traffic densities. As suggested earlier, the flow maps of Minard reached a sophistication of technique never really matched by others since. There was a period of time, however, in the decades of the first half of this century, when a great many quantitative flow maps appeared in college economic geography textbooks, and in many cases with laudable design techniques.

In a study of flow mapping in college geography textbooks by M. Jody Parks, several hundred flow maps were found among 71 books published between 1891 and 1984.[16] Although the author did not consult all books in this category in this time period (an enormous task), the study is remarkable for its breadth and attention to detail, and summarizes carefully the findings of the study and includes numerous examples of this kind of mapping. Although used extensively in this publishing medium, in general flow mapping lagged behind other forms of thematic mapping techniques.[17]

It is interesting to note from the study done by Parks that qualitative flow maps appeared in these books as early as 1891, and that this form outnumbered the quantitative form by three to one.[18] Qualitative flow maps are used to illustrate migration routes, explorer routes, and transportation networks. Desire line maps as a category of qualitative flow maps are used also, especially after 1960. The earliest quantitative maps, from those textbooks studied by Parks, appeared in 1912. Transportation themes are most often illustrated by the quantitative flow maps, especially international import and export of agricultural commodities.[19]

The study by Parks yields a classification of flow map design, including these three distinct patterns: radial, network, and distributive.[20] **Radial flow maps** are easily distinguished by a radial or spokelike pattern, especially when the features and places mapped are nodal in form. Present-day traffic volume maps fit into this category. (See Figure 12.6a.) **Network flow maps** are those used to reveal the interconnectivity of places, especially evidenced by transportation or communication linkages. (See Figure 12.6b.) Airline route maps are good examples.

Flow maps in the third class include those that present the distribution of commodities or migration flows. These are called **distributive flow maps.** Trade flows, such as shipments of wheat among countries, is a good example of this form. (See Figure 12.6c.) Maps that show diffusion of ideas or things are included in this class. Maps illustrating diffusion are often found in textbooks on cultural geography.

Topics mapped by the flow line technique are quite varied, and suggest the innovation often employed by cartographers and geographers in using this method. Railway and airline route maps are common (showing interconnectivity between places). Shipments of natural gas, wheat, animals, migration of people, ore, coal, and cotton are just a few examples of the topics commonly represented on flow maps. Trade flows of marine harvests is yet another. Migration routes, such as those taken by American Indians, and the French, Spanish, and English peoples in settling the New World, and other topics have been mapped by the flow map technique. There are few subjects that contain a from-to relation that cannot be mapped by flow symbolization.

This brief examination of flow maps in textbook cartography is intended to provide only a backdrop to the fascinating study of this form of mapping. Space will not permit a detailed presentation of hundreds of examples and the rich variety of design found among these maps. The student of thematic map design is urged to explore examples from the actual textbooks themselves. A worthwhile design activity can be found in such an experience.

DESIGNING FLOW MAPS

Creating effective flow maps through a careful and thoughtful design plan represents one of the more difficult challenges for the map designer. Three aspects of design must be considered: map organization and figure-ground (including the selection of the projection), line symbolization and data scaling, and legend design.

MAP ORGANIZATION AND FIGURE-GROUND

The map's hierarchical plan must be carefully considered. It seems apparent that the flow lines, or desire lines, are to be the most dominant marks on the map. As with other forms of thematic mapping, they should be placed high in the hierarchy so that they clearly stand out as strong figures. (The topic of the visual hierarchy and map design is dealt with in greater detail in the next section, "Designing Thematic Maps.") Several figures in this chapter, notably Figures 12.2 and 12.3, illustrate this idea. However, achieving this hierarchy is sometimes difficult because the flow lines may stretch over several different levels on the maps. For example, lines may first be over land, then water, then land again, and so on. They may intersect other flow lines, creating confusion for the reader.

As with other thematic symbol types, and as explained in the next section, flow lines should have strong edge gradients and be rendered so that visual conflict with other symbols does not result. (See Figure 12.7.) Interposition, which can be achieved easily by not having transparent symbols, is a technique often helpful in designing strong symbols. In the case here in Figure 12.7, rendering the flow lines black (or nontransparent white) is a way to improve the thematic symbols on this map.

Flow maps are ordinarily quite complex visual graphics. On most thematic maps the organization of the graphic

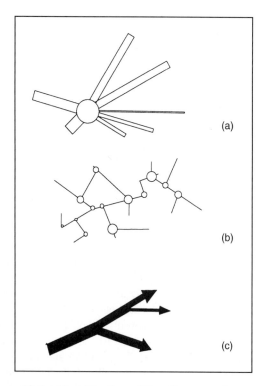

Figure 12.6   Classification of flow line maps in geography textbooks.

A radial type is illustrated in (a), the network type is represented in (b), and the distributive type in (c).

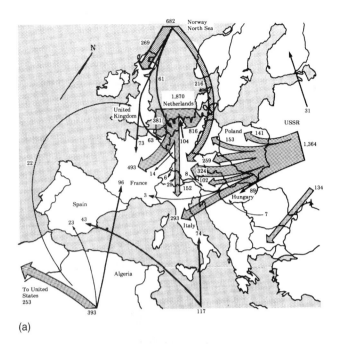

(a)

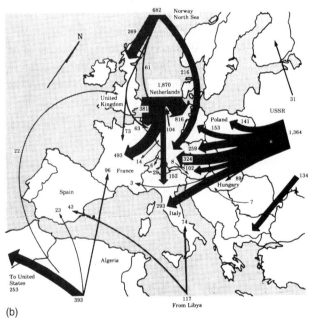

(b)

Figure 12.7 Flow lines that lack strong figure characteristics. The flow lines in (a) could be stronger figures by either eliminating the coastlines and boundaries beneath them or by rendering them as black symbols, as in (b). (*Source: U.S. Department of Energy, Energy Information Administration,* 1981 Annual Report to Congress, *vol. 2* Energy Statistics, *p.112.*)

components, land and water, symbols, titles, legends, and other marks on the maps tends to fall neatly into an easily followed plan. Land and water contrasts are developed by following figure-ground principles, and symbols usually occupy space over land areas. Titles and legends are dealt with similarly. But on many flow maps, especially those illustrating international movements, the thematic symbols are likely to occupy spaces over land or water, or both, and many times the lines themselves are intertwined and appear

to rest in different visual levels. The nature of the visual complexity on many flow maps, then, creates unusual and challenging design problems for the cartographer.

Minimally, the cartographer should provide clear land and water distinctions on flow maps. The flow symbols must be dominant figures in perception, with strong edge gradients and clear continuity. Labeling flow lines with their values, often useful to assist the reader, should not interfere with the symbol's visual integrity. Using patterns or screens on flow symbols should not lead to confusion with other areas on the maps. The scaling of the flow lines should not cause them to be too large for the maps, which can result in too little base-map information showing through. Attention to design details such as these will assist the cartographer in reaching successful results.

Projection Selection

Perhaps of equal importance to that of achieving a good visual hierarchy on the flow map is the necessity of selecting an appropriate projection. Placement of the center of the flow, if there is a center, must be strategically planned and this may require careful consideration of the projection, its center, and aspect. Placement and design of the flow lines should be done so that the map does not become an incomprehensible mesh of confusing lines. (See Figure 12.8.) For flow mapping, the equal-area and conformality attributes of projections may not be as important as other factors, such as continental shapes.[21]

Selecting a projection, adopting its center, choosing an aspect for it, and placing it in the map frame must all be considered when developing a flow map. The final map that results from these choices should be one that connotes organization and control of the image, and deftly satisfies the purpose of the map. It is the responsibility of the cartographer to know the flow pattern he or she wishes to portray, then make decisions regarding the projection that best illustrates this pattern. For example, if the flows are single origin to multiple destinations, or if the pattern reflects multinodal origins and a single destination, the employment of the projection should complement the pattern.

Balance and layout of the map's elements conclude the design activities related to achieving the visual and intellectual plans for the map. Placement of map objects and utilization of space are dealt with so that a pleasing result is reached, which is defined as one in which no other solution seems merited. The map's elements, as with other thematic map forms, are placed in intellectual order and treated graphically to satisfy the plan. Titles, legends, scales, source materials, and other elements are therefore treated accordingly. However, the thematic symbols—the lines themselves—remain the most important features of the map, and all other elements are second in importance.

Essential Design Strategies

A summary of the essential design strategies for flow maps should include these principles:

1. Flow lines are highest in intellectual and therefore highest in visual/graphic importance.
2. Smaller flow lines should appear on top of larger flow lines.

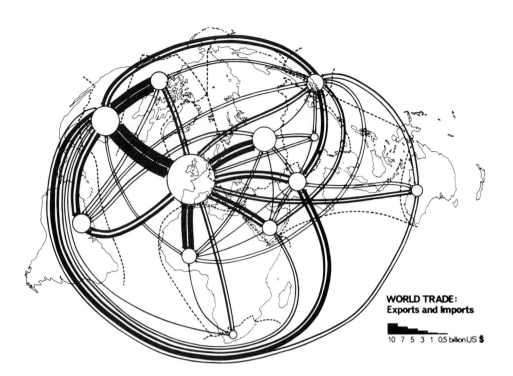

Figure 12.8 World trade: Net exports, 1967. The selection of the projection (uncommon in this case), its orientation in the map frame, and the nature of flow distribution all contribute to a confusing array of lines. (*Reprinted by permission of Van Nostrand Reinhold Company, Ltd.*)

3. Arrows are necessary if direction of flow is critical to map meaning.
4. Land and water contrasts are essential (if the mapped area contains both).
5. Projection, its center and aspect, are used to direct readers' attention to the flow pattern important to the map's purpose.
6. All information should be kept simple, including flow line scaling.
7. Legends should be clear and unambiguous, and include units where necessary.

LINE SCALING AND SYMBOLIZATION

On most quantitative flow maps the widths of the flow lines are proportionally scaled to the quantities they represent. Thus, a line representing 50 units will be five times the width of one symbolizing 10 units. This practice has been employed since the time of Harness, Belpaire, and Minard. Perceptually, the reader is being asked to make this visual judgment, and ordinarily most readers can do this in a linear fashion.[22] Scaling, therefore, appears to be straightforward and should not pose severe problems for the designer.

Chief among the concerns for the designer is the data range that must be accommodated by the widths selected. In many instances this imposes considerable restraint on the project. Ordinarily the best plan is to select the widest line that can be placed on the map (and still preserve the integrity of the base map), and then determine, by looking at the data range, the narrowest line on the map. If the widest line is 1.27 cm (.5 in), and the data range is 5,000 units, then the smallest line would be .000254 cm (.0001 in), obviously too small to draft on the map sheet. Either the widest line would have to be enlarged,

already determined to be unacceptable, or some other solution would need to be reached. It is important to note, too, that as the linear symbols become wider, they cease to appear as such, and may take on the qualities of area symbols. This limit should be avoided. There appear to be three ways out of this dilemma: abandon the map in favor of some other form, provide a standard line to represent values below a certain critical value (and symbolize the remaining values by conventional methods), or range grade the values and symbolize the resulting classes by scaling the flow lines proportionally to the midpoint of the classes. In some cases, both the second and third options can be combined on the same map. (See Figure 12.9.)

Other methods in addition to proportionally scaling line widths have been used to symbolize flows. One method is a dot method in which small dots, each representing a unit of flow

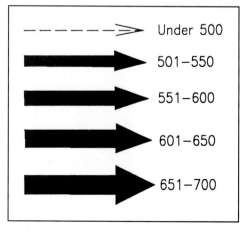

Figure 12.9 Use of a standard line in scaling flow lines. When the data range is too great to scale all flow lines proportionally, a standard line may be selected to symbolize several values.

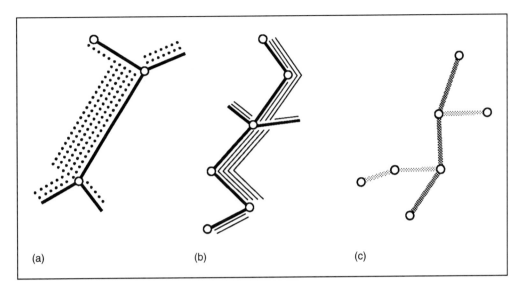

Figure 12.10 Alternative methods of symbolizing quantitative flow lines. In (a), dots symbolizing a unit of flow are placed alongside two actual routes. In (b), parallel lines are used, with each line representing a unit of flow. Finally, in (c), lines of uniform thickness are given area patterns based on numerical classes.

(a)   (b)   (c)

(for example, 100 vehicles), are placed along the route.[23] Visual impression of volume differences is quickly noted, routes are easily followed and, if necessary, the reader can count the dots to retrieve actual magnitudes. (See Figure 12.10a.)

One other method is to render several lines of uniform thickness parallel to the route, and provide the proper number of these lines to represent the volumes. (See Figure 12.10b.) This method yields results similar to the conventional method and, in addition, the individual lines can be counted for greater precision, if necessary. One disadvantage to this method is that compilation is more difficult and time-consuming.

Yet one other scaling solution can be reached by a method analogous to applying patterns to choropleth maps. In this method, broad lines of uniform thickness are drawn connecting points in the mapped space and, depending on the numerical class into which the line segment falls, are given an areal pattern.[24] (See Figure 12.10c.) Areal symbols are selected as they are in choropleth mapping; that is, higher values are represented by darker (higher percent area inked) patterns or screens. Color may be used instead of patterns or screens to represent the different classes. Here different color intensities of the same hue would be selected to represent the different classes. If the segments were classified into different nominal classes, however, different hues may be chosen.

This list of alternative scaling methods for flow lines is not meant to be exhaustive, but rather to be representative. Enterprising designers will no doubt add to these possibilities and newer solutions may be forthcoming. Regardless of the method, however, the scaling method must visually suggest proportional flow and do so easily.

Treatment of Symbols

No rules have emerged that govern how flow lines should be treated graphically in all maps. The only convention appears to be the way distributive flow lines are treated. (See Figure 12.11.) When mapping flows that separate into smaller flows, the widths of the individual branches should add up to the width of the trunk. This makes intuitive sense. The same applies when smaller branches come together to make a larger one.

Figure 12.11   Symbolization of flow lines.
When quantitative flow lines branch or unite, their widths are drawn proportionally.

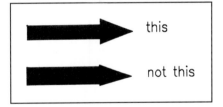

Figure 12.12   Proper arrowheads improve the appearance of flow lines.

If arrowheads are used they should be clear and should be scaled proportionally to the lines to which they belong. Arrowheads with small shoulders should be avoided. (See Figure 12.12.) If flow lines overlap, smaller ones should be made to appear on top of larger ones. (See Figure 12.13.) This is accomplished by breaking the lines of the larger flow line, and having the smaller one continue uninterrupted. This is a simple and effective technique that adds a plastic and fluid dimension to the map. More will be said of interposition in the next section.

LEGEND DESIGN

One of the more difficult tasks in the design of the flow map is the preparation of the legend. The legend is the crucial

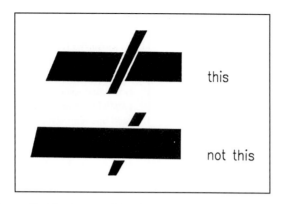

Figure 12.13 Interposition is useful in flow line symbolization.

Interposition is a technique useful in making smaller lines appear to rest on top of larger ones.

link in cartographic communication between cartographer and map reader, as it serves to explain carefully the symbols on the map. The legend must above all be clear and unambiguous. Units of measurements must be prominently displayed and it must be obvious how the lines are scaled, and the flow lines in the legend must appear exactly as they are on the map. If the data have been classed, the class boundaries should be clearly represented.

In one study of flow maps, four general types of legend design were identified: the ruled line, scaled bar or triangle, graduated lines, and key values (including "stairstep" designs).[25] (See Figure 12.14.) There have been no exhaustive studies on flow map legend designs. In the meantime, cartographers must rely on their best judgments. A good idea would be to try out the design solution on representative readers before a final decision is made.

In some cases, the values of the lines on the map may be labeled for greater precision. (See Figure 12.2.) The overall pattern of the lines shows the organization of the movement, and the labeled values provide detailed information for the reader seeking tabular data. Whether or not a legend also is used in these cases depends on the map purpose. In some respects this appears redundant. And in some cases, although the lines are scaled and drawn proportionally, no labels at the lines are provided, and no legend is included. (See Figure 12.15.) Presumably, the cartographer is wishing only to show geographical organization and pattern. It would seem imperative that these drawings be accompanied by written narrative.

UNIQUE SOLUTIONS

A number of innovative and unique solutions for flow mapping have been used by cartographers and geographers. Up to this point in our discussion, the majority of examples have been drawn from a rather conventional pool. Other solutions do exist and can be mentioned to stimulate experimentation. One such example illustrates connectivity by using flow lines of uniform width, and quantity of shipments

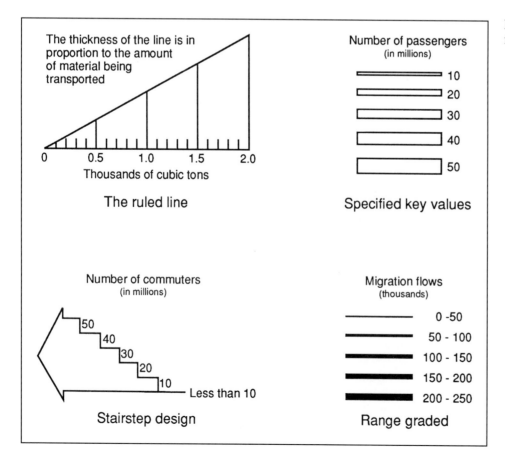

Figure 12.14 Legend designs for quantitative flow maps.

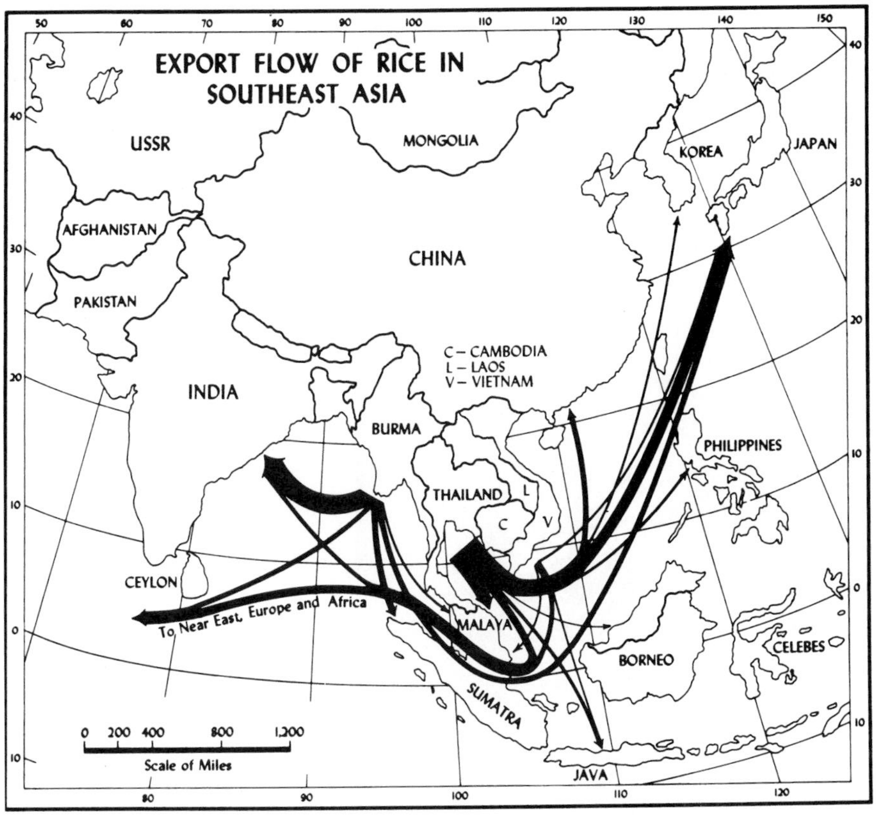

Figure 12.15 Export flow of rice in Southeast Asia.
See the text for an explanation. (*Reprinted by permission of Prentice Hall, Inc.*)

by proportional circle.[26] (See Figure 12.16a.) Another interesting solution, quite different, shows movements of coal by constant-width flow lines connected at their ends by bar graphs scaled to the magnitudes of shipments.[27] (See Figure 12.16b.)

A kind of flow line has been used in centrographic studies. In these examinations, geographical centers of gravity for different time periods are connected by lines of uniform thickness.[28] (See Figure 12.16c.) The visual result shows the geographical trend over time and the flow lines show the connectivity from time to time. These are largely descriptive studies, although centrographic analyses can be employed inferentially also.

Desire lines have been used in very unique ways. Conventional use of desire lines would include lines of uniform thickness (although they may be scaled to proportional widths, and are generally without directional arrows), usually placed on a geographic base map. One unusual use has them placed on a square root–transformed base map.[29] (See Figure 12.17.) In this instance the map projection origin (the "origin" of the desire lines) is located in Toronto, and the lengths of the desire lines are no longer linear. Although these kinds of unique solutions may reveal patterns otherwise hidden, map readers may need, by way of an explanatory legend, some tips on map-reading strategy.

Very heavily generalized and more stylized flow map solutions also have been used. (See Figure 12.18.) In the example used here showing the international trade of mineral fuels (mostly oil), the geographic base map has been abandoned in favor of a block-type organizational chart. The author does not provide any clues about the sizes of the countries and

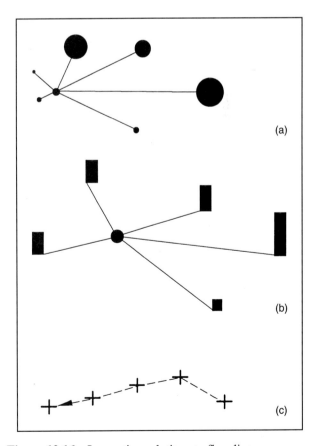

Figure 12.16 Innovative solutions to flow line representation and scaling.
See the text for an explanation.

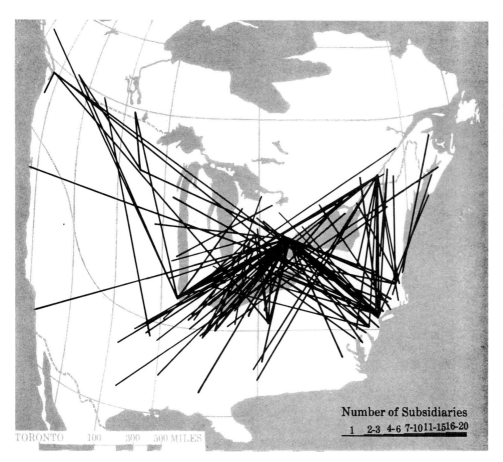

Figure 12.17 Desire lines on a square root-transformed projection, centered on Toronto, Canada. (*Reprinted by permission from D. Michael Ray, "The Location of United States Manufacturing Subsidiaries in Canada,"* Economic Geography 47 [1971]:*392.*)

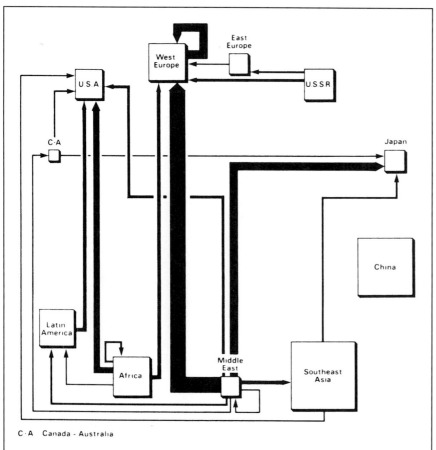

Figure 12.18 Heavily generalized quantitative flow map. See the text for an explanation. (*Reprinted with permission from Butterworths.*)

Table 12.1   Summary of Thematic Mapping Techniques with Examples of Appropriate Data

Technique	When Appropriate	Data Example
Choropleth map	Portraying a geographical theme of continuous or discrete phenomena occurring within areal enumeration units	Derived values, ratios, proportions: violent crime by county
Common dot map	Used to show spatial numerousness of discrete geographic phenomena	Absolute number: dairy cattle
Proportional symbol map	Displaying discrete phenomena at points, or to represent aggregate data at points	Absolute number or derived values: employment in retailing by cities
Isarithmic map	Used when total form of continuous phenomena is warranted	Absolute values, ratios, proportions: population density
Value-by-area cartogram	This map is used for continuous or discrete phenomena when data are available by enumeration unit	Absolute values, ratios, proportions: dollars spent on education per capita
Flow map	Used to show movement of goods or ideas between places; used to show interaction between places	Absolute values: tons of wheat shipped

regional rectangles, and there is no legend describing the widths of the flow lines. However, it is clear that western Europe, the United States, and Japan are the three main destinations in international fuel shipping. This type of presentation illustrates organization in the activity, although at the expense of not showing actual geographical routes.

Computer Solutions

Computer solutions to flow mapping have not gone unnoticed by cartographers. In one cartographic study, internal migration data for the United States, in the form of a 50×50 (states) "from-to" table, was used in the development of a computer program to automatically draw flow lines or bands.[30] Line-width scaling, arrowhead design, flow line overlap, and other typical graphic solutions have to be dealt with as in manual design solutions. The advantage in using the computer, however, is that alternative designs can be examined quickly. A solution is provided here in Figure 12.19.

SUMMARY OF MAPPING TECHNIQUES

Six different mapping techniques have been presented in this part of the book: choropleth mapping, the common dot-mapping method, proportional symbol mapping, isarithmic mapping, value-by-area or cartogram mapping, and flow mapping. This group probably accounts for the majority of thematic *quantitative* maps rendered today, whether by manual or computer means. It is a good idea at this juncture to review *when* each is to be used, and to give examples, before you continue through the remainder of the chapters. The review may be accomplished by examining Table 12.1.

As you review this material it is a good idea to remember that matching map technique with the data mapped continues to be somewhat subjective and will usually vary with the scale of the study or map. For example, at one scale the phe-

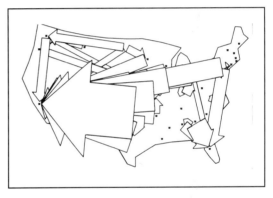

Figure 12.19   Computer solutions to a flow mapping problem.
(Reprinted by permission from Waldo R. Tobler, "Experiments in Migration Mapping by Computer," American Cartographer *14 [1987]: 159.)*

nomena may be considered continuous, but at larger scales, you may find that the phenomena are indeed discrete. At small scales, dairying may be continuous (although clumped perhaps), but at the farm scale, cows are indeed discrete.

After you have a firm grasp of the use of each method, and a feeling for the kind of appropriate data for each, you should resume your work through the next part of the text, which presents methods for better map *design* and *cartographic production.*

NOTES

1. Erwin Raisz, *Principles of Cartography* (New York: McGraw-Hill, 1962), pp. 218–20.

2. Stephen S. Birdsall and John W. Florin, *Regional Landscapes of the United States and Canada* (New York: Wiley, 1981), pp. 50–52.

3. Peter Davis, *Data Description and Presentation* (London: Oxford University Press, 1974), p. 92.

4. Brian J. L. Berry, *Geography of Market Centers and Retail Distribution* (Englewood Cliffs, NJ: Prentice Hall, 1967), pp. 11–21; see also Brian J. L. Berry, Edgar C. Conkling, and D. Michael Ray, *The Geography of Economic Systems* (Englewood Cliffs, NJ: Prentice Hall, 1976), p. 230.

5. H. Gary Funkhauser, "Historical Development of Graphical Representation of Statistical Data," *Osiris* 3 (1937): 269–404.

6. Ibid.

7. Arthur H. Robinson, "The 1837 Maps of Henry Drury Harness," *Geographical Journal* 121 (1955): 440–50.

8. Herman Friis, "Statistical Cartography in the United States Prior to 1870 and the Role of Joseph C. G. Kennedy and the U.S. Census Office," *American Cartographer* 1 (1974): 131–57.

9. Robinson, "The 1837 Maps," pp. 440–50; see also Arthur H. Robinson, *Early Thematic Mapping in the History of Cartography* (Chicago: University of Chicago Press, 1982), pp. 64, 147.

10. Robinson, *Early Thematic Mapping,* p. 147.

11. Robinson, *Early Thematic Mapping,* pp. 147–54; see also Arthur H. Robinson, "The Thematic Maps of Charles Joseph Minard," *Imago Mundi* 21 (1967): 95–108.

12. Robinson, "Thematic Maps of Minard," pp. 95–108.

13. Ibid.

14. Robinson, *Early Thematic Mapping,* p. 150.

15. Edward R. Tufte, *The Visual Display of Quantitative Information* (Cheshire, CT: Graphics Press, 1983) p. 40.

16. M. Jody Parks, *American Flow Mapping: A Survey of the Flow Maps Found in Twentieth Century Geography Textbooks, Including a Classification of the Various Flow Map Designs* (unpublished master's thesis, Department of Geography, Georgia State University, Atlanta, 1987), p. 14.

17. Ibid., p. 34.

18. Ibid., p. 39.

19. Ibid., p. 43.

20. Ibid., pp. 50–66.

21. Borden D. Dent, "Continental Shapes on World Projections: The Design of a Poly-Centered Oblique Orthographic World Projection," *Cartographic Journal* 24 (1987): 117–24.

22. George F. McCleary, Jr., "Beyond Simple Psychophysics: Approaches to the Understanding of Map Perception," *Proceedings of the American Congress on Surveying and Mapping, 1970,* pp. 189–209.

23. Gwen M. Schultz, "Using Dots for Traffic Flow Maps," *Professional Geographer* 8 (1961): 18–19.

24. David E. Christensen, "A Simplified Traffic Flow Map," *Professional Geographer* 8 (1961): 21–22.

25. Parks, "American Flow Mapping," pp. 67–68.

26. George H. Primmer, "United States Flax Industry," *Economic Geography* 17 (1941): 24–30.

27. Walter H. Voskuil, "Bituminous Coal Movements in the United States," *Geographical Review* 32 (1942): 117–27.

28. E. E. Sviatlovsky, "The Centrographical Method and Regional Analysis," *Geographical Review* 27 (1937): 240–54.

29. D. Michael Ray, "The Location of United States Manufacturing Subsidiaries in Canada," *Economic Geography* 47 (1971): 389–400.

30. Waldo R. Tobler, "Experiments in Migration Mapping by Computer," *American Cartographer* 14 (1987): 155–63.

GLOSSARY

desire line map unique flow map in which actual routes between places are not stressed, but interaction is; direction of flow often not shown, p. 223

distributive flow map flow map on which the distribution of commodities or migration is the principal focus, p. 226

dynamic map term often used to describe the ordinary flow map, p. 222

flow map map on which the amount of movement along a linear path is stressed, usually by lines of varying thicknesses, p. 222

network flow map flow map that reveals the interconnectivity of places, p. 226

radial flow map class of flow map that is characterized by a radial or nodal pattern, p. 226

traffic flow map particular kind of flow map in which movement of vehicles past a route point is shown by scaled lines of proportionally different thicknesses, p. 223

READINGS FOR FURTHER UNDERSTANDING

Friis, Herman. "Statistical Cartography in the United States Prior to 1870 and the Role of Joseph C. G. Kennedy and the U.S. Census Office." *American Cartographer* 1 (1974): 131–57.

Funkhauser, Gary H. "Historical Development of Graphical Representation of Statistical Data." *Osiris* 3 (1937): 269–404.

Parks, M. Jody. *American Flow Mapping: A Survey of the Flow Maps Found in Twentieth Century Geography Textbooks, Including a Classification of the Various Flow Map Designs.* Unpublished master's thesis. Atlanta: Georgia State University, Department of Geography, 1987.

Robinson, Arthur H. *Early Thematic Mapping in the History of Cartography.* Chicago: University of Chicago Press, 1982.

Tufte, Edward R. *The Visual Display of Quantitative Information.* Cheshire, CT: Graphics Press, 1983.

PART III

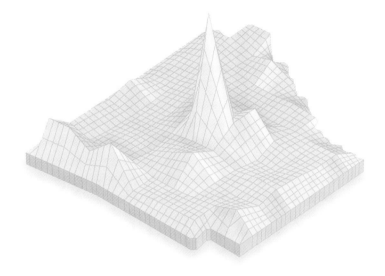

DESIGNING THEMATIC MAPS

The text preceding Part III dealt with forming the base map, techniques of symbolization, data processing, and mapping techniques. All of these subjects are important in developing the final map. Other design considerations are required, however, before one completes the mapping assignment. This part of the book presents many of the techniques regarding the visual design aspects of thematic mapping.

Cartographic designers face two-dimensional design problems much as artists do. Furthermore, the elements of two-dimensional design can be studied carefully and techniques learned. Basic and advanced approaches, and advanced concepts and techniques are presented in Chapter 13, which provides ideas for total map organization and figure-ground relation-ships. Of all the visual design ideas discussed in this book, an approach to a design problem by way of having a visual hierarchical plan is the most fundamental and necessary. This plan steers the page design activity and assures success. Typographics, the study and application of language labels to maps, is a basic part of map design. Chapter 14 provides current ideas about this subject and briefly includes modern techniques of producing type by laser printing. Chief among the design variables for the thematic cartographer is color. Color essentials, specifications, applications, and standards are subjects about color that cartographers need to know. These topics are presented in the last chapter of Part III.

CHAPTER

13

THE MAP DESIGN PROCESS AND THE ELEMENTS OF MAP COMPOSITION

CHAPTER PREVIEW

This chapter is our first venture into the design of maps and describes the design process and the appearance of the final map product. The design process is viewed as a series of stages beginning with problem identification and ending with implementation. Evaluation, creativity, experimentation, and map aesthetics are part of the design activity. The basic concepts of graphic composition as they relate to maps are presented, and they should encourage experimentation. No one best way to a design solution can be predetermined for all maps—only principles and general approaches can guide the cartographer. The ordinary thematic map can be considered to have two or more planes or levels. Thematic map design deals with the arrangement of the map's elements at each level or *between levels. The map's design elements are those marks on the map that must be arranged into a graphic composition suitable to the map's communication purpose. The elements include the map's title, scale, border, symbols, and other marks. By using visualization and experimentation techniques, the composition elements are arranged to satisfy the design goals. These elements include the planar organizational elements of balance, focus, internal organization, figure and ground, and a number of contrast elements. Visual acuity plays an important part in design; cartographers must be aware of the limitations of the human eye as they plan and develop map specifications.*

Chapter 1 presented a brief explanation and several descriptions of map design. There it was established that map design is a complex activity involving both intellectual and visual aspects. This chapter resumes the topic of design by looking in particular and in more detail at the design process and examining the elements of map composition—the first of the visual assignments that must be treated by the designer.

THE DESIGN PROCESS

Most designers would agree that all design takes place in sequential *steps,* ordered in a way that eventually yields the planned result. Identifying these steps or stages in design is helpful in learning how design takes place.

The **map design process,** like any act of designing, includes six essential stages: problem identification, preliminary ideas, design refinement, analysis, decision, and implementation.[1] (See Figure 13.1.)

Needs and design criteria are established in the first stage. Limitations are usually set in this stage. In the case of mapping, this stage includes the identification of map purpose and map reader and such factors as cost and technical considerations. The most creative step in the design process occurs in the second stage, where preliminary ideas are formulated. Brainstorming or having synectic (problem solving based on creative thinking) sessions is helpful here. Many solutions to design problems are found unconsciously, and then "pop" into the mind's eye. Sorting through one's visual memory often takes place in this stage, and is especially helpful in cartographic design.

In the third stage, called design refinements, all preliminary ideas are evaluated—and may be accepted or rejected. Those ideas that are retrieved are refined and sharpened,

and decisions are made that will affect the whole process. For cartographers, this stage usually involves setting down in writing details of the mapping project. For example, critical data needs are reviewed and finalized.

Models are often created in the analysis stage. For product designers, the models serve to make the drawings and sketches come alive and assist in the visualization process. Today, computer modeling in graphic form is preferred, and real models are not made. Nevertheless, this remains a prototype stage. Market research is often conducted on the prototypes and changes can be made. Map designers use this stage to develop detailed drawings and to work out problem areas (such as tricky printing problems or unique symbol schemes). Cartographers may wish to test their preliminary designs on sample readers.

The decision stage is as the name suggests. Changes, if any, are made on the prototype based on research and fact-finding from the previous stage. Ideas are rejected or accepted, and the final stage, implementation, begins. For cartographers this stage signals the beginning of final map production.

Feedback in the design process is continuous. Each design teaches us something about future problems and processes. Feedback is a critical element that helps designers become efficient and recognize that each design process may be unique, and that not every design problem will utilize the design stages in exactly the same manner.

Projection in the process assists the designer in anticipating solutions and problem areas. Projection also involves visualization, especially in cartographic design. Projection allows the designer to "see" in his or her mind's eye the end product, and thereby helps decision making along the way. Being able to project is essential, and comes with experience, regardless of the design area. This is easy enough to say but difficult to do or teach—experience is the only teacher. Nonetheless, the ability to conceive the final solution before it is physically mapped will reduce the number of wrong choices in design.

The designer may cycle through all stages of the design process as many times as required to reach an acceptable solution to the problem. Repetition is to be expected.

DESIGN EVALUATION

It is indeed difficult to evaluate map design, usually because the one doing the evaluating was not intimately involved in the design process. The outsider is never sure what compromises the cartographer had to make to balance the decisions in the design process. We do not know the relationship between map author and designer, and what sacrifices and learning had to take place to get anything in the map. One cartographer does suggest, however, that a map's design should be judged only with regard to the map's purpose and intended audience.[2] This seems fair enough.

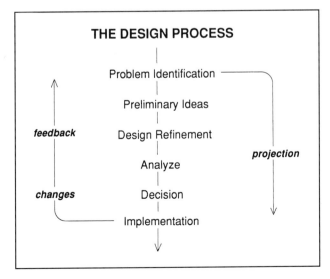

Figure 13.1   The design process.
See text for an explanation.

There are a few guidelines, however, that may be followed. Southworth and Southworth, for example, list, among others, these design characteristics of successful maps:[3]

1. A map should be suited to the needs of its users.
2. A map should be easy to use.
3. Maps should be accurate, presenting information without error, distortions, or misrepresentation.
4. The language of the map should relate to the elements or qualities represented.
5. A map should be clear, legible, and attractive.
6. Many maps would ideally permit interaction with the user, allowing change, updating, or personalization.

CREATIVITY AND VISUALIZATION

Creativity is the ability to see relationships among elements (regardless of the design arena). Although there is no recipe for creativity, certain activities appear to be shared by people considered to be great thinkers, scientists, or artists.[4]

1. Challenging assumptions—daring to question what most people take as truth.
2. Recognizing patterns—perceiving significant similarities or differences in ideas, events, or physical phenomena.
3. Seeing in new ways—looking at the commonplace with new perceptions, transforming the familiar into the strange, and the strange into the familiar.
4. Making connections—bringing together seemingly unrelated ideas, objects, or events in ways that lead to new concepts.
5. Taking risks—daring to try new ways, with no control over the outcome.

6. Using chance—taking advantage of the unexpected.
7. Constructing networks—forming associations for the exchange of ideas, perceptions, questions, and encouragement.

Cartographic designers can learn from this list. Conscious effort to participate in new ways of thinking during the transformation stage should become an integral part of the design procedure.

The **visualization process,** or thinking by incorporating visual images into thought, occurs to the fullest when seeing, imagining, and graphic ideation (see following page) come into active interplay.[5]

For the creative person, *seeing* is integral to all thought processes, and willingness to restructure visual images into new configurations is essential. Looking at a map inside out or upside down may yield solutions never thought possible. **Imagining,** or creating visual images in the mind's eye, is also helpful in developing design experience.[6] These images may or may not be composed of past perceptions. Imagining is the act of "seeing" a map before it is physically produced. The ability to project in the design process calls for the strengthening of imagining talents.

There are many descriptions of how new ideas are formed in the minds of creative people. A composite pattern might include these four stages in the creative process.[7] (See Figure 13.2.)

1. *Preparation.* At this stage, a person consciously "files away" into memory visual images that can be useful for a problem at hand. This may be referred to as an **image pool.**
2. *Incubation.* The person releases all conscious hold on the problem and turns to other tasks. It is theorized that images in a person's mind are rearranged into new alignments and patterns during this truly creative stage.

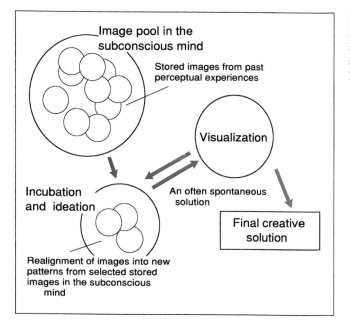

Figure 13.2   The visualization process.
It is theorized that creativity is the result of new arrangements of stored visual images in the subconscious mind. Visualization is the ability to "see" these arrangements, and is an illumination process leading to final solutions.

3. *Illumination.* The solution to the problem appears suddenly, often spontaneously.

4. *Verification* or *revision.* The person consciously works out the details of the solution, bringing all efforts and skills to bear. Formal structures result.

Map designers should try to develop this image pool in the subconscious, so that creative ideas can form during the incubation stage. Some psychologists believe that ideation results from realignment of images stored from earlier perceptual events.[8] *Visualization* is the process of experiencing or seeing these new creations—an illumination stage, or as described earlier, placing visual images into thought.[9] Cartographers should therefore experience as much graphic art, art, and cartography as possible. It is worthwhile to spend time exploring an art museum, seeing an animated film exhibit, or going through old atlases. From such experiences, the designer builds an image inventory that can later provide creative design solutions.

There are two views of visualization. One is the view just presented, where visualization begins from within and leads to new mental formations (creations) from previously stored images. The other view, emerging since the development of sophisticated computer graphic software and data presentation, provides yet another perspective. This latter view, called **scientific visualization,** is a method using computers to transform data into geometric models that can be presented on cathod-ray tubes (CRTs). (See box.) Both operations are instrumental in good cartographic design.

Graphic Ideation

Graphic ideation, or *sketching,* is often practiced by creative persons. The goal of this activity is to bring vague images into clear focus.[10] Designers can improve their creative potential by practicing seeing, imagining, and graphic ideation.

One real contribution of rapidly generated temporary maps (displayed on cathode-ray tubes) is their potential as "what if" images. Hundreds of maps with alternative designs can be created in a short time so that the designer can explore alternatives. Scale, projection, color, typography, classification, and other elements of design can be varied and checked almost instantly. Yet, do not be lulled into thinking this is a panacea for good design; even though this

may be considered a feature in favor of the new technology, it should not replace the designer's own imagining abilities, reducing overall creativity in the long run.

EXPERIMENTATION

Experimentation is necessary in testing the new idea. Sometimes ingenious solutions do not work when put onto the map sheet. Creative solutions must be worked out in detail on paper, to preview their visual effects. Do the elements of the design work effectively together? Do the details, however creative they may be, detract from the map as a whole? Are alternative solutions now apparent that were not noticeable before? A willingness to explore alternative solutions is essential to developing the best graphic design for a map. (See Figure 13.3.)

Unfortunately, many map designs are first solutions. Time and cost are enemies of experimentation but must not inhibit the search for better solutions or refinements of good first ideas. It is altogether too easy to drift into simple solutions. Good map designers take the extra time to explore all possible ideas, even stopping to let incubation begin again. The process from visualization through experimentation to final solution often involves much repetition and backtracking.

MAP AESTHETICS

Although maps may not be objects of aesthetic concern, there is a trend toward giving greater consideration to the quality of appearance of thematic maps. Writing several decades ago, the great map critic John K. Wright addressed this issue:

> *The quality of a map is also in part an aesthetic matter. Maps should have harmony within themselves. An ugly map, with crude colors, careless line work, and disagreeable, poorly arranged lettering may be intrinsically as accurate as a beautiful map, but it is less likely to inspire confidence.*[11]

More recently, John S. Keates, a British cartographer, remarked, "The 'art' of cartography . . . is not simply an anachronism surviving from some prescientific era; it is an integral part of the cartographic process."[12]

Three elements have been identified as forming the basis for the evaluation of map aesthetics: harmony, composition, and clarity.[13] *Harmony* is viewed as the relationship between different map elements (that is, how do the elements look *together?*). *Composition* deals with the arrangement of the elements and the emphasis placed on them. In other words, how does the structural balance of emphasis appear? Finally, *clarity* deals with the ease of recognition of the map's elements by the map user. "A map which lacks one or more of these three main elements, lacks beauty."[14]

The objective of scientific visualization, as it interacts with cartography, is the development of ideas and hypotheses about spatial information, and is embodied in the computational and non-computational tool kit which assists scientists in the exploration of their data so that they can develop ideas (ideation) which lead to the better understanding of the information.

Source: P. Fisher, J. Dykes, and J. Wood, "Map Design and Visualization," *Cartographic Journal* 30 (1993): 136–42.

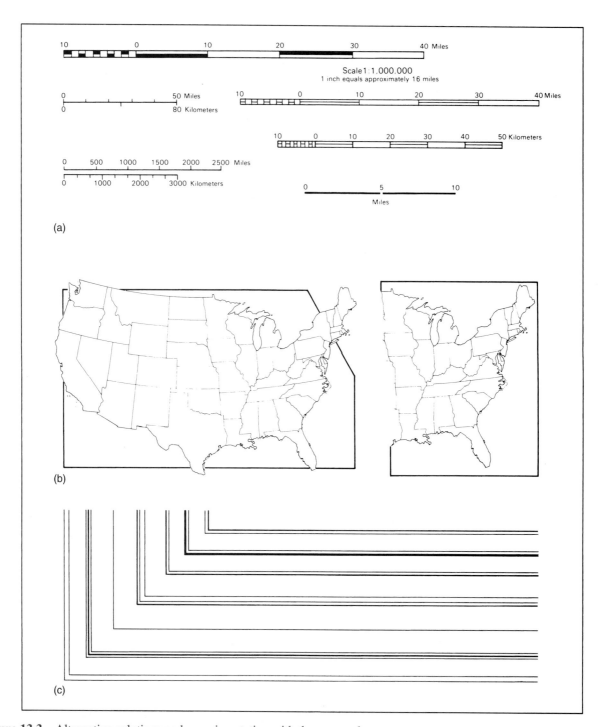

Figure 13.3 Alternative solutions and experimentation with three map elements.

In (a), several different ways of presenting graphic scales are shown. In (b), map borders need not be conventional boxes around the map but can be closely integrated with the graphic design. The illustration in (c) shows different map borders and neatlines. In graphic composition of the map, the designer must be open to many alternative ideas.

Cartographic designers have a certain degree of freedom in the design process. Of course map function is the overriding concern, along with the needs of the user, but beyond that the designer is working in a subjective realm. How well he or she performs in the creative, aesthetic realm will more than likely depend on intuitive judgments, conditioned by fundamental training and experience.

The subjective elements of design have been listed as follows:[15]

- Generalization—beauty of simplified shapes
- Symbolization—beauty of graphic representation
- Color—beauty of color accent and balance
- Layout—beauty of composition
- Typography—beauty of typographic appearance

This brief excursion into map aesthetics is not intended to be exhaustive. It has been included only to suggest to the reader that map design is subjective as well as objective. Beyond the rigid, scientific world of numerical cartography lies the intuitive, artistic, and aesthetic world of maps where designers can exercise their expressive talents.

MAP DESIGN—NEW CHALLENGES

Before we leave this topic of the design process, a few remarks are in order regarding cartographic design in the years ahead. British cartographer Michael Wood, for example, has written that the map user is destined to be much more important to the totality of design in the future.[16] Much of the design research over the past two decades has been aimed at the map—its appearance and symbolization strategies—and has not fully integrated the user and his or her complexity of cognitive abilities. Remarking on this, Wood has said:

> *Early psychological investigations reflected enthusiasm rather than clear direction, but ideas and methods have advanced. A stronger theoretical framework is now available from psychology, although the current focus of cartographic research has moved away from the important subjects of visual efficacy and map information processing. . . . A full understanding of maps as representations will require a multi-directional approach at all cognitive levels from first sensation of the image. As it takes place at a deeper level of cognition, interpretation of symbols relates more to user characteristics and the topics of particular maps, than to their graphic appearance opening up a much wider and deeper area of potential research.*[17]

Indeed the future of cartographic design will be an interesting, if not difficult, one.

THE MAP'S DESIGN ELEMENTS

Thematic maps are instruments of visual communication. The marks that make up a map are visual elements, and transfer of information takes place through them. The map designer arranges the visual elements into a functional composition to facilitate communication. This functional approach to design was first expressed by Arthur Robinson: "Function provides the basis for design."[18] That something is functional means that it has been "designed or developed chiefly from the point of view of use." Design decisions regarding the map's elements should be made on the basis of how each element is to function in the communication. The challenge is to make the map aesthetically pleasing as well as functional.

Most thematic maps contain these **map elements:** titles, legends, scales, credits, mapped areas, graticules, borders, symbols, and place names. (See Table 13.1.) The task of the designer is to arrange these into a meaningful, aesthetically pleasing design—not an easy task. From this perspective, map design is like a series of filters in which selections are made at each filter. (See Figure 13.4.) This analogy directs attention to the fact that map design is a complex affair involving many decisions, each of which affects all the others. Good design is simply the best solution among many, given a set of constraints imposed by the problem. The best design will likely be a simple one that works well with the least amount of trouble. The optimum solution may not be achievable, and what is good design today may be ineffective in the future. We are constantly learning more about the map user, and this will modify our future decisions as designers.

Important design principles include simplicity, appropriateness in a functional context, pleasing appearance, and considerations of economy. The designer's tools of creativity, visualization, ideation, and problem solving are used to sift through the map elements in order to bring these principles into a proper balance.

The graphic solution should seek to cause a change in the reader, as mentioned in Chapter 1. In viewing art, the reader may believe that nothing is required other than mere contemplation. When looking at effective graphic advertising, the reader responds by buying the product. In effective cartographic design, the reader should gain spatial knowledge and understanding.

DESIGN LEVELS ON THE MAP

It is useful at the outset to imagine the thematic map as composed of different planes or levels. (See Figure 13.5.) Usually the levels are differentiated by visual prominence. Each component of the map belongs to a specific level. More than one map element can be placed on a particular

---

The best designs . . . are *intriguing and curiosity-provoking*, drawing the viewer into the wonder of the data, sometimes by narrative power, sometimes by immense detail, and sometimes by elegant presentation of simple but interesting data. But no information, no sense of discovery, no wonder, no substance is generated by chartjunk.

Source: Edward R. Tufte, *The Visual Display of Quantitative Information* (Cheshire, CT: Graphics Press, 1983), p. 121.

Just as some realistically painted cows are full of life while others are deadly mechanical records, so some faithful maps are alive while others leave us untouched.

Source: Rudolf Arnheim, "The Perception of Maps," *American Cartographer* 3 (1976): 5–10.

Table 13.1   Typical Elements of the Thematic Map

Name of Element	Description and Primary Function
Title (and subtitle)	Usually draws attention by virtue of its dominant size; serves to focus attention on the primary content of the map; may be omitted where captions are provided but are not part of the map itself
Map legend	The principal symbol-referent description on the map; subordinate to the title, but a key element in map reading; serves to describe all unknown or unique symbols used
Map scale	Usually included on a thematic map; it provides the reader with important information regarding linear relations on the map; can be graphic, verbal, or expressed as an RF
Credits	Can include the map's data source, an indication of its reliability, dates, and other explanatory material
Mapped and unmapped areas	Objects, land, water, and other geographical features important to the purpose of the map; make the composition a map rather than simply a chart or diagram
Graticule	Often omitted form thematic maps today; should be included if their locational information is crucial to the map's purpose; usually treated as background or secondary forms
Borders and neatlines	Both optional; borders can serve to restrain eye movement; neatlines are finer lines than borders, drawn inside them and often rendered as part of the graticule; used mostly for decoration
Map symbols	Wide variety of forms and functions; the most important elements of the map, along with the geographic areas rendered; designer has little control over their location because geography must be accurate
Place names and labeling	The chief means of communicating with maps; serve to orient the reader on the map and provide important information regarding its purpose

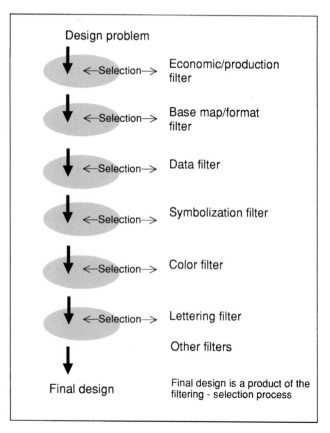

Figure 13.4 Map design as a filtering-selection process. In this view of design, a series of filters must be rotated (selection), allowing design activity to continue until an appropriate final solution is reached.

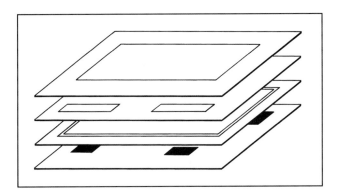

Figure 13.5 The organizational levels of the thematic map. A typical thematic map can be considered to be made of several distinct levels or planes. In planning a map, the designer assigns the various map elements to these levels. This causes the designer to think of each element in its proper role, thus leading to a more organized design.

level, but a *single element should never be assigned to more than one level.* Thinking of the map in this way will facilitate the map's overall design.

Map composition, the arrangement of the map's elements, takes place at each level and between levels. The arrangement at a given level may be called **planar organization,** and that between levels **hierarchical organization.** Hierarchical organization is also referred to as the *visual hierarchy.* The cartographer ordinarily approaches design solutions by simultaneously manipulating all elements *at* and *between* all levels.

ELEMENTS OF MAP COMPOSITION

The map's graphic composition is the arrangement or organization of its elements. The composition principles introduced in this chapter include the purpose of map composition, planar organization, figure and ground organization, contrast, and visual acuity. Knowledge of these principles and their application assists the cartographer in seeking better design solutions.

PURPOSE OF MAP COMPOSITION

Map composition is much more than layout, which is simply a sketch of a proposed piece of art, showing the relationships of its parts. The term *composition* is used here because it indicates the intellectual dimension as well as the visual. Map composition serves these ends:

1. Forces the designer to organize the visual material into a coherent whole to facilitate communication, to develop an intellectual *and* a visual structure
2. Stresses the purpose of the map
3. Directs the map reader's attention

4. Develops an aesthetic approach for the map
5. Coordinates the base and thematic elements of the map—a critical factor in establishing communication
6. Maintains cartographic conventions consistent with good standards
7. Provides a necessary challenge for the designer in seeking creative design solutions

PLANAR ORGANIZATION OF THE VISUAL ELEMENTS

The three aspects of planar visual organization are balance, focus of attention, and internal (intraparallel) organization. Each is important to the designer's language, and their visual possibilities and effects must be explored.

Balance

Balance involves the visual impact of the arrangement of image units in the map frame. Do the units appear all on one side, causing the map to "look heavy" on the right or left, top or bottom? An image space has two centers: a geometric center and an **optical center** (as shown in Figure 13.6). The designer should arrange the elements of the map so that they balance visually around the optical center.

Rudolf Arnheim, a noted author on the psychological principles of art, has suggested in his writings that **visual balance** results from two major factors: weight and direction.[19] Objects in the visual field (e.g., within the borders of a map) take on weight by virtue of their location, size, and shape. Direction is also imposed on objects by their relative location, shape, and subject matter. Arnheim stresses that balance is achieved when everything appears to have come to a standstill, "in such a way that no change seems possible, and the whole assumes the character of 'necessity' in all its parts."[20] In Arnheim's view, unbalanced compositions appear accidental and transitory.

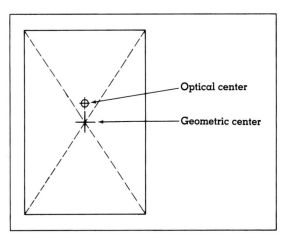

Figure 13.6 The two centers of an image space. The designer should arrange the map's elements around the natural (optical) center, rather than around the geometric center.

Arnheim's observations on balance resulting from visual weight and direction can be summarized as follows:[21]

1. Visual weight depends on location.
   - Elements at the center of a composition pull less weight than those lying off the tracks of the structural net. (See Figure 13.7a.)
   - An object in the upper part of a composition is heavier than one in the lower part.
   - Objects on the right of the composition appear heavier than those on the left.
   - The weight of an object increases in proportion to its distance from the center of the composition.
2. Visual weight depends on size.
   - Large objects appear visually heavier than small objects.
3. Visual weight depends on color, interest, and isolation.
   - Color affects visual weight. Red is heavier than blue. Bright colors appear heavier than dark ones. White seems heavier than black.
   - Objects of intrinsic interest, because of intricacy or peculiarity, seem visually heavier than objects not possessing these features.
   - Isolated objects appear heavier than those surrounded by other elements.
4. Visual weight depends on shape.
   - Objects of regular shape appear heavier than irregularly shaped ones.
   - Objects of compact shape are visually heavier than those not so shaped.
5. Visual direction depends on location.
   - Weight of an element attracts neighborhood objects, imparting direction to them. (See Figure 13.7b.)
6. Visual direction depends on shape.
   - Shapes of objects create axes that impart directional forces in two opposing directions.
7. Visual direction depends on subject matter.
   - Objects possessing intrinsic directional forces can impart visual direction to other elements in the composition.

Of course, Arnheim recognizes that the elements of compositional balance operate together in complex fashion. He also advises not to forsake the content of a composition simply in order to create balance: "The function of balance can be shown only by pointing out the meaning it helps to make visible."[22] Once again, this underscores the idea that map content is more important than the map's design.

It is often difficult to achieve balance on the map. Cartography is not an expressive art form in which the graphic elements may be rearranged at will. Many of the shapes and their locations are imposed by geographical or locational facts. The guidelines presented by Arnheim should nevertheless be applied whenever possible.

Graphic-art professionals who work with two-dimensional design often speak of the **golden section.** This method of devising proportions is attributed to classic Greek architects and sculptors. In the golden section, the proportion of a smaller unit to a larger is the same as that of the larger unit to the whole. This method of sectioning can be duplicated any number of times. (See Figure 13.8a.) Proportion is the relationship of a part of the visual field to the remainder or whole. Balance is achieved when pleasing proportions among the parts are maintained. In fact, one author has noted that the great violin maker Stradivarius employed the golden section in his violin proportions.[23] Cartographers should bear proportion in mind in arranging the different elements of the map. Applying the golden section in cartographic design is an intuitive matter, not subject to rigid quantification.

Visual balance can be looked at from a different point of view. Writing more than 50 years ago, Richard Surrey, an advertising artist, developed important ideas about composition. Surrey observed that layout involves not only the arrangement of units (balance) but also the

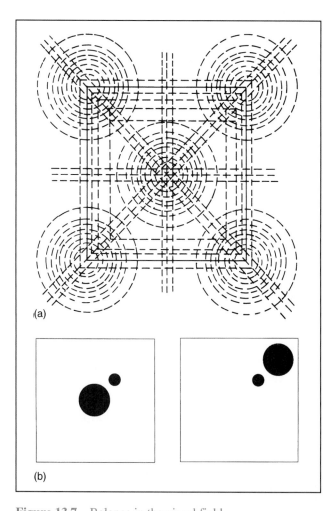

Figure 13.7   Balance in the visual field.
Arnheim stresses that a structural net, as in (a), determines balance. Objects on the main axes or at the centers will be in visual balance. An object is given direction by other objects adjacent to it. In (b), the small disc's directional element is shifted as the large disc's position is changed. Each thematic map will have a unique structural net created by the locational patterns of its elements.

division of space. "In other words, instead of layout being a process of addition (putting together units), it is much more easily grasped when considered as a process of division."[24] (See Figure 13.8b.) His further thoughts on this idea may be summarized:

1. Equal divisions of space are the least interesting. Inequality and the pursuit of equilibrium make layout visually alive.

2. Small spaces struggling against large spaces are visually alive.

3. Variety, for example the division of the image space into four unequal parts, creates interest. Complex designs may be more exciting than simple ones.

To illustrate how balance can affect the impression one has when viewing a map, several different locations of the shape of Africa are included in Figure 13.9. Which appears better balanced within the map frame?

Achieving visual balance, of course, is not always as simple as the case just illustrated. Normally, thematic maps contain most of the elements mentioned earlier, and all must be handled in terms of balance. Visual weight caused by texture, solid black and white areas, and other elements must figure in the planning. Open spaces take up "balance space" and must be used effectively in the overall design. Complex designs require careful planning to use all spaces efficiently while retaining a visually harmonious balance. Acceptable balance is reached when the relocation of any one element would cause visual disturbance. Balance is a state of equilibrium.

In at least one study using thematic maps, the balance of the map's elements is shown to have an initial effect on the way the map reader goes about looking at those elements.[25] However, the longer the reader views the map, the less importance balance seems to have on map-reading behavior. Better balance also leads to less reading difficulty and to somewhat better memory of the map's message. It is not altogether clear exactly what constitutes good and poor balance in such studies because these extremes are subjective at best. Nonetheless, the balance of the map's elements is a vital concern for the cartographic designer.

Focus of Attention

As previously mentioned, the optical center of an image area is a point just above the geometric center. This attracts the viewer's eye, unless other visual stimuli in the field distract attention. Surrey mentioned earlier makes several

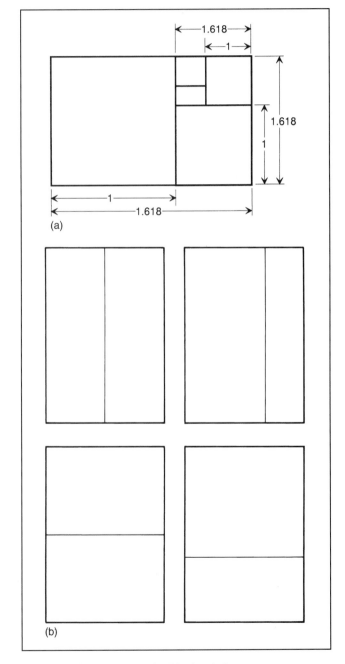

Figure 13.8 Methods of achieving balance.
Balance is dynamic and will result from appropriate proportioning of the image space. The method of arriving at the golden section is illustrated in (a), and several alternatives are pictured in (b). In (b), notice that unequal divisions of space are more interesting.

Every visual pattern is dynamic. Just as a living organism cannot be described by its anatomy, so the essence of a visual experience cannot be expressed by inches of size and distance, degrees of angle, or wave lengths of hue. These static measurements define only the "stimulus," that is, the message sent to the eye by the physical world. But the life of a precept—its expression and meaning—derives entirely from the activity of the kind of forces that have been described. Any line drawn on a sheet of paper, or the simplest form modeled from a piece of clay, is like a rock thrown into a pond. It upsets repose, it mobilizes space. Seeing is the perception of action.

Source: Rudolf Arnheim, *Art and Visual Perception* (Berkeley: University of California Press, 1965), p. 6.

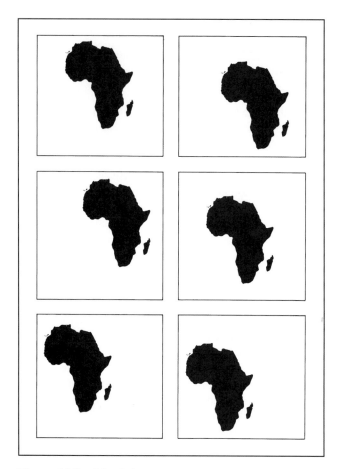

Figure 13.9   Map balance.
Position of map elements in the image space affects the balance of the map. The difference can be visually subtle, as this illustration shows. In which image does a natural visual equilibrium appear to exist?

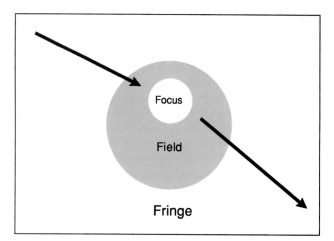

Figure 13.10   Eye movement through the image space. In normal viewing, the reader's eyes enter through the image space at the upper left, proceed through the visual center (focus), and exit the space at the lower right. Cartographic designers may use this pattern when arranging the map's elements, so that the positions of important objects on the map correspond to natural eye movements.

other points that are significant for questions of design. He says that the reader's eye normally follows a path from upper left to lower right in the visual field and passes through the optical center.[26] (See Figure 13.10.) Furthermore, the point of greatest natural emphasis is where a line of space division intersects either the focus or field circles of attention. (See Figure 13.11.)

Surrey's ideas were based on intuitive judgments and personal observations and have not been scientifically proven. Yet they do have appeal for the designer. An examination of recent print advertisements attests to the general applicability of his ideas. We can learn from these and other graphic designs. The map is a visual instrument, so the designer must learn what works in the visual world.

Internal Organization—Intraparallelism

The internal organization of the map's visual field relates to visual or perceptual order. Rudolf Arnheim defines order as "a wealth of meaning and form in an overall structure that clearly defines the place and function of every detail in the

whole."[27] Order implies an underlying structure, graphic or intellectual, that binds the parts of the whole together. In an ordered map, the graphic elements are arranged into a composition that develops a clear visual expression of the meaning of the communication and that shows an underlying structure of the graphic elements.

One technique to give structure to the graphic elements, at least at the planar level, is intraparallelism.[28] Intraparallelism is achieved when the elements of internal structure are aligned with each other. (See Figure 13.12.) Intraparallelism reduces tension in perception; it can be introduced into map designs in subtle ways. (See Figure 13.13.) It is a good idea to experiment with a variety of visual techniques to simplify the graphic design of maps.

CONTRAST AND DESIGN

Closely associated with figure and ground organization, and of nearly equal importance, is the feature called **contrast.** Contrast is fundamental in developing figure and ground but can be considered a design principle in its own right. Visual contrast leads to perceptual differentiation, the ability of the eye to discern differences. A lack of visual contrast detracts from the interest of the image and makes it difficult to distinguish important from unimportant parts of a communication. Map elements that have little contrast with their surroundings are easily lost in the total visual package. Contrast must be a major goal of the designer.

Contrast can be achieved through several mechanisms: line, texture, value, detail, and color. All of these could be

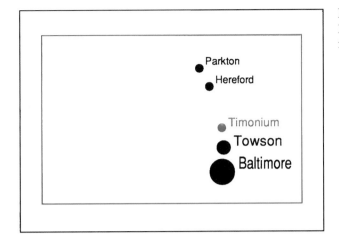

Figure 13.11   Recentering for greater clarity.
If possible, it is a good idea to recenter the map to place the central focus (in this case, the town of Timonium) closer to the optical center.

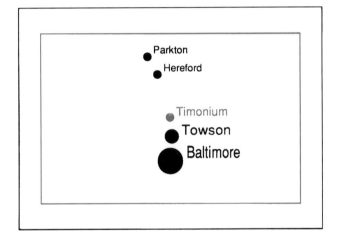

used in one design, but the result might be visual dishar-mony and tension—potentially as unrewarding as having no contrast at all.

Line Contrast

Lines may be put to a variety of uses on maps. They can function as labels, borders, neatlines, political boundaries, quantitative or qualitative symbols, special symbols to divide areas, or graphic devices to achieve other goals. Line contrast can be of two kinds: character and weight. **Line character** derives from the nature of the line and its segments, or its value or color. (See Figure 13.14a.) The order of visual importance of various line characters has not been well established. The subject and purpose of the map very often restrict choice of line character. On some maps, there may be no lines. For example, some recent designs use edges rather than lines to evoke a response.

The thickness of a line is its **line weight,** although no clear-cut relationship exists between thickness and visual or intellectual importance. Although a broader line generally carries more intellectual importance, very fine lines also

can be visually dominant. Strike a balance, keeping the map's purpose firmly in mind.

Contrast of line character and weight introduces visual stimulation to the map. A map having lines of all one weight is boring and lacks potential for figure formation. (See Figure 13.14b.) On the other hand, a map with lines of several weights and characters focuses attention, is lively, and aids the map reader's perceptual organization of the material. (See Figure 13.14c.) Guidelines can assist the designer in choosing lines so that discrimination between weights is possible. Generally, a line-weight difference exceeding .05 in is discernible by more than half of all map readers. A difference of .15 in is easily noticed by practically all readers. Figure 13.15 is a chart that can aid the designer in choosing line width to assure discernible differences.

Texture Contrast

Contrast of texture involves areal patterns and how they are chosen for the map. In this context, texture is a pattern of small symbols (e.g., dots) repeated in such a way that the eye can perceive the individual elements. Texture is often determined by the selection of quantitative or qualitative symbols for the map. Contrast considerations should be part of symbol selection. In some instances, patterns are selected and applied to the map solely to provide graphic contrast (e.g., in the differentiation of land and water). Texture is sometimes applied in order to direct the reader's attention to a particular part of the map.

Another possibility, not often used, is textured lettering. This differentiates labels from other lettering, enabling the designer to use more lettering in the design. Textured lettering is possible only when the letters are geometric, not composed of many thick and thin strokes.

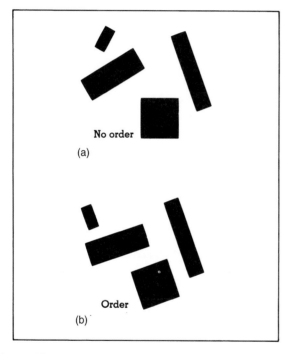

Figure 13.12 Internal structure of graphic elements provided by intraparallelism.
The objects in (a) display a lack of internal structure because each has its own orientation. After realignment in (b), the objects appear to be more of an integrated unit. Lack of intraparallelism leads to tension and disunity.

Value Contrast

Texture is observable because the individual dots or other elements of the pattern are easily seen. Reducing such a pattern to the point where the elements are below the threshold of visual resolution acuity results in the perception of a visual tone or value. Contrast of value is another design technique used by cartographers, although some of the contrast is often dictated by the nature of the data (qualitative or quantitative). In cases not determined by the data, contrast of value can be used in ways similar to contrast of texture. (See Figure 13.16.) Contrast of value leads to light and dark areas on the map. A good place to use this contrast type is in the development of figures and grounds. To stand out strongly, figures should have values considerably different than grounds. Land areas, for example, should be made lighter or darker than water areas.

Variation of Detail

Although designers seldom think of it as a positive design consideration, contrast of detail can be employed effectively, especially in combination with other techniques. Along a continuum ranging from little detail at one end to great detail at the other, the reader's eye will be attracted to those areas of the map with the most detail.

This feature can work against the designer, however. Exquisite detail rendered to an unimportant feature can distract the reader's attention from the communication effort. By judicious use of extra detail in important areas of the map, the designer can subtly lead the reader to them. (See Figure 13.17.) Detail can also be used to strengthen figure formation.

Color Contrast

Employment of color is one of the chief techniques in the development of contrast in design. Color can differentiate areas on the map for a variety of purposes. Color as a major design ingredient is treated in detail in Chapter 15.

VISION ACUITIES

The map designer works in a visual medium, so all elements must be visible to the map reader. If the elements cannot be seen, the map's communicative attempt will be

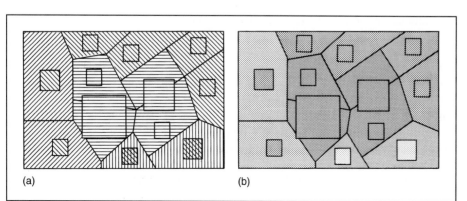

Figure 13.13 Rearranging map elements to achieve greater internal order.
The line patterns of (a), with different orientations, have been replaced in (b) by dot screens having greater internal order. The rows of repeating dots have identical orientations.

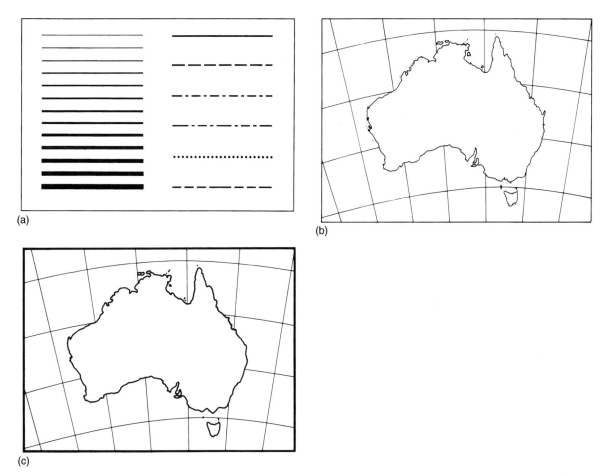

(a)

(b)

(c)

Figure 13.14   Line contrast.
In (a), several different line weights and line characters are shown. The visual effect of varying line weights on a map is illustrated by comparing (b) and (c). More visual interest is achieved with greater contrast of weight. In this case, the figure and ground organization is strengthened.

lost, no matter how well designed the map may be in other respects. There are two important measures of the human ability to see visual elements: **visibility acuity** and **resolution acuity.**

Visibility Acuity

So far it has been assumed that the map reader can see the map and all its design elements, but visibility must not be taken for granted. Fortunately, it is seldom a problem unless the designer overlooks it when preparing art for reduction.

Visibility acuity is a measure of a size threshold.[29] It should not be confused with intensity threshold, which is a measure of sensitivity. The parameter used in visibility is called a subtense. It is an angular measure, because the retinal image size for a 1-inch object 2 feet away is identical to that for a 10-inch object 20 feet away. Strictly speaking, acuity measures the threshold size of the retinal image, not the size of the object. For practical purposes, minimum object sizes can be prescribed so that objects will not fall below the visibility threshold, especially when art is to be reduced.

The subtense varies from about .44 second of arc to about 10 minutes for a black line on a white background, depending on illumination.[30] This means that at a reading distance of 18 in (46 cm), a black line should not be rendered smaller than .006 in (.15 mm) to about .05 in (1.27 mm). The smallest size of nib manufactured by one technical-pen manufacturer is .13 mm (.005 in). A minimum line thickness of .25 mm (.01 in) is a safer all-around specification for most design work.

The visibility acuity threshold for a black dot on a white background is 32 seconds of subtense. This becomes about .04 in (1.0 mm) at an 18-in (46-cm) reading distance.

Resolution Acuity

Resolution acuity is somewhat different from visibility acuity. Resolution is a measure of the detectable separation between objects in a visual field. When two objects are seen apart, the reader is said to resolve them. Again, the threshold subtense is the point at which this occurs accurately. The average threshold separation of two black dots on a

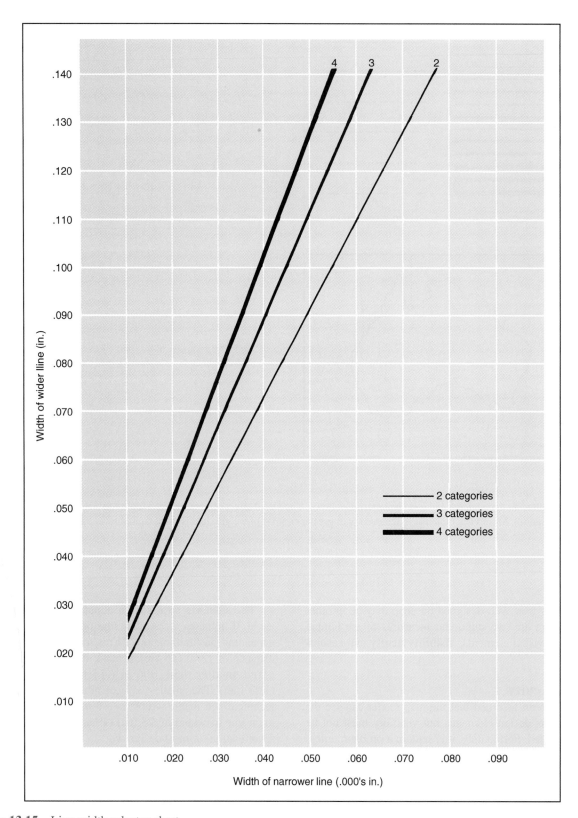

Figure 13.15   Line-width selector chart.
This chart may be used to determine line widths when two, three, or four different widths are used on maps. For the two-line category, simply determine the width of the narrower line and enter the chart along the bottom axis. Extend a perpendicular line to the two-category ordinate, and from this point extend a horizontal line to the vertical axis. Read the width of the wider line directly from the vertical axis. For the three-category case, first determine the wider line of a two-line case, and then use the line width of that determination as the narrower line for the three-line case (this time using the three-category ordinate). The four-line case can be handled in a similar manner. *(Source: Richard D. Wright, "Selection of Line Weights for Solid, Qualitative Line Symbols in Series on Maps" [unpublished Ph.D. dissertation, Department of Geography, University of Kansas, 1967], pp. 91–92.)*

white background is about 1 minute of arc.[31] This is approximately .003 in (.076 mm). This measure can become critical in map design when patterns are specified in design, especially if art is to be reduced. If the elements in a pattern are closer than this, the observer does not see the pattern but begins to perceive only a continuous tone. If the difference in patterns is important for differentiation in such a case, the design has failed in its task of facilitating communication. Map elements are not functional at all if they cannot be seen. The very best map designs will falter if minimum thresholds are not maintained.

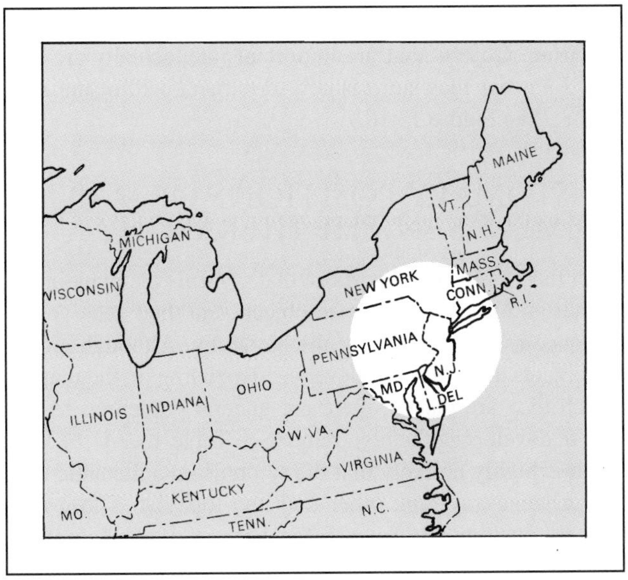

TOTAL MAP ORGANIZATION, THE VISUAL HIERARCHY, AND THE FIGURE-GROUND RELATIONSHIP

There is probably no perceptual tendency more important to cartographic design than **figure and ground organization.** A person's underlying perceptual tendency is to organize the visual field into categories: figures (important objects) and grounds (things less important). This concept was first introduced by Gestalt psychologists early in this century. Figures become objects of attention in perception, standing out from the background. Figures have "thing" qualities; grounds are formless. Figures are remembered better; grounds are often lost in perception.

In the three-dimensional world, we see buildings in front of sky and cars in front of pavements. Likewise, we see some objects in front of others in the two-dimensional world, given wise graphic treatment. In cases such as the words printed on this page, no extraordinary measures are needed to make figures stand out from ground.

In the planar level of design planning, the designer should structure the field in a way that directs the reader's perception along paths commensurate with the communication goals of the map. For example, objects that are important intellectually should be rendered so as to make them appear as figures in perception.[32]

Figure 13.16   Use of differences in value to focus the reader's attention.

In this example, the absence of the pattern over New Jersey forces the eye to that part of the map. Contrast of texture can also be used in this manner.

(a)   (b)

Figure 13.17   Provision of detail can direct the reader's eye.

In (a), the even distribution of place names does not focus the reader's attention. The eye is drawn to that part of the map in (b) that has the most detail—in this case, lettering.

Deborah Sharpe, a color designer not from the cartographic profession, has stated very succinctly the importance of incorporating figure and ground perception into design:

In my own design work, I have found that designating the features that are to represent figure and those that are to represent ground as a first step on the job eliminates the trial and error inherent in the beginning stages of most creative tasks.[33]

VISUAL HIERARCHY DEFINED

Public speakers arrange their material to emphasize certain remarks and subordinate others. Professional photographers often focus the camera to provide precise detail in certain parts of the picture and leave the remainder somewhat blurred. Advertising artists organize ads to accentuate some spaces and play down others. Choreographers arrange the dancers so that some will stand out from the rest on the stage. Professional cartographers must go through similar activities in designing the map.

The **visual hierarchy** (or *organizational hierarchy*) is the intellectual plan for the map and the eventual graphic solution that satisfies the plan.[34] Each design activity should contain such a hierarchy. In this phase of design, the cartographer sorts through the components of the map to determine the relative intellectual importance of each, then seeks a visual solution that will cast each component in a manner compatible with its position along the intellectual spectrum. Objects that are important intellectually are rendered so that they are visually dominant within the map frame. (See Figure 13.18.)

CUSTOMARY POSITIONS OF MAP ELEMENTS IN THE HIERARCHY

Each map has a stated purpose that controls the planning of the visual hierarchy. Mapped objects and their relative importance assume a place in the hierarchy. Although identical objects may vary in relevance, depending on the map on which they are placed, there are general guidelines to follow in developing the hierarchy. (See Table 13.2.)

It is highly unlikely that the symbols on a thematic map will assume any rank other than the topmost. Those map objects customarily toward the bottom of the hierarchy may fluctuate more. For example, water is ordinarily placed beneath the land in the order, but might assume a more dominant role if the purpose of the map has to do with marine or submarine features. The design activity

Figure 13.18 The visual hierarchy.
Objects on the map that are most important intellectually are rendered with the greatest contrast to their surroundings. Less important elements are placed lower in the hierarchy by reducing their edge contrasts. The side view in this drawing further illustrates this hierarchical concept.

Table 13.2 Typical Organization of Mapped Elements in the Visual Hierarchy

Usual Intellectual Level*	Object	Visual Level
1	Thematic symbols	I
1	Title, legend material, symbols and labeling	I
2	Base map—land areas, including political boundaries, significant physical features	II
3–4	Important explanatory materials—map sources and credits	II–III
4	Base map—water features, such as oceans, lakes, bays, rivers	III
5	Other base-map elements—labels, grids, scales	IV

*A map object with a rank of 1 has greater intellectual importance to the map's message than one with a rank of 5.
Visual levels I through IV roughly correspond with the intellectual levels 1 through 5.

calls for a careful examination of each element and its proper placement in the hierarchy.

An interesting activity for students is to analyze thematic maps with a view to their organizational schemes. One noticeable result will be that those maps without a visual plan are the least successful in conveying meaning and are visually confusing. (See Figure 13.19.) Unfortunately, too many such maps exist. On the other hand, a map with a carefully conceived and executed hierarchy pleases the reader, is visually stable, and does not require redesigning.

The oft-stated axiom that the best designs are not even noticed operates in cartography as well as in other disciplines.

ACHIEVING THE VISUAL HIERARCHY

Ordering the intellectual importance of map elements is a relatively simple task, especially when guided by a clearly stated map purpose. Making the hierarchy work visually is another matter, involving knowledge of the perceptual tendencies of map readers. The designer must learn these tendencies if effective results are to be achieved.

FUNDAMENTAL PERCEPTUAL ORGANIZATION OF THE TWO-DIMENSIONAL VISUAL FIELD: FIGURE AND GROUND

The **figure-ground phenomenon** is often considered to be one of the most primitive forms of perceptual organization; it has even been observed in infants.[35] We tend to see objects having form as segregated from their surroundings, which are formless. Objects that stand out against their backgrounds are referred to as *figures* in perception, and their formless backgrounds as *grounds*. The segregation of the visual field into figures and grounds is a kind of automatic perceptual mechanism. Figures will not emerge from homogeneous visual fields, however.

In the everyday three-dimensional environment, we see a table on top of the floor (with the floor continuing beneath the legs of the table), buildings in front of the sky, pictures in front of the walls on which they are hung, and so on. When we lay an eraser on a piece of paper, we see the paper continuing unbroken behind the eraser and the eraser as a complete object on the top of the paper. (See Figure 13.20a.)

Figure formation is possible in two-dimensional spatial organization as well. Figures perceived in this way are seen separately from the remainder of the visual field, have form and shape, appear to be closer to the viewer than the amorphous ground, have more impressive color, and are associated with meaning.[36] The ground usually appears to continue unbroken behind the figure, just as in the three-dimensional case. A simple example of figure from the two-dimensional world is a black disc placed within a frame. (See Figure 13.20b.)

The cartographic designer should use this perceptual tendency of figure-ground segregation as a positive design element in structuring the visual hierarchy, so that the most important map elements appear as figures in perception. With careful attention to graphic detail, all the elements can be organized in the map space so that the emerging figure and ground segregation produces a totally harmonious design. Visual confusion is eliminated, and the intent of the message becomes clear. Fortunately, psychological researchers have examined many of the mechanisms that lead to figure formation, so designers have guidelines for organizing the map's graphic elements.

Perceptual Grouping Principles

Several perceptual grouping principles have been found to be primary mechanisms for figure formation. In **perceptual grouping,** the viewer spontaneously combines elements in the visual field that share similar properties, resulting in new forms or "wholes" in the visual experience. From the designer's point of view, these groupings can act in two opposing ways: as positive mechanisms to use if the map's

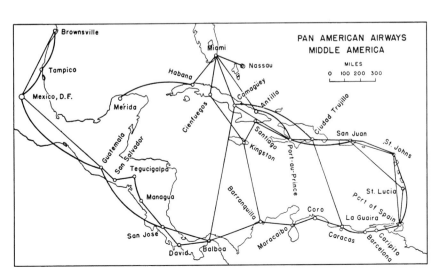

Figure 13.19 Poor organizational map plan.
In this case, it is difficult to see which areas on the map are water and which are land. Further difficulty is introduced by failure to place the thematic symbolization on the most dominant visual level. The map as a whole suffers from lack of contrast. *(Illustration from Fred Carlson,* Geography of Latin America, *p. 480. Copyright Prentice Hall, Inc. 1952. Reprinted by permission.)*

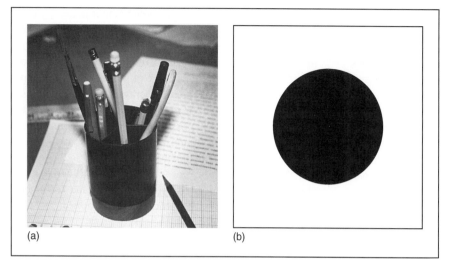

Figure 13.20 Figures and grounds.
In (a), we see familiar objects appearing in front of their backgrounds (the paper in this case). Grounds appear to continue unbroken beneath the figural objects. In the graphic two-dimensional world, as in (b), figures dominate perception and appear to rest on top of seemingly unchanging grounds. The black disc in (b) is usually seen as figure and the surrounding white area as ground.

elements will permit it, or as mechanisms to avoid if the map's elements are arranged in such a fashion that the spontaneous grouping will detract from the planned hierarchy. Their potential as design elements is explained and exemplified below.

Grouping by Similar Shape

Objects in the visual field possessing similar shapes are usually combined into a new group that appears distinct from the remainder. (See Figure 13.21a.) This perceptual feature undoubtedly comes into play when map readers view a map containing several different qualitative map symbols. (See Figure 13.21b.) In fact, if we could not visually combine identical symbols, it would be difficult to "see" the geographical pattern of one symbol type as distinct from others.

Grouping by Similar Size

Viewers tend to group similarly sized objects in the visual field into new perceptual structures. (See Figure 13.22a.) This tendency is especially important to cartographic design in at least two areas: the reading of different type sizes on maps and the visual assimilation of geographical pattern from maps containing range-

graded proportional symbols (explained in Chapter 9). Designers, although they may not be aware of it, depend on this perceptual grouping feature when they work with these cartographic elements. (See Figure 13.22b.)

Research on the perception of graduated point-symbol maps (discussed in Chapter 9) has uncovered a perceptual characteristic that is relevant to the matter of grouping by similar size. It appears that the perceived size of symbols is affected by their immediate environments. Consequently, the perceptual grouping of similarly sized symbols could possibly be affected by this condition. Patricia Gilmartin concluded her study with these findings:[37]

1. When a circle (the point symbol tested) is among circles smaller than itself, it appears larger (an average of 13 percent) than when it is surrounded by circles larger than itself. Isolated circles not surrounded by others (larger or smaller) were judged to be of an intermediate size.
2. The effect of larger or smaller circles on a surrounded circle can be reduced if internal borders (boundaries around the area represented by each graduated symbol) are used on the map.

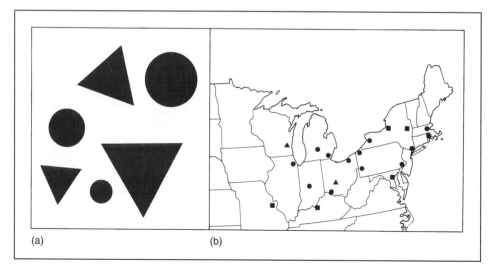

Figure 13.21 Grouping by similarity.
In perception, we tend to group similar objects. In (a), the triangles belong visually to a group distinct from the group formed by the circles. This perceptual tendency is fundamental in some cartographic situations, such as (b). We often use this perceptual grouping phenomenon to communicate thematic messages on maps.

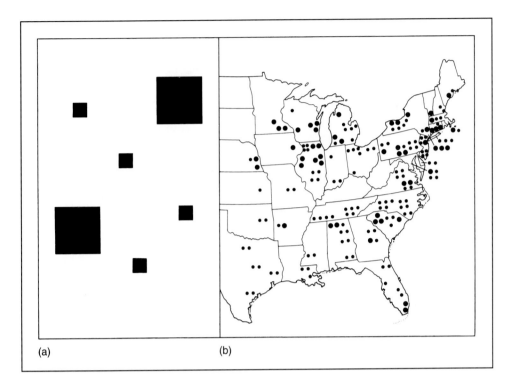

(a)   (b)

Figure 13.22   Grouping by similar size.

In (a), objects in the visual field that are similar in size tend to be grouped together in perception. It is difficult to make the experience take on a different visual structure. Try to place a small square and a large square together into a coherent new group. This perceptual tendency to group similarly sized objects is used frequently by cartographers in the design of maps, as in (b).

3. There is some evidence that smaller circles are more susceptible to being judged differently because of their environments than large circles.

It is clear from such perceptual studies that the designer's task is not easy, or even altogether clear. Perceptual tendencies can be confusing and difficult to manage in cartographic design. However, if the designer strives to create patterns that are unambiguous in perception, the design is likely to succeed.

Grouping by Proximity   Another strong perceptual grouping tendency is that of proximity. Elements in the visual field that are closer to other elements tend to be seen as a unit that stands out from the remainder. (See Figure

13.23a.) In fact, the words on this page are formed by the proximity principle—letters are grouped visually in words. The cartographer relies on this visual tendency when depicting geographical distributions, especially those containing clusters. (See Figure 13.23b.) If the eye did not combine elements in this way, it would be exceedingly difficult for designers to accomplish their task.

FIGURE FORMATION AND CLOSURE

Closure is an important perceptual principle. It refers to the tendency for the perceiver to complete unfinished objects and to see as figures objects that are already completed. A contour or edge is usually associated with closure. (See Figure 13.24a.) Broken contours are spontaneously completed

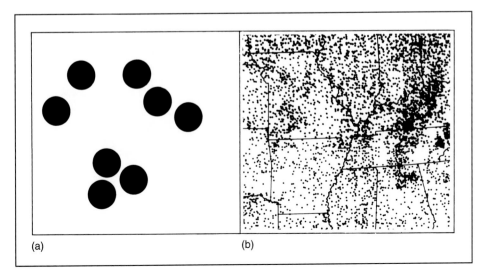

(a)   (b)

Figure 13.23   Perceptual grouping by proximity.

In (a), visual recombination of the groups into a new organization would be difficult. Cartographers use this perceptual grouping tendency in map design, as in (b). (*Dot map in (b) is part of a larger map taken from U.S. Bureau of the Census,* Census of Agriculture, *1969, vol. 5, Special Reports. pt. 15, Graphic Summary [Washington DC: USGPO, 1973], p16.*)

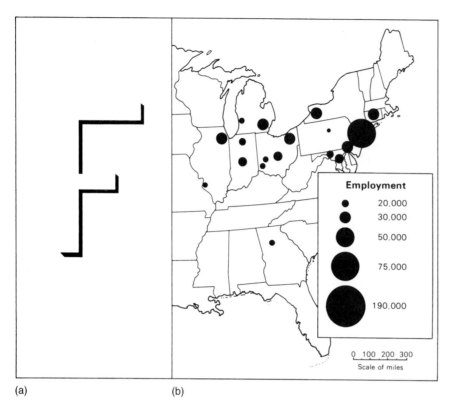

(a)   (b)

Figure 13.24   Closure.
We tend to close figural objects to form more simple structures, as in (a). Our perceptual mechanisms supply missing elements without conscious thought. This occurs in many cartographic situations, and the communication does not suffer. In (b), the person reading the map provides the missing coastlines and is not really bothered by their physical absence.

so that we see the "whole" object. Therefore, map designers need to provide strong, completed contours around figural objects. Sometimes these edges are "broken" by the diagram to provide for lettering or other map elements. (See Figure 13.24b.) Unless skillfully executed, this design solution can have negative effects on figural areas.

USING TEXTURE TO PRODUCE FIGURES

Texture and differences in texture can be used to produce figures in perception.[38] It appears that orientation of the textural elements is more important in figure development than is the positioning of the elements. (See Figure 13.25.) This feature is related to the quality called **intraparallelism,** which is the similar alignment of elements in the visual field to achieve order and harmony in the experience.[39]

The cartographic literature also provides evidence that texture and texture discrimination lead to the emergence of figures. In a study to determine the effects of different graphic representations of land and water, it was found that a representation containing a textured pattern over one surface produced the least amount of ambiguity during perception.[40] Unfortunately, the subjects in the experiment also considered that particular configuration to have a low aesthetic value. This negative response may have been caused by the specific textured patterns selected for the experiment, not the inherent visual differentiation caused by texture as such.

Using texture to evoke figures in perception is closely related to the feature of **articulation** in the visual field: inequality of detail from place to place in the field. Both texture and articulation are ways of achieving heterogeneity in visual experience, a necessary requirement for figure

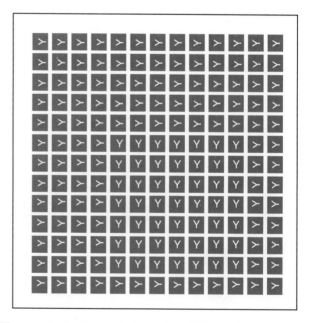

Figure 13.25   Textural elements and figures in perception. Although subtle, as in this case, texture can provide just enough visual contrast to segregate the visual field into figures and grounds.

perception.[41] Adding texture and detail to shapes on the map that are planned as figures is a reliable graphic method of ensuring their visual emergence during map reading. (See Figure 13.26.)

Differential brightness, measured by light reflected from the map sheet, also can be used to cause figural development.[42] A difference of at least 20 to 25 percent in reflected light is required to produce the desired result. Brightness as it relates to color and figure formation will be dealt with in Chapter 15.

STRONG EDGES AND FIGURE DEVELOPMENT

One of the principal ways of producing a strong figure in two-dimensional visual experience is to provide crisp edges to figural objects. Conversely, figural dominance can be weakened by reducing edge definition. Edges result from contrasts of brightness, reflection, or texture. These characteristics have special significance and utility in cartographic design, particularly in coordinating the graphic elements in the planned visual hierarchy. For example, thematic symbols and other elements high in the hierarchy can be rendered in solid hue (black or color), and subordinate features can be screened, thus reducing the sharpness of their edges. Screening also reduces the intensity of the less important elements. (See Figure 13.27.)

Screening to reduce edge sharpness and intensity is a photomechanical technique used to produce **aerial perspective** on a flat map sheet. This phenomenon is part of

our three-dimensional world; it accounts for the haziness and lack of clarity of distant objects. It is often called a *depth cue*. Landscape painters employ this real-world phenomenon by rendering distant objects with less detail and less intense color.[43]

Important work on the contributions that the figure-ground phenomenon can make on design continues. Clifford Wood, for example, remarks on this subject clearly by saying:

> *Cartographic communication can only be successful if sound map design principles are fully understood and used properly. It is obvious that many cartographers still do not understand the fundamental principle of map design, the figure-ground relationship.*[44]

For Clifford Wood, the figure-ground relationship formed the basis of a research effort that has important contributions for designers. Fundamentally, he researched the idea that designs incorporating figure-ground divisions would have an effect on the way readers scan maps. Among others, the results showed that without a figure-ground inspired design, map readers have visual processing difficulties.[45] He further remarks, "It is clear that figure-ground differentiation is a vital element in successful graphic communication."[46]

I believe that the figure-ground phenomenon, as a basic visual and perceptual tendency among map readers, must be accommodated for in *every* thematic map design project. To not do so would be a serious oversight.

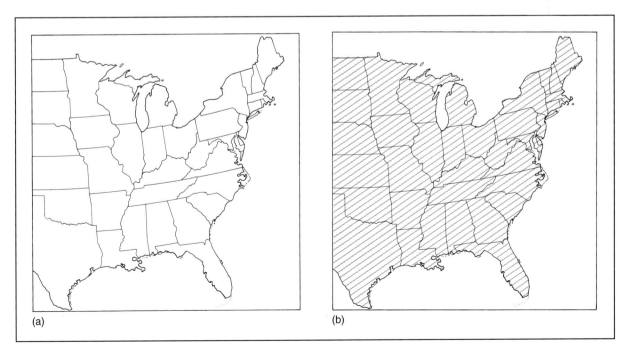

(a)   (b)

Figure 13.26   The map and the use of textural elements.
Textural elements should be applied subtly. Texture can be placed on either figure or ground areas. The illustration here shows screened lines.

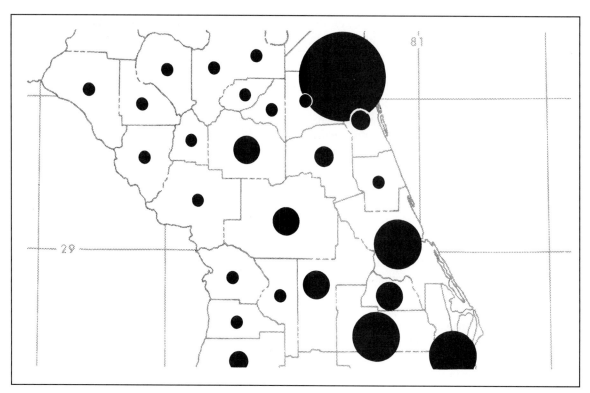

Figure 13.27   The use of screening to develop visual hierarchy.

Less important elements (coastlines, political boundaries) have been screened to reduce the shape of their edges. The graticle has been screened even more, further reducing its contrast and thereby placing it on an even lower visual level. Important elements (symbols) are rendered in total black, which tends to emphasize their importance in the overall organization.

THE INTERPOSITION PHENOMENON

A most useful way of causing one object in the two-dimensional visual field to appear on top of or above another is to interrupt the edge or contour of one of the objects. This phenomenon is usually referred to as **interposition** and is frequently cited as a depth cue in perception.[47] In cartographic design, this technique can be used to strengthen the dominance of certain objects in the visual hierarchy. (See Figure 13.28.) The result is an impression of depth on the map—an interesting, dynamic, and fluid solution.

It is possible to use interposition cues to produce stacking effects of clustered graduated point symbols (these symbols are discussed in Chapter 8). Although this method results in a map that appears three-dimensional (thus adding interest to the design), it makes it difficult for the map reader to see the quantitative differences of the scaled symbols.[48] On the average, errors in symbol reading increase with the amount of the symbol obscured by an overlapping one. Because of this, the use of interposition is not recommended in cases where one quantitative point symbol overlaps another. Otherwise, interposition can be used whenever it enhances the overall hierarchical plan for the map.

FIGURES AND GROUNDS IN THE MAP FRAME

Within a bounded space in the two-dimensional visual field, areas that are smaller and completely enclosed will tend to be viewed as well-defined figures. Cartographic designers have considerable choice in such elements as texture, articulation, edging, and interposition to accentuate objects as figures, but less freedom in the manipulation of figural size and enclosedness. Constraints imposed by location, scale, and map size can preclude any adjustment to enhance figural areas of the map. In cases where these restrictions do not limit design choice, there are guidelines to assist in selecting the proper size ratios of figures and grounds.[49] In one study, acceptable size ratios ranged from 1:2.18 to 1:3.56—nonfigural areas may be from 2.18 to 3.56 times larger than figural areas without interfering with figure formation.

One approach to assure that the figural portion of the map is completely enclosed, in the case of land-water differentiation, is to provide an enclosing contour by continuing the shoreline as the map border.[50] (See Figure 13.29.) This solution, although not preferred by the respondents in the study, did reduce ambiguity somewhat.

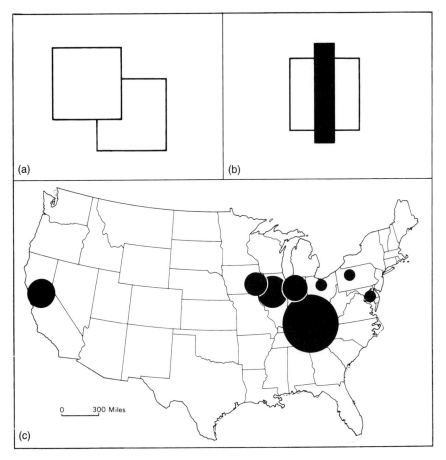

Figure 13.28   Interposition to assist in figural development.

In (a), we tend to see one square in front of the other one. In (b), the back rectangle is seen "over" the square. The objects whose edge contours continue unbroken are the ones seen as being on top. Interposition can be used on maps, as in (c). The back circles break coastlines or state boundaries, and appear "on top" of the land. This enhances their figural properties.

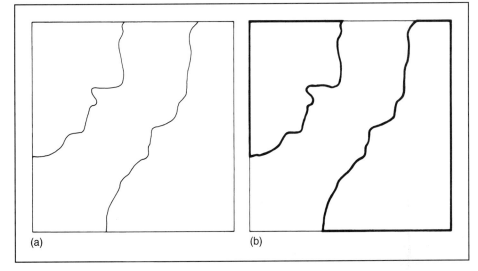

Figure 13.29   Figural development of land areas.

In (a), the configuration of land and water areas and the graphic treatment of the map border hinder the land from becoming figure. In (b), the borders joining the land areas are treated as part of the land. This forms enclosing contours and strengthens the perception of the land areas as wholes. Cartographic designers will need to apply different solutions to varying designs.

THE SPECIAL CASE OF THE LAND-WATER CONTRAST

The principal concern of the thematic map designer is to communicate a spatial message effectively. To a great extent, the success of this effort depends on how the message is presented or arranged for the map reader. If the graphic material is presented in a clear and unambiguous manner, success in communication will probably result. On the other hand, an unclear and confusing graphic picture will make the reader frustrated and unreceptive to the message. (See Figure 13.30.) The graphic elements of the map must therefore be arranged in such a way as to reduce any possible reader conflict. One way of eliminating possible confusion is to provide clues on the map to help the reader in determining geographical location. A worthwhile exercise for a map-design class would be to redesign the map in Figure 13.13 to achieve a more stable and visually clear image.

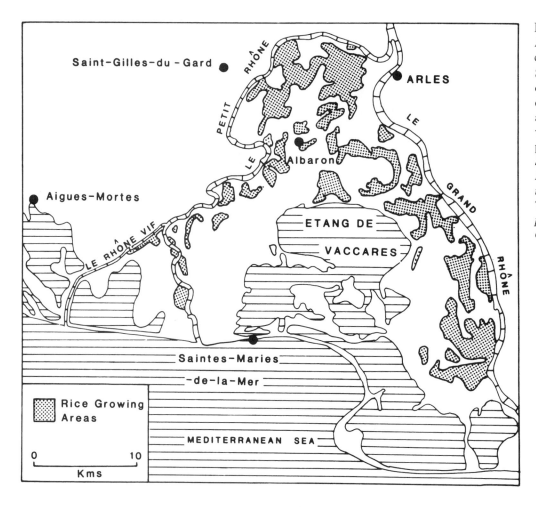

Figure 13.30
Ambiguous land-water contrast.

Some coastlines are very complex, causing difficult design solutions. If looked at long enough, land and water areas alternate in perception. (*Map from A. Scarth, "Rice and the Ecological Environment in the Camargue,"* Geography 71 [1986]:157. *Reprinted by permission of the Geographic Association.*)

A significant geographical clue is the differentiation between land and water, if the mapped area contains both. This distinction has been suggested as the first important process in thematic map reading.[51] Maps that present confusing land-water forms deter the efficient and unambiguous communication of ideas. (See Figure 13.31.) Design solutions should never be visually distracting or lacking in clarity and stability in the intended order. Land-water differentiation usually aims to cause land areas to be perceived as figures and water areas as ground. In unusual cases, water areas are the focal point of the map and would therefore be given graphic treatment to cause them to appear as figures.

VIGNETTING FOR LAND-WATER DIFFERENTIATION

Historically, engravers, drafters, and cartographers solved the land-water differentiation problem by use of **vignetting** at the coastline. Vignetting is any graphic treatment emerging from an edge or border and resulting in a continuous gradient of brightness. Vignetting of the coast is most common, although it has been used on maps at political or other borders for visual differentiation between land areas.

Some of the popular ways of coastal vignetting are stippling, form lines, and continuous tones (both decreas-

ing or increasing in brightness away from the coastline). (See Figure 13.32.) In fact, these methods have come under close scrutiny to determine if they in fact enhance the figure and ground organization of the map. Most research has compared them to other forms of figure enhancement (e.g., heterogeneity, texture, articulation, and graticule) in terms of effectiveness.

One surprising result of these studies, reported by at least two researchers working independently, is that stippling on the water at coastlines is ineffective.[52] In one study, the stippling caused the water to be seen as land! In the other study, stippling on the water side was ranked sixth among eight in figural goodness and also in aesthetic preference. It appears that stippling should be used very cautiously, if at all.

In a study by Lindenberg, a continuous tone of increasing brightness (away from the coastline) was ranked best overall by map-reading judges in developing the figural goodness of land *and* in aesthetic preference. The form line method was not judged very effective or aesthetically preferred. Both studies support the idea that development of heterogeneity in the visual field through differences in surface texture is as effective in developing land-water contrasts as is coastal vignetting.

Summarizing the various methods of achieving land-water differentiation, it would seem that, if vignetting is

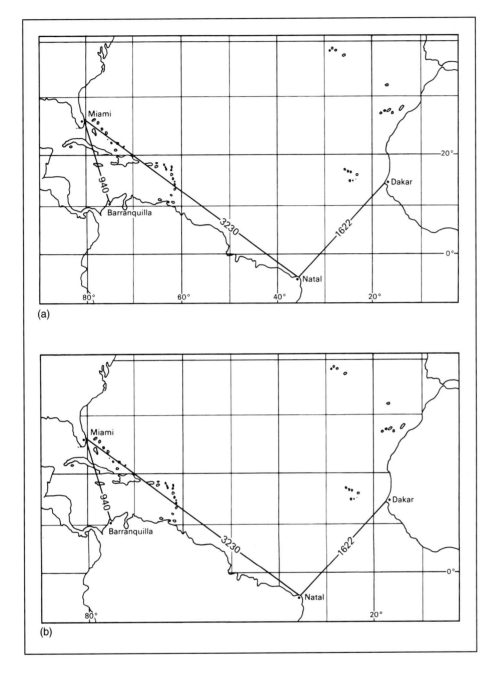

(a)

(b)

Figure 13.31 Land and water
contrasts.
In (a), the contrast between land and
water is not clearly evident. The
graticle creates a similar texture over
the entire map, resisting figure
formation. A very simple correction
can overcome this. In (b), the graticle
has been eliminated over land areas,
causing them to be enhanced as
figures. In many cases, only simple
corrections are necessary to solve
problems of insufficient contrast
between land and water.
(Illustration (a) from Fred Carlson,
Geography of Latin America, *p. 491.*
Copyright Prentice Hall, Inc., 1952.
Reprinted by permission.)

used, a continuous tone of increasing brightness should be applied to the water at the coastal interface. If no vignetting is used, at least surface-texture differences between land and water should be developed. One last suggestion is to arrange the shapes in the map frame to reduce the possibility of ambiguous or reversible figures.

In sum, "The cartographer is responsible for structuring a complete synthesis which will be comprehensible to the user."[53] This can be done by effectively planned visual hierarchy, carried out by incorporating known perceptual principles of the figure-ground dichotomy. This has been attempted here by redesigning Figure 13.30 referred to earlier. (See Figure 13.33.)

DESIGNING THE PAGE-SIZE MAP

An interesting exercise for students is to examine by example how the page-size, or smaller, map is designed, and especially how the various map elements are treated in this process. The following paragraphs discuss how the design of the map in Figure 13.34c was reached.

To begin with, the organization of the land-water was examined. Because there are both elements on the map, a decision about their place in the visual hierarchy had to be determined. (See Figure 13.34a.) As the principal map purpose was to show population, which is on land, the water element needed to be treated lower in the hierarchy. I believed I could achieve this by rendering the land as white and the

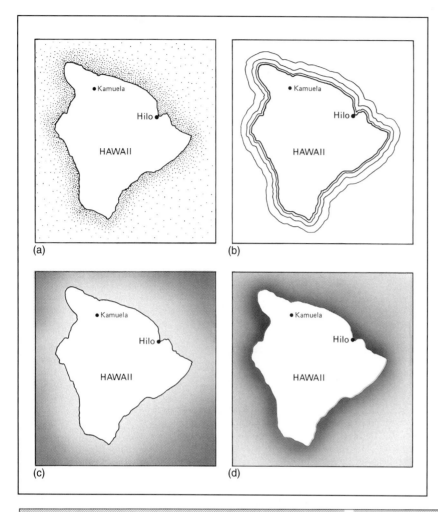

(a)   (b)   (c)   (d)

Figure 13.32 Different methods of coastal vignetting.

In (a), stippling on the water side does not clearly show the land as figure but can be used because it is conventional. In (b), form lines are used; these are not aesthetically preferable. In (c), a continuous tone of decreasing brightness is employed. This is not particularly good because it lacks contrast at the coastline. The continuous tone of oncreasing brightness shown in (d) is the best choice of the ones illustrated, because it enhances land as figure and is aesthetically preferable. *(Source: These figures are redrawn from Richard Lindenberg. Coastal Vignetting and Figure-Ground Perception in Small-Scale Black and White Cartography [unpublished master's thesis, Department of Geography, Georgia State University, 1975] p. 76.)*

Figure 13.33 Redesign for better land-water contrast and overall improvement of the visual hierarchy.

This map is a redesign of the map in Figure 13.33. Appreciation is extended to Elizabeth Pascard, a cartography graduate student at Georgia State University, for her assistance in developing this map.

water a light grey, which would also serve to bring out the distinction between land and water. (See Figure 13.34b.)

The primary purpose of this map was to illustrate the distribution of the population in major population centers in the Canadian Core region. As this data is concentrated at points, it seemed very appropriate to symbolize the data as graduated point symbols. Good contrast along symbol edges would develop the symbols as strong figures, so black was chosen for the symbols (as this was to be a noncolor map).

Effective thematic maps have to have several other elements as part of their design. Map *title, subtitle, legend, scale, sources, and any explanatory elements* were considered and finally placed on this map. A first design organization was attempted, and each element was placed on the map. (See Figure 13.34b.)

After looking at the first map element organization treatment, I decided that the arrangements of the elements could be done to serve map balance better and map space more efficiently. As this map was prepared on a microcomputer using Designer software (a product of Micrografx in Richardson, Texas), it was fairly easy to rearrange the map elements any number of ways.

Different word arrangements were tried for the title and subtitle. (See Figure 13.34b.) I also thought the map needed a scale, always a consideration when mapping. The map's symbols needed an efficient legend to describe their scaling for this map. These elements, too, were placed several ways on the map before the final design was rendered.

The final map design appears here as Figure 13.34c. To me, there is good land-water contrast, and good figure-ground employment in the overall visual hierarchy on the map. The population symbols are strong figures and, because they overlap the coastline edges and other lines on the map, they appear "on top," a feature that strengthens their perception as figures. There is good balance and use of space over the whole map. The title is placed for effective balance and it, too, is positioned for overlap to strengthen its place higher in the overall hierarchy. The

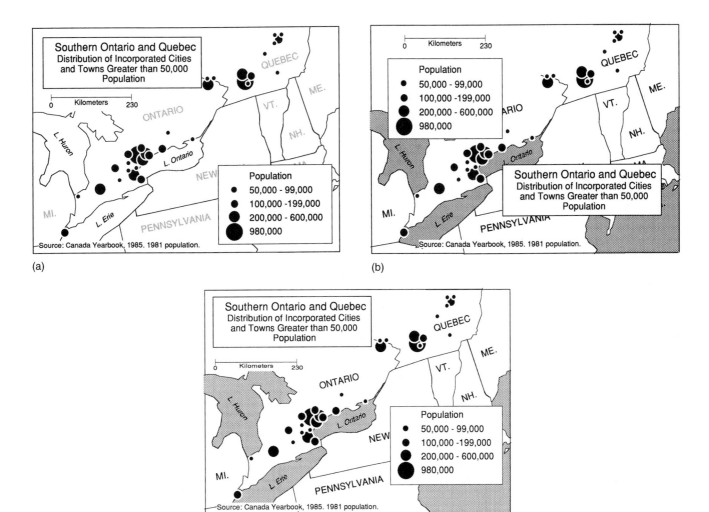

Figure 13.34   Map planning, visual hierarchy, and balance in small-scale map design.
See text for explanation.

scale and explanatory materials are included, but not dominant in the final design.

This exercise was a brief look at how a small thematic map is treated in design. Overall, all the essential map elements were treated and used. The final map appearance is pleasing and efficient.

NOTES

1. Kurt Hanks, Larry Bellistan, and Dave Edwards, *Design Yourself* (Los Altos, CA: William Kaufmann, 1978), pp. 60–61. For additional comment on the design process, see also Eve Downing, "Designing at MIT" (alberti.mit.edu); Charles J. Nuese, *Building the Right Things Right* (New York: Quality Resources, 1995); Victor Papanek, *The Green Imperative: Natural Design for the Real World* (New York: Thames and Hudson, 1995); Milton D. Rosenau, Abbie Griffin, George A. Castellion, and Ned F. Anschultz, *The PDMA Handbook of New Product Design* (New York: Wiley, 1996); and Philip Thiel, "Beyond Design Review: Implications for Design Practice, Education, and Research," *Environment and Behavior* 26 (1994), 363–76.
2. Phillip C. Muehrcke, "An Integrated Approach to Map Design and Production," *American Cartographer* 9 (1982): 109–22.
3. Michael Southworth and Susan Southworth, *Maps: A Visual Survey and Design Guide,* A New York Graphic Society Book (Boston: Little, Brown, 1982), pp. 16–17.
4. The Burdick Group, *Creativity: The Human Resource* (exhibit) (San Francisco: Standard Oil Company of California, 1982).
5. Mike Samuels and Nancy Samuels, *Seeing with the Mind's Eye* (New York: Random House, 1975), p. xi; and Phillip C. Muehrcke, "Maps in Geography," in *Maps in Modern Geography: Geographical Perspectives on the New Geography,* ed. Leonard Guelke (Cartographica, Monograph No. 27, 1981), p. 11.
6. Samuels and Samuels, *Seeing with the Mind's Eye,* p. 43.
7. Samuels and Samuels, *Seeing with the Mind's Eye,* pp. 239–40.
8. Ibid., p. 245.
9. For an interesting discussion of visualization in geography see David DiBiase, "Visualization in the Earth Sciences," *Earth and Mineral Sciences* 59 (1990): 13–18. (This publication is produced by the College of Earth and Mineral Sciences, Pennsylvania State University, University Park.)
10. Muehrcke, "Maps in Geography," p. 13.
11. John K. Wright, "Map Makers Are Human," *Geographical Review* 32 (1944): 527–44.
12. John S. Keates, *Understanding Maps* (New York: Wiley, 1982), p. 127.
13. Aart J. Karssen, "The Artistic Elements in Design," *Cartographic Journal* 17 (1980): 124–27.
14. Ibid.
15. Ibid.
16. Michael Wood, "The Map-Users' Response to Map Design," *Cartographic Journal* 30 (1993): 149–153; see also R. Beard, A. D. Cooper, and K. R. Crossley, "Map Design Education," *Cartographic Journal* 30 (1993): 159–62; P. Fisher, J. Dykes, and J. Wood, "Map Design and Visualization," *Cartographic Journal* 30 (1993): 136–42; J. S. Keates, "Some Reflections on Cartographic Design," *Cartographic Journal* 30 (1993): 199–201.
17. Ibid.
18. Arthur H. Robinson, *The Look of Maps* (Madison: University of Wisconsin Press, 1966), p. 13.
19. Rudolf Arnheim, *Art and Visual Perception* (Berkeley: University of California Press, 1965), p. 14.
20. Ibid., p. 12.
21. Ibid., pp. 14–17.
22. Ibid., p. 27.
23. Kevin Coates, *Geometry, Proportion, and the Art of Lutherie* (Oxford: Oxford University Press, 1985).
24. Richard Surrey, *Layout Techniques in Advertising* (New York: McGraw-Hill, 1929), pp. 11–20.
25. James R. Antes, Kang-tsung Chang, and Chad Mullis, "The Visual Effect of Map Design: An Eye-Movement Analysis," *American Cartographer* 12 (1985): 143–55.
26. Surrey, *Layout Techniques,* p. 15.
27. Arnheim, *Art and Visual Perception,* p. 45.
28. Borden D. Dent, "Simplifying Thematic Maps through Effective Design: Some Postulates for the Measurement of Success," *Proceedings* (American Congress on Surveying and Mapping, Fall Technical Convention, October 1973): 243–51.
29. Albert M. Potts, ed. *The Assessment of Visual Function* (St. Louis: Mosby, 1972), p. 5.
30. Ibid., p. 18.
31. Ibid., p. 21.
32. Borden D. Dent, "Visual Organization and Thematic Map Communication," *Annals* (Association of American Geographers) 62 (1972): 79–93.
33. Deborah T. Sharpe, *The Psychology of Color and Design* (Chicago: Nelson-Hall, 1974), p. 102.
34. Borden D. Dent, "Visual Organization and Thematic Map Communication," *Annals* (Association of American Geographers) 62 (1972): 79–93.
35. William D. Dember and Joel S. Warm, *Psychology of Perception,* 2d ed. (New York: Holt, Rinehart and Winston, 1979), p. 251.
36. Ralph Norman Haber and Maurice Hershenson, *The Psychology of Visual Perception* (New York: Holt, Rinehart and Winston, 1973), p. 184.

37. Patricia P. Gilmartin, "Influences of Map Context in Circle Perception," *Annals* (Association of American Geographers) 71 (1981): 253–58.

38. John P. Frisby, *Seeing: Illusion, Brain, and Mind* (Oxford: Oxford University Press, 1980), pp. 115–16; see also Dent, "Visual Organization and Thematic Map Communication," pp. 79–93.

39. William R. Sickles, "The Theory of Order," *Psychological Review* 49 (1942): 403–21.

40. C. Grant Head, "Land-Water Differentiation in Black and White Cartography," *Canadian Cartographer* 9 (1972): 25–38.

41. E. G. Weaver, "Figure and Ground in the Visual Perception of Form," *American Journal of Psychology* 38 (1927): 194–226.

42. Clifford H. Wood, "Brightness Gradients Operant in the Cartographic Context of Figure-Ground Relationship," *Proceedings* (American Congress on Surveying and Mapping, March 1976): 5–34.

43. For an interesting discussion of psychological depth cues by a cartographer, see M.-J. Kraak, "Three-Dimensional Map Design," *Cartographic Journal* 30 (1993): 188–94.

44. Clifford Harlow Wood, *The Influence of Figure and Ground on Visual Scanning Behavior in Cartographic Context* (unpublished Ph.D. dissertation, Department of Geography, University of Wisconsin–Madison, 1992), p. 128.

45. Ibid., p. 263.

46. Ibid., p. 262.

47. Julian E. Hochberg, *Perception* (Englewood Cliffs, NJ: Prentice Hall, 1964), pp. 87–88; see also Gaetano Kanizsa, *Organization in Vision: Essays on Gestalt Perception* (New York: Praeger, 1979), pp. 94–97.

48. Richard E. Groop and Daniel Cole, "Overlapping Graduated Circles: Magnitude Estimation and Method of Portrayal," *Canadian Cartographer* 15 (1978): 114–22.

49. P. V. Crawford, "Optimum Spatial Design for Thematic Maps," *Cartographic Journal* 13 (1976): 134–44.

50. Head, "Land-Water Differentiation," pp. 25–38.

51. Ibid.

52. Ibid.; and Richard E. Lindenberg, *Coastal Vignetting and Figure-Ground Perception in Small-Scale Black and White Cartography* (unpublished master's thesis, Department of Geography, Georgia State University, Atlanta, 1975).

53. Willis Heath, "Cartographic Perimeters," *International Yearbook of Cartography* 7 (1967): 112–20.

GLOSSARY

aerial perspective   the diminution of detail with increasing distance, p. 257

articulation   providing detail in one part of the visual field; objects that are more articulated are frequently perceived as figures, pp. 256–257

closure   spontaneous perceptual tendency to close contours or edges around objects to make them whole, pp. 255–256

contrast   important element of design; contrasts of line, texture, value, detail, and color are means through which maps become interesting and dynamic, p. 246

creativity   unusual ability to see relationships among elements, p. 238

figure and ground organization   fundamental behavioral tendency to organize perception into figures and grounds; figures are dominant elements, and grounds serve as backgrounds for figures, p. 251

figure-ground phenomenon   a fundamental perceptual tendency in which visual fields are spontaneously divided into outstanding objects (figures) and their surroundings (ground), p. 253

focus of attention   part of the visual field that attracts the reader's eye, pp. 245–246

golden section   method of dividing two-dimensional space so that the proportion of a smaller area to a larger one is identical to that of the larger area to the whole, p. 244

graphic ideation   bringing images into clear focus by sketching, p. 239

hierarchical organization   composition or arrangement of the map's visual elements as they appear between two or more visual or intellectual levels, p. 243

image pool   the collection of mentally stored visual images obtained from previous visual experiences, p. 238

imagining   creating visual images in the mind's eye, p. 238

interposition phenomenon   perceptual tendency for one object to appear behind another because of interrupted contour, p. 258

intraparallelism   similar alignment of elements in a textural surface; can lead to figure formation in perception, pp. 246, 256

line character   the internal elements that make up the distinctive qualities of a line; e.g., dot, dot-dash, dash-dash-dot, etc., p. 247

line weight   the thickness of a line, p. 247

map composition   arrangement of the map's visual and intellectual components, p. 243

map design process   characterized by six stages: problem identification, preliminary ideas, design refinement, analysis, decision, and implementation, p. 237

map elements   marks that make up the total visual image comprising a map, including the title, legend, scale, credits, mapped or unmapped areas, graticule, borders and neatlines, and symbols, pp. 252–253

optical center   the place just above the geometric center in an image space; can be used to create visual balance, p. 243

perceptual groupings   mechanisms identified by psychologists as leading to figure formation; include grouping by shape, size, and proximity, pp. 253–255

planar organization composition or arrangement of the map's visual elements as they appear at one visual or intellectual level, p. 243

resolution acuity ability to discern a separation between objects in the visual field; usually measured by angular subtense, pp. 249–250

scientific visualization transforming symbolic information into geometric forms by representing them on computer display devices, allowing the viewer to see structures not observable before, p. 239

vignetting graphic treatment at an edge or border, resulting in a continuous gradient of brightness, p. 260

visibility acuity size threshold; the ability to discern an object in the visual field; usually measured by angular subtense on the retinal image, as opposed to the physical size of the object, p. 249

visual balance state in which all objects in a visual image appear in equilibrium, p. 243

visual hierarchy the intellectual plan for the map and the subsequent graphic solution to satisfy the plan; may also be called the organization hierarchy, p. 252

visualization mental process in which the designer experiences whole new creations by rearranging previously stored visual images, p. 238

READINGS FOR FURTHER UNDERSTANDING

Adams, Robert. *Creativity and Communications.* London: Studio Vista Limited, 1971.

Antes, James, Kang-tsung Chang, and Chad Mullis. "The Visual Effect of Map Design: An Eye-Movement Analysis." *American Cartographer* 12 (1985): 143–55.

Arnheim, Rudolf. *Art and Visual Perception.* Berkeley: University of California Press, 1965.

———. *Visual Thinking.* Berkeley: University of California Press, 1971.

Beck, Jacob, ed. *Organization and Representation in Perception.* Hillsdale, NJ: Erlbaum, 1982.

The Burdick Group. *Creativity: The Human Resource* (exhibit). San Francisco: Standard Oil Company of California, 1982.

Crawford, P. V. "Optimum Spatial Design for Thematic Maps." *Cartographic Journal* 13 (1976):134–44.

Dember, William N. *The Psychology of Perception.* New York: Holt, Rinehart and Winston, 1961.

———, **and Joel S. Warm.** *Psychology of Perception.* 2d ed. New York: Holt, Reinhart and Winston, 1979.

Dent, Borden D. "Visual Organization and Thematic Map Communication." *Annals* (Association of American Geographers) 62 (1972): 79–93.

———. "Simplifying Thematic Maps through Effective Design: Some Postulates for the Measurement of Success." *Proceedings* (American Congress on Surveying and Mapping, Fall Technical Convention, October 1973): 243–51.

Ferens, Robert J. "Design of Page-Size Maps and Illustrations. "*Surveying and Mapping* 28 (1968):447–55.

Ford, Kathryn. *Perceptual Organization of Complex Atlas Plates.* Unpublished master's thesis. Queen's University, Department of Geography, 1983.

Groop, Richard E., and Daniel Cole. "Overlapping Graduated Circles: Magnitude Estimation and Method of Portrayal. " *Canadian Cartographer* 15 (1978): 114–22.

Haber, Ralph Norman, and Maurice Hershenson. *The Psychology of Visual Perception.* New York: Holt, Rinehart and Winston, 1973.

Hanks, Kurt, Larry Belliston, and Dave Edwards. *Design Yourself.* Los Altos, CA: William Kaufmann, 1978.

Head, Grant. "Land-Water Differentiation in Black and White Cartography." *Canadian Cartography* 9 (1972): 25–38.

Hochberg, Julian E. *Perception.* Englewood Cliffs, NJ: Prentice Hall, 1964.

Jones, Christopher. *Design Methods.* New York: Wiley, 1981.

Kanizsa, Gaetano. *Organization in Vision: Essays on Gestalt Perception.* New York: Praeger, 1979.

Karssen, Aart J. "The Artistic Elements in Map Design." *Cartographic Journal* 17 (1980): 124–27.

Lindbeck, John R. *Designing Today's Manufactured Products.* Bloomington, IL: McKnight and McKnight, 1972.

Lindenberg, Richard E. *Coastal Vignetting and Figure-Ground Perception in Small-Scale Black and White Cartography.* Unpublished masters thesis. Atlanta: Georgia State University, Department of Geography, 1975.

Potts, Albert M., ed. *The Assessment of Visual Function.* St. Louis: Mosby, 1972.

Robinson, Arthur H. *The Look of Maps.* Madison: University of Wisconsin Press, 1966.

Rock, Irwin, ed. *The Perceptual World: Readings from Scientific American Magazine.* New York: Freeman, 1990. An excellent collection of readings on perception, this work is highly recommended for design students.

Samuels, Mike, and Nancy Samuels. *Seeing with the Mind's Eye.* New York: Random House, 1975.

Sharpe, Deborah T. *The Psychology of Color and Design.* Chicago: Nelson-Hall, 1974.

Sickles, William R. "The Theory of Order." *Psychological Review* 49 (1942): 403–21.

Surrey, Richard. *Layout Techniques in Advertising.* New York: McGraw-Hill, 1929.

Tufte, Edward R. *The Visual Display of Quantitative Information.* Cheshire, CT: Graphics Press, 1983.

Weaver, E. G. "Figure and Ground in the Visual Perception of Form." *American Journal of Psychology* 38 (1927): 194–226.

White, Jan V. *Graphic Idea Notebook.* New York: Watson-Guptill Publications, 1980.

Whitfield, P. R. *Creativity in Industry.* Baltimore: Penguin Books, 1975.

Wood, Clifford H. "Brightness Gradients Operant in the Cartographic Context of Figure-Ground Relationship." *Proceedings* (American Congress on Surveying and Mapping, March 1976): 5–34.
———. *The Influence of Figure and Ground on Visual Scanning Behavior in Cartographic Context.* Unpublished Ph.D. dissertation. Madison: University of Wisconsin, Department of Geography, 1992.

Wood, Michael. "Visual Perception and Map Design." *Cartographic Journal* 5 (1968):54–64.
———. "Human Factors in Cartographic Communication." *Cartographic Journal* 9 (1972): 123–32.
Wright, Richard D. *The Selection of Line Weights for Solid, Qualitative Line Symbols in Series on Maps.* Unpublished Ph.D. dissertation. Laurence: University of Kansas, Department of Geography, 1967.

CHAPTER

14

MAKING THE MAP READABLE: THE INTELLIGENT USE OF TYPOGRAPHICS

CHAPTER PREVIEW

Map lettering is an integral part of the total design effort and should not be reduced to a minor role. Lettering on the map functions to bring the cartographer and map reader closer together and makes communication possible. To employ lettering properly, the cartographic designer should be familiar with letterform characteristics, sizes, letterspacing, type personalities, and legibility. Size is a most critical element in design, and lettering style and personality can affect the appearance of the map. Type classification is important for the cartographer in selecting a suitable *typeface. Map typography includes lettering placement; overall lettering harmony can be achieved through adherence to established conventions of placement. Experimental studies expand knowledge of map typography and reveal the need for greater research in this area of cartographic design. Today, most map lettering is generated by cold type methods, primarily by computer software in the form of page-description languages. Digital fonts have taken over at the computer workstation, offering cartographers the most in design flexibility and ease.*

For many cartographic designers, the planning and application of map lettering remains the last task. This is unfortunate—in most instances, the appearance and mood of the entire map can be set by its lettering, so lettering should not be treated lightly in design. Although only a handful of cartographic researchers have examined map lettering in any detail, there is extensive general literature on type, its design, history, application, and production. This chapter deals with the fundamentals of lettering and attempts to illustrate the importance of lettering in thematic map design. Four major topics—the function of map lettering, the elements of type, map typography and design, and the production of map lettering—will be examined. The specific subject of typesetting by computers is addressed in the last pages of this chapter.

FUNCTIONS OF MAP LETTERING

All map lettering, whether on general-reference maps or thematic maps, serves to bring the cartographer and map user together, making communication possible. Only in the highly unusual situation where cartographer and map user discuss the map in person can written language be ignored.

On general-reference maps, lettering serves mainly to *name places* and to identify or *label things* (e.g., scales, mountains, oceans, straits, and graticule elements). On thematic maps, lettering is also provided for titles, legends, and other explanatory marginal materials necessary to make the map content more comprehensible. Because all written language elements (letters and words) are symbols for meaning, they serve the same function on maps.

Map lettering should be viewed first as a *functional symbol* on the map, and only secondarily as an aesthetic object. Nevertheless, map lettering, if not done well, can hinder communication. Therefore, the cartographer should approach the employment of lettering with an appropriate regard for both function and form. Map lettering in this context refers to the *selection* of lettering type and its *placement* on the map. Questions of which words to use in titles, labels, or other marginalia are not addressed here.

Lettering can express the nature of a geographical feature by its *style,* the feature's importance by its *size,* the feature's location by its *placement,* and the feature's extent by its *spacing.*[1] The variables of style and size are most important in the design of titles and legends; size, spacing, and placement are particularly important on the body of the map. These distinctions set apart map lettering from general text lettering and the designer should keep them in mind.

Cartographic writers have provided further convenient classification to help in the understanding of the different uses of map type. Table 14.1 presents one such classification.

THE ELEMENTS OF TYPE

The selection of typeface and the placement of lettering are the two chief concerns of the designer. Proper selection can be made only if the cartographer fully understands the fundamentals of *type design.* As there are hundreds of individual typefaces from which to choose, the designer must recognize the elements and characteristics of their design, how they may be classified to make selection and

---

Table 14.1   Classification of Map Lettering According to Use

Use	Description and Use
Descriptive text	Reflects features that are symbolized on map face by point, line, area
Narrative	Names of objects
Descriptive	Additional property of feature ("scenic route")
Warning	Dangerous nature of feature ("sunken wreck")
Functional information	Locatable ground feature ("Rescue Post")
Regulatory	Legal information (area of land)
Analytical text	Links user with attribute of features
Confirmative	Spatial relations (distance between two towns or bearings on cadastre)
Determinative	Tables placed along map
Interpretive	Difficult to get information from map, so it is provided ("quickest route is . . .")
Reference	Text alongside map
Categorization	Categorization of a theme in codes (soil maps, geologic maps)
Positional Text	Text to describe or confirm location, in space or time
Geocoding	Grid reference notations
Measurement	Relative position (at edge of map, "Twenty miles to . . .")
Temporal Position	Text to give time of events (historic battles)
Metadata	Refer to nature of source data to map as a whole (reference ellipsoid)

Source: D. J. Fairbairn, "On the Nature of Cartographic Text," Cartographic Journal *30 (1993): pp. 104–11.*

use easier, and the fact that different typefaces can make different impressions on the reader. At first, these aspects of type can be bewildering, but familiarity comes easily with practice and experience.

TYPEFACE CHARACTERISTICS

Typeface design, type size, and letterforms are the principal characteristics of type with which the designer works. Letter and word spacing may be added, although strictly speaking these are not typeface elements. They are treated in this section because they can have such a tremendous impact on the perception of individual letters or words.

Letterform Components

All **typefaces** have elements in common regardless of the letter represented. (See Figure 14.1.) Letters, both capital and lowercase, are begun on the **base line.** The height of the body of lowercase letters is referred to as the *x*-height, an important dimension in letter design because it often determines the readability of the type. Letter strokes that are higher than the *x*-height are **ascenders;** all ascenders in a lowercase alphabet will terminate at a common **ascender line. Descenders** are letter strokes that fall beneath the base line and terminate at the **descender line.** In most typefaces, capital letters are shorter than ascenders; there is therefore a **cap line** that defines the vertical dimension of capital letters.

A major element of some letterforms is the **serif.** Serifs are finishing strokes added to the end of the main strokes of the letter. Not all designs have serifs, but it is important to note that in running text such as you are now reading, letterforms with serifs are easier to read. (See Figure 14.2.) Lettering styles that do not contain such finishing strokes are called **sans serif** styles. Serifs have different appearances, depending on the way they are joined to the main strokes. The serif may or may not be supported by a **bracket** (or *fillet*).

Counters, bowls, and loops of letters should be examined carefully when choosing type. **Counters** are the partially or completely enclosed areas of a letter, and **bowls** are the rounded portions of such letters as o, b, d, and the upper part of the lowercase g. The lower part of the lowercase g is referred to as a **loop.** These spaces can close up during the photographic reproduction of the map, possibly because of reduction of the original art, poor photographic negatives, or excessive ink in printing. This reduces their discernibility and can therefore impede communication. Map designers should choose typefaces with letterforms that are open and have few light areas that can lead to such problems.

Another aspect of letterforms is **shading,** a term used to describe the position of the maximum stress in curved letters.[2] (See Figure 14.3.) Shading has a definite impact on the appearance of letters and is especially important in the readability of type for text. Broadly speaking, letters with extreme vertical shading are more difficult to read than those with softer or more rounded shading. Shading also characterizes the historical development of type, a matter addressed later in the chapter, the section on Typeface Classification.

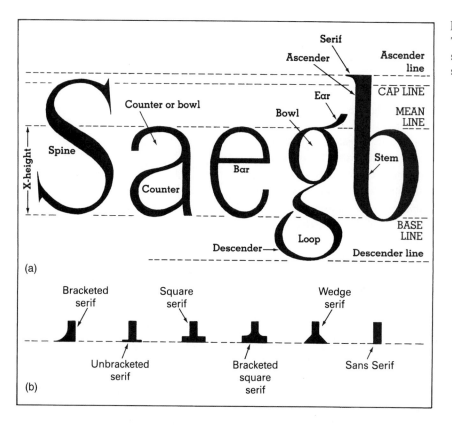

Figure 14.1 Type elements.
The principal elements of letterforms are shown in (a), and (b) illustrates the different serif forms.

TIMES ROMAN

Most geographical analyses involve point patterns (or centers of areas) that have weights attributed to them. Except at the simplest nominal scales, geographical point phenomena do not usually occur everywhere with equal value. An especially interesting study is to plot weighted means over time to discover a spatially dynamic pattern. Weights are most often socioeconomic data such as income, production, sales, or employment data.

HELVETICA REGULAR

Most geographical analyses involve point patterns (or centers of areas) that have weights attributed to them. Except at the simplest nominal scales, geographical point phenomena do not usually occur everywhere with equal value. An especially interesting study is to plot weighted means over time to discover a spatially dynamic pattern. Weights are most often socioeconomic data such as income, production, sales, or employment data.

Figure 14.2   Two lettering text blocks.
Times Roman, a roman, serified type, is easier to read in running text than lettering set in sans serif style, such as Helvetica Regular.

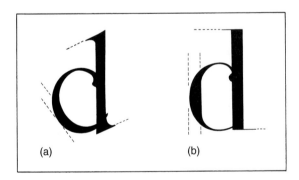

(a)   (b)

Figure 14.3   Shading.
Shading is the term used to describe the main slant of letterforms. Oblique slants, as in (a), characterize most Oldstyle typefaces. In Modern faces, such as (b), the emphasis is on strong verticals and horizontals.

Cartographic Requirements   Map lettering differs considerably from book and text lettering. Type set on lines with identical background is characteristic of book typesetting. Letters are **set solid** (no letterspacing), and the **leading** (spacing) between lines of type is an important concern for the book designer. In cartography, however, letters are often spread out, placed on changing backgrounds (tints, patterns, colors), oriented in a variety of ways, and inter-

rupted by lines or other symbols. The letters themselves may be rendered in different tones, patterns, or colors, thereby causing considerable variation. As a result of these awkward situations, the chief criterion of the cartographic designer in selecting a typeface is *that the individual letters be easily identifiable.*

One cartographer has identified the following considerations in selecting type for maps.[3]

1. The legibility of individual letters is of paramount importance, especially in smaller type sizes. Choose a typeface in which there is little chance of confusion between c and e or i and j.
2. Select a typeface with a relatively large *x*-height relative to lettering width.
3. Avoid extremely bold forms.
4. Choose a typeface that has softer shading; extreme vertical shading is more difficult to read than rounder forms.
5. Do not use decorative typefaces on the map; they are difficult to read.

It might be added that not so many years ago serif letterforms were normally preferable in meeting the unusual requirements of cartographic lettering. This is not the case today as we are seeing more sans serif faces than ever before. However, individual letter recognition continues to be vital on maps. The serif can play an important role in this, because it tends to complete letters optically and ties one letter visually to another.[4] Aesthetic considerations often affect the choice between serif and sans serif forms also.

Type Size

Type size designation is related to the way type was originally produced on metal, or foundry, blocks. (See Figure 14.4.) The body or height of the block specifies the type

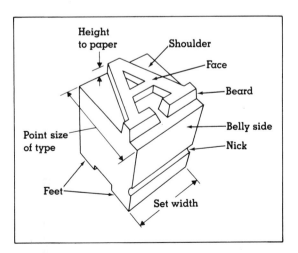

Figure 14.4   Foundry type character and common terminology to describe its parts.
Notice that the point size refers to the size of the body height, not the height of the face on the body; with the exception of point size, none of these terms applies to modern type specification.

size, although the actual letter on the block may not extend to the full height. The system of specifying type size in the United States and Britain is based on division of the inch into 72 parts called **points.** A point equals .0138 in (.351 mm). Thus, 72 points equal .9962 in. All type is measured in point sizes. Because the face of a particular type style may not occupy the whole height of the foundry body, the actual point size of the printed letter may be different from the nominal size given for a particular typeface.

Type size specifications based on foundry production of type are still used today, although the actual production of type may be by photo or computer methods. The important point is that a given style and size of type is standard. The cartographer normally chooses a type based on the actual size of the typeface and specifies the nominal type size as listed in the manufacturer's catalog.

The designer must specify type that is large enough to be read easily. A number of studies have been conducted to determine the minimal size. Professional cartographers rarely use type smaller than 4 or 5 points; safe practice is to set the lower limit at 6 points (.0828 in). Type environment can also play a role in the choice of size. Care must be exercised when original art is to be reduced so that type sizes smaller than 4 to 6 points will not result.

Letterforms and Type Families

Typographic nomenclature includes the word **font,** which is a complete set of all characters of one size and design of a typeface. A font of type normally includes numerals and special characters such as punctuation marks, in addition to the letters of the alphabet. Type designers often design vari-

ations of a basic font, making up the **family** of that font. (See Figure 14.5.) A normal *type family* will include these letterform variants: weight, width, roman, and italic. **Type weight** refers to the relative blackness of a type, although this variant is usually not standard among different typeface designs.[5] It is customary to find three weights: normal, lightface, and boldface. Type may be available in *condensed* or *extended* versions, called **type widths.**

The principal variants in a type family are the **roman** and **italic** forms. Roman is the basic, upright version of the design, and italic is a slanted version of its roman counterpart. Historically, an italic letterform was often slightly different from its roman partner, especially in the serifs. Today's type designers usually make the italic form identical except for its slant. In text, italic forms are frequently used for emphasis, but on maps they are most often used to label hydrographic features: oceans, seas, bays, straits, coves, lagoons, lakes, rivers, and so on.

LETTER AND WORD SPACING

Strictly speaking, letterspacing and word spacing are not elements of typeface design. However, these features are extremely important, especially in the visual perception of letter pairs and words. Entire map projects can be weakened by poor spacing of letters and words; no other feature is so obviously incorrect at first glance. Practice and a critical eye are essential in providing correct spacing.

Letterspacing involves the appropriate distribution of spaces between letters.[6] Letterspacing is usually required for lines of type in all capitals, but not in lines of lowercase

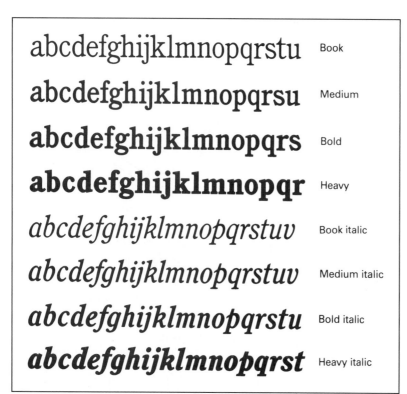

abcdefghijklmnopqrstu	Book
abcdefghijklmnopqrsu	Medium
abcdefghijklmnopqrs	Bold
abcdefghijklmnopqr	Heavy
abcdefghijklmnopqrstuv	Book italic
abcdefghijklmnopqrstuv	Medium italic
abcdefghijklmnopqrstu	Bold italic
abcdefghijklmnopqrst	Heavy italic

Figure 14.5 Variants within a type family. Both roman and italic forms, with different weights in each, may be included in a family. A heavier weight in each form is also availible. (*This type is called ITC Cushing, designed by Vincent Pacella for the International Typeface Corporation. Reproduced with permission.*)

letters. Letters are usually set solid (without letter spacing) in lines of lowercase. **Word spacing** is the proper distribution of spaces between words to achieve harmonious rhythm in the line. Words set in all capitals or all lowercase require word spacing.

Spacing of both letters and words is especially important in cartography because so much of the lettering is stretched across map spaces to occupy geographic areas. Some names cause little difficulty for the designer because they are set solid: town and city names and other labels applied to point phenomena.

The real culprits are capital letters that cause open spaces, such as A, J, L, P, T, U, W, and O. Notice that these contain few vertical strokes and have mostly oblique or rounded strokes. Jan Tschichold, a noted German typographic designer, suggests that spacing of capital letters can be facilitated by the **neutralizing rule.**[7] Figure 14.6 demonstrates this, utilizing the O and the L. If the O is too close to the L (or other letter), a "hole" appears in the word. Letterspacing the O and the L causes the space to be visually equal to the inner part of the O, and the hole in the word disappears. Tschichold's neutralizing rule states that minimum spacing between capitals should always be equal to this optical value (or visually equal space) for all capital letters, not just for words containing an O. Letterspacing can be greater than this minimum value with no disruption of visual harmony, although word recognition will be more difficult.

As a general rule, continues Tschichold, the designer should strive to letterspace the capitals so that each letter's outline is more visually dominant than its inner space. Practice with spacing of capital letters will demonstrate this simple but revealing principle.

For some capital letter combinations, notably AV, AT, AW, AY, LT, LV, LW, and LY, **mortising** may be required. Mortising, also called **kerning,** is fitting letters closer together to achieve proper visual balance in letterspacing. Many typeface manufacturers supply several kern pairs as part of their fonts. According to the International Typeface Corporation, which copywrites and supplies many typefaces to third-party vendors, the 20 most frequently used kern pairs are: Yo We To Tr Ta Wo Tu Tw Ya Te P. Ty Wa yo we T. Y. TA PA and WA. This group suggests these generalizations:[8]

1. Commas, periods, and quotes almost always have to kern.
2. Cap and lowercase letters with outside diagonal strokes require kerning more often than not.
3. T, L, and P generally need to be kerned with nonascending lowercase letters.

Mortising usually requires that other letterspacing rules be applied as well.

Word spacing of names set in all capitals also deserves attention. One rule is to provide a distance between words equal to the letter I, including the letterspacing that is appropriate for it. (See Figure 14.7.) This principle applies to spacing of words set in capitals along a line of text, where it is undesirable for the words to form "islands" in the line. Cartographers need to exercise considerable caution here. Too much word spacing disrupts the continuity of multiple words, yet the cartographer often attempts to stretch words to occupy a given geographical space. Experimentation is required to achieve an acceptable balance in these cases.

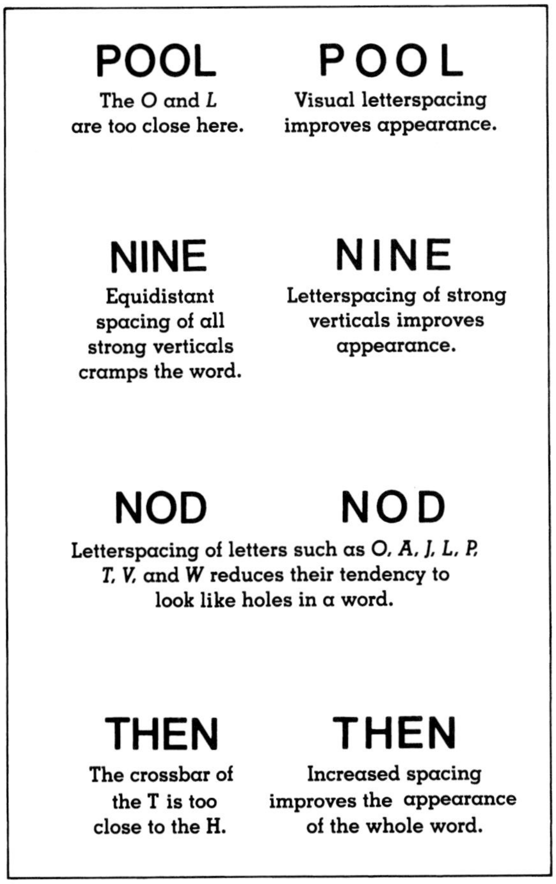

POOL
The O and L
are too close here.

P O O L
Visual letterspacing
improves appearance.

NINE
Equidistant
spacing of all
strong verticals
cramps the word.

N I N E
Letterspacing of strong
verticals improves
appearance.

NOD
Letterspacing of letters such as O, A, J, L, P,
T, V, and W reduces their tendency to
look like holes in a word.

N O D

THEN
The crossbar of
the T is too
close to the H.

T H E N
Increased spacing
improves the appearance
of the whole word.

Figure 14.6 Letterspacing.
Letterspacing is important in cartographic design. These examples point out common mistakes.

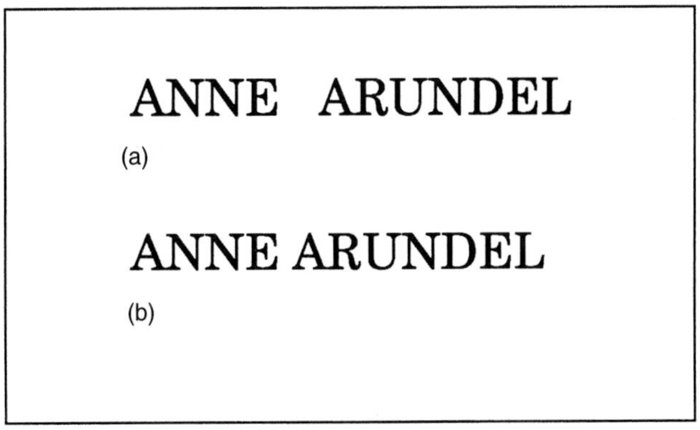

ANNE   ARUNDEL
(a)

ANNE ARUNDEL
(b)

Figure 14.7 Word spacing.
The space between the words in (a) is too great. Improved word spacing, as in (b), is achieved by using approximately the same space as the letter I (plus its normal letterspacing) between words.

> T he first thing to realize is that the rhythm of a well formed word can never be based on equal linear distances between letters. Only the visual space between letters matters. This unmeasurable space must be equal in size. But only the eye can measure it, not the ruler. The eye is the judge of all visual matter, not the brain.
>
> Source: Jan Tschichold, *Treasury of Alphabets and Lettering* (New York: Reinhold, 1966), p. 29.

Table 14.2 Typeface Classes

1. Black Letter
2. Oldstyle
   a. Venetian
   b. Aldine-French
   c. Dutch-English
3. Transitional
4. Modern
5. Square serif
6. Sans serif
7. Script-cursive
8. Display-decorative

Source: Alexander Lawson, Printing Types: An Introduction *(Boston: Beacon Press, 1971), p. 27.*

TYPEFACE CLASSIFICATION

The cartographer who has a knowledge of the history of type design and sees how different typefaces relate to each other will be better prepared to select type for maps. Type classification is one way to begin. The system discussed here is one among many; it is selected for its simplicity and historical emphasis.

Prior to the invention of movable type and the printing of Gutenberg in Germany, writing in Europe was accomplished by hand, usually by a select group of clerical scribes. The alphabets had come to them from Latin alphabets, as modified through time. Manuscript letterforms were called **textura** by the Italians.[9] Early movable type designs simply attempted to replicate the manuscript forms.

A system of type classification begins with the first movable type and places all subsequent typefaces into classes, based on when they were created. A modern typeface is placed in one of these classes according to the match of its style characteristics with those of faces designed earlier. A relatively simple classification, developed by Alexander Lawson, contains eight major classes. (See Table 14.2 and Figure 14.8.) For the most part, **Black Letter** is the term

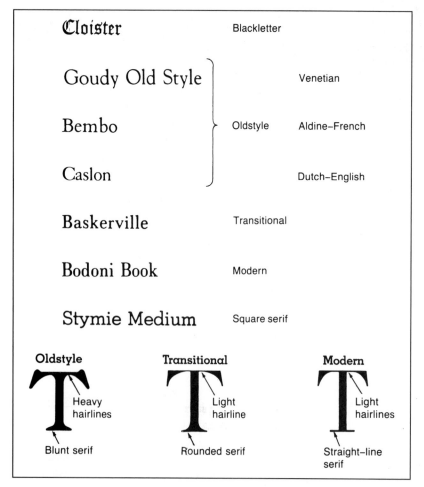

Figure 14.8 Examples of various typefaces for the different classes used in this text. The chief differences in strokes and serifs among the three main classes are illustrated at the bottom.

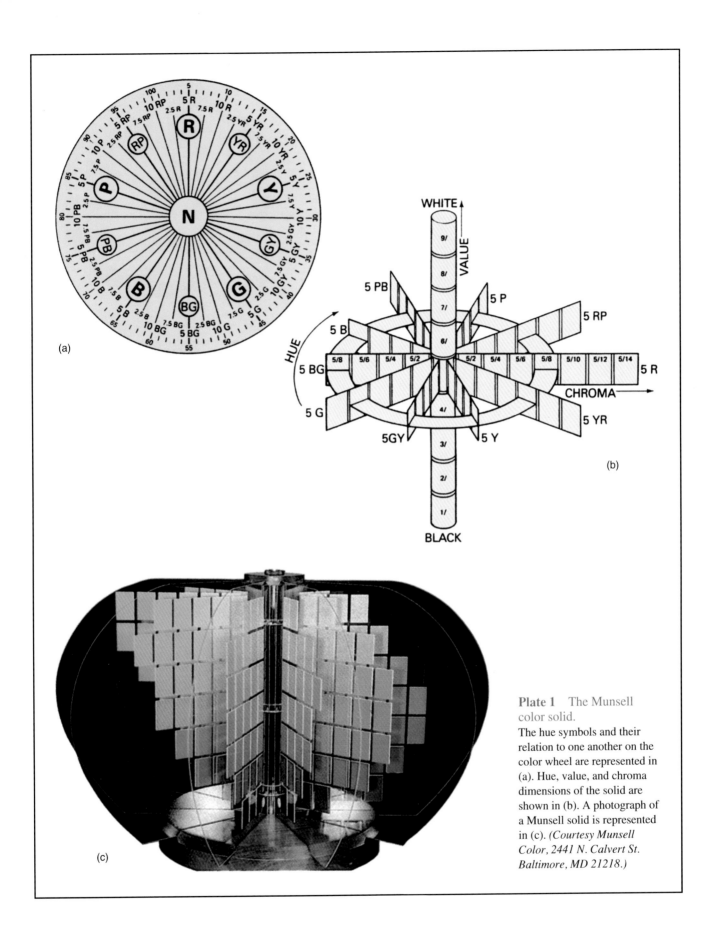

(a)

(b)

(c)

Plate 1 The Munsell color solid.
The hue symbols and their relation to one another on the color wheel are represented in (a). Hue, value, and chroma dimensions of the solid are shown in (b). A photograph of a Munsell solid is represented in (c). *(Courtesy Munsell Color, 2441 N. Calvert St. Baltimore, MD 21218.)*

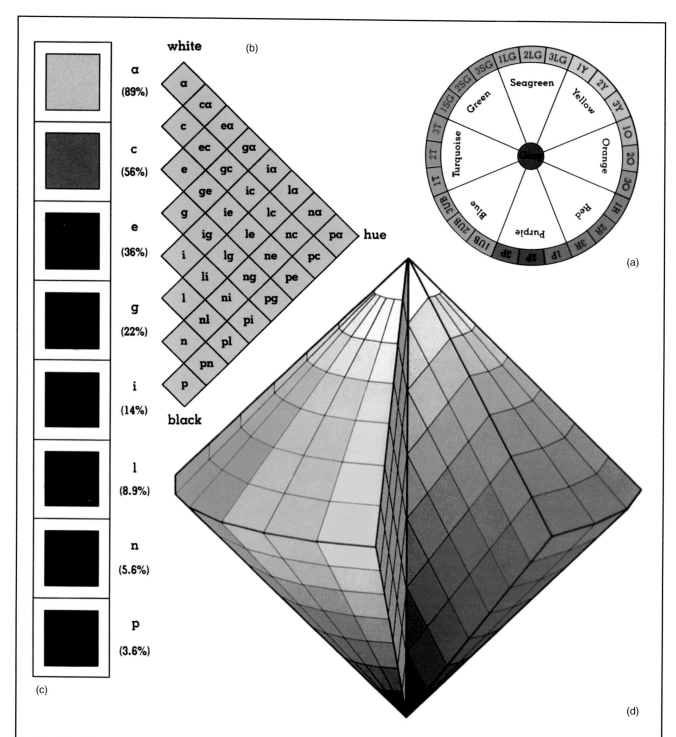

Plate 2 Features of the Ostwald color solid.

The hue circle (a) contains 24 hues. A triangle of the solid (b) contains an arrangement of hues (tints and tones) caused by mixing different amounts of white and black to each hue. The equal value gray scale used by Ostwald determined the appearance of the tints and tones (c). Figures in parentheses show white content; black content in remainder equals 100 percent. An approximation of a completed Ostwald cone is represented in (d). *(The screen tint percentages used here to represent the solid yellow, magenta, and cyan are modeled after those used by Allan Brown in "A New ITC Colour Chart Based on the Ostwald Colour System," ITC Journal [1982] 2:109—118. Used with permission.)*

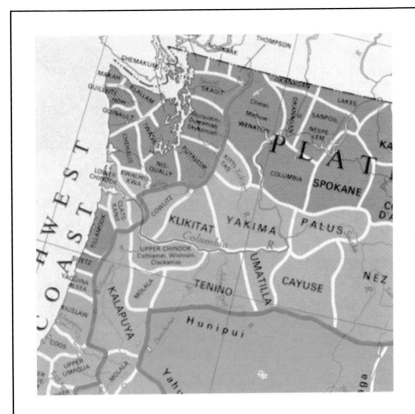

Plate 3 Color used to clarify map elements.

On this map, color is used to separate the linguistic stock of several minor and major Indian tribes and their cultural areas. Color helps the reader differentiate areas, and this is especially critical on this map with so much information. *(Source: United States Department of the Interior, Geological Survey. National Atlas Single Sheet, "Indian Tribes, Cultures, and Languages," 1991.)*

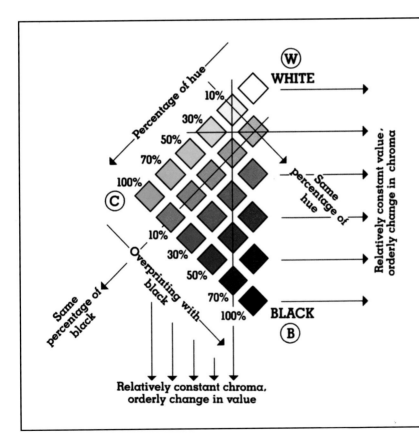

Plate 4 Color scaling for quantitative maps based on the Ostwald color triangle. The percentages shown along the hue and black sides yield perceptual steps of approximate equality. A sequence for scaling may be selected from any vertical, horizontal, or diagonal row, depending on the designer's choice of varying value (select a vertical row), chroma (select a horizontal row), or both (select a diagonal row). No absolute rules exist to assist in this selection. *(This diagram is based on Henry W. Castner, "Printed Color Charts: Some Thoughts on Their Construction and Use in Map Design,"* Proceedings *[Annual Meeting of the ACSM-ASP, St. Louis, 1980], p. 378.)*

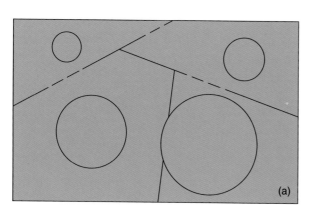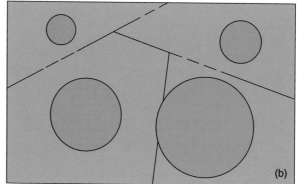

Plate 5 Color and figure-ground development.
Effective use of hue differences can enhance the figure-ground relationship on maps. In the example shown, the proportional symbols in (b) are clearly more visually dominant than those in (a).

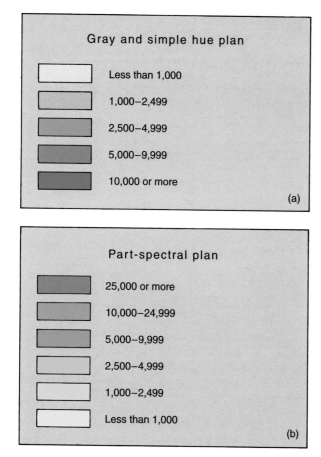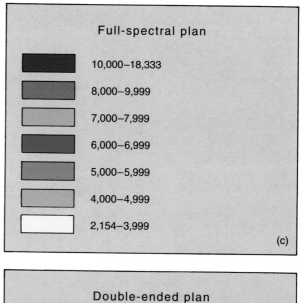

Gray and simple hue plan

Less than 1,000

1,000–2,499

2,500–4,999

5,000–9,999

10,000 or more

(a)

Full-spectral plan

10,000–18,333

8,000–9,999

7,000–7,999

6,000–6,999

5,000–5,999

4,000–4,999

2,154–3,999

(c)

Part-spectral plan

25,000 or more

10,000–24,999

5,000–9,999

2,500–4,999

1,000–2,499

Less than 1,000

(b)

Double-ended plan

50–100

Increase

0–49

0–40

Decrease

50–100

(d)

Plate 6 Different quantitative color plans.
See text for explanation. The color plan for (b) is used on Bureau of the Census GE-50 Series map number 56 (Median Family Income for 1969); the color plan for (c) is used on GE-50 Series map number 60 (Occupied Housing Units, 1970.)

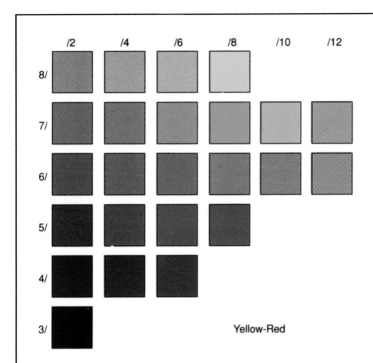

Plate 7 Yellow-red, Munsell-based color chart. Each color box has been printed by using different percentages of yellow, magenta, cyan, and black. The numbers and slashes along the margins refer to Munsell color chip near-equivalents. Refer to text for additional information on how the chart was devised, and for various ways to use the chart for selecting quantitative series from the chart. *(Redrawn from Cynthia A. Brewer, "The Development of Process-Printed Munsell Charts for Selecting Map Colors,"* The American Cartographer *16 [1989] pp. 269-278.)*

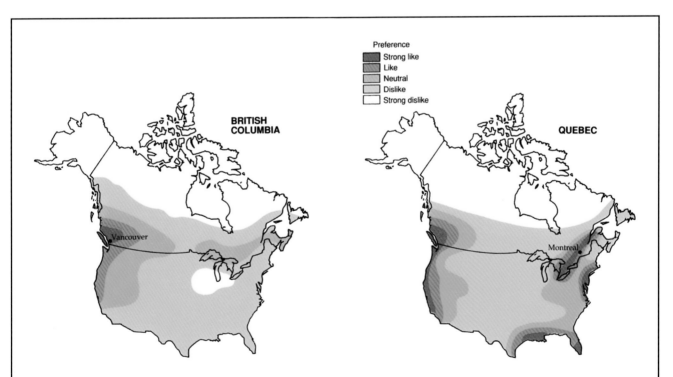

Plate 8 Color and continuous surface mapping.
This map illustrates residential preferences of Canadians, some sampled from British Columbia and some from Quebec. Here the cartographer does not reveal to the reader how much, but rather the rank of, preference. Although not disclosed, it is fairly certain that the color areas are bounded by isarithms, and the areas between the isolines are rendered in different colors, selected to suggest a low to high ranking by the hue and intensity of the selected colors.

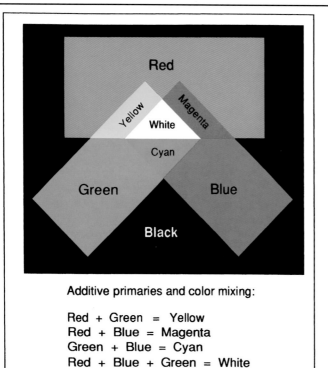

Additive primaries and color mixing:

Red + Green = Yellow
Red + Blue = Magenta
Green + Blue = Cyan
Red + Blue + Green = White
None = Black

Plate 9 Additive and subtractive color mixing.

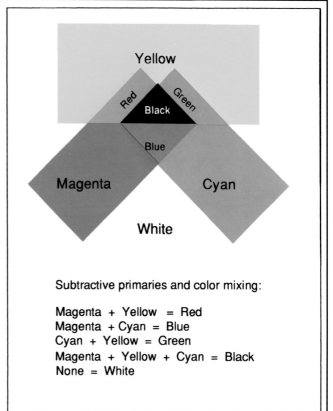

Subtractive primaries and color mixing:

Magenta + Yellow = Red
Magenta + Cyan = Blue
Cyan + Yellow = Green
Magenta + Yellow + Cyan = Black
None = White

Plate 9   Additive and subtractive color mixing.

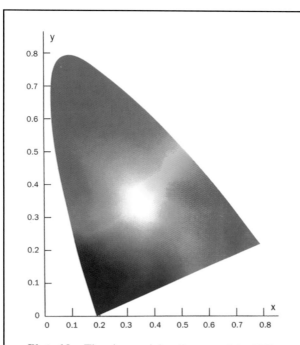

Plate 10   The chromaticity diagram of the CIE color system.
Not all colors of the color area can be printed due to limitations in printing inks. *(M. Nyman,* FOUR COLORS/ONE IMAGE, © *1991, 1993 Mattias Nyman/Peachpit Press. Reprinted by permission of Addison-Wesley Publishing Company, Inc.)*

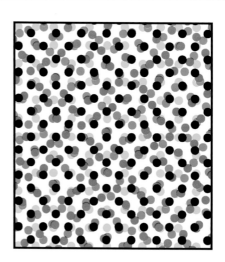

Plate 11   Four-color process screen angles.
Black is angled 45 degrees from the horizontal (beginning on the horizontal and counting counterclockwise), yellow at 90 degrees, cyan at 105 degrees, and magenta at 165 degrees. These angles have been determined to provide the maximum inks of the four colors, thereby giving the purest colors.

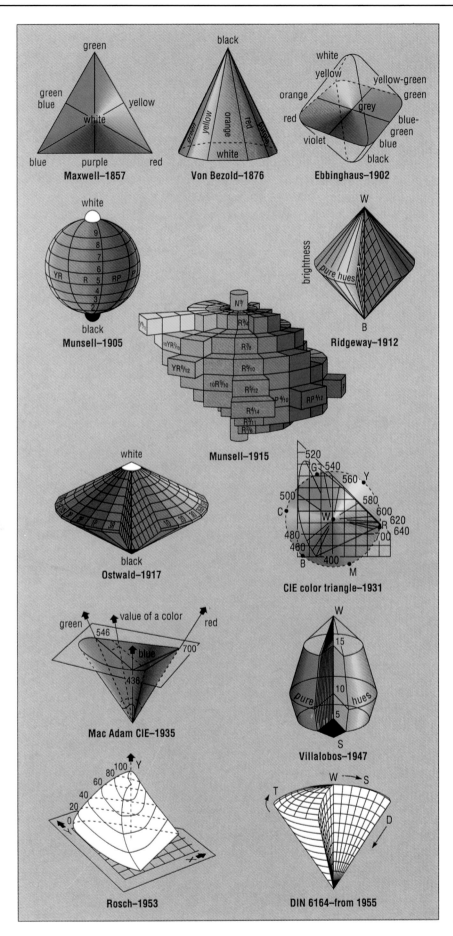

Plate 12 Different color models.
The Munsell (1915), the Ostwald (1917), and the CIE (1931) have become the most useful color models in cartography. *(Reproduced with permission of the Pennwell Publishing Co.)*

Plate 13 The RGB color solid.
This view of the solid is looking directly at the white corner. Maximum red is in th lower left corner, and re\green a the lower right. Not shown is the black corner, directly opposite the white. the view of the model corresponds to figure 17.8. The model may be viewed form any orientation. *(With permission from Xerox Corporation.)*

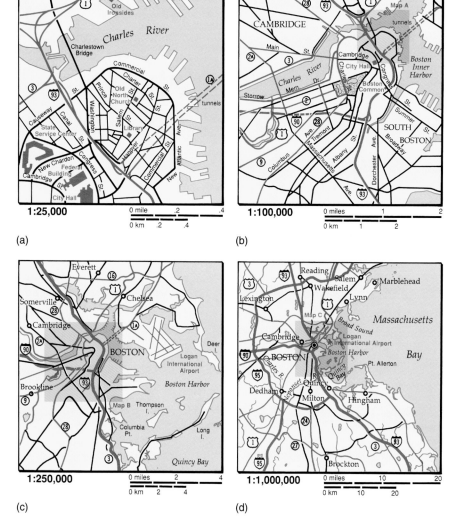

Plate 14 Scale and map detail.
Maps at relatively large scales such as 1:25,000 (a) show considerably more detail than those at small scales, for example, 1:1,000,000 (d). Notice that the large scale individual buildings and streets can be shown, but at smaller scales this is impossible. Maps at intermediate scales show more detail than those at large scales, but less than those at small scales. For geographers, scale sets the limits on the types of questions that can be asked, or answered.

used for at least three or four different styles of early movable type that attempted to replicate manuscript forms. Two principal forms were *lettre de forme* and *lettre de somme*.[10] **Lettre de forme** typifies the design used by Gutenberg in his Bible. It is characterized by angular, sharp strokes with considerable vertical stress. **Lettre de somme** is less formal and more rounded. Another term used to describe Black Letter is *gothic,* but some confusion exists because many in the United States used this name for the early sans serif styles developed in the 1830s. Two styles used today are Cloister Black (Old English) and Goudy Text, which closely resembles the type of Gutenberg's first Bible.

There are few uses of Black Letter in modern cartography. The style is very difficult to read, especially set in text, and is compatible with little else on the map. Perhaps it might find some use on decorative maps with historical content. Diplomas and newspaper logos continue to use Black Letter.

Oldstyle, sometimes referred to as *Old face* types, may be subdivided into three groups. *Venetian* is a style first developed in northern Italy some 20 years after Gutenberg. Venetian Oldstyle is characterized by concave serifs and little contrast between thick and thin strokes. (See Figure 14.8.) A revival of the Venetian Oldstyle in the late nineteenth and early twentieth centuries resulted in the introduction of such type designs as Cloister Oldstyle, Centaur, and Goudy Old Style.[11]

Another class of Oldstyle is *Aldine-French.* This group is modeled after the Venetian but was originally characterized by a greater difference between thick and thin strokes and modifications of the serifs. This style influenced French types of the sixteenth century and dominated style efforts in Europe for nearly 150 years.[12] Modern designs include Bembo and Palatino. (See Figure 14.8.)

Dutch-English forms the third group in the Oldstyle class. The French type forms were popular in Europe but soon gave way to the enterprising Dutch, who in many cases redesigned type to facilitate printing. The contrast of thick and thin strokes was increased, and serifs were straightened and bracketed in the lowercase forms. Early Dutch designs were adopted in England and enjoyed enormous popularity because of William Caslon.[13] (See Figure 14.8.) Caslon's foundry and others in England produced virtually all the type seen in America until the Revolution. Caslon forms abound today; many manufacturers offer the style. The Dutch-English tradition is at the end of the development of the Oldstyle class of letterforms.

Transitional type forms make a historical bridge between the Oldstyle and Modern typefaces. John Baskerville, a printer and type founder in England around 1750, designed the type that bears his name,[14] often thought to be an excellent example of the Transitional style. The style is characterized by less serif bracketing than letterforms of the Oldstyle period. (See Figure 14.8.)

The **Modern** style of letterform began in the late eighteenth century and was considerably influenced by the Transitional forms. Baskerville's designs led Bodoni of Italy and Didot of France to design faces that have become identified with the modern movement. (See Figure 14.8.) The development of typeface designs in the modern period was in response to several other parallel achievements, notably in printing, manufacturing, and advertising. The strokes of the copper engraver led to hairline serifs and thick cross strokes. There was an emphasis on vertical shading. The development of **display types** (those too large for book-text printing) was important because of the Industrial Revolution and the advertising that it brought. Type of the Modern period was still characteristically roman in style, but brought to the extreme in geometric regularity.

Square serif type styles, also called *slab serif* or *Egyptian,* were first introduced around the beginning of the nineteenth century. These type forms are characterized by even stroke widths and usually have unbracketed slablike serifs. The Clarendon typeface appeared in the mid-nineteenth century and was a more graceful form of the square serif. Stymie was produced in the early part of this century and carries on the tradition of this class of type. (See Figure 14.8.)

The most notable letterform to appear in the history of type design is **sans serif,** a letter having no serifs. Previous innovations in design had at least carried along the Latin alphabet's tradition of serifs, but the new sans serif styles marked a radical departure. They have remained in use since their introduction and have enjoyed widespread popularity. A Swiss type designer named Adrian Frutiger styled a complete family of sans serif alphabets called **Univers,** one of the most popular such forms ever designed. Unique to this family was Frutiger's use of a number series, rather than descriptive names, to designate weights. Those in the 40s are lightface, in the 50s medium, in the 60s bold, and in the 70s extra bold. Both roman and italic forms are included in each number series. (See Figure 14.9.)

Two other classes of lettering styles round out this classification. **Script-cursive** and **display-decorative** are similar in that they are not useful for ordinary text typography but find their place in decorative or advertising display. They are difficult to read in most circumstances and have had little use in ordinary cartography. (See Figure 14.10.)

Nearly any type class may be used effectively on the thematic map except Black Letter, script, or decorative forms. Certain designs may be more effective than others or may impart a more appropriate mood to the map. Certain principles regarding lettering class combinations and use of families are recommended and outlined later in the chapter. First, however, let us look at one important singular quality of letter designs—their personality.

THE PERSONALITY OF TYPEFACE

Most graphic artists and designers seem to agree that typefaces are capable of creating moods, and that there is such a thing as **type personality.** There is no question that the selection of type can alter the appearance of a map, but it is

Figure 14.9 The complete univers palette.
A unique feature of this type family is the use of numbers instead of descriptive terms to designate weight. (*The idea for representing the Univers family this way comes from Alexander Lawson,* Printing Types [*Boston: Beacon Press, 1971*], p. 100. Redrawn with modifications from same source.)

exceedingly difficult to categorize each type design by the kind of response it is likely to produce. Certainly, little is known about typeface and moods in cartographic design; more detailed knowledge is needed.

In general, Oldstyle designs, with their emphasis on diagonal shading, create more mellow and restful designs.[15] Bibles are more likely to be printed in Caslon or Bembo than Bodini. On the other hand, **Modern** faces, characterized by strong vertical and horizontal strokes, create abrupt, dynamic, and unsettling lettering. Lettering set in a sans serif face is dazzling, which makes it difficult for the eye to settle down. These observations are especially true in text typography but are also observable in display settings, such as in most cartographic applications.

As a starting point, the student might wish to consider the contents of Table 14.3. These are only generalizations; the appearance of type in a cartographic environment may be significantly altered in the map's content and other style features. Nonetheless, an awareness of a type's influence on mood is important for the designer.

LEGIBILITY OF TYPE

Although the designer may be interested in the aesthetic qualities of letters and alphabets, the real concern is readability. If type on the map cannot be read easily, its employment has been a wasted effort. **Type legibility,** the

Figure 14.10 Script and decorative typefaces.
Thematic map designs do not employ these typefaces.

term most often used in research dealing with readability, has been defined this way:

> *Legibility deals with the coordination of those typographical factors inherent in letters and other symbols, words, and connected textural material which affect ease and speed of reading.*[16]

In cartography, **type discernibility**—the perception and comprehension of individual words not set in text lines—is more critical than legibility.

There is extensive literature dealing with text-type legibility, but little that deals specifically with display-type discernibility. Nonetheless, certain important facts are known to apply to cartographic lettering.

The individual lowercase s, q, c, and x are difficult to distinguish, and the letters f, i, j, l, and t are frequently mistaken for each other.[17] The lowercase e, although a frequent letter in print, is often mistaken. In general, these legibility patterns have been recorded:

dmpqw	high legibility
jrvxy	medium legibility
ceinl	low legibility

The legibility of individual letters becomes important in cartography because the map may be embedded with different backgrounds, textures, or colors that can make letters difficult to distinguish. In addition, legibility may be

Table 14.3	**Personality of Typefaces**
Dignity	Oldstyle typefaces
Power	Bold sans serif
Grace	Italics or scripts
Precision	New sans serifs and slab serifs
Excitement	Mixtures of typefaces

Source: Roy Paul Nelson, Publication Design *(Dubuque, IA: William C. Brown, 1972), pp. 34–35.*

Some types are versatile enough to be appropriate for almost any job. Others are more limited in what they can do. But all types have some special qualities that set them apart. Art directors are not in agreement about these qualities, but here are a few familiar faces, along with descriptions of the moods they seem to cover.

1. Baskerville—beauty, quality, urbanity.
2. Bodoni—formality, aristocracy, modernity.
3. Caslon—dignity, character, maturity.
4. Century—elegance, clarity.
5. Cheltenham—honesty, reliability, awkwardness.
6. Franklin Gothic—urgency, bluntness.
7. Futura—severity, utility.
8. Garamond—grace, worth, fragility.
9. Standard—order, newness.
10. Stymie—precision, construction.
11. Times Roman—tradition, efficiency.

These qualities, if indeed they come across at all to readers, come across only vaguely. Furthermore, a single face can have qualities that tend to cancel out each other. (Can a type be both tradition-oriented and efficient?) While the art director should be conscious of these qualities and make whatever use he/she can of them, he should not feel bound to any one type because of a mood he wants to convey.

Assume he is designing a radical, militant magazine like *Ramparts.* Baskerville seems an unlikely choice for such a magazine, and yet it has been used effectively by *Ramparts,* as it has been by *Reader's Digest!*

Sometimes the best answer to the question, "What type to use in title display?" is to go with a stately, readable type—like Baskerville—and rely upon the words in the title to express the mood of the piece.

Source: Roy Paul Nelson, *Publication Design* (Dubuque, IA: William C. Brown, 1972), p. 94.

affected by the way the letters are shown; they can be screened or rendered in color.

Lettering in all capitals, more than any other factor, slows reading. Apparently this is caused by a larger number of eye fixation pauses; because capital letters occupy more space, the result is a reduction in the number of words perceived. Lowercase letterforms lead to quicker recognition. "Lowercase words impress the mind with their total silhouette while capitals are mentally spelled out by letter."[18] (See Figure 14.11.) Also, the upper halves of lowercase letters contribute more recognition cues than the lower halves. Unfortunately, many of the tops of the letters in the sans serif forms are identical or nearly so, creating problems of legibility.

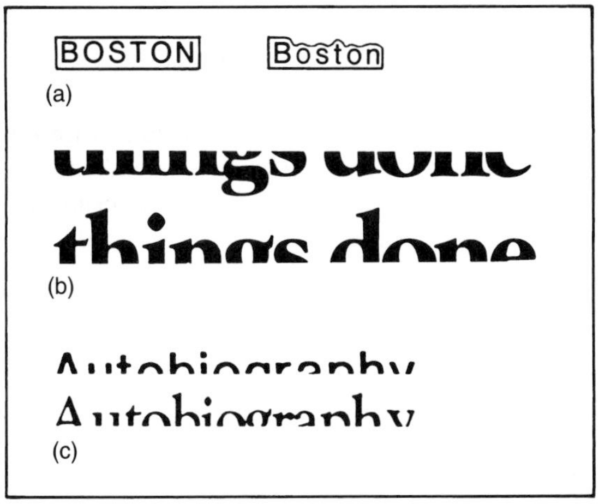

Figure 14.11 Reading type.
(a) Lowercase letters lead to easier word recognition because of the silhouette they produce. (b) It is easier to distinguish words by the tops of letters than by their bottoms. (c) Serif letterforms are frequently easier to read, because the letters are more distinct, especially along the upper halves of lowercase letters.

In the cartographic context, placement of letters relative to their environments and the selection of letterforms can either add to or detract from overall legibility.

In the styles commonly used in text printing, legibility is not influenced by typeface style. This also applies to sans serif styles. Aesthetic preference might lead the designer away from sans serif, even though it might be equivalent in legibility.[19] It is interesting to note that the United States Geological Survey selected a sans serif typeface (Univers) to appear on many of the 1:100,000 series maps, as well as others they produce.[20]

A significant research finding is that lettering in all italics slows reading, or at least is not preferred by most readers.[21] Italic letterforms have historically been used for labeling hydrographic features, and this practice is not likely to change. Of course, the legibility studies of italic forms were in a text format. Similar findings might not apply in cartographic contexts.

THEMATIC MAP TYPOGRAPHICS AND DESIGN

The selection of map type and its proper placement have been developed mostly by tradition. There are few experimental studies that pertain solely to map design. This section presents the highlights of conventional practices and the few experimental results available.

LETTERING PRACTICES

General guidelines, typeface selection, the use of capitals and lowercase, and lettering placement have all developed through empirical use. For the most part, the trained eye and aesthetic judgment of the cartographer have produced the design strategies that are now standard.

General Guidelines

There are four major goals in approaching the lettering of a map.[22]

1. Legibility
2. Harmony
3. Suitability of reproduction
4. Economy and ease of execution

Legibility research traditionally deals with the characteristics of letter design and how they affect the speed and accuracy of reading. Because lettering on maps usually occurs in a much more complex environment (such as confusing background, textures, and linework), legibility is achieved not only by choosing the appropriate typeface, but also by attending to good placement and adequate spacing of letters and words. Care and advanced planning help avoid problems.

Lettering harmony involves several features related to the selection of a typeface. This goal of typographic design will be treated in detail below. The third aim, suitability for reproduction, pertains to the character of the typeface and how well it stands up to reduction (if any) and printing. Typefaces that contain small bowls and counters tend to "close up" during printing, especially if excessive ink is used. Styles with unusually thin strokes and serifs may not photograph well; these letter features may disappear during printing. The designer should evaluate how well each letter in a proposed style will reproduce in the job at hand.

Finally, the designer needs to evaluate the cost of lettering. The various means by which cold type can be produced introduce options that should be investigated. If possible, unit costs (cost per word) should be compared for the different means. For many mapping projects, the cost of the lettering exceeds any other material costs.

Selecting Typefaces and Lettering Harmony

Many professional cartographers maintain that only one typeface should be used on the map. To achieve contrast and harmony, the cartographer may, however, select several variants of a single type family.[23] The preferred way of mixing type is to vary roman and italic forms in the same family and to vary weight (light, medium, or bold). The least preferred is to mix different typefaces; if different typeface designs are used, do not select those that contrast markedly.

The following combinations have also been suggested if more than one typeface is used: a strong geometric Modern with a sans serif; and a square serif with a sans serif.[24] It is also good practice to choose different types (from the same period) that are not too similar; each face should be distinct in style to avoid confusion.

Using only different weights in one family of type should not be considered a limitation in design. The different weights, if used effectively, can be very expressive. Weights should be chosen in accordance with the importance of the feature the lettering identifies. Name and label

weights may be lightened to deemphasize lettering that is large for other design reasons.

Typographic harmony is achieved by choosing a typeface that is compatible with the map's content. An example of an incompatible choice would be Black Letter on a map depicting a twentieth-century geographical theme. Harmony means using a typeface with a personality that befits the map, in type sizes that correspond to the map's intellectual hierarchy. If the type size is carefully planned, the hierarchy is apparent to the map reader.

Very small lettering (4 or 5 points) is difficult to read at a normal reading distance of 30 to 46 cm (12–18 in). As a general rule, the larger or more important the feature, the larger its label should be.[25] This does not suggest that label sizes are proportional to the sizes of features, but rather that there should be an ordinal association. Rarely should there be more than four to six different sizes.

The selection of the typeface for a map requires careful inspection of each letter in the font. Typefaces with great contrast in thick and thin strokes, large x-heights, open bowls and loops, and dissimilar top halves of lowercase letters should be favored over others. Serif forms are good when letters are spread out, because they help tie the letters together. Display types should be investigated for their usefulness, as these can be more appropriate for a given map than ordinary text faces.[26]

The Use of Capital and Lowercase Letters

Good lettering design on the map can be achieved by contrast of capitals and lowercase. A map that contains only one form or the other is exceptionally dull and usually indicates a lack of planning. In general, capitals are used to label larger features such as countries, oceans, and continents, and important items such as large cities, national capitals, and perhaps mountain ranges.[27] Smaller towns and less important features may be labeled in lowercase with initial capitals.

One practice is to label features represented by point symbols with lowercase lettering. In contrast, features that have large areas should be labeled in capitals. (See Figure 14.12a.) Scale determines the choice. At small scales, the city of Baltimore is represented by a dot and labeled in lowercase; at a much larger scale, it is represented in capitals. An exception is made for country names, normally set in capitals, which are still set in capitals when placed outside the country at smaller scales. (See Figure 14.12b.) The general rule is, however: inside the feature, capitals; outside the feature, lowercase with initial capital.

The Placement of Lettering

Careful lettering placement enhances the appearance of the map. There are several conventions, supported by a few experimental studies. Most decisions regarding lettering placement fall into one of the following categories: labeling point symbols, designating linear features, naming open areas, and placement and design of titles and legend materials.

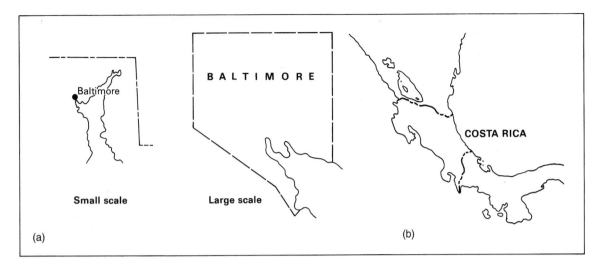

Figure 14.12   The use of capital and lowercase letterforms.
Point phenomena (a) are customarily labeled with lowercase. At larger scales where inside lettering can be accommodated, capitals are suggested. Because of space problems, areal phenomena such as countries may have their labels outside, as in (b). In these cases, the convention of using capitals is retained.

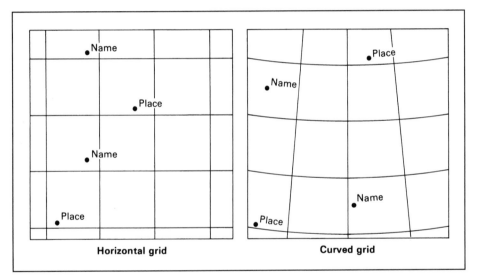

Figure 14.13   Lettering placement conventions and the map graticule.

Point-Symbol Labeling   Most professional cartographers agree that point symbols should be labeled with letters set solid (no letterspacing). On small-scale maps, these should follow the line of the parallels (if apparent). They should be set horizontally on large-scale maps.[28] (See Figure 14.13.) Several years ago Yoeli established a priority for the placement of labels adjacent to point symbols.[29] (See Figure 14.14.) His system of rules was not based on empirical work, but on his own preferred judgments. Imhof's rules, though less strict, agree with Yoeli's, especially regarding the most preferred location (up and to the right). Others have developed their own priority schemes, some similar and some dissimilar to both Yoeli and Imhof, but none universally adopted. It seems apparent, though, that one should never locate the label on the same horizontal line with the symbol, and if possible should not locate it to the left of the symbol.

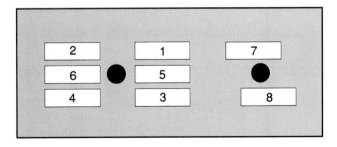

Figure 14.14   Point feature label location priorities, after Yoeli.

Several cartographers have looked closely into point label locations now that automation in map production is practiced routinely. One study in particular is interesting because it examines three commercially produced road maps (containing many label names) to determine what placement scheme they follow (especially to determine if they follow the priority scheme of Yoeli).[30] The work done by Wu and Buttenfield found that none of the three maps followed the priority scheme of Yoeli. After sampling from the three maps, these researchers grouped several possible relative locations into the eight outlined by Yoeli and this revealed the results tabulated in Table 14.4. A majority of the names fall into four positions—upper right, upper left, lower right, and lower left. One surprising finding was the relative strength of Yoeli's priority seven. On two of the maps, priority positions one, two, and three were at the top, though not in the same order.

Research has shown that distractors (other graphic marks) can impede the finding of names on a map when located anywhere near the beginning of a word.[31] Consequently, never use open point symbols that can be confused with letters of the label. Other marks, such as lines and texture, can also distract the reader. With careful planning, these situations can be avoided.

Names of ports and harbor towns should be placed seaward, if possible.[32] Names at coastlines or rivers should never be placed so they overlap the coastline. Likewise, town names should be placed on the side of the river on which the town is located.

Designating Linear Features Linear features on thematic maps include rivers, streams, roads, railroads, streets, paths, airlines, and many linear quantitative symbols. The general rule is that their labels should be set solid (no letterspacing) and repeated as many times along the feature as necessary to facilitate its identification.

The ideal location of the label for a linear feature is above it, along a horizontal stretch if possible. (See Figure 14.15.) Do not crowd the label into the feature. Room must be reserved for lowercase descenders, if any. If at all possible,

place a river's name so that its slant is in the direction of the river's flow (assuming the label is italic, which is preferred).

Names in Areal Features If a space on the map is large enough to accommodate lettering completely inside it, it may be designated an areal feature. Examples include oceans and parts of oceans, large bays, lakes, continents, countries, states, forests, and geographic subdivisions. The general rule is to letterspace the words so as to reach the boundaries of the feature.[33] The extent of the feature is then obvious from the letterspacing of its label. (See Figure 14.16.)

Each map presents unique problems in the placement of areal labels, but the following simple rules are helpful:

1. Curved lines of letters should be gentle and smooth, and the curve should be constant for the entire word. (See Figure 14.17a.)
2. Do not hyphenate names and labels.
3. If a line of lettering is not horizontal, make certain it deviates significantly from the horizontal so that its placement will not look like a mistake.
4. Do not locate names and labels in such a way that the beginning and ending letters are too close to the feature's borders.
5. Choose a plan for the lettering placement of the whole map in accordance with the normal left-to-right reading pattern.[34] (See Figure 14.17b.) Never place labels and names vertically.
6. Never position names so that parts of them are upside down. (See Figure 14.17c.)

Placement and Design of Titles and Legends
Map titles, legends, and other explanatory information, like any other graphic elements, need to fit into the whole plan for the map. Titles are generally the most important

Table 14.4 Percent of Names Falling into Aggregated Positions Around Point Symbols*

Rand Map		AAA Map		Gousha Map	
Position	Percent	Position	Percent	Position	Percent
2	18.5	1	24.9	1	17.8
7	17.4	3	18.5	3	17.4
1	15.9	2	13.5	2	16.7
3	15.4	4	12.0	7	14.1
8	13.2	5	8.5	8	13.9
4	12.3	7	8.3	4	11.3
6	4.4	6	8.1	5	4.9
5	2.9	8	6.2	6	4.1

*The aggregated positions refer to the relative positions as identified by Yoeli.
Source: Adapted from Chyon Victor Wu and Barbara P. Buttenfield, "Reconsidering Rules for Point-Feature Name Placement," Cartographica 28 (1991): 10–27.

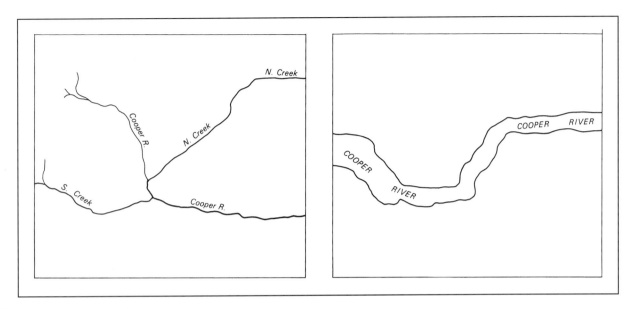

Figure 14.15   Labeling linear features.
Names outside a feature should be set solid, never letterspaced (a). Labels may be repeated for clarity. Inside a feature such as a river at large scale, the name may be set in all capitals as in (b), but not letterspaced.

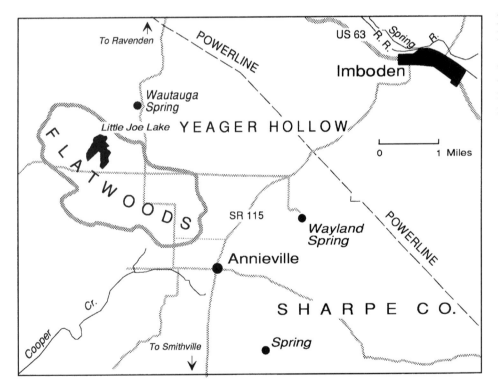

Figure 14.16   Areal labels.
Names set inside areal features should be letterspaced to occupy the entire feature. Such labels may be curved to suggest the shape of the feature. Names outside features are not usually letterspaced.

intellectually and should therefore be largest in type size. Any subordinate titles should have somewhat smaller point sizes. Legend materials are important elements on the thematic map but are subordinate intellectually and visually. Their lettering size should reflect their position in the hierarchy. Map sources, explanatory notes, and the like are the smallest in point size.

Spacing in title or legend boxes requires attention from the designer. When three lines of type are used and the middle line is to be largest, the lower line must be a bit smaller than the top line if it is to appear as the same size. (See Figure 14.18a.) One type designer notes that "lines consisting of letters of the same size in the same arrangement must, under all circumstances, have the same letterspacing."[35] (See Figure 14.18b.) This has particular significance when placing titles within boxes or title frames. Short words should never be letterspaced to achieve a real or imagined rectangle.

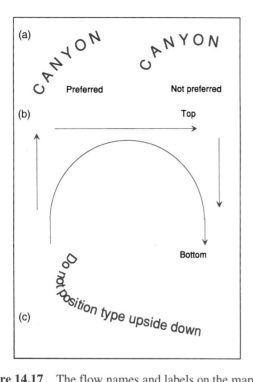

Figure 14.17 The flow names and labels on the map. See the text for an explanation.

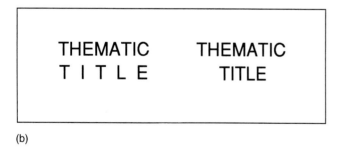

Figure 14.18 Planning title designs.
If the bottom line of a three-line title is to appear nearly equal to the top line, it should be in a smaller point size, as on the right in (a). Never spread letters to fill in a legend box or imaginary rectangle, as on the left in (b). It is better to use consistent letterspacing and center the titles.

Because of their importance, map titles usually should be set larger than any subtitles, and in uppercase and lowercase. Subtitles can be made to appear subordinate by using smaller point sizes. Remember that words set in all capitals must be letterspaced for aesthetic reasons.

The placement of titles, legends, and other written information will be dictated in part by the overall layout of the map. These elements must be worked into the total design to achieve proper balance and proportion. Their treatment must be subtle so that they command the attention they deserve without overpowering the map. Title lettering or title boxes can be located so that they overlap other map features. (See Figure 14.19.) This helps to position them in the visual hierarchy and establish their intellectual rank.

Experimental Studies

Few experimental studies have dealt with lettering in thematic map design, but some significant findings have come to light. In one study, researchers investigated the effect of map typography on the speed of searching for place names on a map—a task similar to finding a city or town name on

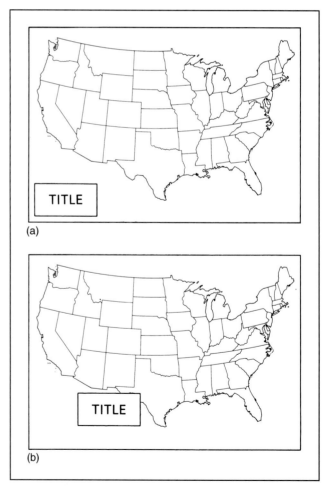

Figure 14.19 Using titles and title boxes in map design. Positioning titles and title boxes over map areas may make them more integral to the overall visual and hierarchical plan of the map.

a general-reference map. The findings that have direct bearing on typographic design for the map are as follows:[36]

1. Names set in lowercase with an initial capital are easier to find than names set in all capitals of the same point size. Lowercase names are recommended.
2. Boldface type is no more legible than type of normal weight. It should be avoided because it has a "cluttering" effect on the map.
3. Choice of typeface has little effect on legibility in the map context.
4. Type should be placed in as clear a space as possible. Avoid clutter close to the initial letter of a word.
5. Typographic coding by color or point sizes can reduce search time; irrelevant coding can increase it.

Another researcher studied the map-search abilities of seventh- and eighth-grade students, reaching the following conclusions:[37]

1. Ordinary type styles are about equal in "searchability." A 4-point size variation had no significant effect on search times.
2. On a map containing three type styles, search is slowed if the reader does not know beforehand in which typeface the target will be printed. On the other hand, search time is reduced if the reader knows beforehand the target's typeface. This supports the design practice of replicating map elements (including symbols and lettering) in legend materials.
3. Text reading is not similar to map reading, at least in terms of type.

A common practice in thematic map design is to suggest an ordinal classification of phenomena with labels of different size. Thus, cities having small populations are labeled with lettering of relatively small point size, and larger cities are identified by labels in larger point sizes. Barbara Shortridge investigated this design practice to determine what lettering sizes and differences between them would best serve to indicate this kind of ordinal ranking. Her general conclusions were that[38]

1. Lettering size differences of 34 percent are easily discriminated. Thus, a difference of 2 to 2½ points in the range of 5½ to 15 points is above this threshold.
2. Lettering size differences of less than 22 percent should not be used.

THE PRODUCTION OF COLD TYPE MAP LETTERING

Cold type production should be distinguished from hot type methods. **Hot type** is any type production process that involves a raised (relief) surface.[39] The term is a carryover from earlier periods in printing when individual type was cast from molten metal. Some graphic artists will use the method, but few others. **Cold type** composition, on the other hand, involves the preparation of letter images for some form of pho-

tographic reproduction. Practically all cartographic lettering today is produced by cold type methods. The most popular method used today is type production using microcomputers and some form of map, graphic, or desktop mapping software, to produce the entire map including its type. Final output from this production method is often sent on discs to a service bureau for the image to be made into negatives for printing. More will be said of these techniques in Chapter 17.

DIGITAL TYPOGRAPHY

Typesetting was originally done hundreds of years ago by wooden, raised relief blocks, set together to form words. When maps were prepared, each letter block was often dropped into the larger wood block forming the map image and whole words for map lettering were thus created. As copperplate engraving took over, the engraver actually engraved the words onto copperplate. Phototypesetting was the next major technological breakthrough, emerging after World War II (although experimented with as early as the late 1800s). Cartographers made use of this technology by ordering whole words set by the phototypesetter, which were then printed on adhesive-backed vinyl sheets and applied to the cartographic art before final negatives were made. Today, digital typesetting has replaced the phototypesetter, and digital type is now available at the microcomputer workstation.

Mapping and graphic application software incorporate some form of type handling within the program. Type can be inserted directly in the map image and stored along with the file. So-called "typeset quality" lettering is easily achieved, depending on the quality and resolution of the output device. (See Figure 14.20.)

Digital type fonts are of two kinds: bit maps of fixed sizes, or scalable (also called *outline*) typefaces. (See Figure 14.21.) Bit-map fonts were actually the first kind of digital fonts, and were only seen on computer screens. A character in a *bit-map* format is composed of an arrangement of pixels, and each typeface style, for each character,

Figure 14.20 Low-cost laser printer typographics. Helvetica typeface on the left is printed actual size at 300 dpi, and on the right reduced by 25 percent.

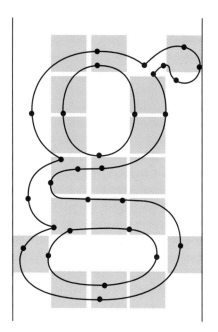

Figure 14.21 Bit-mapped versus scalable (outline) font characters.
Characters defined in bit-mapped format are formed by pixels, and are not as precise as those described by mathematical formula, called scalable, or outline fonts.

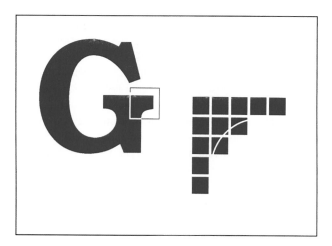

Figure 14.22 Bit-mapped fonts.
If bit-mapped fonts are enlarged, the irregularity caused by the fixed pixel resolution causes the edges of the letterform to appear jagged (the "jaggies").

and for each point size has its own bit map. These type of fonts require an enormous amount of computer storage memory, and are inflexible in that they cannot be displayed other than in the original pixel pattern. If they are enlarged, for example, the irregularity caused by the fixed pixel resolution causes the edges of the letterface to appear jagged. (See Figure 14.22.)

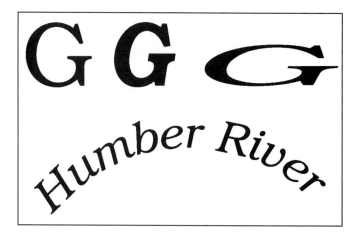

Figure 14.23 The flexibility of scalable fonts.
Scalable fonts can be turnned, twisted, skewed, and be made to conform to curves without losing their sharpness.

Bit-mapped fonts are referred to as **screen fonts,** because they are generated in the pixel format of the display screen. A pixel stands for "picture element" and is the smallest addressable unit on the computer screen of a given resolution. More will be said of these in Chapter 17. The ordinary type of the display device uses bit-mapped fonts. It is possible to display outline fonts on the screen through a **font rasterizer,** which is also necessary for printing outline type.[40] Font rasterizers may be in the form of software read into the CPU, or stored into the printer's ROM, in the case of a PostScript, or other PDL (page description language) printer.

The other form of digital type is called **scalable,** or *outline,* **type.** In this form, each character is defined as a mathematical formula, the formula defining the lines and curves of the character's outline. One advantage of this kind of type is that it can be output to any size, and its accuracy is only dependent upon the resolution of the output device. For example, a type can be generated and its point size specified and printed at 300 dpi (dots per inch), 600 dpi, 1,200 dpi or higher resolutions of the imagesetter, or output to a laser printer. Digital type output to film at 2,450 dpi results in a crispness that rivals the best phototypesetter of a few years ago.

Another advantage of scalable type characters is that they can be turned, skewed, twisted, and most important to cartographers, can be made to conform to curves without losing their sharpness. (See Figure 14.23.)

One caution about scalable fonts must be expressed. Although the same typeface may be specified, the actual *design* of that face may be different among vendors.[41] Often the differences are very subtle, perhaps in only the smallest of the finishing strokes. Some typeface designs differ also in the way the rasterizer deals with fitting the outline to the underlying pixel grid. And because the way a given design behaves with the underlying grid when the font is reduced or enlarged causes its computer design to differ somewhat from the original version of that style. Cartographic designers need to be aware of these nuances of digital type.

abcdefghijklmnopqrstuvwxyz
ABCDEFGHIJKLMNOPQRSTUVWXYZ

a A e E k K r R t T z Z

Figure 14.24   Digital type.
Digital type, in caps or lowercase of any size, can be generated
with just a few computer keystrokes. This is Times New Roman,
printed with a low-resolution printer.

With the graphics and mapping programs available
today, and with the output devices at hand, the cartographer
is no longer faced with significant type production and reproduction problems. The cartographic designer has literally
hundreds of digital type styles from which to choose. (See
Figure 14.24.) Of all the advances made in the technology
of cartographic production, I have found that among them,
those offered by digital type are at the very top.

NOTES

1. Erwin Raisz, *Principles of Cartography* (New York:
   McGraw-Hill, 1962), pp. 54–55.
2. John R. Biggs, *The Use of Type* (London: Blandford
   Press, 1954), p. 18.
3. John S. Keates, "The Use of Type in Cartography,"
   Surveying and Mapping 18 (1958): 75–76.
4. R. A. Gardiner, "Typographic Requirements of
   Cartography," *Cartographic Journal* 1 (1964): 42–44.
5. Alexander Lawson, *Printing Types: An Introduction*
   (Boston: Beacon Press, 1971), p. 29.
6. Jan Tschichold, *Treasury of Alphabets and Lettering*
   (New York: Reinhold, 1966), p. 29.
7. Ibid.
8. Allen Haley, "Kerning: Fine Typography or Marketing
   Hype?" *Upper and Lower Case* 18, no. 1 (Spring
   1991): pp. 22–24; see also Biggs, *The Use of Type,*
   p. 40.
9. Lawson, *Printing Types,* p. 47.
10. Daniel Berkeley Updike, *Printing Types, Their History,
    Forms, and Use: A Study in Survivals,* 3d ed., 2 vols.
    (Cambridge: Harvard University Press, 1966),
    pp. 61–64. The reader is encouraged also to read W. S.
    Shirreffs, "Typography and the Alphabet,"
    Cartographic Journal 30 (1993), pp. 97–103.
11. Lawson, *Printing Types,* p. 59.
12. Ibid., p. 61.
13. Updike, *Printing Types,* pp. 101–6.
14. Ibid., pp. 107–16.
15. Biggs, *The Use of Type,* pp. 18–20.
16. Miles A. Tinker, *Legibility of Print* (Ames: Iowa State
    University Press, 1963), p. 8.
17. Herbert Spencer, *The Visible Word* (New York:
    Hastings House, 1968), p. 25.
18. Tschichold, *Treasury of Alphabets,* p. 35.
19. Tinker, *Legibility of Print,* p. 64.
20. Clarence R. Gilman, "1:100,000 Map Series," in "Map
    Contemporary," *American Cartographer* 9 (1982):
    173–77; see also David Woodward, "Map Design and
    the National Consciousness: Typography and the Look
    of Topographic Maps," Technical Papers of the
    American Congress on Surveying and Mapping (Falls
    Church, VA: American Congress on Surveying and
    Mapping, 1982), pp. 339–47.
21. Spencer, *The Visible Word,* p. 31.
22. A. G. Hodgkiss, "Lettering Maps for Book Illustration,"
    Cartographer 3 (1966): 42–47.
23. Arthur H. Robinson, *The Look of Maps: An
    Examination of Cartographic Design* (Madison:
    University of Wisconsin Press, 1966), p. 39.
24. Ibid.
25. Raisz, *Principles of Cartography,* p. 54; also Hodgkiss,
    "Lettering Maps for Book Illustration," pp. 42–47.
26. Gardiner, "Typographic Requirements," pp. 42–44.
27. Robinson, *The Look of Maps,* p. 37.
28. Eduard Imhof, "Positioning Names on Maps,"
    American Cartographer 2 (1975): 128–44; see also
    Robinson, *The Look of Maps,* p. 47.
29. P. Yoeli, "The Logic of Automated Map Lettering,"
    Cartographic Journal 9 (1972): 99–108.
30. Chyan Victor Wu and Barbara P. Buttenfield,
    "Reconsidering Rules for Point-Feature Name
    Placement," *Cartographica* 28 (1991): 10–27.
31. Liza Noyes, "The Positioning of Type on Maps: The
    Effect of Surrounding Material on Word Recognition
    Time," *Human Factors* 22 (1980): 353–60.
32. Imhof, "Positioning Names on Maps," pp. 128–44.
33. Ibid.
34. Ibid.
35. Tschichold, *Treasury of Alphabets,* p. 41.
36. Richard J. Phillips, Elizabeth Noyes, and R. J. Audley,
    "Searching for Names on Maps," *Cartographic Journal*
    15 (1978): 72–76.
37. Barbara S. Bartz, "Experimental Use of the Search Task
    in an Analysis of Type Legibility in Cartography,"
    Cartographic Journal 7 (1970): 103–12.
38. Barbara Gimla Shortridge, "Map Reader Discrimination
    of Lettering Size," *American Cartographer* 6 (1979):
    13–20.
39. J. Michael Adams and David D. Faux, *Printing
    Technology: A Medium of Visual Communication*
    (North Scituate, MA: Duxbury Press, 1977), p. 66.
40. Barrie Sosinsky, *Beyond the Desktop: Tools and
    Technology for Computer Publishing,* Bantam ITC
    Series (New York: Bantam Books, 1991), pp. 346–50;

see also James Felici and Ted Nace, *Desktop Publishing Skills: A Primer for Typesetting with Computers and Laser Printers* (Reading, MA: Addison-Wesley, 1987), p. 9; see also Bob Carter, Ben Day, and Philip Meggs, *Typographic Design: Form and Communication* (New York: Van Nostrand, 1985), pp. 96–102.

41. Students are encouraged to read further from numerous articles found in professional computer books and magazines that deal with the basics of digital typography.

GLOSSARY

ascender letter stroke extending above the *x*-height or body of a lowercase letter, as in h, d, b, p. 270

ascender line horizontal line marking the maximum extent of all ascenders, p. 270

base line bottom horizontal line from which all capital letters and lowercase bodies rise, p. 270

Black Letter name of the type style first used in movable type; attempted to replicate textura or manuscript forms, pp. 274–275

bowls rounded portions of such letters as o, b, and d, p. 270

bracket a fillet that helps join serifs to the main stroke, p. 270

cap line horizontal line marking the height of all capital letters; usually lower than the ascender line, p. 270

cold type preparation of letter images for photographic reproduction, p. 283

counters partially or completely enclosed areas of a letter, p. 270

descender letter stroke extending below the *x*-height or body of a lowercase letter, as in y, g, q, p. 270

descender line horizontal line marking the maximum extent of all descenders, p. 270

display-decorative special typefaces used in art and advertising; not ordinarily useful in cartography, p. 275

display types type sizes not ordinarily used in running text; usually 12 points or larger, p. 275

family a series of lettering styles all related to the basic style; a family contains variations of the basic style in weights, widths, roman, and italic, p. 272

font a complete set of all characters of one size and design of a typeface, p. 272

font rasterizer software that converts scalable type to raster format for screen display, p. 284

hot type any type production process that involves a raised (relief) surface; traditionally refers to type cast from molten metal, p. 283

italics slanted version of a type style; minor letter form differences may exist between an italic and its roman counterpart, p. 272

kerning same as mortising, p. 273

leading spacing between horizontal rows of continuously running text, p. 271

letterspacing additional horizontal spacing placed between letters; required when all capital letters are to be set, but optional in lowercase, pp. 272–273

lettre de forme angular shaped Black Letter with extreme vertical stress; very difficult for modern readers, p. 275

lettre de somme more rounded version of Black Letter; less formal, easier to read than lettre de forme, p. 275

loop rounded portions of lowercase descenders such as j, g, and y, p. 270

Modern type design developed in the early eighteenth century; these styles have a pronounced vertical shading that distinguishes them from earlier forms, pp. 275–276

mortising fitting letters closer than in the ordinary (geometric) arrangement in order to achieve a good visual result, p. 273

neutralizing rule a letterspacing principle based on visual rather than geometric assessment of letterspacing, p. 273

Oldstyle early type designs that did not try to replicate manuscript styles; Venetian (Italy), Aldine-French (mostly French), and Dutch-English are three variations of Oldstyle, p. 275

point unit of measurement for type; 72 points to the inch (0.138 in per point), p. 272

roman basic, upright version of a type style, p. 272

sans serif a letterform containing no serifs, pp. 270, 275

scalable type each character in the font is defined by mathematical formula, and can be displayed in infinite point sizes (limited by resolution of output device), slanted or skewed, p. 284

screen font font characters generated in pixel format for display on a screen, p. 284

script-cursive type style that replicates handwriting; not ordinarily useful in cartography, p. 275

serif finishing stroke added to the end of the main strokes of a letter, p. 270

set solid no letterspacing, p. 271

shading the position of maximum stress or slant of a letterform; Oldstyle type has greater shading than Modern faces, p. 270

square serif bold and squared off, with thickness similar to that of main letter strokes; also called *slab serif* or *Egyptian*, p. 275

textura early manuscript alphabet form common at the time of Gutenberg's first printing of the Bible, p. 274

Transitional type design that bridged the gap between Oldstyle and Modern; Baskerville (English) is the best example of period, p. 275

type discernibility perception and comprehension of individual words not set in running text; perhaps of greater theoretical use in cartographic design than type legibility, p. 276

typeface   particular style or design of letterforms for a whole alphabet, p. 270

type legibility   type design characteristics that can affect ease and speed of reading, p. 276

type personality   informal way of describing the impressions that type can elicit, e.g., dignity, power, grace, precision, and excitement, pp. 275–276

type weight   relative blackness of a type: normal, lightface, boldface, p. 272

type width   different width of a letter design, such as condensed or extended, p. 272

Univers   a popular sans serif typeface designed by Adrian Frutiger in the 1930s, p. 275

word spacing   space between words set horizontally, p. 273

x-height   height of the body of lowercase letters, p. 270

READINGS FOR FURTHER UNDERSTANDING

Adams, J. Michael, and David D. Faux. *Printing Technology: A Medium of Visual Communication.* North Scituate, MA: Duxbury Press, 1977.

Bartz, Barbara S. "An Analysis of the Typographic Legibility Literature: An Assessment of Its Applicability to Cartography." *Cartographic Journal* 7 (1970): 10–16.

———. "Experimental Use of the Search Task in an Analysis of Type Legibility in Cartography." *Cartographic Journal* 7 (1970): 103–12.

Biggs, John R. *The Use of Type.* London: Blandford Press, 1954.

Brook, Valerie Francene. "Information Architect." *Print,* September/October 1990, pp. 120–29, 157–62.

Cook, Rick. "Page Printers." *Byte* 12, September 1987, pp. 187–97.

Dailey, Terrence, ed. *Illustration and Design: Techniques and Materials.* Secaucus, NJ: Chartwell Books, 1980.

Felici, James, and Ted Nace. *Desktop Publishing Skills: A Primer for Typesetting with Computers and Laser Printers.* Reading, MA: Addison-Wesley, 1987.

Gardiner, R. A. "Typographic Requirements of Cartography." *Cartographic Journal* 1 (1964): 42–44.

Gilman, Clarence R. "1:100,000 Map Series." In "Map Commentary," *American Cartographer* 9 (1982): 173–77.

Hodgkiss, A. G. "Lettering Maps for Book Illustration." *Cartographer* 3 (1966): 42–47.

Imhof, Eduard. "Positioning Names on Maps." *American Cartographer* 2 (1975): 128–44.

Keates, John S. "The Use of Type in Cartography." *Surveying and Mapping* 18 (1958): 75–76.

———. *Cartographic Design and Production.* New York: Wiley, 1973.

Lang, Kathy. *The Writer's Guide to Desktop Publishing.* London: Academic Press, 1987.

Lawson, Alexander. *Printing Types: An Introduction.* Boston: Beacon Press, 1971.

Lewis, John. *Typography: Design and Practice.* New York: Taplinger, 1978.

Monmonier, Mark Stephen. *Technological Transition in Cartography.* Madison: University of Wisconsin Press, 1985.

Nelson, Roy Paul. *Publication Design.* Dubuque, IA: William C. Brown, 1972.

Noyes, Liza. "The Positioning of Type on Maps: The Effect of Surrounding Material on Word Recognition Time." *Human Factors* 22 (1980): 353–60.

Phillips, Richard J., Elizabeth Noyes, and R. J. Audley. "Searching for Names on Maps." *Cartographic Journal* 15 (1978): 72–76.

Poulton, E. C. "A Note on Printing to Make Comprehension Easier." *Ergonomics* 3 (1960): 245–48.

Raisz, Erwin. *Principles of Cartography.* New York: McGraw-Hill, 1962.

Robinson, Arthur H. *The Look of Maps: An Examination of Cartographic Design.* Madison: University of Wisconsin Press, 1966.

———, **Randall Sale, and Joel Morrison.** *Elements of Cartography.* 4th ed. New York: Wiley, 1978.

Schlemmer, Richard M. *Handbook of Advertising Art Production.* Englewood Cliffs, NJ: Prentice-Hall, 1976.

Shortridge, Barbara Gimla. *Map Lettering as a Quantitative Symbol: A Preliminary Investigation.* Unpublished Ph.D. dissertation, Lawrence: University of Kansas, Department of Geography, 1979.

———. "Map Reader Discrimination of Lettering Size." *American Cartographer* 6 (1979): 13–20.

Spencer, Herbert. *The Visible Word.* New York: Hastings House, 1968.

Tinker, Miles A. *Legibility of Print.* Ames: Iowa State University Press, 1963.

Tschichold, Jan. *Treasury of Alphabets and Lettering.* New York: Reinhold, 1966.

Updike, Daniel Berkeley. *Printing Types, Their History, Forms, and Use: A Study in Survivals.* 3d ed. 2 vol. Cambridge: Harvard University Press, 1966.

Woodward, David. "Map Design and the National Consciousness: Typography and the Look of Topographic Maps." Technical Papers of the American Congress on Surveying and Mapping. Falls Church, VA: American Congress on Surveying and Mapping, 1982, pp. 339–47.

Wu, Chyan Victor, and Barbara P. Buttenfield. "Reconsidering Rules for Point-Feature Name Placement." *Cartographica* 28 (1991): 10–27.

Zachrisson, Bror. *Studies in the Legibility of Printed Text.* Stockholm: Almquist and Wiksell, 1965.

CHAPTER

15

PRINCIPLES FOR COLOR THEMATIC MAPS

CHAPTER PREVIEW

The application of color to the thematic map is one of the most exciting aspects of cartographic design, yet perhaps the least studied by cartographers. Color is produced by physical energy, but our reaction to it is psychological. Color has been found to have three dimensions: hue, value, and chroma. Our perception of color is influenced by its environment and by the subjective connotations we attach to colors. Map readers have preference for certain colors, but these change throughout life. Our understanding and use of color, *especially in color research, are enhanced by learning how color can be specified. Color theorists may use the Munsell, Ostwald, ISCC-NBS, CIE specification, or other systems. Cartographic designers are aware that colors can function in certain ways in design. Certain color conventions guide the cartographer, especially in quantitative mapping such as on choropleth maps. Design strategies to achieve figure and ground, the proper degree of contrast, and color harmony improve the use of color on the thematic map.*

Introducing color into the design of a thematic map can be both exciting and troublesome. On the one hand, color provides so many more design options that designers often quickly seize the opportunity to include it. Yet the inclusion of color invites many potential problems. Costs increase, design solutions require greater precision, and production problems can begin to mount. The designer is plagued by uncertainty as to how the reader will respond to the map. However, if allowed to choose, with no budgetary restraints, most map designers choose color mapping because of its inherent advantage of greater design freedom.

Color in thematic mapping is perhaps the most fascinating and least understood of the design elements. Color is subjective rather than objective:

> *The concept of color harmony by the color wheel is an objective conclusion arrived at by intellectual activity. The response to color, on the other hand, is emotional; thus there is no guarantee that what is produced in a purely intellectual manner will be pleasing to the emotions. Man responds to form with his intellect and to color with his emotions; he can be said to survive by form and to live by color.*[1]

Another problem in dealing with color is that it is difficult, if not impossible, to set color rules. Certain standards for color use have been adopted for some forms of mapping, notably on USGS topographic maps and maps of other national mapping programs. No standards or rules for color use, except for a few conventions, exist for thematic maps. This situation is getting greater attention as the production of temporary color maps, such as those generated on color CRT (cathode-ray tube) screens, can be designed from a palette of thousands of colors. On the other hand, rigid rules can also be restrictive in design.

Color is a complex subject and can be studied in many different ways. Physicists, chemists, physiologists, psychologists, philosophers, musicians, writers, architects, and artists approach color from different perspectives, with distinct purposes.[2] The physicist looks at the electromagnetic spectrum of energy and how it relates to color production. Chemists examine the physical and molecular structures of **colorants,** the elements in substances that cause color through reflection or absorption. The physiologist treats the mechanisms of color reception by the eye-brain pathway, and the psychologist deals with the meaning of color to human beings. Finally, the artist works with aesthetic qualities of color, using information gained from the physiologist and psychologist.

This chapter introduces the characteristics of color that are important in thematic mapping: color perception, color specification systems, cartographic conventions, and design strategies. The focus throughout is on the psychological and aesthetic aspects of color for the printed map (as opposed to the color we see when viewing a CRT image). Color as it applies to printing is dealt with in Chapter 16.

The student is encouraged to read further from the fascinating literature on color listed at the end of this chapter.

COLOR PERCEPTION

Reading thematic maps is a process involving the eyes and brain of the map reader. Light reflected off the map is sensed by the eyes, which report sensations to the brain, where cognitive processes begin. The sensing and cognitive processing of color is called **color perception.** Aspects of color perception treated in this section are the physiology of the human eye, the physical properties of color production, and the dimensions of color.

THE HUMAN EYE

The human eyes are wondrous organs, often considered to be external linkages to the brain itself. The processing of light that falls on the human eye is like a vast, complex data-analyzing system. Each eye is a spherical body about 1 in. in diameter. (See Figure 15.1.) Light enters through the **cornea,** a transparent outer protective membrane. The amount of light entering the eye is controlled by the **iris,** a diaphragm-like muscle at the center of the eye. Light is then focused by the transparent **lens** and passed onto the back wall of the eye chamber. The inside of the back of the eye chamber is covered by a thin tissue called the **retina.** A fluid (vitreous humor) fills the eye and serves to keep the distance between lens and retina nearly constant.

Light-sensitive cells make up the retina. **Rod cells,** numbering about 120 million, provide for only achromatic sensations and have no color discrimination. **Cone cells,** numbering about 6 million, are of three types. Some have peak sensitivity to blue, some to red, and others to green portions of the electromagnetic spectrum. As these rod and cone cells are triggered, electric impulses pass through a complex system of specialized cells to the **optic nerve,** which is a bundle of nerve cells that transmits the electrical

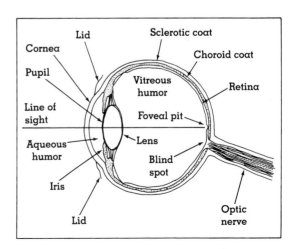

Figure 15.1   The parts of the human eye.

impulses to the brain. Color perception takes place in the brain and involves cognitive processes that add meaning and substance to the light patterns sensed by the eye. Color does not exist in the environment, or in objects. Color is psychological, a product of the mind.

The cartographic designer does not ordinarily need to pay particular attention to the details of the color-sensing apparatus of the map reader, except to make sure that graphic marks do not exceed the lower limits of visual acuity. Illumination is very important. At low levels, only achromatic sensing is possible; color detection becomes possible as illumination increases. Most map-reading situations exceed the minimal illumination threshold for color discrimination. When objects subtend an angle on the retina of less than about 12 minutes of arc, discrimination of color becomes difficult. Absolute color identification is generally poor for objects subtending less than 10 minutes of arc.[3] It is advisable to limit color use to map objects that subtend an angle on the retina of 20 minutes of arc or more. At a reading distance of 30 cm (12 in), this would be about .06 in (.15 cm).

Color blindness is a facet of color perception that affects about 7 percent of males and less than 1 percent of females. This is usually a hereditary condition, although not always. The manifestation is that the subject may have difficulty in perceiving distinctions between red/green, blue/yellow, pale green/yellow, or some of all three. Although certainly not a major concern in map design, some mapping projects may call for the designer to accommodate color-blind readers, especially those maps that use hue differences to define symbol categories such as way-finding road maps, rapid-rail system maps, bus route maps and others. There is more research being done now on the design of color maps for the color blind, principally by using CRT displays.[4] This is not altogether unreasonable in this age of scientific and cartographic visualization, where much analysis takes place while viewing color CRT displays using maps, graphs, and satellite color images.

According to two recent studies in molecular biology, a difference in a single amino acid—the minimum genetic difference between two people—can cause a perceptible difference in color vision. Research teams at the University of Washington in Seattle and at Johns Hopkins University tracked the genetic basis of red photopigments, a type of protein, and show that the newly discovered amino acid affects the part of the cone cells where color perception begins. The studies show that there is nearly an infinite number of ways to see red alone. The variations are caused by subtle differences in genetic makeup, offering a biological explanation for the extreme subjectivity of chromatic responses. . . . With research pointing to complete individuality in perception, what theoretical models have a chance?

Source: Charles A. Riley, II, *Color Codes* (Hanover, NH: University Press of New England, 1995), pp. 1–2.

PHYSICAL PROPERTIES OF COLOR PRODUCTION

The production of color requires three elements: a light source, an object, and the eye-brain system of the viewer. This defines the **object mode of viewing.** Color production in cartography is normally the object mode, that is, when light strikes the map some of that light is reflected back to the eye of the map reader.

Visible light is that part of the electromagnetic energy spectrum to which our eyes respond. Sir Isaac Newton showed that the visible portion of the energy spectrum is composed of the various colors, each having a different wavelength. (See Figure 15.2.) Moreover, different light sources, such as the sun, a tungsten lightbulb, or a fluorescent bulb, generate different spectral energy patterns. When we attempt to describe the physical properties (physical dimensions) of color, we must specify which source is used. Each produces a characteristic **spectral energy distribution curve,** so that the colors of the viewed objects vary. In a paint store, color chips are often displayed in such a way that the customer can switch on either incandescent or fluorescent lights, to see how the colors behave in different light. It is no wonder that color specification and color

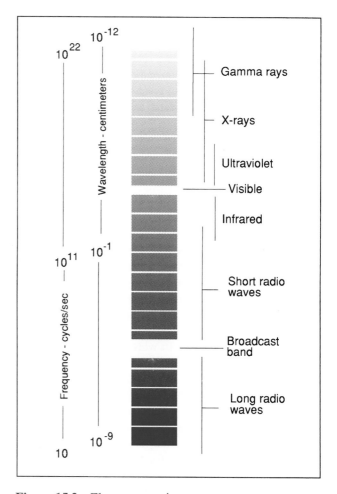

Figure 15.2 Electromagnetic energy spectrum.

choice are so difficult in cartography; color maps may be designed in one environment and viewed in another.

The physical characteristics of color are also affected by the quality of the object's surface. Some surfaces permit all light to pass through them. These are called **transparent objects.** Most objects are **reflective** of portions of the visible spectrum (such as paper maps); that part of the spectrum that is reflected defines the color of the surface. Surfaces that absorb all light are **opaque** and appear black. The amount of light that is reflected from surfaces can be plotted on a diagram that is called a **spectral reflectance curve.** This type of diagram defines the physical dimensions of reflected light from objects. (See Figure 15.3.)

Cartographers usually do not delve into the spectral reflectance curves of the papers and inks of their printed maps. The combinations of printed color inks and different papers are so numerous that measurement of this kind is usually cost prohibitive. Research cartographers, however, need to be especially mindful of these characteristics, so that their work can be replicated and extended. The complexities of this kind of research have had a negative impact on color experimentation in cartography.

COLOR DIMENSIONS

Cartographic designers are directly concerned with the psychological dimensions of color—those that describe what we see in the form of reflected light. These dimensions are much more difficult to explain because of the range of variation in human perception, the difficulty of reporting visual perception objectively, and the fact that we often associate colors with behavior moods. No set of exactly tailored rules is available to the designer, although a few standards regarding color use for quantitative maps have emerged. These are discussed later in the Color Conventions in Mapping section of this chapter.

The Desert Island Experiment

A subject is asked to imagine that he or she is on a desert island that is covered by many large pebbles of different colors.[5] The task is to arrange (classify) the pebbles by color, using any system. The subject first decides to divide the pebbles into two groups—those with color (**chromatic**) and those that are white, gray, or black, or without color (**achromatic**).

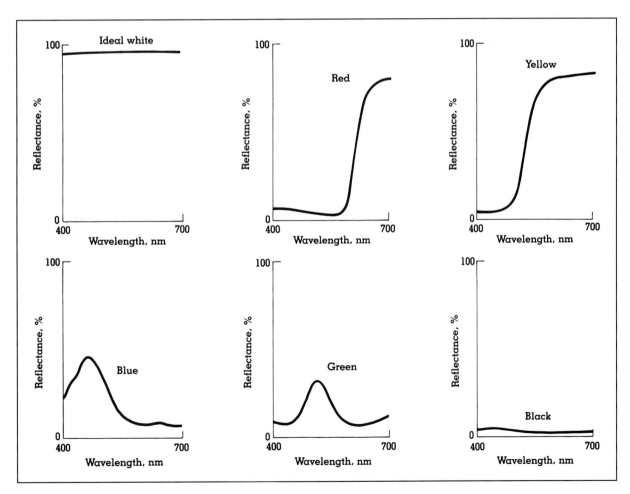

Figure 15.3 Spectral reflectance *(R)* curves of several colored materials.
Notice that white reflects the most light and black the least. Also, the similarity in the curves of colors close to each other (red and yellow or blue and green) is quite remarkable.

The next step in classification by this subject might be to arrange the achromatic pebbles (the whites, grays, and blacks) into an array from white to black, with the grays in between. This array, based on the single quality *lightness,* is called a **value scale.**

To deal with the sorting of the chromatic pebbles, the subject decides to make piles of red stones, blue stones, green stones, and so on. If greater precision is needed, there could be groups of blue-green pebbles, red-orange, and so forth. The chromatic pebbles are now divided into **hue** categories.

Within each pile, the subject now decides to array the colored pebbles by lightness or value. Thus, for example, the red stones could be arrayed from light red to dark red, from pinks to the deepest cherry reds.

In this experiment, a last judgment is made by the subject. After careful examination of, say, all the red pebbles, it may be noticed that even though two stones are similar in value (lightness), they still *look* different. A red radish looks somehow different from a tomato whose red is equally dark. After careful study, the subject decides that the two reds can be compared in terms of their neutral (or middle) gray content; all reds of the same value can be further divided into yet another spectrum. This is sometimes referred to as *saturation,* or **chroma.** A hue having less gray in it is more saturated.

This completes the desert island experiment. The subject has classified colors into groups based on appearance or **psychological dimensions.** Because these dimensions are based on perception, they have particular significance in thematic map design.

Hue

Hue is the name we give to the various colors we perceive: the reds, greens, blues, browns, red-oranges, and the like. Each hue has its own wavelength in the visible spectrum. Now, although hues and their relationships to other hues may be illustrated in many ways, a customary way, especially by the artist, is on a **color wheel,** as in Figure 15.4 and Plate 2. Theoretically, a color wheel can contain an almost infinite number of hues, but most include no more than 24. Eight or 12 are more common, 12 being especially useful for artists.[6] The organization of hue on the color wheel is not based on the physical relationships of the hues, but is simply a conventional way of showing visual hue relationships. It may be added here that some writers about color suggest that the color wheel is simply the human's "impulse" to simplify color into a tabular schema, much as the chemist places the elements into a periodic table.[7]

Although the color wheel normally contains 12 hues, the human eye can theoretically distinguish millions of different colors. In ordinary situations, however, we are not called on to do so; in cartographic design, our assortment of hues is certainly far less. In fact, it would be difficult to imagine any case where a map reader would have to iden-

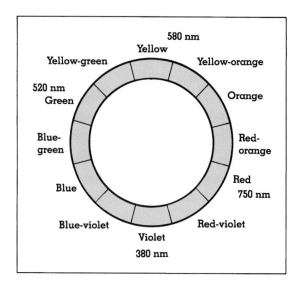

Figure 15.4 A 12-part color wheel, showing approximate wavelengths of the various hues.
Color wheels are typically used by artists, but can be useful in describing certain color associations (such as color complementaries and opposites). Typical color wheels often contain 8 or 12 hues. The abbreviation nm stands for nanometers; one nanometer is one-billionth of a meter.

tify more than a dozen (this is common on soil or geologic maps). The United States census has done 16-color two-variable choropleth maps,[8] but these have not found wide use because of the complexities inherent in reading them.

Value

Value is the quality of *lightness* or *darkness* of achromatic and chromatic colors. Conceptually easy to grasp, value is difficult to deal with in practice. Sensitivity to value is easily influenced by environment, and apparent lightness or darkness is not proportional to the reflected light of achromatic surfaces. A given gray, or hue of a specified value, looks one way when examined individually against a white background but differs when viewed in an array of grays against different backgrounds.

People distinguish better among grays if all are similar to their backgrounds. Against a light background, light grays are more easily distinguished from each other; against a dark background, dark grays are better distinguished. (See Figure 15.5.) Also, relative to the whole value spectrum, greater differences at each end (both white and black) are required for better discrimination.[9] These conditions are compounded when the value scales are chromatic. People cannot easily discern values from an array containing a number as large as 10, unless they have training. Five is preferable for cartographic purposes.

In art, value is controlled by the addition of white or black pigment to a hue. If white is added to a hue, a **tint** results. When black is added, a **shade** is produced. A **tone** results from adding equal amounts of a hue, white, and

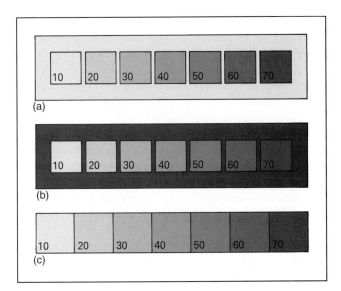

Figure 15.5   The effect of the environment on the perception of value.

The surrounding environments of (a) and (b) modify the appearance of the group (otherwise identical) in the strips they surround. Induction results when color patches of different values are closely juxtaposed, as in (c). Notice also that induction does not occur in (a) and (b) because the gray patches are separated by open spaces or black strips.

black.[10] Tint, shade, and tone are discussed also in the section on harmony later in the chapter. These terms are mentioned in this chapter because they are used so frequently in informal discussions of color. *Cartographers in specifying colors for printing use different terms.*

Controlling value on printed color maps is difficult. One suggested method is discussed below, but it will be better understood if the chroma dimension is first explained.

Chroma

Chroma is also called *saturation, intensity,* or *purity.* This color dimension can best be understood by comparing a color to a neutral gray. With the addition of more and more pigment of a color, it will begin to appear less and less gray, finally achieving a full saturation or brilliance. For any given hue, chroma varies from 0 percent (neutral gray) to 100 percent (maximum color). At the maximum level, the color appears pure and contains no gray. Chroma levels vary with hues; the most intense yellow appears brighter than the most intense blue-green. Achromatic colors are said to have zero chroma.

COLOR INTERACTION

When we look around us we see a world of many colors, patches of color surrounded by other colors, and colors adjacent to other colors. In many cases, maps are composed of more than one color. Reaction to a color patch is always modified by its environment. Seldom is the reader asked to look at a single color—even simple color maps usually contain variations of one hue, combined with black. There are at least three **color interactions**—simultaneous contrast, successive contrast, and color constancy—that must be considered in cartographic design.

Simultaneous Contrast

When the eye spontaneously produces the complementary color (opposite on the color wheel) of a viewed color, the effect is called **simultaneous contrast.**[11] A color, if surrounded by another color, begins to appear tinged by the complementary color of the surrounding color. A simple experiment demonstrates this effect. Place a small gray patch inside a box of pure color. After you gaze at the surrounding color for a brief period, the gray square begins to look like the surrounding color's complementary color.

Simultaneous contrast (Figure 15.5) causes adjacent colors to be lighter in the direction of the darker adjacent colors, and darker in the direction of the lighter colors. This is sometimes referred to as **induction,** a special case of simultaneous contrast.[12] Induction is particularly bothersome in map design when several different values of the same color are juxtaposed on the map or in the legend. The effects can be reduced by separating color areas by white or black outlines.[13] An interesting feature of induction, or any simultaneous contrast, is that it cannot be photographed; it is produced by the eye alone.

Successive Contrast

Successive contrast results when a color is viewed in one environment and then in another in quick succession.[14] The color will be modified relative to these new surroundings. It may appear darker or lighter, more or less brilliant compared to each new environment. For example, a gray will look darker against a lighter background and lighter against a darker background. An orange will appear darker and more red on a yellow background and lighter and more yellow on a red background. New contrasts are therefore established in each new situation. Cartographers cannot control all color environments, but positive design attempts to reduce the number of successive contrasts on the map will help.

Color Constancy

Although it is not of central concern to the map designer, **color constancy** is a feature of color perception that should at least be noted. We tend to judge colors based on presumed illumination. For example, the shadow areas of the folds of a red drape are gray rather than red, but because we assume the drapes to be lit by a common source (e.g., sunlight) we perceive these gray areas to be red also. This kind of judgment is made during map reading when color areas fall in shadows. Although the shadows cause actual color areas to change, we perceive them in constancy. If it were not for this, reading color maps would be almost impossible.

SUBJECTIVE REACTIONS TO COLOR

Human beings react to color for a number of reasons. Simultaneous and successive contrast are caused by our physiological system (eye-brain); most people have nearly identical response mechanisms to these. More variable, and much more difficult to control in design, are psychological reactions to color: color preferences, the meanings of color, and behavioral moods and color. Taken together, these are called the **subjective reactions to color.**

Color Preferences

An enormous amount of research has dealt with color preferences. Most of the formal research has come from the fields of psychology and advertising. Little research has been conducted by cartographers on color preference, so our understanding must be borrowed from the other disciplines.

There is abundant evidence that color preference is somewhat developmental. We must first define terms. **Warm colors** are those of the longer wavelengths (red, orange, and yellow), and **cool colors,** at the opposite end of the spectrum, have shorter wavelengths (violet, blue, and green). Warm and cool, of course, are psychological descriptions of these color ranges. Children about the age of four or five years prefer warm colors. Red is most popular, with blues and greens next.[15] Young children also prefer highly saturated colors, but this preference begins to drop off after about the sixth grade.

The findings regarding children's color choices have prompted one cartographic researcher to comment on color design for children's maps:[16]

1. As children are aware only of small ranges in hue, colors should be chosen from within the basic spectrum colors—blue, green, yellow, orange, and red.
2. Because school-age children begin to reject fully saturated colors, a step or two down the saturation range is more desirable.
3. Children appear to dislike dull unattractive colors, so color choice should avoid these. Stay close to the spectral hues.
4. Children generally reject achromatic color schemes—the gray scale. These reduce the attractiveness of the map.
5. Choosing colors that have greater compatibility with the *expected* yields greater comprehension. Thus, for example, blue is better than red for water.

As we mature and leave childhood, our color preferences change. Generally, we tend to favor colors at the shorter wavelengths. Color preferences among North American adults are blue, red, green, violet, orange, and yellow, in that order. The greenish-yellow hues are the least liked by both men and women. Women show a slight preference for red over blue and yellow over orange, whereas men slightly prefer blue over red and orange over yellow. Both sexes choose saturated colors over unsaturated ones. Although color preference yields some insight into the broader realms of color psychology, a simple list of preferences may not be enough. Color preference tests performed by psychologists are conducted for the sake of color choice only and are not product related. Most color experts would agree that other variables, such as color environment, product name, packaging, context, and merchandising schemes, are also important in color choice.[17]

Parallels may be drawn for color use in cartographic design. Color conventions should overrule broader color preferences; cartographic design should otherwise seek to follow the preferences of the marketplace. Until detailed studies are provided, map designers need to follow the leads provided by merchandising, art, and related psychological literature.

Colors in Combination   In one particular study,[18] 10 different hues, each with three different values and chromas, were selected as object colors. A random sample of 25 background colors was selected, and the subjects were asked color combination preferences. The general findings were these (see also Table 15.1):

1. The most pleasant combinations result from large differences in lightness. Lightness (or value) contrast is necessary in pleasant object-background combinations.
2. A good background color must be either light or dark; being intermediate in lightness makes it poor.
3. Consistently pleasant object colors are hues in the green to blue range, or other hues containing little gray.
4. Consistently unpleasant object colors are in the yellow to yellowish-green range, or other hues containing considerable gray.
5. To be pleasant, an object color must stand out from its background color by being definitely lighter or darker. This is the single most important finding of the study for the cartographer.
6. Vivid colors combined with grayish colors tend to be judged as pleasant.
7. Good and poor combinations are found for all sizes of hue differences (that is, distances apart on the color wheel).

These findings were not the result of a cartographic inquiry, so they can be used only as general guidelines. Nonetheless, they can serve the cartographer as a starting point for color selection.

Connotative Meaning and Color

Perhaps the most interesting design aspect of color is that people react differently to spectral energies. Two people look at the same red wavelength, but their responses may be entirely different. This presents the cartographic designer with a considerable challenge, so the subjective and connotative meanings of color should be studied.

Connotative responses to colors vary considerably. The literature, both in psychology and advertising, is sometimes vague and contradictory. Nonetheless, some generalization may be made. (See Table 15.2.) Psychologists suggest that

Table 15.1   Object-Background Preferred Color Combinations

Option Hues			Highest Ranked Favorable Background Hues*			Highest Ranked Unfavorable Background Hues†		
Munsell Notation		ISCC-NBS Color Names††	Munsell Notation		ISCC-NBS Color Names	Munsell Notation		ISCC-NBS Color Names
5R	5/8	Moderate red	5.0G	2/2	Very dark green	10.0RP	4/12	Vivid purplish red
5YR	5/8	Brownish orange	5.0G	2/2	Very dark green	10.0RP	4/12	Vivid purplish red
5Y	5/6	Light olive	5.0G	2/2	Very dark green	5.0R	4/14	Vivid red
5GY	5/8	Strong yellow-green	5.0G	2/2	Very dark green	5.0G	5/4	Moderate green
5G	5/6	Moderate green	5.0R	8/2	Pale to grayish pink	2.5PB	5/4	Grayish blue
5BG	5/6	Moderate bluish green	N	10/	Black	5.0G	5/4	Moderate green
5B	5/6	Moderate greenish blue	5.0Y	9/12	Vivid yellow	5.0G	5/4	Moderate green
5PB	8/2	Very pale blue	5.0R	2/2	Dark grayish red-very dark red	5.0GY	7/10	Strong yellow-green
5P	5/2	Grayish purple	N	10/	Black	5.0Y	5/4	Light olive
5RP	5/10	Moderate purplish red	N	10/	Black	2.5YR	5/14	Vivid orange
N	5/	Medium gray	5.0R	8/2	Pale pink	2.5PB	5/2	Grayish blue

*The background color shown for each object color represents only the top-ranked of five favorable ones selected by test subjects.
†The background color shown for each object color represents only the top-ranked of five unfavorable ones selected by test subjects.
††ISCC-NBS (Inter-Society Color Council—National Bureau of Standards). Color names are identified for each Munsell Color Notation.
Source: H. Helson and T. Lansford, "The Role of Spectral Energy of Source and Background Color in the Pleasantness of Object Colors," Applied Optics *9 (1970): 1513–62; color names from National Bureau of Standards,* The ISCC-NBS Method of Designating Colors and a Dictionary of Color Names. *Circular 553 (Washington, DC: USGPO, 1955).*

Table 15.2 Connotative Meanings and Color

Color	Connotations
Yellow	Cheerfulness, dishonesty, youth, light, hate, cowardice, joyousness, optimism, spring, brightness; strong yellow—warning
Red	Action, life, blood, fire, heat, passion, danger, power, loyalty, bravery, anger, excitement; strong red—warning
Blue	Coldness, serenity, depression, melancholy, truth, purity, formality, depth, restraint; deep blue—silence, loneliness
Orange	Harvest, fall, middle life, tastiness, abundance, fire, attention, action; strong orange—warning
Reddish browns, russets, and ochres (earth colors)	Warmth, cheer, deep worth, and elemental root qualities; can be friendly, cozy or dull, reassuring or depressing
Green	Immaturity, youth, spring, nature, envy, greed, jealousy, cheapness, ignorance, peace; midgreens—subdued
Purple (violet)	Dignity, royalty, sorrow, despair, richness; maroon—elegant, and painful
White	Cleanliness, faith, purity, sickness
Black	Mystery, strength, mourning, heaviness
Grays	Quiet and reserved, controlled emotions; can be used to create sophisticated atmosphere

Source: James E. Littlefield and C. A. Kirkpatrick, Advertising: Mass Communication in Marketing *(Boston: Houghton Mifflin, 1970), p. 204; Al Hackl, "Hidden Meanings of Color,"* Graphic Arts Buyer *12 (1981): 41–44; and Marshall Editions,* Color *(New York: Viking Press, 1980), pp. 138–41; see also Frank H. Mahnke and Rudolf H. Mahnke,* Color and Light in Man-Made Environments *(New York: Van Nostrand Reinhold, 1995), pp. 11–16.*

reds, yellows, and oranges are usually associated with excitement, stimulation, and aggression; blues and greens with calm, security, and peace; black, browns, and grays with melancholy, sadness, and depression; yellow with cheer, gaiety, and fun; and purple with dignity, royalty, and sadness.[19] Advertising people are quick to point out that color context and copy also suggest the connotative meaning of color.

The implications of this in cartographic design are not altogether clear. There is a paucity of cartographic research into color meaning and map design. Until adequate research can lead the way, it appears that we must borrow from the psychologists and advertising people. In general, the map designer should attempt to select colors by carefully taking into account their connotative meanings in association with the map's message.

ADVANCING AND RETREATING COLORS

The eye of the reader seems to perceive colors as **advancing** or **retreating.** The colors with the longer wavelengths, notably red, appear closer to the viewer when seen along with a color of shorter wavelength. There is some evidence of a physiological basis for this claim. The lens of the eye "bulges" when it refracts red rays. This same convex shaping of the lens takes place when we view objects close up. For this reason, red may have come to be associated with proximity. In general, warm hues (long wavelengths) advance, cools recede. For brightness or value, high values advance, and low values recede. In terms of saturation, deep or highly saturated colors advance, and less saturated colors recede.

There is at least one case in which this aspect of color has an effect on color selection. The development of figures and grounds on maps can be enhanced by choosing colors with their advancing and retreating characteristics in mind. Advancing colors should be applied to figural objects.

COLOR SPECIFICATION SYSTEMS

Color specification—the exact naming of colors—has been examined by many artists, scientists, and others ever since Newton showed that color could be generated by passing light through a prism. Yet because of the large number of colors available to the senses, it was not easy to agree upon such a naming system. It was not until the early part of this century that standards began to appear. The Munsell system, named after its American inventor, Albert H. Munsell (1858–1918), has become the standard in the United States. Wilhelm Ostwald (1853–1932), a Latvian by birth, invented a similar system that has become commonly used throughout Europe.

Two other systems, the ISCC-NBS (the Inter-Society Color Council and the National Bureau of Standards), and the CIE (Commission International de l'Eclairage, or the International Commission on Illumination) have evolved since the work of Munsell and Ostwald. ISCC-NBS uses a unique naming system applied to the Munsell color solid, and CIE specifications are based on the physical spectrometry of reflected light. The RGB color model, specific to CRT displays and increasingly more important in map production, is discussed in Chapter 17, which deals with digital compilation and desktop mapping.

Cartographers, especially researchers, are called on more and more to be specific in their color designations for papers and inks used in printing. Understanding the terminology of color specification has therefore become increasingly important to map designers as they attempt to communicate about color. The value of these systems, especially Munsell's and Ostwald's, is in the way they help us conceptualize interrelationships of the three dimensions of hue, value, and saturation of colors. Practical color specification deals mainly with ink selection.

MUNSELL COLOR SOLID

Munsell recognized the three dimensions of color to be hue, value, and chroma (or saturation) and incorporated these into the **Munsell color solid.**[20] A color solid is a representation of **color space,** a way of integrating and showing the relationships of the three color dimensions. (See Plate 1.) The solid is only roughly a sphere, because of the unevenness of chroma.

The color solid of Munsell and the quantitative specifications derived from it, are based on the whole array of colors available for human sensation. Steps or divisions between colors are based on psychological intervals, measured through extensive testing. As new color pigments are devised by science and industry, new color chips can be added to the model. This is a principal advantage of this color solid.

Ten hues make up the color band around the solid; each hue has a letter designation. Ten number divisions are associated with each hue—the "5" position designating the pure color of each hue. There are therefore 100 different hues on the color wheel; for example, 7R, 2Y, and 8B each specify a particular hue on the circle.

Psychophysical testing has determined that people do not perceive color value in an equal-step manner. Munsell's value scale, called an *equal-value scale,* records the reflectance percentages of what appear to be equal steps from white to black. Value on the solid is represented along a vertical scale (the axis) of the color solid, perpendicular to the plane of the color wheel. The top of this value scale is white (10) and the bottom position is black (0). Between these two "poles" are nine divisions, graduating from white to black, that yield different grays. The middle gray has a value of 5. Notation of value is by the position on this vertical scale, followed by a slash (7/, 2/, etc.).

With these two dimensions, dark red or middle red, for example, can be exactly specified. A designation of 10BG 2/ would indicate a blue-green closer to blue and dark (nearly black) in value.

Munsell also incorporated chroma designation in his color solid. Chroma or saturation is represented by steps going inward toward the value axis for each hue band. The chroma scale is also psychologically stepped. As a color gets closer to the axis, it becomes weaker (more gray); conversely the farther out on the chroma scale a color is, the stronger and more saturated it becomes. Chroma designation is a number that follows the value slash, such as /7. A complete Munsell color designation would thus be, for example, 8P 7/5.

Maximum chroma differs for the variety of hues and values. For example, for 5R, maximum chroma at a value of 8/ is 4, but for a value of 4/ it is 14. For 5PB, maximum chroma at value 8/ is 2, and for 2/ it is 6. These give the color solid an uneven and unbalanced appearance, causing it to deviate significantly from a perfect sphere. (See Plate 1.)

OSTWALD COLOR SOLID

Like the model of Munsell, the **Ostwald color solid** displays an orderly arrangement of the colors humans can sense. The Ostwald solid differs in several important respects, however. It is shaped like a double cone, with the bases touching. (See Plate 2.) The color wheel, which forms the outer ring or circumference of the cones, unlike Munsell's, is composed of 24 separate hues, 3 in each of 8 primary hues. Ostwald's solid does not recognize the dimensions of hue, value, and chroma, but achieves distinct colors by hue, white, and black.[21]

The color solid is formed by the introduction of a monochromatic triangle of 28 *tones* (distinct colors achieved by adding percentages of white and black to a given hue). A neutral gray scale becomes the axis of the solid and is the innermost tier of tones in a given hue's triangle. (See Plate 2.) The neutral gray scale is a psychological gray scale—its divisions or percentages of gray are based on equally appearing amounts of gray. This scale differs from Munsell's.

A whole range of *tints* (hue plus white), *shades* (hues plus black), and *tones* (hues plus white and black) are achieved simply on the Ostwald solid. Specifications follow letters that indicate certain percentages of white and black, and hue is designated by its number from the color wheel. For example, 8 pa designates a relatively pure red (p indicates an absence of black and a an absence of white). A soft, grayish red would be 8 le.

Ostwald's color solid has one major limitation: its inflexibility. A new color having stronger brilliance (because of the invention of new dyes and colorants) cannot be added without changing the whole solid. The Munsell solid is not limited in this way, because new chromas may be added at the peripheries as they are introduced.

The Ostwald system of color specification finds its greatest utility among printers, artists, ink manufacturers, and others who develop colors by the mixing of colorants, white, and black pigments. Cartographers likewise find this system appealing.

> For a variety of information about color, color articles, color institutes, color trends, and color products, visit Web site *www.Pantone.com*. This is an especially interesting site for cartographers.

THE ISCC-NBS METHOD

One interesting approach to color designation deserves comment. The ISCC (Inter-Society Color Council) and the NBS (National Bureau of Standards) color name charts were originally devised as color standards to be used by the United States Pharmacopoeia.[22] In this system, all hues are preceded by modifiers that describe the color. (See Figure 15.6.) Such adjectives as pale, light, brilliant, very pale, weak, or dusky are used to designate such colors. The adjectives are used throughout all hues, but their locations vary relative to the value and chroma scales of the Munsell system, to which they are matched.

CIE COLOR SPECIFICATION

The CIE system of color designation describes the psychophysical properties of color as judged by a standard observer and has become an international standard for this kind of color description. This system is based on additive color mixing, generated by reflecting red, green, and blue colored lights from a white screen in an attempt to match the color of a test lamp.[23] These colored lights are considered *primary* lights, and the amounts of the lights necessary to produce a test color are called the test color's **tristimulus values.** Because this system also considers the reflective characteristics of the surface and the quality of the source(s) of illumination, color specification becomes quite complex. Sophisticated recording machines are required, such as a spectrophotometer that measures reflectance from a colored surface at different wavelengths.

Color charts, or physical samples such as color chips, are not produced by the CIE system. For this reason, the practicing cartographer will find little direct utility in this color system. The cartographic researcher pursuing objectives in color investigation, however, will find it useful because of its specificity in describing the physical attributes of colors.

OTHER SPECIFICATION SYSTEMS

There have been other color specification systems not specifically dealt with here. Some, indeed, predated the 1915 Munsell. The interested student is encouraged to read *Color Desktop Publishing Fundamentals* (Pennwell Graphics Group, 1994) for a brief look at other systems. (See Plate 12.) Some systems are used in printing specifications, especially the *Pantone Matching System,* which is discussed in detail in the next chapter.

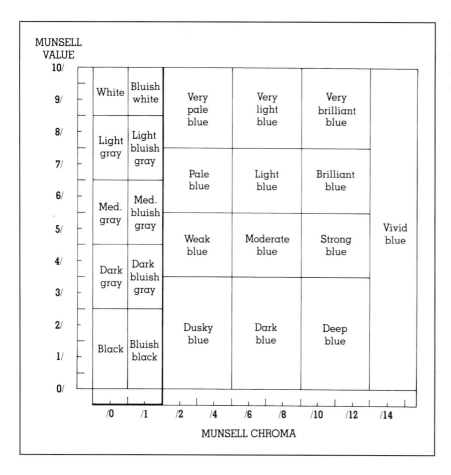

Figure 15.6 An example of the ISCC-NBS color specifications system. Each color is described by adjective identifiers and related to the chroma and value scales of the Munsell solid.

COLOR IN CARTOGRAPHIC DESIGN

Through the use of a variety of design strategies, several functional uses of color can be achieved on the map, as discussed in this final section of our introduction to the use of color in cartographic design.

THE FUNCTIONS OF COLOR IN DESIGN

Arthur Robinson of the University of Wisconsin–Madison has written succinctly on the various functions of color in mapping,[24] summarized as follows:

1. Color functions as a *simplifying* and *clarifying* agent. In this regard, the use of color can be useful in the development of figure and ground organization on the map. Color can unify various map elements to serve the total organization of the planned communication. (See Plate 3.)
2. Color affects the general perceptibility of the map. Legibility, visual acuity, and clarity (of distinctiveness and difference) are especially important functional results of the use of color.
3. Color elicits subjective reactions to the map. People respond to colors, especially the hue dimension, with

connotative and subjective overtones. Moods can be created with the use of many colors.

Thus, *structure, readability,* and the reader's *psychological reactions* can be affected by the use of color. It might be added that color functions to clarify thematic symbolization.

Lessons learned by the advertising industry regarding the function of color in printed media are also instructive for the cartographic designer. The use of color advertisements is not hit or miss, as it once was. Marketing studies have shown that color is an effective attention-getter and mood-setter and can be used as a strong product identifier. Some advertising experts estimate that readership can be increased by 15 percent by using color.[25] The goals of thematic mapping are vastly different from those of most print advertising, but the two disciplines do share common ground in the way they use color.

The functions of color in print advertising have been listed as follows:[26]

1. To attract buyer's attention
2. To stimulate interest
3. To identify products
4. To relate successive advertising
5. To provide emphasis
6. To illustrate, interpret, and prove

7. To contribute to structural motion; lead the reader's attention
8. To embody prestige and a certain mood or atmosphere
9. To relate components in the ad
10. To show structure, design, and installation of product

The functions of color in mapping are indeed similar to these.

DESIGN STRATEGIES FOR THE USE OF COLOR

Map designers employ several strategies to use color to its fullest potential in map communication. Five will be treated here, with figure and ground development first.

Developing Figures and Grounds

The organization of the map into figures and grounds can be enhanced by the use of color. Color provides contrast—a necessary component in figure formation. Perceptual grouping by similarity is also strengthened by the use of color. For example, similar hues are grouped in perception (although they may in fact be of different wavelengths). On a world map, for example, continents are more easily grouped as landmasses if they are rendered in similar hues. Colors of similar brightness or dullness are also grouped, as are warm colors, cool colors, or other like tints and shades.[27] Tints and shades are grouped with their primary colors. Perceptual grouping of colors is a strong tendency and should be a positive design element.

Generally, warm colors (reds, oranges, and yellows) tend to take on figural qualities better than cool colors (greens, blues, and purples), which tend to make good grounds. (The map "Per Capita Retail Sales: 1963" on page 221 of *The National Atlas of the United States of America* is a good example of the use of colors to bring out figures and grounds.) This may be partially explained by the tendency of the warm colors to advance and the cool colors to recede. In one set of United States census maps, *Race and Hispanic Origin Population Density of the United States, 1990,* United States Maps, GE-90 No. 6, the bright red in the highest density class stands out clearly from the remaining colors on the map.

The **power factor** concept can be applied in the context of developing figures and grounds. The power factor of a color is defined as the product of its Munsell value and chroma numbers.[28] Small areas of higher power factor are more likely to be seen as figures.

Color combinations also affect figure and ground development. (See Plate 5 and Table 15.3.) Yellow on black is noted to be the most visible color combination—yellow tends to be perceived as figure. The least visible and therefore worst combination, in terms of figure and ground development, is red on green. An example of the weakness of red on green in developing figures and grounds can be found on page 161 of *The National Atlas of the United States of America.* The colors listed in Table 15.3 are to be used only as starting points for selection, because any modification

Table 15.3 Color Combinations Useful in Developing Figure and Ground Organization on the Thematic Map

Figure Colors		Ground Colors
Yellow	Best	Black
White		Blue
Black		Orange
Black		Yellow
Orange		Black
Black		White
White		Red
Red		Yellow
Green		White
Orange		White
Red	Worst	Green

Source: Deborah Sharpe, The Psychology of Color and Design *(Chicago: Nelson-Hall, 1974), p. 107.*

of chroma and value will affect the results. The cartographic designer needs to balance other design elements with these color combinations to achieve an overall solution.

The Use of Color Contrast

Contrast is the most important design element in thematic mapping. Contrast in the employment of color can lead to clarity, legibility, and better figure-ground development. A map rendered in color with little contrast is dull and lifeless and does not demand attention. Even in black-and-white mapping (or one color other than black), contrasts of line, pattern, value, and size are possible. With color, additional possibilities exist. One colorist lists seven possible color contrasts:[29]

1. Contrast of hue
2. Contrast of value (light-dark)
3. Contrast of cold and warm colors
4. Complementary contrasts
5. Simultaneous contrast
6. Contrast of saturation (intensity or chroma)
7. Contrast of extension

Hue contrast can be used in cartographic design as a way to affect clarity and legibility and to generate different visual hierarchical levels in map structure. Some cartographers believe that hue is the most interesting dimension in color application in mapping, more so than value or chroma.[30] Contrast of hue demands an overall color plan for the map and requires that some thought be given to color balance and harmony, topics not very well researched by cartographers. It is also important to note that identical hues appear differently, depending on their color environment.

Saturation and value are two contrasts that provide visual interest and, depending on the nature of the map, can carry quantitative information. Contrast of value is a fundamental necessity in structuring the color map's visual field into figures and grounds. Objects that are high in value (relatively light) tend to emerge as figures, provided other components in the field do not impede figure formation. It is difficult to talk of saturation and value in isolation; they are usually a part of any hue selection. Both are essential in the design of quantitative maps.

Contrast of cold and warm colors can also be used to enhance figure and ground formation on the map. Artists use this contrast to achieve the impression of distance; faraway objects are rendered in cold (blue/green) colors and nearby objects in warmer tones. Figures on maps should be rendered in colors of the warmer wavelengths, relative to the ground hues.

In most mapping cases the designer should avoid simultaneous contrast, because this effect is visually troublesome and distracting. For some, simultaneous contrast can be minimized, if not eliminated altogether, by separating color areas with black or white lines.

Many artists believe that providing complementary colors in a composition establishes stability. Complementary colors are opposite on the color wheel. The primary complementaries are yellow-violet, blue-orange, and red-green. The eye will spontaneously produce the complementary of a color fixated on if it is not present. Because of this, it is said that harmony can be achieved in a composition by a balance of complementary colors. The eye does not then need to produce complementaries on its own, so the image is more stable. The contrast of complementaries appears to be a feature of color use that needs further investigation.

Contrast of extension relates to the much broader topic of color balance and harmony, dealt with in more detail later in the chapter.

Developing Legibility

The legibility of colored objects, especially lettering, is greatly influenced by their colored surroundings. Symbols in color must be placed on color backgrounds that do not affect their legibility. Black lettering on yellow (object on background) is a very high legibility combination, and green lettering on red is the least. (See Table 15.4.) The difficulty is exacerbated because the lettering is usually spread over several different background colors. Black lettering may become illegible as it crosses dull or gray colors. The designer must pay careful attention to lettering and all color environments in which it is placed. An otherwise good design can fail if caution is not exercised in this respect.

Color Conventions in Mapping

Conventional uses of color in mapping may be separated into qualitative and quantitative conventions. On most color thematic maps, these conventions must be observed because breaking a convention can be extremely disconcerting to the map reader.

Table 15.4 Legibility of Colored Lettering on Colored Backgrounds

Color Combination (Object on Background)

Black on yellow	Most legible
Green on white	
Blue on white	
White on blue	
Black on white	
Yellow on black	
White on red	
White on orange	
White on black	
Red on yellow	
Green on red	
Red on green	Least

Source: Al Hackl, "Hidden Meanings of Color," Graphic Arts Buyer *12 (1981): 41–44.*

Qualitative Conventions   Colors used on maps in a qualitative manner are those applied to lines, areas, or symbols that show kind or quality, not amount. Qualitative color conventions use color hue and chroma to show nominal (and in some cases ordinal) classifications. Many conventions are quite old—such as showing water areas as blue—and the logic of their use is well established.

Arthur Robinson has itemized many of the conventional uses, as follows:[31]

1. Blue for water
2. Red with warm and blue with cool temperature, as in climatic and ocean representations
3. Yellow and tans for dry and little vegetation
4. Brown for land surfaces (representation of uplands and contours)
5. Green for lush and thick vegetation

The color dimension most logically used on the qualitative map is hue. Hue shows nominal classification well and is especially appropriate because hue is difficult to associate psychologically with varying amounts or quantity of data. Caution must be exercised in using contrast of value to show nominal characteristics, because people tend to assign quantitative meanings to value differences.

Quantitative Conventions   Several International Statistical Congresses were held in Europe between 1853 and 1876. The question of statistical graphics (including maps) became an issue at several of these. In particular at the Congress of 1857 in Vienna, the sixth section of one report contained several recommendations on the use of color for statistical maps. Among the most important to us were as follows:[32]

When one wishes to make use of the method [graphic], so often useful, of applying colors in various shades to geographic maps, he should have regard for the following points:

a. Twelve classes or divisions permit the establishment of a gradation of tints easy to distinguish from the lightest to the darkest shades and present the further advantage of allowing the comparison to be limited to six, four, or three classes, when there is a desire to consider only the principal relations.

b. The strongest proportions should be indicated by the darkest shades.

c. When it is a question of showing opposite extremes, the intermediate shades should be omitted from the scale of colors used.

It is indeed interesting to note that as long ago as 140 years, those in the emerging discipline of geographic mapping should have been so intuitive about how to use color value on the quantitative maps.

Conventions associated with color use on quantitative thematic maps today operate in terms of color choice or color plan. No conventions exist for color choice on quantitative maps (e.g., population density maps are always blue, income maps are always green, and so on). **Color plan** is the way the designer chooses to use the color dimensions of hue, value, and chroma to symbolize varying amounts of data on the map. (See Plate 6.)

There are a number of cartographic researchers who have examined basic color plans for quantitative maps. We will look at three, Cuff, Mersey, and Brewer, in particular.

The four color plans identified by David Cuff are:[33]

1. *Gray and simple-hue plan.* More of a single ink is applied to areas that represent greater amounts. This is achieved by *screening.* An excellent example of this method can be seen at the top of page 166 of *The National Atlas of the United States of America.*

2. *Part-spectral plan.* Colors adjacent on the color wheel show variations in amount: blue, green, and yellow (blue represents high amounts, yellow low); red, orange, and yellow (red represents high amounts, yellow low). Usually, the colors are produced by two inks, with the middle color achieved by overprinting.

3. *Full-spectral plan.* A separate hue is used to represent different amounts of data on the map. Red, orange, yellow, green, and blue may be used, with red usually chosen to represent the higher amounts. Colors are achieved by overprinting or by separate inks. This plan is the most commonly used for hypsometric layer tints on maps that show elevation.

4. *Double-ended plans.* In schemes that illustrate both positive and negative aspects on the same map, dark red at the positive end may grade to light red, and light red to light blue to dark blue at the negative end, or some other hue choices.

In cartographic research since Huff, Janet Mersey labeled six different color series (very similar to Cuff's and in some cases identical), ranging from one that displays the least order (hue-based) to one that is most ordered (value-based):[34]

1. Hue series (with practically no difference in value and intensity).

2. Double-ended series (usually used when a mapped variable deviates either above or below zero) and the class symbols go from lighter to darker values of different hues on either side of zero or middle class.

3. Spectral series (when the classes range through the spectrum and there are differences in value and intensity, such as is customary on hypsometrically tinted maps). Part-spectral series only use part of the spectrum. (See Plate 6.)

4. Hue-value series (this describes a series of colors that range from a light value of one hue to a dark value of another hue).

5. PMS value series (one colored ink plus black). In this series, one hue is selected and the series is produced by screening the hue so that different values of the hue result. This series also may be produced by taking one hue (ink) and overprinting different black screens, thus changing the values of the hue.

6. Black-and-white series (a true value series lacking in any hue) and producing only steps of gray. Often white and black anchor the ends of the spectrum.

An ordered color scheme, says Mersey, borrowed from the work of Bertin, is one in which the color symbols are spontaneously and universally ordered from high to low according to the magnitudes they represent. An unordered scheme is one in which the color symbols have no logical order (such as a set of different hues with no differences in value, that is, lightness and darkness).

Until recently I have held fast to the philosophy that quantitative color plans adhere to the simple hue plan of Cuff or the PMS series described by Mersey. This seemed the most *reasonable* and was consistent with my understanding of color and numerical value. However, recent work by cartographic researcher Cynthia Brewer tends to dispel this notion. In an exhaustive research project with the National Center of Health Statistics, she and her colleagues found spectral schemes "most pleasant and easy to read," and they go on to say:

Contrary to our expectations, spectral schemes are effective and preferred (the spectral scheme we tested included diverging lightness steps suited to the quintile-based classification of the mapped mortality data).[35]

The Munsell color specifications for the hues used by Brewer in this study are provided in Table 15.5 and, for examples of the different plans that have been used by the Census Bureau in its mapping, see Table 15.6.

Table 15.5 Spectral Color Schemes used by Brewer and Colleagues in Mortality Study

| Color Label* | Munsell Specification | |
	H	**V/C**
7-class scheme		
Rd-7F	5R	5/14
Or-6F	2.5YR	6/14
OY-5F	10YR	7/14
Yl-4F	7.5Y	8.5/12
Gn-3F	10GY	7/10
BG-2F	5BG	6/10
Bu-1F	10B	5/10
5-class scheme		
Rd-5Q	5R	4/14
Or-4Q	5YR	6/14
Yl-3Q	7.5Y	8.5/12
Gn-2Q	10GY	6/10
Bu-1Q	10B	4/8

*Labels of colors are those used by Brewer and colleagues in the study.

Source: Abbreviated from Cynthia Brewer et al., "Mapping Mortality: Evaluating Color Schemes for Choropleth Maps," Annals (Association of American Geographers) 87 (1997): 411–38.

When you do plan to use the single hue scheme, there are three ways that can be pursued in which gradations of value or chroma (or both) are achieved through screening or multiple screening. (See Figure 15.7.)

1. Gradation is achieved by screening solid ink with successively higher or lower screen values. For example, 10-, 25-, 50-, and 75-percent screened areas (with solid ink as the highest) may be used. Experimental results show that the higher the value level of the initial hue, the more the screened gradations vary in chroma than in value.[36]

2. Gradation is achieved by *overprinting* a solid color ink with successively higher (or lower) screened black ink. Thus, for example, 10-, 25-, 50-, and 75-percent black screens are overprinted on the selected color ink. In terms of sensation, the color's value or chroma (or both) are altered by this method.

3. The two above methods are combined.

Cartographic designers need guidelines in the production of such graded series. Fortunately a solution has been suggested by Henry Castner using the Ostwald color solid and the concepts of tint and shade.[37] The plan is patterned after a triangular slice from the Ostwald solid and shows the relationship of screened hues, overprinting black screens, and their effects on saturation and value. (See Plate 4.) Although not based on experimental psychological testing, this model is very helpful in devising the color plan for symbolization on quantitative maps.

Table 15.6 Color Plans Used on Several GE-50 Series Choropleth Maps

GE-50 Series No.	Title	Color Plan (1)	Hues (2)	Number of Classes	Printing Method (3)
29	Size of Farms, 1964	Part spectral	Light beige to dark blue	7	A
42	Population Trends, 1940–1970	Double-ended	Dark blue to dark red	6	B
6	Families with Incomes under $3,000 in 1959	Full spectral	Red, brown, orange, yellow-green, blue	6	C
38	Population Density, 1970	Part spectral	Beige to red	4	D
47	Number of Negro Persons, 1970	Part spectral	Light yellow to dark blue	6	A
56	Median Family Income for 1969	Full spectral	Yellow-beige, light green, green, blue, orange, red, purple	7	A

Notes:

(1) A part spectral plan may be considered an ordered hue array; the descriptions in this column follow those of Cuff.

(2) In each case, the first color listed represents the lowest data magnitude, and the last color listed, the highest data magnitude.

(3) Method A = 3 solid inks, and screened inks overprinted

Method B = 2 solid inks, each screened successively

Method C = 6 solid inks

Method D = 2 solid inks, and screened inks overprinted

No black screen tints are used in any of these color plans.

Source: U.S. Census Bureau, GE-50 Series Maps listed.

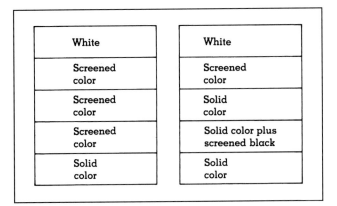

Figure 15.7 Two commonly used systems of screening and overprinting to achieve a graded series in color on quantitative maps.

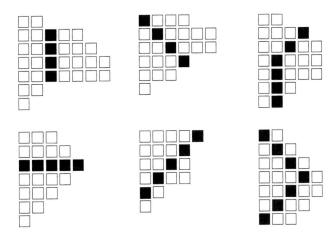

Figure 15.8 Representtaive color series for quantitative thematic maps from Munsell color charts.
Either value or chroma alone, or together, may be selected from the chart. *(Redrawn from Cynthia A. Brewer, "The Development of Process-Printed Munsell Charts for Selecting Map Colors,"* American Cartographer *16 [1989]: 269-78, Figure 6.)*

Mersey's work with color specification for choropleth maps indicates that the color plan should be selected based on the task placed before the map reader.[38] If the task is to extract specific information (that is, to look for an area containing a given data amount), the color plan should be based on hue variations, with few data classes. If the task is to gain overall knowledge of the spatial variation of the data mapped, the color plan should be a simple hue plan having regular value increments ordered to the data amounts (larger magnitudes are symbolized with a hue of darker value and smaller magnitudes with the same hue of lighter value). Map readers in this latter case appear to do equally well with choropleth maps having either a few data classes, or many. *If the designer is developing a choropleth map that encompasses multiple purposes, the best compromise is a hue-value plan in which regular value amounts are combined with a corresponding ordered-hue array.* Again, this appears to work equally well with choropleth maps containing a few or many data classes.

Cartographer Cynthia Brewer states, "The selection of thematic [color] schemes is assisted by the perceptual organization of the Munsell-based charts. Quantitative relationships between map categories are commonly represented by a progressive change in values accompanied by a systematic treatment of chroma."[39] This is consistent with Mersey and many other cartographers on the use of color for such thematic maps. In her work, Brewer provides representative paths through the Munsell chips to achieve several appropriate color series; these are included in Figure 15.8. A good student exercise in design would be to reproduce these on a CRT color monitor that can set colors in a CMYK palette, although a perfect representation would not result because of the limitations of producing color on the monitor.

Little study has been conducted on color plan selection for other types of quantitative maps. A notable exception is one done by Lindenberg on the perception of color graduated symbols. His conclusions show that estimation of size differences of graduated circles is not affected significantly by different hues.[40]

The Brewer Color Schemata

Reflecting considerable work of assimilation and amalgamation from many writers and color researchers by cartographers and others, and from her own expertise and experience, cartographer Cynthia Brewer provides a very useful organizational schemata for the use of color on thematic maps.[41] This color schemata is provided by using hue (the name of a color, such as red, orange, and so forth), and "lightness" (a term she prefers instead of value, darkness, intensity, and so forth), and, only secondarily, saturation. These schemata are captured in Table 15.7.

The essential strength of Brewer's organization is that it is consistent with the overall approach that color use must align logically with the data being shown on the map. (See box.)

As Brewer notes in her work, color application for quantitative mapping must take into consideration the data being displayed and the logical/perceptual organization of color. She notes that this is especially important for dynamic (animated) displays because events happen quickly and the logic between data and color must be clear and easily grasped. The coordination of these variables is essential in the use of color and map design.

In the more recent work of Brewer and her associates, they found that diverging schemes produce "better rate

When color is used "appropriately" on a map, the organization of the perceptual dimensions of color corresponds to the logical organization in the mapped data.

Source: Cynthia A. Brewer, "Color Use Guidelines for Mapping and Visualization," in *Visualization in Modern Cartography,* eds. Alan M. MacEachren and D. R. Fraser Taylor. (New York: Elsevier, 1994), pp. 123–47.

Table 15.7   Brewer Color Schemata for Qualitative and Quantitative Maps

Schemata Type	Color Organization
One-variable map	
Qualitative	Hues of similar lightness.
Binary	One hue and lightness step.
Sequential—no hue	Lightness steps of neutral grays.
Sequential—one hue	Lightness steps of a single hue.
Sequential—hue transition	Lightness steps with a part-spectral transition in hue.
Sequential—hue steps	Lightness steps with hue steps that progress through all spectral hues, but beginning in the middle of the spectrum, with low values represented by light yellow.
Spectral	Not appropriate.
Diverging	Two hues differentiate increase from decrease and lightness steps within each hue, colors lighter show minimal change.
Diverging	Two hues with lightness steps diverging from the midpoint.
Two-variable map	
Qualitative/binary	Different hues for the qualitative variable with a lightness step for the binary variable.
Qualitative/sequential	Different hues for the qualitative variable crossed with lightness steps for the sequential variable.
Sequential/sequential	Sequential/sequential scheme with cross of lightness steps of two complementary hues with mixtures producing a neutral diagonal.
Combination of sequential schemes	Combinations of hue differences resulting in hue and lightness steps.

retrieval than spectral or sequential schemes. . . ." This is really a remarkable finding and bears close scrutiny in future quantitative color work and map application.[42]

Color Harmony in Map Design

Color harmony traditionally is believed to belong to the realm of the artist, and is therefore not considered a matter of concern to the cartographic designer. This is far from the truth, but this element of design has been overlooked in cartographic research. Some aspects of this important subject are introduced in this section.

Color harmony, as it applies to maps, is more than the suitable and pleasing relationship of hues. Color harmony relates to the overall color architecture for the entire map. Harmony includes these components:

1. Effectiveness of the *functional uses* of color on the map
2. Appropriateness of the *conventional uses* of color on the map
3. Overall appropriateness of *color selection* relative to map content
4. Effective use of the *quantitative color plan*
5. Effective employment of the *relationship of hues*

Effectiveness of functional uses relates to how well the designer has employed color as a simplifying and clarifying agent. A harmonious design can be judged also on its appropriate use of color convention. Has every care been taken to provide conventional color wherever possible? Do departures from convention restrict the ease of communication?

Another important component of overall color harmony is the proper selection of color relative to map content. Maps illustrating January temperatures should not be rendered in warm hues, because of convention and connotative aspects, and deserts having sparse vegetation should not be shown in green.

The quantitative color plan (graded series of colors to show varying amounts) should be designed so that either color value or chroma differences correspond with numerical gradations. On maps of nominal data classes, symbols should be rendered in different hues.

Achievement of an effective relationship of hue balance of all colors on the map requires great skill and careful planning. Any multicolor map has its different colors occupying areas of varying sizes. **Color balance** is the result of an artful blending of colors, their dimensions, and their areas so that dominant colors occupying large areas do not overpower the remainder of the map. Dominant colors are those that contrast greatly with their environment; any color can be dominant, depending on its surrounding colors. *Contrast of extension* is another expression for color balance.

Brilliance (or light value) and extent (occupied area) are the two elements of **color force**.[43] This concept is similar to the power factor of a color, mentioned earlier. The light values of colors have been measured by colorists: yellow–9, orange–8, red–6, violet–3, blue–4, and green–6. For example, if a composition contains yellow and violet, the proportionality would be 9:3. To compute the areas and to maintain equal dominance, the reciprocal must be used. As yellow is three times stronger than violet, it should occupy one-third the space. Several examples of how yellow can dominate a map are formed in the yellow-blue color series of choropleth maps in the U.S. Department of Commerce *1987 Census of Agriculture, Agricultural Atlas of the United States,* listed in the readings at the end of this chapter.

Cartographic designers have their hands full when dealing with balance. In most instances, areas are fixed by geography, so only choice of color is possible when planning balance. Consideration must be given to the other elements of color harmony as well. The designer needs experience with color and a knowledge of how colors behave in different environments to solve balance problems.

Color harmony, as artists and colorists use the term, refers to the pleasant combination of two or more colors in a composition. Harmony also refers to order—the arrangement of colors. According to Ostwald, the simpler the order, the better.[44] However, what is pleasant to one map reader may not be to another. Experimentation with the ideas of color harmony as developed by Munsell, Ostwald, Birren, Itten, and others would reveal differences.

In considering harmony, perceived color can be categorized into six forms: pure hue, white, black, tint, shade, and tone. (See Figure 15.9.) White, black, and the three primaries produce the secondaries, tint, shade, and tone:

hue + white = tint
hue + black = shade
hue + black + white = tone

The following thoughts regarding these relationships of harmony are derived from Birren:[45]

1. Most people like pure hues rather than modifications (reds, not purples).
2. Every color should be a good example in its category. If a red is chosen, it should not be possible to mistake it for a purple. Pure colors, if selected, must be brilliant and saturated.
3. Whites should be white, and blacks should be black.
4. Tints, shades, and tones should be easily seen as such, not confused with pure hues, blacks, or whites.

To these may be added the following general harmony rules:

5. Colors opposite on the color wheel tend to be harmonious.
6. Harmony can also be achieved by using colors adjacent on the color wheel.
7. Harmony can also be achieved by combining colors of the same hue but with different tints, shades, and tones.

The words *tint, shade,* and *tone* to treat harmony are not often used by cartographers, but merit inclusion here because cartographers often use them in informal conversation. Professional cartographers must explore in considerable depth the relationships of colors. The task is made more difficult because cartography is not a purely expressive art. Color selection and employment are constrained by the map's other elements.

The next chapter is devoted to printing technology as it relates to the cartographer, and in that realm the discussion of color is extended to yet another level.

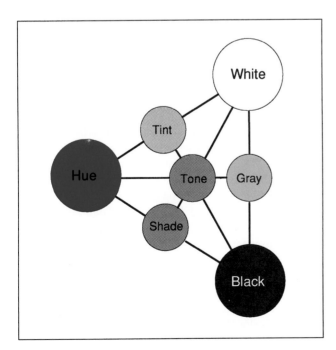

Figure 15.9   A color triangle devised by Faber Birren. As sensation, all colors fall into one of these seven forms. *Source: Faber Birren,* Selling Color to People *(New York: University Books, 1956), p.170.*

NOTES

1. Deborah Sharpe, *The Psychology of Color and Design* (Chicago: Nelson-Hall, 1974), p. 123.
2. Johannes Itten, *The Elements of Color* (New York: Van Nostrand Reinhold, 1970), p. 12; see also Charles A. Riley, II, *Color Codes* (Hanover, NH: University Press of New England, 1995).
3. Mari Riess Jones, "Color Coding," *Human Factors* 4 (1962): 355–65.
4. Judy M. Olson and Cynthia A. Brewer, "An Examination of Color Selections to Accommodate Map Users with Color-Vision Impairment," *Annals* (Association of American Geographers) 87 (1997): 103–34.
5. This experiment is retold in many books on color. A good version can be found in Fred W. Billmeyer, Jr., and Max Saltzman, *Principles of Color Technology* (New York: Wiley, 1966), pp. 15–17.
6. Howard T. Fisher, "An Introduction to Color," in *Color in Art,* ed. James M. Carpenter (Cambridge: Fogg Art Museum, Harvard University, 1974), p. 25.

7. Riley, *Color Codes,* pp. 7–8.

8. Judy M. Olson, "Spectrally Encoded Two-Variable Maps," *Annals* (Association of American Geographers) 71 (1981): 259–76.

9. Arthur H. Robinson, *The Look of Maps: An Examination of Cartographic Design* (Madison: University of Wisconsin Press, 1966), p. 90.

10. Faber Birren, *Selling Color to People* (New York: University Books, 1956), p. 171.

11. Itten, *The Elements of Color,* p. 52; the reader also is encouraged to look at Josef Albers, *Interaction of Color* (New Haven: Yale University Press, 1975); and Jerry Whitely and Joseph Roberts, "Josef Albers' Interaction of Color: From Print to Interactive Media," *Academic Computing,* January 1990, pp. 6–11.

12. William N. Dember, *Psychology of Perception* (New York: Holt, Rinehart and Winston, 1960), pp. 204–5.

13. Robinson, *The Look of Maps,* p. 94; good examples of simultaneous contrast can be found in Jan V. White, *Color for the Electronic Age* (New York: Watson-Guptill, 1990), pp. 16–18.

14. Jacob Beck, *Surface Color Perception* (Ithaca, NY: Cornell University Press, 1972), p. 187.

15. Sharpe, *The Psychology of Color and Design,* p. 18; see also Faber Birren, *Selling with Color* (New York: McGraw-Hill, 1945), p. 21.

16. Patrick Sorrell, "Map Design—With the Young in Mind," *Cartographic Journal* 11 (1974): 82–91.

17. Watson S. Dunn and Arnold M. Barbau, *Advertising: Its Role in Modern Marketing* (Hinsdale, IL: Dryden Press, 1978), pp. 429–32; see also White, *Color for the Electronic Age,* pp. 22–23.

18. H. Helson and J. Lansford, "The Role of Spectral Energy of Source and Background Color in the Pleasantness of Object Colors," *Applied Optics* 9 (1970): 1513–62. See also Deane B. Judd, "Choosing Pleasant Color Combinations," in *Contributions of Color Science by Deane B. Judd,* ed. David L. MacAdams (U.S. Department of Commerce, National Bureau of Standards—Washington, DC: USGPO, 1979).

19. Sharpe, *The Psychology of Color and Design,* p. 55.

20. Faber Birren, *Munsell: A Grammar of Color* (New York: Von Nostrand Reinhold, 1969), pp. 17–27.

21. Wilhelm Ostwald, *The Color Primer* (New York: Van Nostrand Reinhold, 1969), pp. 17–18.

22. U.S. Department of Commerce, National Bureau of Standards, *The ISCC-NBS Method of Designating Colors and a Dictionary of Color Names,* Circular 553—(Washington, DC: USGPO, 1955), p. 1.

23. Fred W. Billmeyer, Jr., and Max Saltzman, *Principles of Color Technology* (New York: Wiley, 1966), p. 31.

24. Arthur H. Robinson, "Psychological Aspects of Color in Cartography," *International Yearbook of Cartography* 7 (1967): 50–59; see also Robinson, *The Look of Maps,* pp. 75–97.

25. Jack Engel, *Advertising: The Process and Practice* (New York: McGraw-Hill, 1980), p. 503.

26. James E. Littlefield and C. A. Kirkpatrick, *Advertising: Mass Communication in Marketing* (Boston: Houghton Mifflin, 1970), p. 205.

27. Sharpe, *The Psychology of Color and Design,* p. 105.

28. A. Jon Kimerling, "Color in Map Design" (paper delivered at the workshop, Map Perception and Design, at the annual meeting of ACSM, Cartography Division, St. Louis, 1980).

29. Itten, *The Elements of Color,* pp. 33–64.

30. Robinson, "Psychological Aspects," pp. 50–59.

31. Ibid.

32. Gray Funkhouser, "Historical Development of the Graphical Representations of Statistical Data," *Osiris* 3 (1937): 268–404.

33. David J. Cuff, *The Magnitude Message: A Study of the Effectiveness of Color Sequences on Quantitative Maps* (unpublished Ph.D. dissertation, Department of Geography, Pennsylvania State University, University Park, 1972), p. 21.

34. Janet E. Mersey, *The Effects of Color Scheme and Number of Classes on Choropleth Map Communication* (unpublished Ph.D. dissertation, Department of Geography, University of Wisconsin–Madison, 1984), pp. 212–16; see also Janet E. Mersey, "Color and Thematic Map Design," *Cartographica* 27 (1991, Monograph, No. 41): 1–157.

35. Cynthia A. Brewer, Alan M. MacEachren, Linda W. Pickle, and Douglas Herrmann, "Mapping Mortality: Evaluating Color Schemes for Choropleth Maps," *Annals* (Association of American Georgaphers) 87 (1997): 411–38.

36. David J. Cuff, "Value versus Chroma in Color Schemes on Quantitative Maps," *Canadian Cartographer* 9 (1972): 134–40.

37. Henry W. Castner, "Printed Color Charts: Some Thoughts on Their Construction and Use in Map Design," *Proceedings* (American Congress on Surveying and Mapping, March, 1980): 370–78.

38. Mersey, "Color and Thematic Map Design," pp. 1–157.

39. Brewer, "The Development of Process-Printed Munsell Charts for Selecting Map Colors," (*American Cartographer* 16 (1989), pp. 269–78.

40. Richard E. Lindenberg, *The Effect of Color on Quantitative Map Symbol Estimation* (unpublished Ph.D. dissertation, Department of Geography, University of Kansas, Lawrence, 1986), p. 121.

41. Cynthia Brewer, "Color Use Guidelines for Mapping and Visualization," in *Visualization in Modern Cartography,* eds. Alan M. MacEachren and D. R. Fraser Taylor (New York: Elsevier, 1994), pp. 123–47.

42. Brewer et al., "Mapping Mortality," pp. 411–38.

43. Itten, *The Elements of Color,* p. 29.

44. Ostwald, *The Color Primer,* p. 5.

45. Birren, *Selling Color to People,* pp. 169–85.

GLOSSARY

achromatic having no hue (such as red, green, blue), but having characteristics (such as white, gray, or black), p. 291

advancing colors a hue of higher wavelength (notably red), a color of high value, or a highly saturated hue appears closer to the viewer; apparently caused by both physiological and learned mechanisms, p. 296

chroma the saturation, intensity, or purity of a color; one of the three color dimensions, p. 292

chromatic having the quality of hue, such as red, green, and blue, p. 291

colorant the elements in substances that cause color, either through absorption or reflection, p. 289

color balance the result of artful blending of colors and their areas so that dominant colors occupying large areas do not overpower the remainder of the composition; sometimes referred to as contrast of extension, p. 304

color constancy the tendency to judge colors as being identical under different viewing conditions, such as different illumination, p. 293

color force light values given colors by artists; brilliance, p. 304

color harmony the pleasant combinations of two or more colors in a composition; can be developed by functional, conventional, and color selection plans, p. 304

color interaction the way we perceive colors together; always modified by their environment, p. 293

color perception cognitive process involving the brain, where meaning and substance are added to light sensation, p. 289

color plan choosing the color dimensions of hue, value, and chroma to symbolize varying amounts of data, p. 301

color space a conceptual three-dimensional space used to illustrate the color dimensions of hue, value, and chroma, p. 296

color wheel the organization of hues of the visible spectrum into a circle; many distinct hues can be shown, but 12 are customary, especially among artists, p. 292

cone cells cells in the retina with peak sensitivity to blue, red, or green light energy, p. 289

cool colors colors of the shorter wavelengths, such as violet, blue, and green, p. 294

cornea transparent outer protective membrane over the lens of the eye, p. 289

hue the quality in light that gives it a color name such as red, green, or blue; a way of naming wavelength; one of the three color dimensions, p. 292

induction a special case of simultaneous contrast; causes adjacent colors to be lighter in the direction of the darker adjacent colors, and darker in the direction of the lighter colors, p. 293

iris diaphragm-like muscle that controls the amount of light coming into the eye, p. 289

lens crystal-like tissue that focuses light onto the back of the inside of the eye, p. 289

Munsell color solid a three-dimensional geometric figure, roughly equivalent to a sphere, designed to show the interrelationships of hue, value, and chroma, pp. 296–297

object mode of viewing the production of color in which a light source, an object, and the eye-brain system of the viewer are present, p. 290

opaque objects absorb all light striking them; appear black, p. 291

optic nerve bundle of nerve cells connecting the retina to the brain; conveys electrical impulses that carry light information, p. 289

Ostwald color solid a three-dimensional geometrical figure roughly equivalent to two cones whose bases touch; designed to show the relationships of color dimensions; particularly useful in showing how colors can be achieved by blending hue, white, and black, p. 297

power factor the product of a color's Munsell value and chroma numbers; related to color force, p. 299

psychological dimensions of color hue, value, and chroma, p. 292

reflective objects reflect some or all of the light striking them, p. 291

retina membrane that lines the back of the inside of the eye; contains light-sensitive cells, p. 289

retreating colors a hue of lower wavelength (blues), a color of low value, or a poorly saturated hue appears farther away to the viewer; apparently caused by both physiological and learned mechanisms, p. 296

rod cells cells in the retina sensitive to achromatic light, p. 289

shade the result of mixing a hue with black, p. 292

simultaneous contrast a color interaction; the eye produces the complementary color of the one being viewed, p. 293

spectral energy distribution curve a graphic plot of the energy and wavelength of a light source; different sources produce different curve characteristics, p. 290

spectral reflectance curves graphic plots of light reflected from objects or surfaces, p. 291

subjective reactions to color involve color preferences, meanings, and behavioral moods produced by colors, p. 294

successive contrast a color appears different to the eye on different backgrounds, especially when viewed successively, p. 293

tint the result of mixing a hue with white, p. 292

tone the result of mixing a hue, white, and black, pp. 292–293

transparent objects transmit all light passing through them, p. 291

tristimulus values amounts of red, green, and blue light necessary to match a test lamp color; used in CIE color specification, p. 297

value scale an array of color based on lightness and darkness qualities; one of the three color dimensions, p. 292

warm colors colors of the longer wavelengths such as red, orange, and yellow, p. 294

READINGS FOR FURTHER UNDERSTANDING

Albers, Josef. *Interaction of Color.* New Haven: Yale University Press, 1975.

Arnheim, Rudolf. *Art and Visual Perception.* Berkeley: University of California Press, 1974.

Billmeyer, Fred W., Jr., and Max Saltzman. *Principles of Color Technology.* New York: Wiley, 1966.

Birren, Faber. *Selling with Color.* New York: McGraw-Hill, 1945.

———. *Selling Color to People.* New York: University Books, 1956.

———. *Munsell: A Grammar of Color.* New York: Van Nostrand Reinhold, 1969.

Brewer, Cynthia A. "The Development of Process-Printed Munsell Charts for Selecting Map Colors." *American Cartographer* 16 (1989): 269–78.

———. "Color Chart Use in Map Design." *Cartographic Perspective,* Winter 1989–90: 3–10.

Burton, Philip Ward, and William Ryan. *Advertising Fundamentals.* Columbus, OH: Grid Publishing, 1980.

Carpenter, James M. *Color in Art.* Cambridge: Fogg Art Museum, Harvard University, 1974.

Castner, Henry W. "Printed Color Charts: Some Thoughts on Their Construction and Use in Map Design." *Proceedings* (American Congress on Surveying and Mapping, March, 1980): 370–78.

Color. New York: Viking Press, 1980.

Color Compass: An Illustrated Guide for Color Mixing and Selection, Color Theory and Harmony. New York: Grumbacher, 1972.

Cuff, David J. *The Magnitude Message: A Study of the Effectiveness of Color Sequences on Quantitative Maps.* Unpublished Ph.D. dissertation. University Park: Pennsylvania State University, Department of Geography, 1972.

———. "Value versus Chroma in Color Schemes on Quantitative Maps." *Canadian Cartographer* 9 (1972): 134–40.

———. "Color on Temperature Maps." *Cartographic Journal* 10 (1973): 17–21.

———. "Shading on Choropleth Maps: Some Suspicions Confirmed." *Proceedings* (Association of American Geographers, April, 1973): 50–54.

———. "Impending Conflict in Color Guidelines for Maps of Statistical Surfaces." *Canadian Cartographer* 11 (1974): 54–58.

Deluca, James P. *Pantone Matching System and Pantone Matching System Formula Guide.* Moonachie, NJ: Pantone, 1980.

Doslak, William, Jr., and Paul V. Crawford. "Color Influence on the Perception of Spatial Structure." *Canadian Cartographer* 14 (1977): 120–29.

Dunn, Watson S., and Arnold M. Barban. *Advertising: Its Role in Modern Marketing.* Hinsdale, IL: Dryden Press, 1978.

Engel, Jack. *Advertising: The Process and Practice.* New York: McGraw-Hill, 1980.

Fisher, Howard T. "An Introduction to Color." In *Color in Art,* ed. James M. Carpenter. Cambridge: Fogg Art Museum, Harvard University, 1974.

Hackl, Al. "Hidden Meanings of Color." *Graphic Arts Buyer* 12 (1981): 41–44.

Hardin, C. L., and Luisa Maffi, eds. *Color Categories in Thought and Language.* Cambridge, England: 1997.

Helson, H., and T. Lansford. "The Role of Spectral Energy of Source and Background Color in the Pleasantness of Object Colors." *Applied Optics* 9 (1970): 1513–62.

Itten, Johannes. *The Elements of Color.* New York: Van Nostrand Reinhold, 1970.

Jones, Mari Riess. "Color Coding." *Human Factors* 4 (1962): 355–65.

Judd, Deane B. "Choosing Pleasant Color Combinations." In *Contributions of Color Science by Deane B. Judd,* ed. David L. MacAdams. U.S. Department of Commerce, National Bureau of Standards. Washington, DC: USGPO, 1979.

Keates, J. S. *Cartographic Design and Production.* New York: Halsted Press, 1973.

Kelley, Charles A., II. *Color Codes.* Hanover, NH: University Press of New England, 1995.

Kimerling, A. Jon. "Color in Map Design." Paper delivered at the workshop, Map Perception and Design, at the annual meeting of ACSM, Cartography Division, St. Louis, 1980.

———. "Color Specifications in Cartography." *American Cartographer* 7 (1980): 139–53.

Kueppers, Harold. *Color Atlas: A Practical Guide for Color Mixing.* Woodbury, NY: Barron's, 1982.

"The Language of Color." 16-mm film produced by Pantone, Inc., Moonachie, NJ, 1983. Available through Modern Talking Picture Service, St. Petersburg, FL.

Lindenberg, Richard E. *The Effect of Color on Quantitative Map Symbol Estimation.* Unpublished Ph.D. dissertation. Lawrence: University of Kansas, Department of Geography, 1986.

Littlefield, James E., and C. A. Kirkpatrick. *Advertising: Mass Communication in Marketing.* Boston: Houghton Mifflin, 1970.

Mahnke, Frank H., and Rudolf H. Mahnake. *Color and Light in Man-Made Environments.* New York: Van Nostrand Reinhold, 1987.

McCormick, Ernest J., and Mark S. Sanders. *Human Factors in Engineering and Design.* New York: McGraw-Hill, 1982.

Mersey, Janet E. *The Effects of Color Scheme and Number of Classes on Choropleth Map Communication.* Unpublished Ph.D. dissertation. Madison: University of Wisconsin, Department of Geography, 1984.

Meyer, Morton A., Frederick R. Broome, and Richard H. Schieweitzer, Jr. "Color Statistical Mapping by the U.S. Bureau of the Census." *American Cartographer* 2 (1975): 100–117.

Olson, Judy M. "Spectrally Encoded Two-Variable Maps." *Annals* (Association of American Geographers) 71 (1981): 259–76.

Ostwald, Wilhelm. *The Color Primer.* New York: Van Nostrand Reinhold, 1969.

PennWell Graphics Group. *Color Desktop Publishing Fundamentals.* New York: PennWell Publishing, 1994.

Robinson, Arthur H. *The Look of Maps: An Examination of Cartographic Design.* Madison: University of Wisconsin Press, 1966.

Sargent, Walter. *The Enjoyment and Use of Color.* New York: Cover Publications, 1964.

Sharpe, Deborah. *The Psychology of Color and Design.* Chicago: Nelson-Hall, 1974.

Sorrell, Patrick. "Map Design—With the Young in Mind." *Cartographic Journal* 11 (1974): 82–91.

U.S. Department of Commerce, National Bureau of Standards. *The ISCC-NBS Method of Designating Colors and a Dictionary of Color Names.* Circular 553. Washington, DC: USGPO, 1955.

———. *1987 Census of Agriculture.* Vol. 2 Subject Series, pt. 1, *Agricultural Atlas of the United States.* Washington DC: USGPO, 1990. A useful publication for the examination of several different choropleth maps containing a variety of color plans.

U.S. Department of the Interior, Geological Survey. *The National Atlas of the United States of America.* Washington, DC: USGPO, 1970.

White, Jan. *Color for the Electronic Age.* New York: Watson-Guptill, 1990.

PART IV

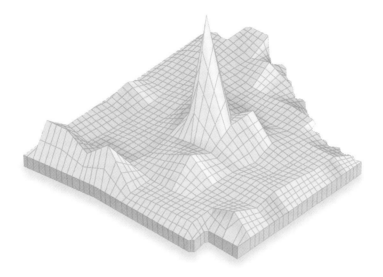

ELECTRONIC MAP PRODUCTION

This last part of the book contains two chapters that treat various subjects pertaining to the production of thematic maps. The first chapter focuses on printing fundamentals for the cartographer, and begins with a brief history of printing technology. Today, techniques of final map preparation are rapidly changing because of newer electronic digital processes, and cartographers need this background, especially one that provides a good grasp of the photographic methods used in prepress operations, also discussed in the first chapter, because these continue to be used even in the electronic publishing world. Knowledge of how screens are used in color printing and how color is obtained on the map sheet is critical for the designer. Setting

forth printing specifications, even if the map is stored on electronic media, requires an understanding of many of the concepts and methods in this first chapter.

The second chapter provides an overview of the complex and technical world of desktop publishing, the method of the 1990s. Whole books are written on this fascinating technology, and it is impossible to delve into much detail in the limited number of pages provided here. Nonetheless, an attempt is made to deliver a broad overview of the most important organizational subjects of the electronic workstation. The material provides at least an underpinning for this kind of thematic mapmaking. Both chapters are helpful in the study of map production and reproduction

CHAPTER

16

PRINTING FUNDAMENTALS AND PREPRESS OPERATIONS FOR THE CARTOGRAPHER

CHAPTER PREVIEW

Cartographic production and reproduction techniques are integral to the design planning of a thematic map. From the outset the designer develops the artwork specifications based on a knowledge of how the map will be reproduced. A working knowledge of the technology of printing is essential for the well-trained cartographic designer. An understanding of color reproduction is especially important, because this form of printing requires specific detail and is considerably more complex than black-and-white duplication. All phases of map production—executing the design concept into final art form ready for printing—are referred to as "prepress operations."

Printing technology has caused revolutionary changes in the printing industry in the 1990s with more computer control, newer printers, new platemaking devices and preparations, and better—or even no—chemicals, and will continue to change the face of printing. Nonimpact printing, now limited to relatively small sheet printing, will likely make an impact on large-map reproduction in the future. Some cartographers continue to prepare maps by scribing and photoetching on special films, especially in large firms, but this is dying out rapidly. Design work requires careful planning to accommodate sophisticated printing technology, and cartographers need to work closely with service bureaus to successfully reach design goals.

This chapter covers several topics essential to the practice of thematic map production and reproduction. These topics play an important role in the overall design process. One cannot adequately approach a design task without at least an elementary knowledge of how maps are made after they leave the designer's office, and this requires some background into printing technology. Even today when the cartographer puts together a final map on a cathode-ray tube (CRT) screen, if the map is to be reproduced on other than the small office printer, it must be prepared most often for the photolithographer. This chapter includes first a brief discussion of the history of printing, and then of the photographic and other techniques that bring the final map to life. The next chapter examines the computer production of small-scale maps, although many of the techniques learned in this chapter are applied to the map designed on the microcomputer and later produced by standard printing techniques.

Throughout this chapter the term *cartographic technician* refers to the individual who may be making the map and preparing the media for reproduction but who did not function as the originator or designer of the map. In some cases, of course, he or she may be performing both roles.

Regardless of whether the designers actually make the map or not, they should be familiar with all techniques because they may come to instruct others in this aspect of cartography. Through knowledge and experience of these processes, better design will be possible. Although many professional cartographers wish to isolate themselves from that part of cartography dealing with the final preparation stages of the map, really good cartographic designers do not completely divorce themselves from the execution phase. They keep informed of new techniques and recognize potential improvements.

PRINTING TECHNOLOGY FOR CARTOGRAPHERS

A distinction should be made between production of manuscript maps and of maps for reproduction, because the approaches are different for each. In **manuscript map** preparation, only one copy of the map is usually being made. The art in final form is the completed map and will ordinarily not be duplicated. Most maps today are constructed for production of multiple copies, in one form or another. This chapter deals primarily with the methods of making maps for reproduction.

BRIEF HISTORY OF MAP PRINTING

Map printing has paralleled the history of all printing. The development of **letterpress,** the process used in making the Gutenberg Bible in 1452, led to the first **woodcut map** in 1472.[1] From that time to the present, there has been a close working relationship between cartographers and printers. In fact, during the first 300 years of printing, the printer probably had as much to say about cartography as the cartographer did. This was no doubt a result of the strong influence of the printing craft guilds. It was not until the present century that cartographers began to control their own products. Designers now have access to a broad range of printing technology and can write knowledgeable specifications to implement their designs.

During the history of printing, three major printing methods have been used by cartographers: relief, intaglio, and planar. Each has its own merits and weaknesses for map reproduction. Most large sheet maps are reproduced today by planar methods, which will be examined in detail later in this chapter. Some small sheet maps may be produced by noncontact printing and this topic will be dealt with in the next chapter.

Relief—Letterpress

Letterpress, also called *relief* printing, is the oldest printing method. Although popularly thought to have been invented by Johann Gutenberg about 1450, the method was actually used by the Chinese hundreds of years earlier.[2] What Gutenberg actually invented was relief printing with movable type. In relief printing, ink is applied to a raised surface and pressed onto the paper. (See Figure 16.1a.) The relief blocks

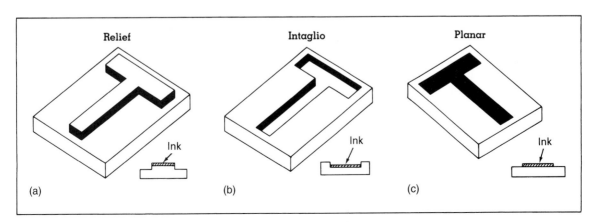

Figure 16.1   The three principal ways of non-electronic printing: (a) relief, (b) intaglio, and (c) planar.

(letters) are reversed so that they will be right-reading after printing. For reproducing art and cartography products, this relief method is referred to as *woodcut*. End-grain wood was chiefly used. Portions of the image that would not be printed were chiseled out, leaving the print portions raised in relief.

Early maps reproduced from woodcuts suffered from a variety of problems inherent in the method. Lines were necessarily thick. Images were often smeared because each impression required a new inking. The paper had to be nearly smooth and free of imperfections. In addition, it was difficult to add lettering to maps—in some cases, holes were cut in the woodcut and letter blocks dropped in. Also, large pieces of wood could not be used because of warping; large maps were often printed by piecing together the impressions made from smaller blocks. Gradation of tone was not really possible with the woodcut method, and the image had to be chiseled backward.

Letterpress printing today is accomplished by first coating a metal plate with a light-sensitive chemical. After drying, the plate is exposed to a very bright light through a negative of the image to be printed. Those areas of the coated plate that are exposed will harden, and the other areas can be washed away by repeated acid baths. This process leaves the image areas raised in relief; they receive and transfer the ink during printing.

Intaglio—Engraving

Intaglio (pronounced in-tal′yo) printing is also called *etching, engraving*, or *gravure*. The elements of the image are first cut out by hand or incised by acid (etching). (See Figure 16.1b.) Ink is then applied and the plate cleaned. Some ink remains in the depressed portions; deeper depressions contain and transfer more ink than shallower depressions. When the plate is pressed onto the paper, the ink is transferred. Best results are achieved when the paper is dampened and considerable pressure applied during transfer. Actually, the paper is pressed into the depressions during printing, and afterward there is a slight raised relief on the paper where the ink adheres. This is one way of identifying materials prepared by the intaglio process.

This form of printing was first used in the early fifteenth century and had achieved prominence by 1700. The engraved copperplate map became standard in map printing for about 150 years. In copperplate engraving, much finer lines are possible than in woodcut, which was then its chief competitor. As intaglio became popular, such techniques as *mezzotint* (working the surface to create tonal image), **stipple engraving** (creating tonal effects by specialized tools), and **aquatint** (special etching to create tones) were introduced. These all had wide appeal to cartographers and were especially useful for vignetting (gradation of tone or texture) at coastlines.

Modern intaglio methods include various forms of gravure, and printing on high-speed web (continuous paper) presses is possible. Techniques of platemaking for gravure include electromechanical or laser scanning of the image in order to break it into a pattern of dots used to control the mechanical engraving of small depressions (cups) in the plate. Similar techniques are used in conventional platemaking, which involves the mechanical transfer, through a carbon tissue medium, of the image onto the plate. Original art is photographed through a gravure screen. The negative is placed in contact with a photographically sensitive gravure plate and exposed. After exposure, the image is chemically etched onto the plate. The image is created on the plate by a pattern of small depressions of varying size and depth. Plates are usually of polished copper, often chromium-plated for protection. Gravure is expensive and is generally economical only for large press runs. Most maps for reproduction are not directly prepared for intaglio or gravure printing.

Planar—Lithography

Planar printing was introduced in 1796 by Alois Senefelder in Germany. Planar printing is usually referred to as lithography. The Greek word *lithos* means stone, and limestone was first used to make the plates. Today it is often called **photolithography, offset lithography,** or photo-offset lithography. This printing method relies on the fact that water and oil (grease) do not mix well. On the printing stone or plate, the image area receives the greasy ink and the non-image areas do not. When paper is pressed to the plate, only the inked areas will transfer to the paper. Unlike the relief or intaglio methods, planar printing creates no relief differences on the plate. (See Figure 16.1c.) Lithography no longer uses stones for plates except in rare circumstances—perhaps only in the graphic arts. This form of printing has come into wide use in practically all commercial applications, and most maps today are printed by this method.

As lithography was introduced to the printing industry, cartographers gradually began to see its advantages. Old copper engravings could easily be updated, transferred to stone, and then printed. Lithography also meant faster preparation of plates than copper engraving; most engraving craftspeople learned the new lithographic techniques easily. Modern lithographic techniques are examined in more detail below.

Nonimpact Printing

Nonimpact printing, or that kind of printing in which ink is applied to the page without an intermediate (such as a plate), and often referred to overall as electronic or electrographic printing, may have a decided impact on cartographic printing in the future. This form of technology is also called *intelligent copier/printer (IC/P) printing*. Cartographers are familiar with this kind of printing when they do small sheet printing on the desktop laser printer. A study recently commissioned by the Graphic Arts Marketing Information Service sought to look at the trends in the industry for nonimpact printing.[3]

Overall, the study found that although IC/P printing technology represented less than 1 percent of the installed, electronic printing equipment base, it accounted for nearly

26 percent of the paper volume of this equipment category. Thus it is an important component of the printing industry. The question here is, How will this form of printing affect the way we make cartographic art in the future? The answer seems to be that it already has, at least for the production of small sheet, black-and-white maps, and to some degree color versions as well (this topic is dealt with in more detail in Chapter 17).

The outlook according to the study, at least for the foreseeable future, appears to be that this form of technology will have little effect on the traditional preparation and final printing for large sheet, multicolor maps. The report states:

> End-user printing needs that require traditional equipment capabilities (print quality above 1,200 dpi [dots per inch], print speeds over 200 ppm, large print sizes, color and run lengths over 10,000 pages) currently are not addressed by electronic printing equipment.[4]

It appears likely, though, that with the rapid pace of technological change in printing, we will see major breakthroughs in the next decade with nonimpact printing, and these changes will affect cartography significantly. After all, in one short decade we have gone from ink on paper to laser-jet printing and negatives prepared from art displayed on a CRT screen.

Cartographic Design and the Printer

The relationship between the cartographer and the printer has gone through different stages since 1450. Arthur Robinson of the University of Wisconsin at Madison, identifies the following periods:[5]

1. During the period when woodcut maps predominated, most cartographers did not do their own woodworking. This usually led to better map designs, because woodcutters were better craftspeople than cartographers.
2. For the most part, this relationship remained unchanged during the time when intaglio was preeminent. A few cartographers—such as Mercator—were also engravers. When national map surveys first flourished, cartographers and in-house engravers worked more closely than ever. Map design was still mainly shaped by engravers' styles.
3. When the transfer process (making printing plates from right-reading material) and lithography were introduced, anything could be reproduced. Specialists such as engravers and cartographers could be bypassed entirely, replaced by others who did not possess any cartographic knowledge. Cartographic design suffered.
4. Today, the situation is reversing. Since about 1950, printers and cartographers have once again been working closely together. Although it is increasingly difficult for cartographers to master the full range of modern printing technology, design is surely improving in the attempt.

As the printing industry became more and more dominated by lithography (especially photolithography), the printer became primarily a duplicator. There were no craftspeople intervening to influence the look of maps. Cartographic education did not stress the aesthetic aspects of maps, except in rare cases. Today, the cartographer must assume the craftsperson's role, handling the aesthetic aspect of map design. Knowledge of the printing industry and its varied techniques gives the cartographer greater freedom of choice in design.

MODERN PHOTOLITHOGRAPHY

Lithography, as noted above, had its beginnings around 1796. By the middle of the nineteenth century, the development of photography made it possible to put the image onto a thin metal plate that could be attached to a rotating cylinder. The impressions were made directly onto paper, requiring the image to be backward until printed. It was not until 1905, when Ira Rubel accidentally printed an impression on a blanket cylinder, that the *offset principle* was discovered. (See Figure 16.2.) Print, art, and map materials could be right-reading through the entire preparation process, making this phase much less risky.

In producing the photolithographic plate, a light-sensitive chemical coating is applied to a thin metal, plastic, or paper plate. (See Figure 16.3.) Most plates today are purchased with the coating already applied. A **copy negative** of the original art is placed in contact with the plate, and both are exposed to an intense light source. After exposure, the plate is chemically washed with a developing fluid that washes away the coating where the plate has not been exposed to light. The image area, that part exposed to light, hardens and becomes ink-receptive (water-repellent) during printing; the plate is first dampened in a water bath, making the nonprinting areas ink-repellent. The plate receives ink only on the printing or image areas. The ink is transferred to a rubber blanket cylinder and then to incoming paper.

Photolithography requires several steps in sequence:

1. Preparation of original art (which may be done by computer to produce positive artwork), or
2. Preparation of plate-quality negatives (if not done in step one, and which may be by imagesetter)
3. Proofing
4. Platemaking
5. Printing

Today the "artwork" prepared for the photolithographer is more likely than not a digital image on computer diskette. Although compiling the image is the first step along the way toward completion of the final printed map, we will look at the other steps first. An understanding of lithographic principles directs how the art is rendered in the computer map production software.

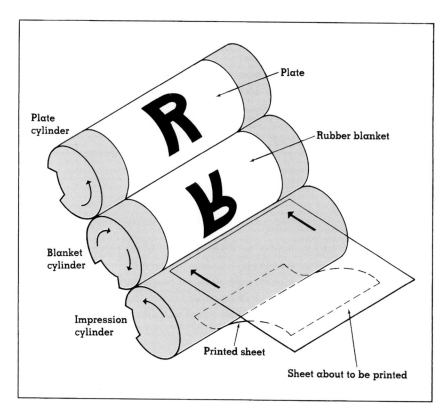

Figure 16.2 The principle of offset printing.
Notice that the image on the plate is right-reading, a feature that has made this form of printing very popular.

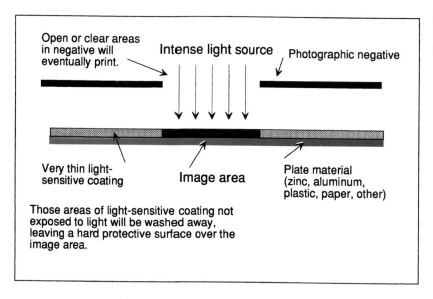

Figure 16.3 Photolithographic negative-working plate process.
See text for explanation.

Screen Specifications

With the exception of only black-and-white (or one color and white) maps, map images usually are made up of area and lines that are solid ink, or composed of very small dots. Printing is a binary process—either you have ink or you do not. To get the areas that appear as gray, or those that look like continuous tones, the ink placed on the paper has to be broken into smaller dots—called **screen tints.**

Not so long ago, the photolithographer would place special dot-screen negatives between the copy negative and the printing plate to break up the image areas into different dot patterns (this is still done by some printers). Today, the cartographer composes and specifies these screens directly

at the computer when designing the map, and screen instructions become part of the electronic image file. It remains useful, however, for the cartographer to be able to visualize how these screens will appear when printed.

When cartographers worked with actual film screens they were referred to as *mechanical* or *flat screens*. As mentioned previously, these screens are placed between the line negative and the light source when the plate is made. Their effect is to break up the solid-line or black areas so that, when printed, they will appear gray. Customarily film screens are designated by two variables, the line designation (number of rows of dots per inch, or lpi, and sometimes referred to as screen frequency) and **gray tone** (approximate

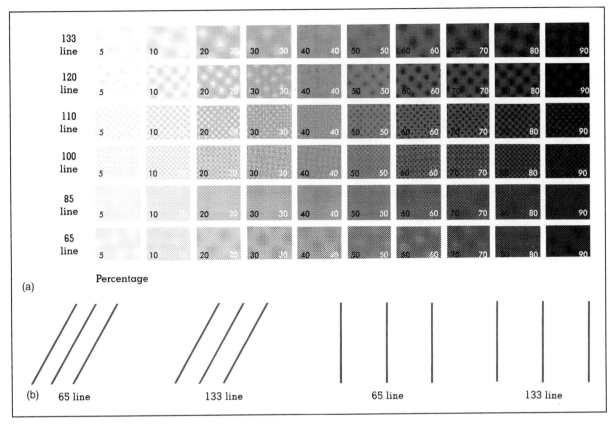

Figure 16.4 Screen tints and the effects of screening lines.
A sample of various tints is illustrated in (a). In (b), lines have been screened to show the effect of lines on at least two orientations. Note the ragged edges when coarse screens are used. (*Screen tints courtesy of the ByChrome Company, Box 1077, Columbus, Ohio 43216.*)

percentage of area covered with ink after printing). Common line designations are 65, 85, 100, 110, 120, 133, and 150 lines per inch. Coarse lines, such as 65 and 85, are not used in most map work because the individual dots can be seen by the human eye. This detracts from the desired flat tonal quality. The 133 line is recommended as a minimum for most map work; 150 line is preferable when fine lines or lettering are to be screened. (See Figure 16.4.)

Gray tones usually range from 5 to 95 percent, in 5 percent gradations, although with computer-generated screens increments of 1 percent are common. The cartographer chooses to specify screen tints when contrast of tone is sought over the map or when a quantitative map is being symbolized to achieve a gray spectrum. Under normal conditions, no less than a 15 percent difference between two tints on the map should be specified (20 percent is preferred). The human eye finds it difficult to differentiate two tints closer in percentage than 15 percent.

When a continuous tone image (such as a photograph or a hill-shaded relief map) is to be printed, another kind of screen is used—the halftone screen. The **halftone negative** is really an ordinary copy negative that has been broken into a pattern of very small dots by exposing it to the art through a *halftone screen.* The halftone screen is a special **contact film screen** marked with small vignetted dots

arranged in a grid pattern that produce a pattern of very small dots of varying sizes when exposed to light and the halftone negative. The resulting halftone negative can be exposed to the lithographic plate in the usual fashion.

The halftone method works because the dots are too small to be resolved by the human eye at normal reading distances. They tend to merge together, creating the impression of gray tones. The dots actually vary in size, depending on the amount of light that was reflected from the original art during exposure. Most thematic map applications do not require the use of the halftone technique. Exceptions are vertical hill-shaded terrain maps and symbols rendered for plastic effect. (See Figure 16.5.)

In electronic or digital preparation of halftone negatives the image is divided into a pattern of dots of varying sizes by altering the sizes of the dots within a regular array of halftone cells. (See Figure 16.6a.) The lines per inch (lpi) designation—or resolution—is selected by the cartographer within the limits imposed by the software, and remains constant over the image. The sizes of the dots are controlled by the software and are governed, too, by the resolution of the output device's pixel sizes (see next chapter for more details about pixels).

Relatively new to the digital production of halftone negatives is **stochastic screening.** Stochastic screening is a

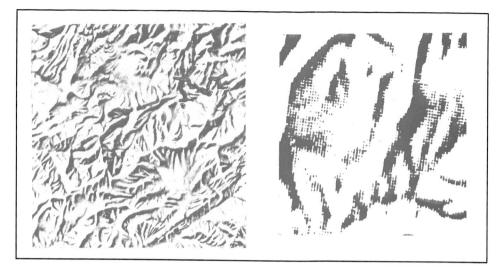

Figure 16.5 The effect of halftoning.
The halftone screen has produced this hill-shaded map, which is composed of a pattern of dots of varying size and density. A continuous-tone effect is produced because we cannot normally resolve the individual dots. The enlarged section on the right reveals the dots.

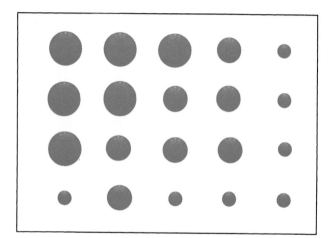

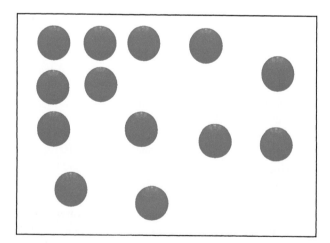

Figure 16.6 Conventional and stochastic halftone screening.
In conventional screening, dots vary in size along evenly spaced rows (a). In stochastic screening, dots are of uniform size, but distances between them vary, as in (b).

method that varies the *distance* between the dots on a halftone, but the sizes of the dots remain constant. (See Figure 16.6b.) This is in contrast to conventional screening that holds the distance constant, but varies the size of the dots. "That's why stochastic screening is also called *FM* (for frequency modulation) *screening*—the dots vary in frequency."[6] The method is being tested by several printers and may hold promise in the near future especially for color reproduction.

As suggested earlier, the cartographer today will compose and design a map and simply select a proper screen percent as he or she completes the map at the computer monitor. (See Figure 16.7.) Most mapping software today allows for this, or for the selection of ordinary or custom line screens. These screens are part of the instructions on the diskette sent to the service bureau that will produce the map on a special machine called an **imagesetter.** The imagesetter will produce plate-quality negatives. (See Figure 16.8.) These machines are also discussed in the next chapter.

Plate Preparation

There are a number of ways to prepare plates. Most modern photolithography shops have a *platemaking machine*, which is basically a large box with a flat surface covered by a glass plate, where the plate and negatives (taped onto paper sheets called **flats**) are placed. These are subjected to a vacuum so that firm contact between the two is assured. On larger machines, this bed is rotated so that it can be exposed to a very intense light source from within the box. The duration of exposure is varied by means of timing controls on the machine.

Photo-direct platemakers are machines that incorporate a camera for making the final printing plate. Light from the original art (in nonnegative form) is reflected to the camera which redirects the image onto the light-sensitive plate material. This kind of platemaker is usually contained in one housing. Most cartographic products are not done in this fashion, primarily because most map specifications call for several fine-textured screens that cannot be handled on original positive artwork, and because imagesetters are more frequently used today.

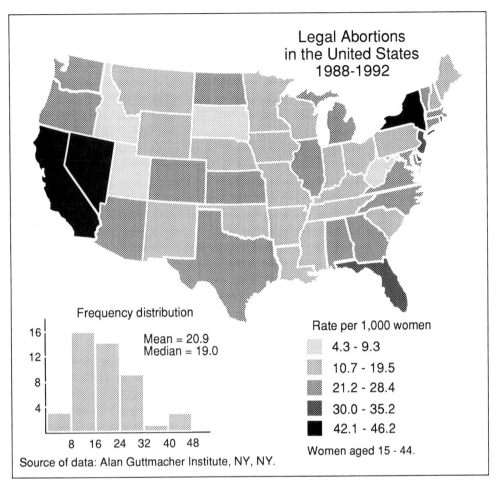

Figure 16.7   A complete map from an electronic file. This map was designed and composed at a computer monitor and printed directly from the electronic file. The screens were generated by the computer software.

Figure 16.8   Modern digital imagesetter.
(Photo courtesy of Dupont Printing and Publishing.)

It is possible also to produce offset printing plates by **laser scanning** original art. This is usually done at the same size as the original. The map artwork is scanned by special laser sensors, processed by computers into digital form, then sent to a device that directs a small laser light source onto the sensitized plate to reproduce the original exactly. The plate is then prepared in the usual way for printing. The scanner and laser platemaker need not be in the same location, that is, the digitally scanned image can be sent by telephone to a remote laser platemaker. In this way, many plates can be made at different locations (even worldwide) at the same time. Laser platemaking eliminates negative preparation (and opaquing and stripping) and can be done for color separations as well.

Two other innovations have emerged in recent times for the direct, computer-controlled production of either plates or negatives (flats) ready for plate preparation. In the former, a digital laser platemaker produced by Polychrome Americas (a division of Dainippon Ink and Chemicals, Inc., and used here as an example) develops either aluminum or paper plates directly from the computer image files with imaging software. (See Figure 16.9.) This machine has a resolution of 1,200 by 1,200 dpi (dots per inch), and can produce fully developed and ready to run plates up to 16 by 27 inches. It interfaces directly with standard computer

Figure 16.9   Modern digital platemaker.
This digital platemaker, driven by computer imaging software, produces either aluminum or paper plates with resolutions of 1,200 by 1,200 dpi, and makes plates as large as 16 by 27 inches. *(Photograph courtesy of Polychrome America.)*

ports. The company claims that its aluminum plates can make from 50,000 to 100,000 impressions. Color separation plates would have to be generated by the designer using the imaging software at the computer.

The other recent development is the production of film negatives (flats) by direct computer control using imaging software (such as page compositors). One is the Sprint 110 Imposetter (a product of Optrotech, Inc.). This machine develops plate-ready negatives coming from such computers as the Apple Macintosh IIci and accepts any PostScript page data (PostScript is a trademark of Adobe Systems, Inc.). Negative sizes are possible up to 32 by 40 inches. Other manufacturers offer different devices and specifications. Here again the cartographer would need to develop separation art if multicolor maps are planned. In the printing industry, going from computer software to plate (and skipping the negative preparation step) is called **direct-to-plate.**

It is not likely that most small cartography operations will have these highly specialized devices for the production of maps. Nonetheless, the cartographic technician who is looking after a whole production job should be aware that these devices are available and might wish to plan the whole job around the printing company that can integrate this printing technology with his or her computer cartography imaging software. This could reduce costs and eliminate many steps in the production process.

Color Printing and the Four-Color Process

Color printing can be divided into two kinds: **flat color printing** and **process color printing.** Flat color printing is accomplished in a manner similar to black-and-white printing except that ink colors other than black are used. Process

color printing is more complex, requiring considerably more expertise and detailed planning. It is difficult to speculate on which kind of printing is mostly used in cartographic work, but it is suspected that flat color is more common.

Discussion of color as it relates to cartographic design was provided in the last chapter, but further aspects of it are presented here. Most of the present material deals with the objective aspects of color, having already looked at the more subjective and perceptual aspects earlier. Of course, cartographers need to know both for effective map design.

Light and the Color Spectrum

Light is that part of the electromagnetic energy spectrum that is visible to the human eye. (See Figure 16.10.) The radiation spectrum is characterized by energy falling on us at different wavelengths. These wavelengths vary from very short (10^{-12} cm) to very long (10^5 cm, or 1 km). All visible light varies from 7.5×10^{-5} cm to about 3.5×10^{-5} cm. **Color** is simply light energy at different places along this spectrum. When our eyes detect light energy at approximately 7.5×10^{-5} cm in wavelength, we see red; when we detect wavelengths at 3.5×10^{-5} cm, we see violet. Other colors and their distinct wavelengths are in between these two.

Additive Primary Colors

Color can be *refracted* from white light by passing light rays through a different medium. For example, we can see the different colors after passing a ray of light through a prism. Although visible light is composed of a myriad of different colors at various wavelengths, we consider white light to be made up of *three primary colors*—red, green, and blue—because these cannot be made from combinations of other colors. If we take each of these three and project them on a wall so that they partially overlap, two things will happen. (See Figure 16.11 and Plate 9.) First, in that area where all three colors overlap, we will see white. Second, in those areas where two colors overlap, we will see a combination called magenta, cyan, or yellow. Red, green, and blue are called the **additive primary colors** because they can, in various combinations, produce any other hue in the visible part of the energy spectrum. Colors produced on television screens are the result of additive primaries (more will be said of this in the Chapter 17).

Color Printing and Subtractive Primary Colors

Color produced by printing is not based on the additive primaries of projective light, but on inks or pigments laid down on paper. Normally, flat color printing uses opaque printing inks; process color printing requires the use of transparent inks.

Opaque printing ink reflects its color, and absorbs the remaining colors at its surface. (See Figure 16.12.) For example, red opaque ink absorbs the blues and greens and reflects the red to the reader's eye. **Transparent inks** behave somewhat differently. The pigments of red transparent ink

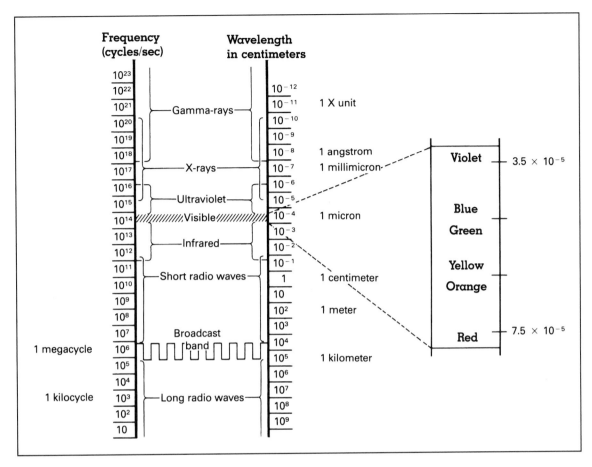

Figure 16.10   The electromagnetic spectrum.
Visible light occupies only a small portion of the entire spectrum.

absorb the blues and greens, but in this case the red light is first transmitted to the paper surface before it is reflected back to the reader's eye. Thus the color of the paper or *other ink* is perceived by the reader. Process color printing uses the transparent inks magenta, cyan, and yellow, which together can create any hue or recreate a continuous-tone color image. These pigment colors are called the **subtractive primary colors.** (See Plate 9.)

Cartographic designers have two choices in color printing. They can select the desired color of opaque ink from literally hundreds of choices.[7] The alternative is to develop virtually any color by using transparent magenta, cyan, and yellow inks. The number of colors planned for the final map determines the choice. Printers normally charge by the number of different inks (regardless of whether they are process or opaque) and press time. The visual quality of a flat opaque ink color may be somewhat better, but competent printers can achieve excellent results using transparent ink overprinting. The designer must look closely at other costs in order to determine the best alternative. Regardless of what choice is made, the decision must come early on because art preparation is dependent on it.

Process color printing in map work can begin in two ways. In one method, the cartographer produces a color manuscript map. (See Figure 16.13.) That is, the map is completely finished and looks exactly as it will appear after printing. The completed art is photographed four times, once for each of the primary colors and black.

Black is usually used (but not absolutely necessary) to overcome the limitations in faithfully reproducing the original color by the transparent inks (hence the term four-color process printing). The reason for this is that, in practice, mixing magenta, cyan, and yellow never produces pure black, but a dark muddy brown, although theoretically black should be produced. Different filters are used to take out all colors except the one being worked. A negative and a plate are made for each of the process colors, plus black. Four runs through the press are required, once for each ink, unless a multicolor press requiring a single press run is used. In this way, the original map is reproduced. This is the same general procedure followed in the color reproduction of continuous-tone color images such as photographs, illustrations, or similar art.

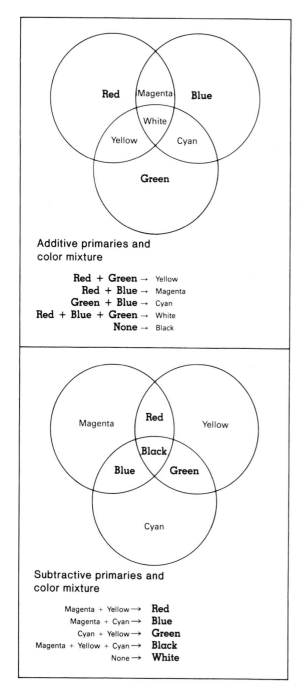

Figure 16.11   Additive and subtractive colors and mixtures.

Process color printing techniques use the principle of subtractive mixture to produce colors on the map. Different percentages of the process inks are used to create various hues.

A second way to reproduce a process color map is to prepare the art originally into **color separations.** (See Figure 16.14.) In this method, the original art is produced in pieces (separations), usually four—one for each process ink and black—and negatives are made of each separation.

In four-color process printing the images are developed by adding four separate screen images to the map sheet, one for each of the primary subtractive color inks. (See Plate 9.) In this way almost any color can be achieved on the printed sheet. In practice, the screens are angled to maximize the amount of ink placed on the sheet for each color, and to avoid **moiré** (a troublesome pattern effect that interferes with the image).[8] Printers create different colors by specifying different screen percentages for each of the four process colors. You are encouraged to look at the color sheet provided with this book.

You may recall from the last chapter that the CIE color specification system was introduced. In that system each hue has a **tristimulus value** as determined by a standard observer. The tristimulus values correspond to the amounts of red, green, and blue lights to match a test color. It has been the practice to represent colors on a **chromaticity diagram** by using these graphing formulas:

$$X = \frac{X}{X + Y + Z}$$

$$Y = \frac{Y}{X + Y + Z}$$

$$Z = \frac{Z}{X + Y + Z}$$

and, therefore, for all values of X, Y, and Z, $x + y + z = 1$. Customarily, any hue would be represented on the diagram by specifying its x and y values. (See Plate 10.)

Actual color space varies—that is, the space that can be perceived is greater than that which can be printed using the additive color primaries (the four process colors), and those that can be generated on a color monitor. This especially causes problems for cartographers and others when movement from one system to another is necessary, such as attempting to replicate a color from the computer monitor to the printed page. This requires some sort of **color calibration** among all devices in the production process. Fortunately, in desktop mapping, most versatile graphics and mapping software have capabilities for this.

Most computer software used by cartographers is capable of producing the art into four separations (and properly registered), once again easing the art preparation stage of map making. Separate plates are again made from each negative, usually by an imagesetter, and the map is then printed in the usual fashion. (See Figure 16.15.) In unusual cases where the cartographic art is provided in finished color form, it may be scanned electronically to produce individual color-separation negatives ready for printing. This often is done in book publishing when a reproduction of a color map is sought for inclusion in a new work.

Hi-Fi Color   An innovative approach to extending the color gamut of printed color images is to add a fifth, or "bump" color to the four colors CMYK. This has been

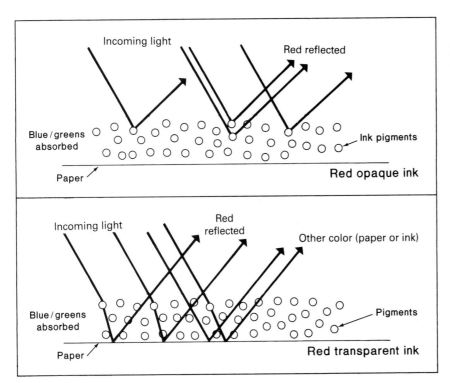

Figure 16.12 The reflecting qualities of opaque and transparent inks.

Light striking opaque inks is immediately reflected. When light strikes transparent inks, it is first transmitted to the paper surface and then reflected.

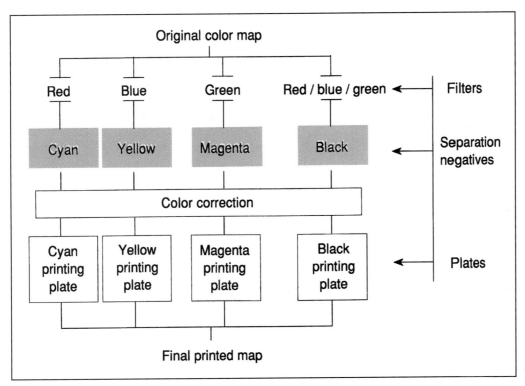

Figure 16.13 Steps in reproducing a color original map.

Color correction is the process of removing from the separation negatives some unwanted dots (thereby reducing final ink) to compensate for certain printing deficiencies caused by overprinting of some process inks.

done for some time. However, now printers are exploring ways to extend color further by adding as many as three and four additional colors. Several test sites around the country are involved in the project, and what appears to be emerging is the addition of orange, green, and violet to the original CMYK colors to achieve what the industry is calling **Hi-Fi color.**[9]

The process of introducing a new standard such as Hi-Fi color is difficult because specifications must be written into computer software, and to be fully integrated most vendors must agree to the process. In the past cartographers have been content with the four-color process, and how any new Hi-Fi color standards will affect them is not yet understood. Nonetheless, the cartographer who wishes to control the

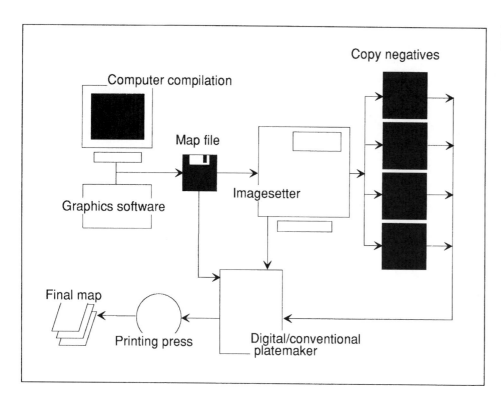

Figure 16.14 Map production in the electronic age.

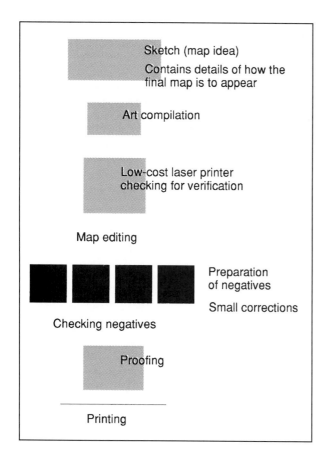

Figure 16.15 Cartographic prepress operations.

color output of a large mapping project should at least explore these new color methods.

There was a time in cartography not so long ago when it was fairly simple to describe how map makers went about their business of producing maps. Much has changed with the development of desktop mapping and it is not entirely clear if there is a single best way that cartographers should work. Regardless of the approach taken, cartographers should have a *plan* for the project. The next section describes the elements that go into this prepress plan.

COLOR PRINTING SPECIFICATION

In printing practice the cartographer specifies color with a *color-matching system* adopted more or less as a standard through consensus by the printing industry. One such system, the **Pantone Matching System (PMS),** has become common today, so much so that cartographers dealing in color should become familiar with it. This system has become so widespread that computer color model (especially RGB) specification can be matched to it.

The new print technologies are to printing and publishing what morphing images and creating special effects are to Steven Spielberg.

Source: David Rosquist in William C. Lamparter, "Hi-Fi Color: Nonsense or Niche?" *American Printer* (May 1994), pp. 52–54.

Table 16.1   Sample Colors from Pantone Mixing Guide

Color:		
Pantone 165C	8 parts Pantone yellow	50.0%
	8 parts Pantone warm red	50.0%
		(Ratio 1:1)
Color:	4 parts Pantone yellow	25.0%
Pantone 164C	4 parts Pantone warm red	25.0%
	8 parts Pantone white	50.0%
		(Ratio 1:1:2)
Color:	8 parts Pantone yellow	40.0%
Pantone 168C	8 parts Pantone warm red	40.0%
	4 parts Pantone black	20.0%
		(Ratio 2:2:1)
Color:	2 parts Pantone yellow	12.5%
Pantone 163C	2 parts Pantone warm red	12.5%
	12 parts Pantone white	75.0%
		(Ratio 1:1:6)

Source: James P. DeLuca, *Pantone Matching System and Pantone Matching System Formula Guide* (Moonachie, NJ: Pantone, Inc., 1980), p. 4.
The Pantone Matching System is a registered trademark of PANTONE, Inc., Moonachie, NJ 07074.

Eight basic colors, plus black and white, form the elementary colors of the PMS: yellow, warm red, rubine red, rhodamine red, purple, process blue, reflex blue, and green. By specifying different mixtures of these colors, over 500 different printing ink hues are possible. The cartographer selects a color from color charts produced by Pantone and provides the printer with the number of the color. The printer mixes the ink by using a *Formula Guide* that shows the percentages of the elementary colors needed to achieve the desired color. (See Table 16.1.) Mixing is by weight. Four-color process printing is done with Pantone process colors. Colors are usually specified in percentages of C (cyan), M (magenta), Y (yellow), and K (black). Process printing is covered in detail in Chapter 17.

A variety of helpful materials is provided by the manufacturer: a *Color Tint Selector* (which shows the printed effects of a variety of screens on nearly 500 Pantone colors), a *Color and Black Selector* (which shows the effects of mixing screened black with about 81 different colors), and a *Four-Color Process Guide* (which shows the different hues achievable by four-color process printing with Pantone process inks). There are other products not mentioned here. Some desktop publishing software now comes equipped with color palettes in which screen colors (RGB) can be specified in Pantone numbers.

Although other color-matching systems are available, most printers use the PMS. Other manufacturers are licensed by Pantone to produce products matched to its colors. Examples include self-adhesive overlay sheets, paper, colored markers, and the like. These materials can assist the cartographer in visualizing the final color map during various stages of production.

Color specification is possible by using such a system as Pantone for ordinary color use in cartography. Because the colorants used are relatively standard (the eight elementary colors) as are mixing guides, mixed hues are reasonably consistent. Variability comes from printing presses, impressions, and paper. These sources of inconsistency will probably always plague the cartographer, making exactly accurate color matches difficult. To make life easier for the cartographer, it is now possible to obtain charts that show the CMYK percentages that best replicate the various Pantone color chips. In fact, many new computer-graphics software packages include such CMYK values for many of the Pantone colors.

Sometimes cartographic researchers like to specify color in the Munsell system, especially if they are conducting color research. However, until very recently it has not been easy to get CMYK percentages of the Munsell color system. A very useful research project carried out by Cynthia Brewer has now produced CMYK percentages for some of the color areas of the Munsell color solid.[10] Brewer searched through a color atlas that had been spectrophotometrically scanned to determine the Munsell color specifications to find those *closest in appearance* to the Munsell student set of 238 color chips. A detailed description of the entire procedure followed by Brewer is too lengthy for inclusion here, but the results are helpful for anyone attempting to replicate some of the colors of the Munsell color solid with the four-process color specification system. The student is encouraged to pursue these studies further by looking at the Brewer readings at the end of this chapter.

The results of the research described by Brewer include the CMYK values closest to those chips from the Munsell student set in 10 color groups: red, yellow-red, yellow, green-yellow, green, blue-green, blue, purple-blue, purple, and red-purple. A table of values is reproduced here in Table 16.2, and one color set is included here in Plate 7.

CARTOGRAPHY AND PREPRESS

The production of a thematic map requires detailed planning so that the printed map will look in every way as the designer intended. The cartographer necessarily deals with content, generalization and symbolization, scale, and other mapping components, but the final map will not emerge as envisioned without careful attention to all phases of the production process. The purpose of this section is to present ways of organizing and managing this plan.

GETTING STARTED

Map production requires putting the sketch map (the preliminary map idea) into press form, which calls for considerable detailed planning by the designer. The sketch map is the documentation of the designer's authoritative plan for the final appearance of the map. Detailed specifications for

Table 16.2 Four-Process Color Specifications for Selected Munsell Colors

Red

	/2	/4	/6	/8	/10	/12	/14
8/	00 05 00 20	00 10 00 10					
7/	00 10 00 30	00 20 00 20	05 30 00 10	10 40 00 05	20 50 00 00		
6/	00 20 00 40	05 30 00 30	10 40 00 20	20 50 00 10	30 60 00 05	40 70 00 00	
5/	05 30 00 50	10 40 00 40	20 50 00 30	30 60 00 20	40 70 00 10	50 80 00 05	60 90 00 00
4/	10 40 00 60	20 50 00 50	30 60 00 40	40 70 00 30	50 80 00 20	60 90 00 10	70 99 00 05
3/	20 50 00 70	30 60 00 60					
2/	30 60 00 80						

Blue

	/2	/4	/6	/8	/10	/12
8/	00 00 10 10	00 00 20 05				
7/	00 00 20 20	00 00 30 10	00 00 40 05	00 00 50 00		
6/	00 00 30 30	00 00 40 20	00 00 50 10	05 00 60 05		
5/	00 00 40 40	00 00 50 30	05 00 60 20	10 00 70 10		
4/	00 00 50 50	05 00 60 40	10 00 70 30			
3/	05 00 60 60	10 00 70 50				
2/	10 00 70 70					

Green

	/2	/4	/6	/8	/10	/12
8/	10 00 10 10	20 00 20 05	30 00 30 00			
7/	20 00 20 20	30 00 30 10	40 00 40 05	50 00 50 00		
6/	30 00 30 30	40 00 40 20	50 00 50 10	60 00 60 05	70 00 70 00	
5/	40 00 40 40	50 00 50 30	60 00 60 20	70 00 70 10		
4/	50 00 50 50	60 00 60 40	70 00 70 30			
3/	60 00 60 60	70 00 70 50				
2/	70 00 70 70					

Purple-Blue

	/2	/4	/6	/8	/10	/12
8/	00 00 10 10	00 05 20 05				
7/	00 05 20 20	00 10 30 10	00 15 40 05	00 20 50 00		
6/	00 10 30 30	00 15 40 20	00 20 50 10	00 25 60 05	00 30 70 00	
5/	00 15 40 40	00 20 50 30	00 25 60 20	00 30 70 10	00 40 80 05	
4/	00 20 50 50	00 25 60 40	00 30 70 30	00 40 80 20	00 50 90 10	
3/	00 25 60 60	00 30 70 50	00 40 80 40	00 50 90 30		
2/	00 30 70 70	00 40 80 60				

Yellow-Red

	/2	/4	/6	/8	/10	/12
8/	10 05 00 20	20 10 00 10	30 15 00 05	40 20 00 00		
7/	20 10 00 30	30 15 00 20	40 20 00 10	50 30 00 05	60 40 00 00	70 50 00 00
6/	30 15 00 40	40 20 00 30	50 30 00 20	60 40 00 10	70 50 00 05	80 60 00 00
5/	40 20 00 50	50 30 00 40	60 40 00 30	70 50 00 20		
4/	50 30 00 60	60 40 00 50	70 50 00 40			
3/	60 40 00 70					

Red-Purple

	/2	/4	/6	/8	/10	/12
8/	00 10 00 10	00 20 00 05	00 30 00 00			
7/	00 20 00 20	00 30 00 10	00 40 00 05	00 50 00 00		
6/	00 30 00 30	00 40 00 20	00 50 00 10	00 60 05 05	00 70 10 00	
5/	00 40 00 40	00 50 00 30	00 60 05 20	00 70 10 10	00 80 15 05	00 90 20 00
4/	00 50 00 50	00 60 05 40	00 70 10 30	00 80 15 20	00 90 20 10	
3/	00 60 05 60	00 70 10 50	00 80 15 40			
2/	00 70 10 70	00 80 15 60				

Notes: The numbers with slashes on the vertical columns refer to Munsell values, and those arrayed horizontally refer to Munsell chromas. For each color specification, the four-process color ink percentages are in this order: top left for yellow (Y), top right for magenta (M), bottom left for cyan (C), and bottom right for black (K). A solid (100-percent) ink is represented by the number 99.

Source: Selected from a larger group from Cynthia A. Brewer, "The Development of Process-Printed Munsell Charts for Selecting Map Colors," American Cartographer *16(1989): 269–78.*

production must appear directly on the sketch or in instructions accompanying it. The appearance of lines and any screening of them, all colors, lettering location and styles, the completed base map, and all other parts of the map must be carefully specified. Often the sketch map contains information about where the map sources are to come from—especially the base map and thematic data. In many instances the sketch map contains actual art that is to be scanned if it is to be electronically produced. In other cases, the sketch map may simply be a set of detailed instructions on how to make the final art. Whichever procedure is used, a sketch map is completed first.

SUPERVISING PRODUCTION

In many instances the cartographer is not the one who does the actual map compilation at the computer. These cases call for the cartographer to supervise the map's production by making certain that the computer images are correct when compared to the original sketch or specifications. This often requires the map to be printed periodically on low-cost laser printers for verification (if the art is sized so that this can be done). Whatever strategy is used, it is critical that some form of checking be done as the map is being compiled. This will reduce problems in the later stages of production. One author has commented on this stage, and suggests these seven steps ". . . to help navigate the terrain of the digital landscape:"[11]

1. The map is more important than the vehicle—have a detailed workflow map of all operations in the project.
2. Know your destination and locations at all times— know exactly your specifications and how to achieve them, both by the hardware and service bureaus.
3. Always communicate—in the digital world of art production, be specific in communicating with others; send along laser proofs to help others see what you mean.
4. Don't be afraid to ask questions—as we are not experts in all phases, ask questions about details of production techniques to be certain you can get where you want to go.
5. Abandon your vehicle and walk if necessary—it is all right to mix computer techniques and manual ones if the computer cannot do what you want it to do.
6. Change drivers as needed—plan to make changes in those specialists you first identified, if others are better.
7. Don't base your workflow off a price list—do not become infatuated with digital gimmicks, and keep the goal in sight at all times.

MAP EDITING

Map editing, which consists of proofreading the finished art for mistakes or omissions, is a crucial step in map production and reproduction. Editing is especially important when the map contains hundreds of place names or when many complex political boundaries are included. Even the simplest map must be checked for errors.

No single method is used in editing, and the activity varies depending on the size of the office and the complexity of its products. Regardless of the procedure used, some editing must be done. The benefits usually outweigh costs in terms of avoiding embarrassment or having to redo the artwork or negatives. Editing should be done before the extensive and costly negative work and again before negatives are sent to the platemaker.

Editing can be done in a systematic way at the computer monitor, or better, from black-and-white proofs from a low-cost laser printer. Whatever method is used, it is critical that the map be *compared in detail with the original sketch map containing the set of specifications.*

One method of editing from black-and-white proofs is to grid the surface of the map and then systematically go through each grid cell and check for spelling or other errors. This has a tendency to distract the editor from the map's contents, and to ensure that all parts of the map have been checked.

THE SERVICE BUREAU

After the production art of the map has been thoroughly edited it is then sent usually on a diskette to a **service bureau** for negative preparation (if the printer does not produce its own copy negatives). It is very important that a good relationship is maintained with this service provider. Normally, the workflow plan specifies the service bureau because some design decisions are made on the abilities of this vendor. For example, some bureaus can work only with software done on Apple Macintosh platforms, whereas others may or may not also work on IBM platforms. Some vendors can work with both, but often they will tell you beforehand that they cannot guarantee their work on a certain platform. The bureau will tell you what kind of computer files they prefer to work with. This the cartographer must know before any production is begun. Direct, careful communication is the key when working with those that prepare your negatives.

PREPRESS PROOFING

The map designer always should strive to have proofs made of the cartographic art before it is printed, especially for color maps. These are called **prepress proofs**—a means of representing the final map, in color or black and white, before printing. The two principal reasons for proofing at this stage are to evaluate the design (especially color quality) before the expensive process of printing begins, and to check on the quality of the photoengraver's work (all the mechanical steps of assembling the work). Printers appreciate, and most require, prepress proofing as a way of assuring approval by their clients before printing. The designer can often visualize the overall final design more easily with

Table 16.3 Methods of Color Proofing and Representative Systems

Product/Manufacturer	Resolution	Maximum Size (in)
Electrophotographic/digital		
Kodak Approval	Dot-for-dot	12 × 19
3M Matchprint	2,400 dpi	18.2 × 26.8
Ink-jet/digital		
Iris 3024 printer	1,800 dpi	24 × 24
Lamination/analog		
DuPont Chromalin	150–200 line ruling	25 × 38
3M Matchprint II	200 line scr.	25 × 38
Agfaproof	400 line scr.	26 × 40
Overlay/analog		
Kodak Color Accord	Dot-for-dot	25 × 38
Dye-sublimation/digital		
3M Rainbow	300 dpi	11 × 17
Thermal transfer/digital		
QMS ColorScript 100	300 dpi	11.3 × 17.2
Linotype-Hell Color Printer 30	300 dpi	11 × 17

Source: "Mirror Images," American Printer (October, 1992), pp. 31–36.

these composite proofs than with negatives pasted onto masking sheets and liberally marked with notations.

Black-and-white art is not usually proofed unless it contains complex screen tints. In these instances, *silverprints* (blueline or brownline) or *diazo* proofs can be made by exposing special papers (or films in the case of diazo) through the copy negative and chemically processing them.

For process color or complex multicolor printing, there are two commonly used methods of prepress color proofing: transparent (or overlay) and opaque (or laminate). In the former, the copy negatives are used to contact-expose films chemically processed to yield one of the process colors. Proofing is done by sandwiching these together, in correct register, and examining them over a sample of the paper that will be used for printing, in bright light. Reasonably good results can be achieved, although colors will not be exact and glare is a problem. Separations can be checked and errors in the planning of the job easily detected. One trademarked product name in this category is Kodak Color Accord. (See Table 16.3.) This is a wet process and can accommodate art sizes up to 25 × 28 inches. By sandwiching the individual overlays, the final image is ready to preview.

Opaque proofing, also called laminate proofing, is similar to transparent proofing except that the proof is on only one sheet of paper. In this system separate, colored layer films are processed (laminated) onto the one sheet of white paper. One problem is a glossy appearance to the paper, although many

cartographers like to work with this method because good results are achieved. One such system is Du Pont's Chromalin (trademarked by Du Pont). This process can handle image sizes of 25 × 38 inches also.

As the printing industry moves steadily to the processes of direct-to-plate and direct-to-press, it appears likely that digital proofing will continue to make further inroads among the proofing world. This process is not completely new, but is increasingly improved. Two of these systems are the Kodak Approval and the 3M Digital Matchprint (both trademarked by their respective companies). These are referred to as direct digital color proofing (or **DDCP**). Digital proofing especially is finding its way into the smaller format desktop publishing arena. Very large printers doing very large press runs will often be using conventional (nondigital) proofing. As with any method of proofing, the cartographer needs to develop a good relationship with the bureau or printer doing the proofing to completely understand the process and know its limitations.

Two other methods of digital proofing include ink-jet and dye-sublimation systems. These are finding increased use. With ink-jet systems, inks are sprayed onto the substrate carrier (paper), and with dye-sublimation systems, a heat source is used to cause ink to flow to the substrate. Digital systems are usually controlled by PostScript software (a trademark of Adobe Systems, Inc.).[12]

Press Inspection Press inspection, or *press proofing*, is generally done only on very large press runs involving large printing costs. A press proof is a printed copy of the map on the final paper. Press inspections are specified in printing contracts and are usually very expensive. The presses must be stopped and the cartographer summoned for the inspection, which is time-consuming. Prepress proofing, if done carefully, is generally adequate.

Registration A critical factor in successful map printing is **registration**—the preplanned alignment of multiple images on the paper. In printing involving several line tints, pattern overlays, or separations, it becomes even more important that proper registration be achieved. Nothing is more troublesome and distracting on a map than improper registration.

The accurate positioning of multiple images on a single piece of paper is the responsibility of the designer, the photoengraver, and the printing-press operator. The designer ensures that all artwork is properly aligned; the photoengraver (or the cartographer, if doing his or her own photography) maintains the alignment during photography (especially while making composite negatives with complex screening); and the press operator controls registration during the actual printing, especially in multicolor press runs. It is difficult to determine the source of poor registration unless one is involved in a specific job. Cartographic designers must make sure that all art is properly registered when it leaves the office.

If the cartographer is producing art with computer software, most such large applications today provide automatic registration for multiple-separated art. Position or registration marks are placed on all artwork overlays as they are prepared for the job. The marks are placed outside the map borders but are part of the image area, so that they will appear when negatives are prepared. The printer may further register the negatives with holes punched on the negatives. Two will be used for small map sheets; on larger sheets several holes, customarily used, are called pin-bar registration. Whatever system is used, it is imperative that some form of registration be provided for jobs having multiple pieces of art.

The next chapter introduces digital map compilation, or as it has come to be known by some, desktop mapping. The background on printing just presented provides you with the material needed to appreciate more fully this world of electronic map production.

NOTES

1. Arthur H. Robinson, "Mapmaking and Map Printing: The Evolution of a Working Relationship," in *Five Centuries of Map Printing*, ed. David Woodward, (Chicago: University of Chicago Press, 1975), pp. 1–23.

2. Michael J. Adams and David D. Faux, *Printing Technology, A Medium of Visual Communication* (North Scituate, MA: Duxbury Press, 1977), pp. 213–14; another useful reference for offset printing is John E. Cogoli, *Photo-Offset Fundamentals* (Mission Hills, CA: Glencoe, 1986).

3. Jill Roth, "Poised for Takeoff," *American Printer* 207 (1991): 31–33.

4. Ibid.

5. Robinson, "Mapmaking and Map Printing," pp. 16–23.

6. Jake Wideman, "Stochastic Screen Test," *Publish,* June 1994, pp. 35–39.

7. Cartographers usually choose colors based on the Pantone Matching System, a color-mixing system that has nearly become standard in the printing industry. It is a product of Pantone, Inc., of Moonachie, NJ.

8. Students are encouraged to read more on these subjects by looking at, for example: Mattias Nyman, *Four Colors/One Image* (Berkeley, CA: Peachpit Press, 1993), pp. 3–4; Kirty Wilson-Davies, Joseph St. John Bate, and Michael Barnard (revised by Ron Strutt), *Desktop Publishing* (London: Blueprint, 1991); and Miles Southworth and Donna Southworth, *Color Separation on the Desktop* (Livonia, NY: Graphics Arts Publishing, 1993).

9. William C. Lamparter, "Hi-Fi Color: Nonsense or Niche?" *American Printer,* May 1994, pp. 52–54.

10. Cynthia A. Brewer, "The Development of Process-Printed Munsell Charts for Selected Map Colors," *American Cartographer* 16 (1989), pp. 269–78.

11. Vincent D. Freeman, "A Roadmap for Digital Design and Production," *Color Publishing*, November/December 1994, pp. 32–34; see also Southworth and Southworth, *Color Separation on the Desktop*, various pages.

12. Students are encouraged to read a variety of trade publications to keep pace with the many changes taking place in the printing world. Three good ones are *American Printer, Color Publishing*, and *Publish*. The latter two deal specifically with desktop publishing.

GLOSSARY

additive primary colors   red, green, and blue; additive colors are those that, when combined with other colors, yield new ones, p. 320

chromaticity diagram   a chart for plotting the tristimulus values of a color, p. 322

color   light energy at different wavelengths along the visible electromagnetic spectrum; red is about 7.5×10^{-5} cm in wavelength, and violet is about 3.5×10^{-5} cm, p. 320

color calibration   a method of matching colors between a computer monitor and the printed page, p. 322

color separation   a process of creating a printed color image that looks like a continuous-tone image but is actually formed by patterns of small dots printed in black plus the three transparent inks of magenta, cyan, and yellow, p. 322

contact film screen   a screen containing vignetted dot patterns arranged at 45 degrees to produce halftone effects on artwork, p. 317

copy negative   a piece of photographic film containing the reverse of the image to be printed; black areas of the original are open on the negative, and white areas of the original are black on the negative; line negatives have only black or open areas, unlike continuous-tone negatives, p. 315

DDCP   otherwise known as direct digital color proofing, a method of going directly from the computer image files to a completed proof in prepress, p. 328

direct-to-plate   expression used in prepress when plates are made directly from computer image files, p. 320

flat color printing   accomplished by using different colored opaque inks, in contrast to four-color process printing, p. 320

flats   gridded sheets, often orange, onto which negatives are taped for plate preparation, p. 318

gray tone   in black-and-white cartography, the amount of ink printed per unit area on paper surface; e.g., 10 percent or 20 percent, pp. 316–317

halftone negative   photographic film containing a precision-ruled pattern of lines; used to photograph a continuous-tone negative; breaks the original image into a pattern of very small dots of uneven size and density, p. 317

Hi-Fi color process of extending the color gamut in printing by adding orange, green, and violet to the four process colors of cyan, magenta, yellow, and black, p. 323

imagesetter a computer-driven machine that produces positives or plate-quality negative film from digital files, p. 318

intaglio forms of printing in which depressions are engraved in the block or plate; the depressions receive ink, which is transferred when pressed onto the paper; copperplate was a popular method of intaglio map printing during the seventeenth and eighteenth centuries, p. 314

laser scanning technique used to make printing plates directly from original art, p. 319

letterpress form of printing in which the printing surfaces are raised in relief from the printing block or plate; raised portions receive the ink, p. 313

light that part of the electromagnetic spectrum that creates sensations in human eyes; light occupies wavelengths from 7.5×10^{-5} to 3.5×10^{-5} cm, p. 320

manuscript map a single map not made for reproduction, or the working compilation map that will be used for making scribe sheets, p. 313

moiré a troublesome pattern effect that interferes with an image, p. 322

offset lithography an intermediate drum or roller on the press allows the plate image to be right-reading; the image on the blanket or offset roller is backward, p. 314

opaque printing ink reflects back its color at the ink's top surface but absorbs all other colors, p. 320

Pantone Matching System (PMS) a color specification system used widely throughout the printing industry, pp. 324–325

photo-direct platemaker a machine incorporating a camera for direct platemaking, p. 318

photolithography modern lithographic printing; so called because of the photographic preparation of the printing plate, p. 314

planar printing relies on the fact that water and grease do not mix well; areas on the printing block or plate are at the same elevation, with ink either adhering or not, depending on the surface preparation between water and grease; a popular form is lithography, p. 314

prepress proof system to develop black-and-white composites of a printing project before it is printed; used for editing and checking specifications and for visual approval of a printing job, p. 327

process color printing achieved by using a subtractive color combination and transparent inks; often called four-color process printing, p. 320

registration any system that ensures the proper alignment of all pieces of art, p. 328

screen tint photographic film used to break a line negative into a regular pattern of very small dots; after printing, the image appears as a percentage of solid color, pp. 316–317

service bureau company that does data processing for other companies and organizations, p. 327

stipple and aquatint engraving special methods of engraving a copperplate in order to create the impression of tints when printed, p. 314

stochastic screening a digital process method of halftone screening involving changing the distance between halftone dots, rather than the size of dots as in conventional screening, pp. 317–318

subtractive primary colors combinations of colors resulting from the absorption of a primary color or colors, p. 321

transparent inks first reflect their color to the paper or other ink on which they are printed, then reflect it back to the reader's eye; process printing requires the use of the transparent inks magenta, cyan, and yellow, pp. 320–321

tristimulus value amount of red, green, and blue lights used to match a test color, p. 322

woodcut map method used to print maps during the fifteenth and sixteenth centuries; based on the relief method; similar to early letterpress, p. 313

READINGS FOR FURTHER UNDERSTANDING

Adams, Michael J., and David D. Faux. *Printing Technology, A Medium of Visual Communication.* North Scituate, MA: Duxbury Press, 1977.

Brannon, Gary R. *An Introduction to Photomechanical Techniques in Cartography.* Waterloo, Ontario: Cartographic Centre, University of Waterloo, 1986. Useful ideas for manual preparation of films and still useful in many cartography prepress shops.

Cardamone, Tom. *Color Separation Skills.* New York: Van Nostrand Reinhold, 1980.

Carter, James R. *Computer Mapping, Progress in the 80s.* Resource Publications in Geography. Washington, DC: Association of American Geographers, 1984.

"Computer Aided Designers." *Designer*, Spring 1988, pp. 8–9.

Dalley, Terence, ed. *The Complete Guide to Illustration and Design, Techniques and Materials.* Secaucus, NJ: Chartwell Books, 1980.

Earle, James H. *Engineering Design Graphics.* Reading, MA: Addison-Wesley, 1977.

Felici, James, and Ted Nace. *Desktop Publishing Skills.* Reading, MA: Addison-Wesley, 1987. A very useful primer incorporating design, typography, graphics, and hardware associated with all aspects of desktop publishing. Highly recommended for the beginning cartographer.

International Paper Company. *Pocket Pal: A Graphic Arts Production Handbook.* New York: International Paper Company, latest edition.

Johnston, Peter. "From Computer to Page." *Computer Graphics World,* February 1989; also reprinted in Pennwell Graphics Group. *Graphic Arts.* Nashua, NH: Pennwell Graphics Group, 1994.

Lang, Kathy. *The Writer's Guide to Desktop Publishing.* London: Academic Press, 1987.

Monmonier, Mark S. *Computer-Assisted Cartography: Principles and Prospects.* Englewood Cliffs, NJ: Prentice Hall, 1982.

———. "Geographic Information and Cartography." *Progress in Human Geography* 8 (1984): 381–91.

———. *Technological Transition in Cartography.* Madison: University of Wisconsin Press, 1985.

Morrison, Joel. "Computer Technology and Cartographic Change." In *The Computer in Contemporary Cartography,* ed. D. R. Fraser. New York: Wiley, 1980.

Nyman, Mattias. *Four Colors/One Image.* Berkeley, CA: Peachpit Press, 1993.

Olson, Judy M. "Component-Color and Final-Color Separation in Mapping." *Cartographica* 22 (1985): 61–69.

Peucker, Thomas K. *Computer Cartography.* Commission on College Geography, Resource Paper No. 17. Washington, DC: Association of American Geographers, 1972.

Robinson, Arthur H. "Mapmaking and Map Printing: The Evolution of a Working Relationship." In *Five Centuries of Map Printing*, ed. David Woodward. Chicago: University of Chicago Press, 1975.

Sena, Michael L. "Back to Camp—Advances in Computer-Aided Map Publishing Drive New Industry." *Computer Graphics World,* March 1982, pp. 119–24.

Southworth, Miles, and Donna Southworth. *Color Separation on the Desktop.* Livonia, NY: Graphics Arts Publishing, 1993.

U.S. Department of the Army. *Offset Photolithography and Map Reproduction.* Technical Manual TM 5–245. Washington, DC: Headquarters, Department of the Army, 1970.

CHAPTER

17

DIGITAL MAP COMPILATION AND DESKTOP MAPPING

CHAPTER PREVIEW

Today most small-scale thematic maps are compiled from digital source materials and therefore are not prepared by manual means. The cartographic designer works at a microcomputer workstation, or directs those who do, and guides the flow of digital data and base maps, creating the final design on a visual display device. In this kind of map preparation, the cartographer works with the RGB (red/green/blue) color model and uses computer technology for incor- *porating data and sending them to various output devices. Selection and knowledge of different software programs are important in this kind of map production, as is an understanding of the electronic environment. The cartographer also relies on a knowledge foundation of printing technology because final printing requires this understanding. In the foreseeable future, electronic production of small-scale thematic maps will be the only kind.*

In the previous chapter a variety of subjects dealing with printing techniques were presented, things that cartographers need to know in order to deal adequately with map production. The purpose of this chapter is to provide you with an introduction to the techniques necessary for the preparation and digital compilation of thematic maps, especially small-scale maps useful in geographic cartography. The focus here is not the computer manufacture of large-scale topographic maps or an in-depth look at digital images received from remote sources (satellites or aircraft). These subjects are covered in much more detail in texts written for those purposes.

This chapter is an extension of the previous chapter. The primary goal is to introduce the newer (digital) *tools* for the production of well-designed thematic maps on *paper* (or photographic film). The emphasis is not on how to make maps look good on a computer screen. Although the literature includes some contributions on designing maps for the cathode-ray tube (CRT), these are not the focus here. Furthermore, this is only an introduction to the subject; more advanced and specialized texts treat these techniques in far greater detail than is permitted in the limited space provided here.

This chapter includes descriptions of components—including hardware, software, and interfaces—of the systems that make up a desktop mapping environment, and addresses how each one can facilitate the digital compilation of a small-scale thematic map. Discussions include the advantages and disadvantages of each, and how the constraints of the components may or may not affect good map design. The material here is not exhaustive. Rather it is an attempt to describe the systems and component types that will remain intact for some time, rather than discuss individual products that may soon be replaced by newer versions. In the computer world, what we may say today may be different from what we know tomorrow.

A BRIEF HISTORY OF COMPUTERS AND MAPPING

A somewhat confusing array of terms has been applied to the use of computers for the generation of maps. At first, of course, there was **computer mapping,** which simply meant that computers were somehow used in the generation of maps. Some cartographers applied the term **automated mapping** to define the same operations. A bit later, the term **computer-assisted cartography** emerged on the scene. As the possibilities grew for the production of maps on the microcomputer, and with the advent of *desktop publishing,* we now hear the term **desktop mapping (DTM)** being applied to the activity of producing small-scale thematic maps.[1] This term should not be confused with a DTM-digital terrain model—also used today, in preparing three-dimensional map views.

Desktop mapping as it is used in this chapter refers to all the computer operations necessary to digitally assemble the map's component parts (base maps, data, lettering, symbols, and so forth) in order to produce final negatives (or even printing plates) for the eventual printing of a final map in the traditional form of ink on paper. The time may come when we can go from the computer screen to final map without negatives or even printing plates, but this sophistication must await improvement in nonimpact printing technology.

The history of computer mapping probably began in the late 1950s and early 1960s. One of the first completely contained mapping programs was SYMAP, developed by the Harvard Laboratory for Computer Graphics and Spatial Analysis in 1965, and first made available to geographers by a correspondence course[2] which, by 1968, was generally available to universities that had large mainframe IBM computers. (This author participated in that course.) The output of the SYMAP program was on-line printers that displayed the maps by printing alphanumeric characters (and overstrikes of those characters to give different densities). The maps were coarse and resolutions restricted to the sizes of printed character spaces. Nonetheless, SYMAP became the standard for a number of years and many cartographers became trained in computer mapping with this program.

From the very beginning, computer cartographers simply attempted to write computer programs that instructed machines to replicate what human cartographers were doing. Soon after SYMAP, the pen plotter was called on to do much of the cartographic drawing. Here again, this was simply a machine doing the work of the cartographer. Software was written containing commands that would drive the plotter. Good lines could be drawn, with different thicknesses depending on the pen used, and area fills could be achieved with fairly good results. Even color was possible by changing pens. This was a great improvement over the output of the line printer, but still lacked the finish required by most cartographers. Many firms, now mostly engineering companies, still rely heavily on plotters, especially because

---

The phrase "desktop publishing" or DTP, was coined on January 28, 1985, by Paul Brainard, president of Aldus Corporation, at an Apple Computer annual stockholder meeting. Aldus was the developer of the PageMaker page layout program which, combined with a Macintosh computer, Aldus Pagemaker, Adobe's PostScript page description language, digital type licensed from the International Typeface Corporation, and an Apple LaserWriter, created a new medium-resolution publishing system. The original Apple LaserWriter was a printer that created images at a resolution of 300 dots per inch (DPI). It used the Canon LCB-CX print engine, and was built around the technology used for photocopiers, in particular, the Canon Personal Copier cartridge line.

Source: Barrie Sosinsky, *Beyond the Desktop: Tools and Technology for Computer Publishing,* Bantam ITC Series (New York: Bantam Books, 1991), p. 15.

they can handle very large format drawings inexpensively. Lettering was another matter, and had to wait for the microcomputer and DTM. Even so, the automatic placement of lettering remains one of the chief problems with DTM mapping software.

Several textbooks on computer use in cartography deal with its history in much more detail than is allowed here.[3] It is sufficient to say, however, that the employment of computers for the generation and display of maps has occupied only a brief time in the long history of the discipline. At no previous time has cartography changed as rapidly as today, which has led one cartographer to remark, "Since the advent of the digital computer, however, we have become used to cartography as being in a state of almost constant technological revolution."[4] This situation will likely continue in the years ahead. In fact, many professional papers of the past 5 or 10 years have dealt with the implementation of computer technology, hardware and software, into the cartographic environment.

When SYMAP was developed, no one had the remotest idea that today we would be generating maps with microcomputers or that laser devices could generate film negatives and printing plates, or that we could go from computer screen to printing plates without ever "touching" the map or its component parts. A cartographer writing on the subject of desktop mapping in the United Kingdom has captured the remarkable changes in the field this way:[5]

> I always give a wry grin when the word "traditional" is used, as a person who was trained at the Ordinance Survey in the early 1960s, drawing on enamel, progressing to scribing on glass and then, of course plastics. The question begs to be asked; what is traditional? I suggest it is transitional—depending on one's age and entry point into the profession. I bet some of you engraved on stone tablets: I just remember them being used as door stops!

One cartographic specialist in computer mapping has said that one of the main benefits of computer mapping, aside from ridding us of the manual tasks of making the map, is that now "we can worry about other things, for example, design—ultimately what the cartographer is trying to perfect."[6] He goes on to say that it has allowed us to develop a design loop—another way of saying that we can operate in a "what-if" state. If we don't like what we see, we can experiment with another design quickly and efficiently, cycling through a variety of loops until we get a product we are satisfied with.

It is probably fair to say that DTM was born of desktop publishing. (See box.) As gains were made in the 1980s with the introduction of better computer software that could create maps, and with the introduction of special computer and printer interface languages, maps could be designed easily with a microcomputer and saved to electronic media in digital form; then negatives could be made and the final map printed in traditional ways. In all but the very unique cases, both large- and small-scale mapping is now done electronically, in digital form. A few agencies, both public and private, continue manual production techniques, but these methods are rapidly disappearing. (Even so, the cartographic designer must remain well informed about the technology of printing.) The era of DTM is with us, and with it map designers have greater freedom to explore more fascinating designs.

The best place to start our discussion of digital compilation of maps is with the typical microcomputer "workstation"—a term that needs some clarification.

THE MICROCOMPUTER CARTOGRAPHY WORKSTATION

Not so long ago a cartographer's workstation was comprised of a drawing or drafting table and included a variety of drawing pencils, pens, compasses, triangles, and so on. Although this accurately represented a manual production workstation, it does not hold true for the microcomputer workstation. The **microcomputer workstation** is better defined as a "configuration of computer equipment designed for use by one person at a time."[7] In practice, this assemblage of equipment will include the computer itself, and some or all of the input and output devices required to produce the map.

The associated equipment tied to the computer is usually referred to as **peripherals.** These are physically tied to the computer by way of cabling and include such input and output devices as keyboards, graphics tablets, digitizers, scanners, CD-ROM readers, visual display terminals (CRTs), printers, and plotters, and may also include external data-storage devices such as hard or soft disk drives. (See Figure 17.1.) These devices are explained more fully later in the chapter.

In some cases the workstation may include such devices as **imagesetters,** machines that transform images in Post-Script languages (explained below) into high-resolution forms that can develop the image onto photosensitive films and papers by laser devices.[8] Because imagesetters are expensive, only cartographic production shops that prepare many maps will have them. Most cartographers compiling maps not requiring very high resolution output save their images to a storage device (diskette) and send them to "service" bureaus that do the actual photographic work. Service bureaus are companies that do data processing for other companies or organizations. Some bureaus may specialize in processing graphic images onto film or paper, or make copies from slides that have been digitally scanned.

THE COMPUTER PLATFORM

In the world of digitally produced maps, terms often heard include *platforms, operating systems,* and *interfaces.* Sometimes people use the term *environment,* and in a computer context this does *not* refer to such physical conditions as

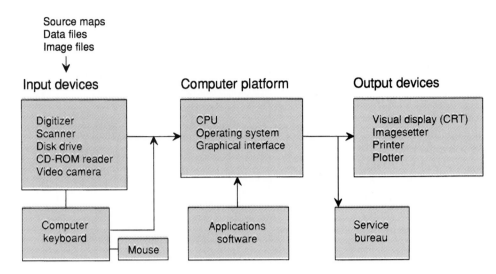

Source maps
Data files
Image files

Figure 17.1   The desktop mapping (DTM) workstation. A typical configuration includes input and output devices coupled to the CPU. In some cases a service bureau may be used to produce final map products.

temperature, humidity, and the like. A computer environment usually refers to a mode of operation—such as "stand alone" or "time sharing." **Platform** usually refers to the complete package of a computer, an operating system, and an interface. We often hear that people are working in an "IBM environment." Some might argue that this is not correct, but what matters most is that there be no misunderstood communication.

The Computer and the Central Processing Unit (CPU)

Microcomputers contain a variety of internal components necessary to do the job of processing information—graphical, nongraphical (words, as in word processing), and strictly numerical processing (mathematical). At the heart of the computer is the **central processing unit (CPU),** a miniaturized electronic component capable of controlling the interpretation and execution of instructions.[9] It is also the central logic unit of the machine. Throughout the short history of the microcomputer, several things have happened to CPUs that have affected the computer. First, they have become smaller; second, they have become much faster; third, they have become much less expensive; and fourth, they are capable of more operations.

Microcomputers have reached their importance in desktop mapping only as their speed has increased over the years. In the world of digitally produced maps, the speed of the CPU is of utmost importance. There are a variety of ways to measure computer speed, but usually for the microcomputer, the speed of the CPU will determine the final effective speed of the machine. Microcomputers do not process information (throughput) as quickly as mainframe computers.[10]

The microcomputer, of course, contains other components necessary for it to produce the dazzling graphics and maps that we see today. In addition to the CPU, it contains circuit boards that house the permanent memory in the system (so-called ROM or read-only memory), volatile memory (so-called random-access memory that is necessary for computational processing), a circuit board that controls the disk drives (both hard and soft drive units), and a circuit board that controls the video output devices. Strictly speaking, the keyboard is an input device, but may be considered part of the computer itself. Instructions that tell the computer what to do come from the keyboard.

The Operating System

The operating system of a microcomputer has a very specific meaning. An **operating system** is a program of instructions that controls and supervises all other programs running in a computer, handling all input and output functions of the computer, disk operations, and printer commands. The two most popular operating systems used by cartographers for designing maps are the **Macintosh system,** used by a family of computers manufactured by Apple, Inc., and the **MS-DOS** (Microsoft Disk Operating System), used by the IBM family of microcomputers and compatibles. In each, the operator can manage files (a record or related records similar in concept to the term in a noncomputer setting), control what the visual display will look like, and organize, send, and receive instructions from peripheral devices. Cartographers who design maps with microcomputers usually become fairly efficient with the operating systems of the computers on which they work, although seldom-used commands may need verification in the system's manuals.

The MS-DOS operating system is provided on a disk or disks read into the computer's memory, which takes up some of the valuable computational memory space. There are ways to circumvent this shortcoming, and most MS-DOS applications software do so, especially the graphics programs. The operating system of the Macintosh is included mostly in the computer's ROM (read-only memory). The operating system of the Macintosh is proprietary.

There are other operating systems (UNIX and OS/2) that work with microcomputers, but at this time most of the mapping and graphics programs make use of the MS-DOS or Macintosh systems.

The Interface

A computer's **interface** is a program that runs in addition to the operating system, and specifically has the function of interacting with the user. The purpose of the interface is to set up a set of commands that all programs will use in the computer. Some interfaces are simple, such as the command lines in DOS, and some are quite complex such as the **graphical user interfaces (GUIs).** Graphical user interfaces have become quite popular, probably because of the early success of the Macintosh line of computers. The Macintosh's graphical user interface is called Finder. The PC line of GUIs include the popular Windows program; others for the PCs are Microsoft NT and OS/2 Presentation Manager. All graphical user interfaces are menu- and icon-based and require much more computer memory than the nongraphic interfaces. Today many software programs are written to work in conjunction with a certain interface. Users therefore benefit, and interacting with the computer is simplified considerably.

THE VISUAL DISPLAY

As Figure 17.1 suggests, a number of peripheral devices can be used to digitally generate a thematic map. These are usually separated into either input or output devices depending on their functions. One device, the visual display, actually functions in a somewhat different manner. It functions as a way to *display* the instructions that have been selected (thereby behaving as if it were part of the computer), *and* it can be used as an output device. It can also function as an editor, to "test" map designs before they are stored or sent to an output destination. This latter function manipulates "virtual" maps. **Virtual maps,** according to Moellering, are digital preliminary maps, maps that are not yet completed or permanent.[11]

The discussion here of the visual display device is limited to the color raster refresh type, primarily because it has become the de facto standard for most microcomputer workstations used for digitally compiling thematic maps. (See Figure 17.2.) There are two fundamental reasons why a basic understanding is required, namely (1) knowing how the display devices function and their specifications assists in selecting them for purchase, and (2) knowing how they function helps understand one other color model—the RGB color solid.

The raster-type color visual display device—also called a monitor or **CRT** (from cathode-ray tube)—gets its name because the image is created on the screen by an electron beam that *scans* in a row-by-row movement called **raster scanning.** (See Figure 17.3.) The focused electron beam (negatively charged) is drawn to the positively charged

Figure 17.2   NEC MultiSync E900 monitor, a typical DTM visual display (CRT) device.
Most visual displays are set on adjustable platforms for comfortable viewing. Large screen displays are becoming standard. (*Illustration courtesy of NEC Technologies, Inc.*)

phosphor-coated screen, which glows. The phosphors are arranged in a small dot pattern.[12]

With color raster scanning, there are three electron beams, each emitting at certain energies that excite one of *three* phosphor dots—red, green, or blue—arranged in a triangular pattern called a triad (although on some monitors the patterns are vertical stripes). These phosphor dots glow when their corresponding beams strike the dots. These are thus called **RGB color monitors.** The three beams must pass through a **shadow mask** before they strike the color phosphor dots. (See Figure 17.4.) The shadow mask contains a pattern of holes that control how the beams strike the dots. After passing through the hole, the blue beam, for example, is prevented (masked) from intersecting the red or green dots. When a CRT needs a convergence adjustment, it simply means that an adjustment must be made to control focus of the beams through the shadow mask.

It is possible to vary the strength of each of the different electron beams so that different shades of the different colors (red, green, blue) result. This is one way to produce so many different colors on the color display.

The spacing of the dots in the shadow mask is referred to as the **pitch** and this roughly determines how far apart the phosphor dots are on the screen. The finer the pitch, the finer the "grain" of the image. Today, many in DTP and DTM would suggest that a pitch of .31 mm (which produces a grain of about 80 dots per inch) is maximum for good quality color display work, at least on a 14-inch (diagonal measurement) monitor. For a monitor that has, say, a viewing area of

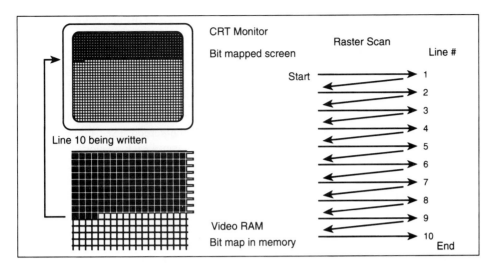

Figure 17.3 Raster scanning CRT.

The video memory (RAM) contains a bit map of the image, and the CRT guns write (raster scan) to the screen to create the image, in a line-to-line pattern.

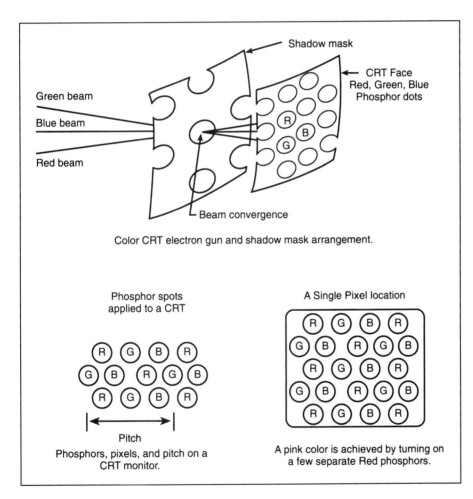

Figure 17.4 CRT shadow mask.

Electron beams of different energy (for red, green, blue) pass through the shadow mask and excite the R, G, B phosphor dots on the screen. A pixel can be composed of just one triad of R, G, B dots, or several triads.

about 7 by 9 inches, there are over 400,000 phosphor dots on the screen (too many for the human eye to resolve). It is not uncommon today to find displays with a .26-mm pitch, especially on 17-inch monitors.

The electronics of the computer video "drivers" and the display device standards used today create the smallest "resolvable" element of the screen as a **picture element,** called *pixels* by most but sometimes referred to as *pels.*

Each pixel can be made up of just one triad of color dots, or by several. So-called high-resolution monitors have two to three color triads per pixel.

The electron guns of the CRT write to the screen in a pattern of *scan lines,* starting in the upper left-hand corner and moving to the right, then passing to each successive lower line, then across the screen again, until the entire screen is filled with the image. For the Macintosh computer, the entire image is scanned 60 or more times per second (the scan rate). Scanning can be done in interlaced or noninterlaced modes. When the image is scanned **interlaced,** the scan skips a row, then comes back and fills in the rows it skipped. (See Figure 17.5.) If the scan is *noninterlaced,* each row is scanned while forming the image. Noninterlaced is better because it reduces flicker on the screen, although it is more expensive.

Image resolution is determined by the number of pixels displayed on the screen, the size of the screen, and of course the dot pitch, which has already been described. (See Figure 17.6.) The Apple color RGB monitor has a 640 × 400 pixel resolution, and an 80 dpi pitch. A 256 color video board is standard for the Macintosh II. The VGA standard in the IBM platform is 16 colors at a pixel resolution of 640 × 480 pixels, and 256 colors can be obtained at a lower pixel resolution of 320 × 200 pixels. Pitch is determined by the monitor selected for the workstation.

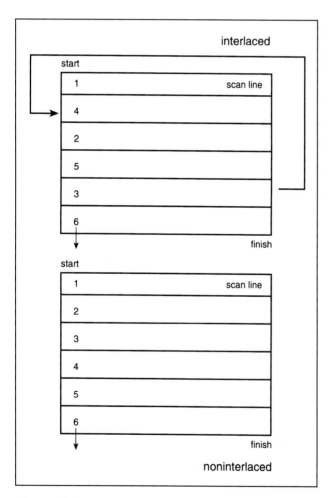

Figure 17.5   Interlaced and noninterlaced CRT scanning. Interlaced scans take two passes to write the image—completely doing every other line on the first pass, then filling in the missed lines on the second pass. Noninterlaced scans complete each row during the first pass. Noninterlaced monitors are more expensive but reduce flicker.

One last topic concerning color display devices is interesting, especially because it helps illustrate how color is achieved on the display, *and* the internal complexity of these devices, as well as why so much computing "power" is necessary to achieve color. On black-and-white displays, a one-plane "frame buffer"—a memory area—is set aside on the computer video board that contains one memory bit for each of the pixels on the screen. (See Figure 17.7a.) This is called a *bit plane.* Because each memory bit has two states—On or Off (0 or 1)—a display containing 512 × 512 pixels must have (2 × 512) × (2 × 512), or 262,144 memory bits for each bit plane.

Simple color is achieved by having one bit plane for each of the RGB colors, which can generate eight colors. (See Figure 17.7b.) Multiple bit planes for each of the primary colors can be used to generate varying intensities of the three colors. Memory requirements escalate quickly, as do speed requirements, to achieve acceptable refresh rates. The student who is interested in reading further might wish to consult *Procedural Elements for Computer Graphics,* a well-written resource on this subject.

The RGB Color Model

In the chapter on color, several models that describe color space were discussed. One, the RGB model, was not discussed. Because it is most applicable to the color generated by computer display devices, we will turn our attention to it here.

You may recall the discussion of the CIE color specification system. This system is based on additive mixture characteristics of color, and on the values of the three primaries, red, green, blue. Each hue is specified in terms of its tristimulus values (of red, green, and blue). This feature explains why it is possible to generate many hues on the RGB color display. It is important to note that additive color mixture is the system used in the RGB color display.

In the **RGB color model** and specification system, a color is located in three-dimensional space based on the following formula:

$$C = xX + yY + zZ$$

where the capital letters *X, Y,* and *Z* refer to red, green, and blue respectively, and the small *x, y,* and *z* refer to the amount of light generating each reflective color.[13] In the model of color space, black is at the origin of the space, and the three colors red, green, blue "are calibrated so that equal amounts give grays along a scale from black to white. The amounts of the primaries that are added to give white usually are normalized such that they have a value of 1. Therefore, the white point is at (1, 1, 1) in the coordinate system. The line joining the black and white points in color space is the neutral or gray axis."[14] (See Figure 17.8 and Plate 13.)

This type of color works because the red, green, and blue phosphor dots on the screen are so small that the human eye does not see them individually, but "blends"

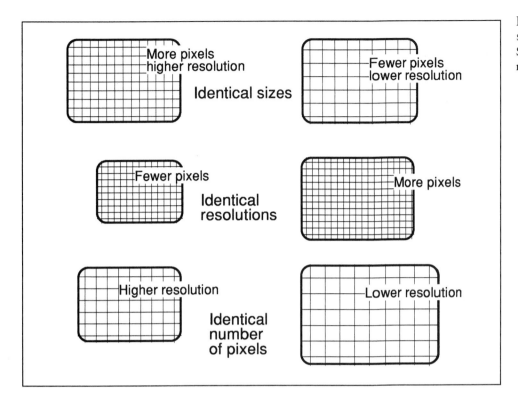

Figure 17.6 Pixel size and screen resolution.
Screen size, pixel number, and resolution are interrelated.

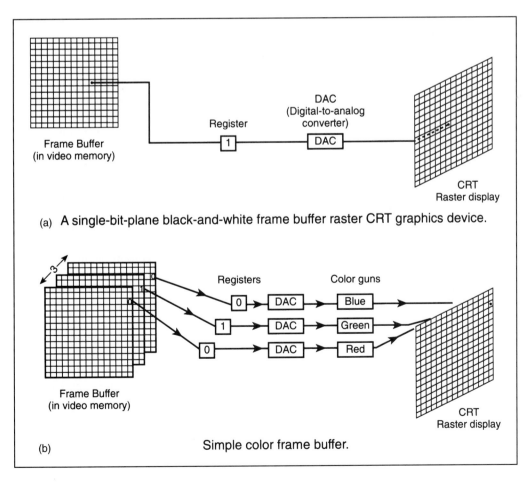

(a) A single-bit-plane black-and-white frame buffer raster CRT graphics device.

(b)   Simple color frame buffer.

Figure 17.7 CRT display and bit-plane buffers.
Buffers are in the computer's video memory. Simple black-and-white (pixel is On or Off) images require only one frame buffer (to show shades of gray at each pixel, more frame buffers are needed), as in (a). Simple color needs one frame buffer for each color, R, G, B, as in (b). Video processors that can control color displays of different intensities require more frame buffers and much more memory.

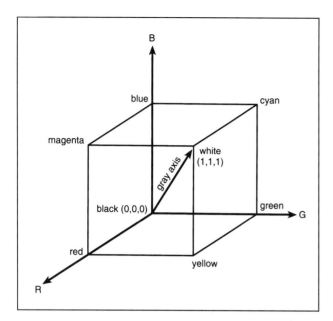

Figure 17.8 The RGB color model.
This model is based on the subtractive colors of red, green, and blue (equal amounts yield black). By varying the intensity of each hue (on the screen at each triad), a separate color is perceived. See text and also Plate 10.

them into full colors. The reds, greens, and blues, each having different intensities, create the myriad of colors possible on the display.

HSB Color Specification

It is possible in many color graphics programs to select colors based on hue, saturation, and brightness. The **HSB color system** is based on the properties of a color, rather than the components of red, green, and blue. Hue of course is the color's position in the energy spectrum, indicating its frequency. Saturation describes the intensity of the color (100 percent being the most intense and 0 percent the least). In this system, brightness is the degree to which the color is approaching white (pure white is 100 percent, black is 0 percent). In some respects this system is similar to the Ostwald color solid.

The HSB system is also called HLS, for hue, lightness, and saturation. No specific color models to describe these color spaces have been generated—based on the parameters of hue, saturation, and brightness—for color display devices.

DIGITAL CARTOGRAPHIC COMPILATION AND ASSOCIATED PERIPHERAL DEVICES

At this point in our discussion it seems pertinent to outline the various stages in the preparation of a digital map and then provide a somewhat more detailed examination of the individual peripheral devices common in the digital workroom or workstation. The process is diagrammed in Figure 17.9.

Base-Map Compilation from Nondigital Sources

In many cases the digital map begins by converting an already-formed map from several conventional or nondigital media formats. This original map may be a simple pencil or ink drawing, a drawing on scribed media, or a film negative of an existing map. Any of these forms may be called **original map data**.[15] This operation is typical when an original digital map is being created.

The original map must be converted to digital form, either in a vector or raster format. Here is a clear and simple way of stating the process:

> *Whenever it is necessary to process graphic material (such as points, lines, areas, conventional symbols and lettering) in a computer, it must be expressed digitally, i.e., with numbers. To this end we make use of a basic graphic concept: plane coordinates. All graphics are reduced to points (which can later be connected by straight or curved lines) and each of these can be expressed by a pair of coordinates . . . which can be fed into, and processed by computer.*[16]

Converting original map data into vector digital format is called **digitizing.** As stated above, capturing map data and storing them into vector format means that the map components in the file are recorded at discrete points in *x–y* locations in Cartesian, two-dimensional space. (See Figure 17.10.) Points, lines, and polygons are possible with vector file manipulation, which is most suitable for maps containing lines (such as coastlines, rivers, political boundaries, and the like). This form of map data capture has been popular in computer-assisted mapping from the very beginning.

Another way to convert original map data into digital form is by raster scanning. **Raster scanning** places the map's component parts into discrete cells, which can emulate the precision of the vector process to varying degrees, depending on how small the cells are. This operation is done by a **raster scanner.** The smaller the cells the closer to a vector image a final map might appear. Raster digitizing has become much more popular, just as the refresh raster display has increasingly replaced other forms of display tubes.

For a map that is to be converted to digital form, the decision regarding which form of capture (vector or raster) may well depend on the mapping software one is using, although today it is possible to *convert* from one format to another (more will be said about this later in the chapter). Also, the scale of the map may determine which type of data capture is required. Raster methods are usually more useful for mapping large amounts of data at small scales, and the vector method is better for detailed information at large scales.[17]

Digitizers **Digitizers** are devices that convert analog images (original maps, drawings, and other images) into a digital representation of the original image. (See Figure 17.11.) Large digitizers may be 36 × 48 inches in size.

Figure 17.9 Work flow of digital thematic map compilation. See the text for an explanation.

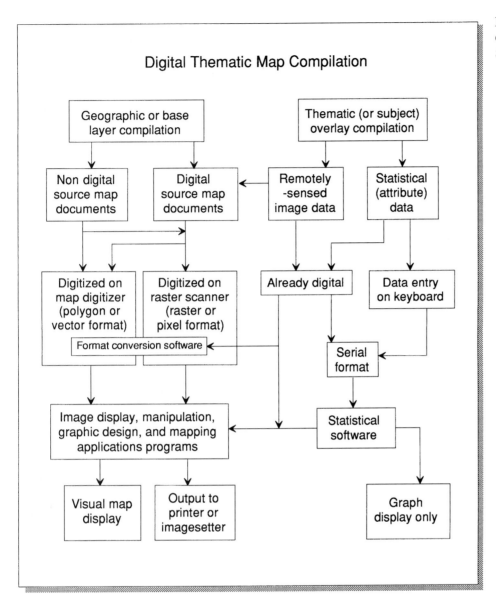

Most have boards embedded with a fine mesh of gridded wires that transmit electric pulses. A handheld **cursor** moved over the surface can detect these pulses to within the resolution of the gridded surface. Many digitizers have resolutions as close as .001 of an inch (1,000 lines per inch). This means that there are 1,000,000 addressable points *per square inch*. The digitizers are linked to the computer, and software processes the information as it is received from the movable cursor. Often there is a menu on the digitizer surface that can be activated by the cursor and which is integral to the software. Small, or tabletop, digitizers for small map sheets are called tablets.

A line can be digitized by selecting points along the line by pressing a button on the cursor, or by foot pedal. Continuous digitizing may be done, and in this case the machine is told to record a point at certain intervals along the line. Points by themselves, of course, can be recorded. The vector digitizing software allows for a variety of options, such as polygon development, area measurement, and others.

The digitized information can be saved to computer files, and later processed by graphics and mapping software.

Scanners   Raster scanning is quite popular in map production; for example, as one cartographer has noted:

The grid is an integral part of cartography and has long been used to structure geographic information. More recently, cartographic data have been acquired in grid formats directly, for example, satellite and scanned air photo data. This aspect is becoming increasingly important as remote sensing becomes a source of new data. Using the grid structure, analytical operations are easier, such as computation of variograms, some autocorrelation statistics, interpolation, and filtering. Resolution on raster devices is improving steadily, and the cost of the graphic memory boards which support display in raster mode is falling, whereas vector technology has not changed significantly in cost or capability in 10 years.[18]

Vector image

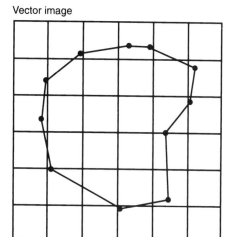

Raster image

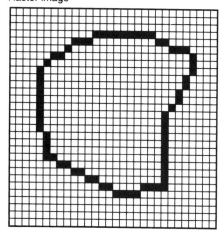

Figure 17.10   Vector and raster images. Somewhat different images result from the two formats, with oblique lines created by the raster format resulting in a "jagged" appearance (the severity is dependent upon display resolution).

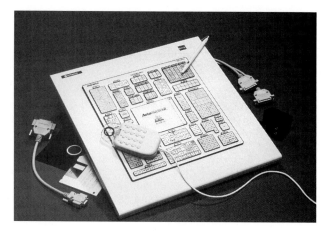

Figure 17.11   A small-format digitizing tablet useful in DTM.
(*Photo courtesy of Hitachi America Ltd.*)

Scanners are of several types, but all are optical devices built to recognize visual images and, like digitizers, convert original map images into digital form. The most expensive devices use laser technology—the so-called flying-spot scanners. These scanners employ a laser light source that quickly scans the image and detects whether the light is reflected or absorbed (for simple black-and-white scanning); in color scanning, they can measure the amount of reflected light and its frequency. (See Figure 17.12.) The analog image is then converted to signals stored as digital information, and at very high resolutions. Large sheet sizes can be scanned.

The desktop mapping workstation normally does not use this type of scanner because of cost, but the cartographer may send cartographic films to a service bureau for scanning. The image can be returned in the form of a digital file that can then be processed with the mapping or graphics software at the desktop workstation. The files returned to the cartographer most likely will be in the Tagged Image File Format (*TIFF*) because most DTM software can easily import (transfer) these into the DTM programs, or JPEG and GIF formats because these are used often on the World Wide Web for image transfer. The TIFF format is practically a de facto standard in the DTP and DTM worlds.

Page scanners often found at the DTM workstation work a bit differently. (See Figure 17.13.) These are generally considered to be low-resolution scanners, and data delivery at resolutions of 300 dpi is virtually standard; but resolutions of 400, 600, or even more dpi are becoming increasingly affordable. *Their primary use in desktop mapping is to import to the mapping software an image that can be used as a tracing template for final map preparation.* This is an economical way of importing "working" images, but should not be confused as a way of importing final images from which plate-quality negatives can be made. It should be pointed out also that this type of scanned image is to be used only when "ground truth" (accurate planimetric) location is not a primary concern.

One possible use of the desktop scanner, though the product is not perfect, is to scan a shaded relief map into the graphic software that will serve as a base map for thematic overlay added by the cartographer. One must be cautious in determining whether the scanned material is in the public domain so that copyright infringement is not a problem, or one must obtain permission for its use. Great care must be exercised, however, because scanned images of previously halftoned images may result in undesirable moiré patterns.

An alternative to the page scanner is the *handheld scanner*. Technologically this scanner is the same as the page scanner, except that it is drawn across the original image manually, and is usually limited to a little over 4 inches in width. The length of the scan depends on the resolution set and the amount of free memory in the CPU. Although capable of scanning in continuous tone images, in association

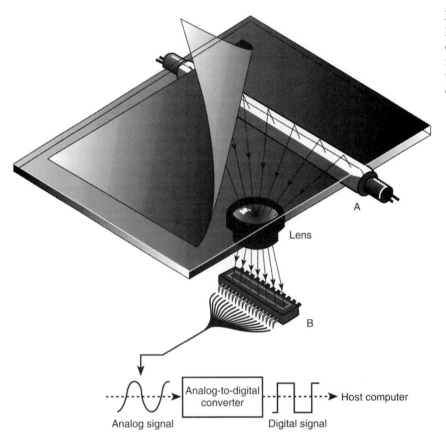

Figure 17.12   The scanning operation. Light is reflected off the source image, collected through a lens, and detected by a myriad of photoelectric cells. The energy at the cells is converted into digital signals that can be processed by special image software.

Figure 17.13   A DTM page scanner. (*Photo courtesy of Microtek, Inc.*)

with **image editors,** these scanners are really most useful for scanning in small original line art—again, only for use as tracing templates for the preparation of final art in the mapping software.

Using Digital Base-Map Products

One of the most useful programs for the cartographer in digital map compilation is the program that offers a myriad of map projections, digitized coastlines, and political boundaries. The cartographer needs only to select the projection, its scale, aspect, and extent, and decide whether coastlines or boundaries are needed, and bring the projection into the

image display to be blended with thematic information for final map compilation. This has been a possibility for the DTM environment for only the past several years.

The MicroCAM Projection Programs   The MicroCAM[19] program and its different databases (world coastlines, islands, lakes, country boundaries, rivers, and states of the United States), were developed from CAM (Cartographic Automatic Mapping), a mainframe program developed by the Central Intelligence Agency. The program can send images directly to output devices (printers and plotters) or to computer files that can be imported later into graphics programs to be enhanced and to have thematic information layers applied to them. With the latest version of MicroCAM it is possible to create some thematic maps directly in the program. Area fills, point symbols, and text may be added to the base maps. Of course the image is displayed on the computer monitor. An example of a projection created by MicroCAM is illustrated in Figure 17.14. Table 17.1 lists the different projections available in the program.

With MicroCAM it is possible to display and save the images to files in a format called DXF (explained more fully later in the chapter), which makes the projections and other features of the program very useful in DTM.

TIGER Files   The United States Bureau of the Census, building on experience with its popular **DIME** (Dual Independent Map Encoding) georeferencing system used in

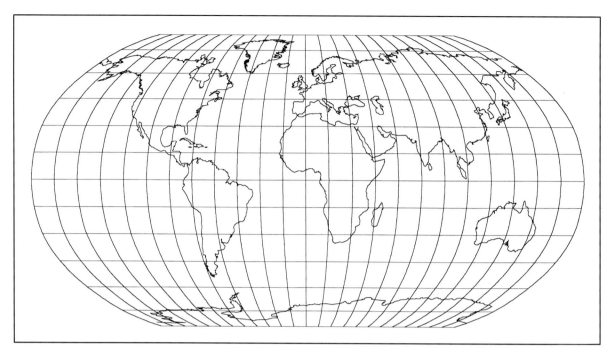

Figure 17.14   A MicroCAM projection
This projection (world Robinson) was created by the MicroCAM program. The image was reduced from the original one done on a 300 dpi laser printer on paper and copy negatives were made in a conventional manner. The MicroCam program is available through the microcomputer users speciality group in the Association of American Geographers. Contact this professional academic organization at 1710 16th Street NW, Wash. DC 20009-3198.

Table 17.1   Projections in the MicroCAM Program

Azimuthal Equi-Distant	Oblique Mercator
Azimuthal Equal Areal	Orthographic
Bonne Equal Area	Perspective
Albers Conic Equal Area	Polyconic
Kavraiskiy IV Conic Equal Interval	Polar Stereographic
	Equi-Rectangular
Ptolemy Conic Equal Interval	Robinson
Gnomonic	Sinusoidal Equal Area
Hammer Equal Area	Stereographic
Lambert Conformal Conic Mercator	Universal Transverse Mercator
Miller Cylindrical	Van der Grinten
Mollweide Equal Area	

At the University of Utrecht we have developed a map projection package. When one wants a world map or a continental map for a specific purpose, one is asked by the computer for the geographical coordinates of the projection center, and for the properties wanted, and the projection that answers these specifications is generated on the screen in 20 seconds with a simple world map, or 80 seconds when World Data Bank is used. . . . The point is, that students can work with them, can experiment with them and are able to generate the world at will, according to their specifications, and learn a lot more a lot quicker than by the time-consuming manual plotting of projections of only 20 years ago, or even a decade ago when they had to immerse themselves first in operating systems.

Source: Ferjan Ormeling, "Requirements for Computer-Assisted Map Design Courses," *International Yearbook of Cartography* 28 (1988): 265–73.

the 1970 census, developed a geocoding system for the 1990 census called **TIGER** (Topologically Integrated Geographic Encoding and Referencing) system. Objects in this geographic referencing system include points (nodes), lines (segments), and areas (blocks, census tracts, or enumeration districts). For many users, this system may become as popular for planners and marketing professionals as did the older GBF/DIME files (still used). Not all planners find

TIGER to be a panacea, however, especially in some uses in geographic information systems (GISs) that require resolution down to the parcel, or land lot, level (TIGER's area definitions go down only to the block level).

Many cartographers will find the TIGER/Line files especially useful. They can be used in the microcomputer environment, coupled with the software necessary to manipulate them. In 1991 the Census Bureau began to release the

TIGER/Boundary extracts from the TIGER database, which includes state, county, numbered census tracts and block boundaries (along with all their codes).[20] Most of these products related to TIGER are available on floppy disk or CD-ROM storage media. TIGER files are also available on Internet Web site *www.cenus.gov*. Thematic cartographers find these digital databases of tremendous value.

United States Geological Survey Products

As part of early planning, and as a direct result of interagency cooperation, the United States Geological Survey assisted with the digitizing of many of their products used in many of the TIGER base files. The **DLG** (digital line graphs) are digital records of the USGS's 71/2- and 15-minute series topographic maps, and some of the smaller-scale map sheets of *The National Atlas of the United States.*

At the 1:24,000 scale, boundaries, rivers, standing water, transportation features (roads and pipelines), and some other prominent features are included in the files. These are vector files. The 1:100,000 series include similar data, as well as the maps digitized from *The National Atlas of the United States.* The United States Geological Survey also produces digital elevation models, called **DEMs.** They are described this way:

The Digital Elevation Models are digital maps of the surface elevation of topography. Two scales are involved, 1:24,000, and 1:250,000. The 1:250,000 data are 3-arc-second increments of latitude and longitude for 1-degree blocks covering most of the United States. . . . Each 1-degree block contains 1,201 elevation profiles, with 1,201 samples in each profile. Elevations are in meters relative to mean sea level, and latitude and longitude use the 1972 World Geodetic System datum.[21]

Both DLGs and DEMs are available on the Internet for downloading at *www.usgs.gov*. Space does not allow a close examination of other digital products of the United States Geological Survey. Several are described in detail in books that specialize in digital (computer) mapping, such as Clarke's *Analytical and Computer Cartography.*[22] The interested student might want to pursue this very interesting book.

Scanned Topographic Maps and Air Photos

The United States Geological Survey (USGS) is providing scanned raster images of its 1:24,000, 1:25,000, and 1:63,500 topographic map sheets, along with the marginal information on each. These are called **DRGs** (digital raster images) and use .tif file extensions. These have been georeferenced for wide application. Scanned black-and-white, color, or color infrared images of aerial photos are also available from this agency, in JPEG format. These are called **DOQ** (digital orthophoto quads). These cover 3.75 minutes latitude and longitude and are also georeferenced, and can be combined (mosaicked) to cover a standard 7.5-minute quad. The **EROS** (The Earth Resources Observational Satellite) Data Center in North Dakota is also a widely used source for air photos, especially useful in

GIS for historical imagery. Geomatics Canada (on the Internet at *www.geocan.nrcan.gc.ca*) and the *National Atlas of Canada* (on the Internet at *www.nais.ccm.emr.ca*) are two sources for digital data and maps in Canada.

The Digital Chart of the World (VMap)

The United States National Imagery and Mapping Agency (NIMA) (in a multinational effort with Canada, Australia, and the United Kingdom) developed a digital chart of the world, called the VMap (Vector Map Level 0, formerly the DCW). This product is in vector format, and more than 250 operational and jet navigation charts covering most of the world were digitized for the project. These source maps are at the 1:1,000,000 scale, and contain elevation contours at 1,000-foot intervals, major transportation routes, railroads, canals, and populated places having outlines, coastlines, and international boundaries.

The files are to be available on CD-ROM when completed, and will be user-friendly. In addition to the primary use (defense systems), the data base is to be available for civilian uses, and can be employed by cartographers and users of geographic information systems to provide basic base-map locational data. Information about the DCW is available from the Defense Mapping Agency.

Private-Party Digital Base-Map Files

Numerous private companies provide their own digital base-map files that are useful for microcomputer thematic mapping. Some vendors have simply obtained the files from the government and market them in their mapping programs. Others use **proprietary files** and/or programs. These files are owned by companies or individuals and are copyrighted or not yet released to the public. These are to be compared with **public domain** documents, which are not protected by copyright, and therefore can be copied or used without fear of legal prosecution.

Some of the base-map files that are particularly useful are those of ZIP code boundaries. Several companies provide vector files of three-digit and five-digit ZIP code areas. These are most useful for marketing studies and display of marketing data. Many cartographers employ these digital base-map files. In some cases, local governments have done the digitizing and share these files with others, as they are in the public domain.

Compiling Thematic Layer Data

As the diagram in Figure 17.9 illustrates, the electronic or digital production of a thematic map requires the addition of thematic information to the base map, just as it does in manual production. The cartographer may wish to use remotely sensed data, statistical data, or both, depending on the map's purpose.

Image Data   It is possible to obtain remotely sensed data in digital form; the most easily obtainable are from **EOSAT** (Earth Observation Satellite Company), which manages the distribution of the images from the Landsat satellite launched by the United States (discussed in Chapter 4) or **SPOT** satellite images (for Systeme Probatorie d'Observation de la Terra). In the case of Landsat, for example, the images of a particular place on the earth may be obtained in digital form on floppy disk media, although these must be processed with appropriate software (usually at a site away from the microcomputer workstation) to make them usable. Photographic products may be purchased from EOSAT as well, and these can be sequentially scanned and thematic data added in the graphic image software at the cartographic workstation to compile a final thematic map.

Statistical Data   The cartographer will most often use statistical data for the compilation of digital thematic maps. These may or may not be brought to the task already in digital form. The cartographer may need to transform the data into digital format by entering them into a statistical or spreadsheet program from the computer's keyboard. This is called **data entry.** A **spreadsheet** is "any number of programs that arranges data and formulas in a matrix of cells."[23] Many of these programs are available for the microcomputer, and cartographers now have enormous power to manipulate numerical data for processing before mapping. Whenever possible, and if the mapping or graphics programs will accept the digital data, they may be transferred directly for final map compilation.

This method was employed by Professor Richard Pillsbury when he produced *American Cornucopia: An Atlas of American Agriculture* (Macmillan, 1995). He imported United States Census of Agriculture CD-ROM data directly into Map Viewer's (a product of Golden Software, Inc.) worksheet. He experimented with different maps until he achieved the ones he wanted for final display. This process saved many hours of labor and was an efficient way to handle a large number of maps.

Today, the cartographer often has the luxury of obtaining numerical data already in digital format. For example, many of the United States census products are either on floppy diskettes or CD-ROM, discussed more fully below. It is also possible to obtain the United States Census Bureau's *County and City Data Book,* as well as the popular *State and Metropolitan Area Data Book,* entirely on floppy diskette or compact disc. These data are in formats that can be transferred directly into some mapping programs. Compatibility among formats remains the only possible stumbling block, but this problem is largely being overcome.

CD-ROM Technology

Compact disc read-only memory, or **CD-ROM,** discs are the same storage devices that are now commonly found in the music store, except that data, in the form of words, symbols, boundary files, attribute data, and images, can now be stored on them for use in thematic mapping.

There is no doubt that, in the very near future, CD-ROM technology will be the storage device used most often by the cartographer. One compact disc can hold as much information as 1,600 floppy diskettes. Because compact discs are only storage devices (at reasonable cost levels, but these are falling), information can only be transferred from them into the computer workstation, and not written to them.

To use CD-ROM discs at the microcomputer workstation, the cartographer needs several items to make them usable. A 4.75-inch reader is necessary, as is a software program to operate the reader (such as Microsoft Extensions) in a compatible operating system. The data discs distributed by the Census Bureau use MS-DOS software at the present time, but are expected to become available for Apple, Inc., systems in the near future. The United States Census Bureau provides many of its census records on CD-ROM discs. (See Table 17.2.)

The Census Bureau CD-ROM files are in dBase III format (a spreadsheet format developed by the Ashton-Tate Company). Several other spreadsheet programs use the dBase file format, making compatibility easier. If a mapping program accepts this format, then the development of quantitative thematic maps can proceed without much delay from digital data input to image display manipulation to final map.

In addition to the Census Bureau, several companies now offer TIGER/Line boundary files on compact discs, and it is possible to obtain DLG files on these media as well. This provides the cartographer with base-map files, and with the statistical data in this format it is possible to more easily compile a thematic map completely in electronic or digital form.

This map was created entirely with Apple Macintosh computers and compatible software. Data were acquired from the Los Angeles office of the United States Bureau of the Census on a CD-ROM disc. The digital files manufactured in dBASE format were read and manipulated using FoxBASE+Mac. Spreadsheet compilation was accomplished with Microsoft Excel, and statistical analysis was undertaken with Strategic Mapping's Exstatix software. The map was produced using MapMaker software and geographic files were also provided by Strategic Mapping, Inc., of San Jose, California. Final map editing was done in Canvas 3.0, manufactured by Deneba Software Miami, Florida. Except for final reproduction, the process took less than a day.

Source: The above narrative appears as a credit at the bottom of a map titled "Non-Hispanic White Population in Metropolitan Los Angeles County, California 1990" published by the Center for Geographical Studies, Department of Geography, California State University, Northridge, CA 91330, Professor William Bowen.

Table 17.2   Abbreviated List of Census Products on CD-ROM

Subjects	Geography
1990 Census of Poupulation	
1. Total population; 18 years and older; Hispanic origin; by race; total housing units	States, counties, county subdivisions, census tracts/blocks, block numbering areas (BNAs), block groups (entire country on 10 discs)
2. Age, sex, race, Hispanic origin, marital status, other	States and their subdivisions down to block-group level (entire country on 17 discs)
3. Digital map data for census geography areas, basic map features (streets, rivers, railroads), names, address ranges, and ZIP codes	Entire country on 44 discs
1987 Economic Censuses	
1. Census of Retail Trade, Wholesale Trade, Service Industries, Manufacturers, Mineral Industries, others	All geographic areas, varies depending on census, including ZIP codes

Source: United States Census documents.

DESKTOP MAPPING AND GRAPHIC-IMAGE MANIPULATION

In the world of desktop mapping the cartographer usually sits before the display monitor and creates a map by using an **applications mapping** or graphics program suitable to the design task. These programs are like any other computer program, that is, they are written in computer languages that instruct the computer to perform certain operations. There are, of course, several on the market that have found widespread use in the cartographic arena area. Some are used strictly in Apple Macintosh applications (such as MacDraw), some strictly for IBM and compatible applications (such as Golden Software's Map Viewer). Others are really graphics or drawing programs that work in only one environment (such as Micrografx Designer for IBM), and some have versions for both environments (for example, Adobe Illustrator).

The chief advantage of programs that are purely mapping programs is that they use and process numerical data much as the cartographer would in the various mapping techniques. For example, in the Map Viewer program the cartographer either imports data or enters data that are referenced to geographic areas into a spreadsheet-like table, specifies a particular type map (choropleth,

proportional symbol, or prism), selects data classes and area symbols (if choroplethic), and sets circle size for the proportional symbol map. The program then performs the operations as specified by the selected commands and draws the map on the screen (and can then send it to an output device).

The drawing and graphics programs, on the other hand, do not offer these advantages. With these each map is built in a custom fashion with the cartographer doing all the data and numerical calculations before designing the map. However, the graphics and drawing programs offer more design flexibility in handling type and line specification and layout options, and patterns and colors are usually more generously available in these programs. In many respects these programs give the cartographer greater control over the design process than do those that are purely mapping programs (at least at the present time).

The individual needs and money resources of the cartographer will probably dictate the choice of the specific applications program or programs used. In many cartography production offices that do digital map compilation, it is not unlikely to see several programs installed on the computers.

GRAPHIC FILE FORMATS AND CONVERSIONS

The true flexibility in preparing digital cartographic products is embedded in file compatibility. That there is any problem at all is a legacy of the early days of microcomputer applications in creating graphics. A **file** is a collection of records that is handled as a basic unit of storage in the computer. The **format** of the file is the specific way that the information has been stored in the file, along with certain codes on how that information is to be read. In the very early days of microcomputer work, many manufacturers of programs established their own formats for their files, so compatibility was not common among the different programs. There is still a plethora of formats, but fortunately many programs now accept files with different formats. There are also **format conversion programs** that can change the format of one file into that of another, allowing a file to be used easily by different applications programs.

There are two fundamental file formats: raster-based, and vector-based.[24] **Raster formats** are also called bit-mapped or pixel file formats. These kinds of files contain descriptions of the On/Off state of each of the picture elements of the graphic image (and are synonymous with the pixels of the display monitor). The chief advantages of this type of format are that it is processed quickly and handles color easily. The primary disadvantages are that it produces jagged impressions of curved lines, consumes much storage space in the computer's memory, and is limited to the resolution technology that existed when it was created. Files with these formats are most often used by paint programs. (See Table 17.3.)

Table 17.3 Abbreviated List of Popular File Formats

Format	Name/Principal Application
Bit-Mapped Files	
BMP	MicroSoft Windows bit-map file
GIF	CompuServe Graphics interchange format
JPG	JPEG graphics format
MAC	MacPaint; used for various graphics programs
PCX	PC-Paintbrush graphics format
TGA	Targa; programs using video capture
TIFF	Tagged Image Format file; used by many graphics programs
Vector-Based Files	
CGM	Various DTP programs
BNA	Atlas GIS export file format
DRW	Micrografx native format file
DS4	Micrografx native file format
DXF	Data Exchange Format; developed by and used in AutoCAD
E00	ARC info export file format
GEM	Graphics Environment Manager; used by Digital Research Products, and Ventura Publisher
HPGL	Hewlett-Packard Graphics Language; used by many engineering application programs especially with plotter output
PLT	Plot; used in AutoCAD and many other drawing programs
Combination Bit-Mapped and Vector-Based Files	
EPS	Encapsulated PostScript; used by many DTP and graphics programs
PCT	Picture; used by most Macintosh graphics programs
Text Files	
TXT, AST, CSV	ASCII text files
DBF	dBASE
DIF	Document Interchange Format; used by spreadsheet and database programs
WKS	Used by Lotus 1-2-3
WP	WordPerfect; native format
Combined Text and Graphic Files	
PDF	Adobe text and graphic file

Source: Compiled from a variety of sources, but see Barrie Sosinsky, Beyond the Desktop: Tools and Technology for Computer Publishing, *Bantam ITC Series (New York: Bantam Books, 1991), pp. 309–402, for a more detailed list and a further discussion of file formats.*

The other common file format is the **vector format.** This is also called an object-oriented file. The files use mathematical instructions to identify the beginning and ending points (the vectors) of the graphic shapes in the image. The instructions also tell how the points are connected, that is, by lines or by normal or other curves. The advantages of vector-based files are that they take up far less storage space, represent curves without jagged impressions and, most important, the resolution is limited only to the output device—thus, as resolutions improve, the files are equally suitable. Disadvantages include difficulty of editing and a longer processing time. Many drafting, drawing, and engineering programs use files with these formats. (See Table 17.3.)

An important point is that a simple inspection of the CRT display will not reveal the file's format. Consider a black line drawn across a display. In bit-mapped format, the condition (On/Off) of each dot on the screen (even the white or nonblack ones) is stored in the file. In vector format, only the instructions for drawing the line are in the file.

A special file format is the TIFF (Tagged Image File Format), a pixel-based file that also contains gray values for each pixel, and can be used at many different resolutions. This file format has become virtually standard by scanning software, and files with this format can be used by both IBM and Apple applications programs. Another special file format is the **EPS** (Encapsulated PostScript) file. This kind of file is a blend of both bit-mapped and vector-based formats, and is used by the PostScript page-description language (described more fully below). The primary advantage of EPS is that, like the TIFF files, it can be used by both IBM and Apple applications. The only disadvantage is that a PostScript output device is required.

Most large applications programs accept graphic files in different formats.[25] Each applications program will list the kinds of formats that can be imported. It is general practice to name files with extension names that identify the format, especially if it has become common. For example, a TIFF file will have as its extension .tif. Some files have become so common that their extension names identify which program generated them. For example, files developed by AutoCad will have an extension called .dxf, and some other program manufacturers state that DXF can be imported directly.

The preparation of a quantitative map, of course, depends on other files as well. **Boundary files** (called spatial data files in GIS) contain the information used to create outlines (polygons) and are in the graphic file formats. **Data files** (called attribute files in GIS) on the other hand, contain the IDs associated with the specific map areas (boundary files), and the numerical data associated with them. Compatibility is a problem here, too, with many applications programs containing proprietary files. Some of the larger mapping programs are written to accept data files in different formats, which makes them more versatile. The United States Census Bureau, for example, provides data in a format that is used by a large database program (dBase III), and this format can be imported into some mapping programs. Again, file-conversion programs may be necessary to make the data files compatible with the mapping software. Many files come to the cartographer in *compressed* format. A computer software utility compresses a file to reduce its storage space or transmission time. These files can later be *expanded* before use.

CREATING THE MAP IMAGE

The thematic map image is created on the computer display, using either mapping or graphic software. The design activity, however, continues to include those subjects discussed in earlier chapters, and should adhere to the plan of the sketch map as presented in Chapter 16. In desktop mapping,

the plan requires attention to completely different forms of technology available for the final map product. In this regard, some thought must be paid to both computer software and hardware, as these are integrally related in desktop mapping. Probably the most important element for consideration is the adoption of a page description language (or PDL) and, because they are so closely tied, the output device.

Page Description Languages

A **page description language** may be defined as a "graphics programming language that allows you to place text, objects, and images on a page, and modify them through the use of various commands."[26] There are probably three such PDLs in common use: PostScript (a product of Adobe Systems, Inc.), Macintosh Quickdraw (a PDL from Apple Computer, Inc.), and those for the Hewlett-Packard laser printers, called PCL (these are products of the Hewlett-Packard Company); others emulate PostScript.

PostScript is perhaps the best known, and many graphics software programs, as well as many output devices, support this language. For example, a PostScript language file can be sent to laser printers (of various resolutions), imagesetters (described below), slide and film recorders, or any device that supports the language. One of the chief benefits of this language, in fact, is that it is device-independent. Our discussion here deals primarily with this language.

Any page-description language simply tells the printer (or other device) what will appear on the printed page, and how. A page-description language is one of the most important considerations in developing a desktop mapping workstation. Any page-description language can describe these things:[27]

1. Type description and modification
2. Objects (i.e., defining graphic primitives such as object fill, line widths, and positions)
3. Object groupings (i.e., placing objects in groups such as layers)
4. Color sets and gray values
5. Special operators (i.e., setting special lines, arcs, miter joints, and so forth)

PostScript   PostScript is a very advanced device-independent page-description graphics language. Since its development in 1985, it has become virtually a de facto standard throughout the desktop publishing world. PostScript, like other PDL's, is simply a set of commands written in a high-level computer language that instructs how a page—composed of both graphic marks and lettering—is to be printed. Actually PostScript tells how a raster image processor (RIP) is to be driven. In the chain of commands from the computer to output device (printer), the PDL instructs the RIP to convert mathematical formulas into bit maps that the laser graphics printer or image setter can interpret for printing. Although PostScript is widely used, there are other competitive PDL's as mentioned above, notably PCL used by Hewlett-Packard and TrueImage from Apple and Microsoft.

Output Devices

The types of output devices used in digital mapping have gone through many technological changes in the past 30 years. **Character-based printers** (similar to the standard typewriter) were used early, such as in SYMAP in the late 1960s. Today, these are seldom used for quality output in DTM.

Plotters are still used in many applications. A plotter is an output device that develops a map by automatically controlling pen movements over a piece of media (papers or films). Some large ones can drive laser cutters to etch scribe films. Plotters may be a flatbed or drum variety. Many small plotters create the image by having the head of the pen move in the X-direction, with the paper moving in the Y-direction. Most plotters are driven by software languages that contain special commands to drive the device (such as the Hewlett-Packard Graphics Language, or HPGL). Today many laser printers have interpreters in them that convert HPGL codes into ones that can drive the printer.

The standard today in output devices, especially in the DTM workstation, are the graphics laser printers or imagesetters. Although other output devices may be used, such as **ink-jet printers** (Figure 17.15), or **thermal transfer printers,** the ones of choice are the laser printers or imagesetters. Color laser printers are also available, although today they are used primarily for proofing rather than for final output.

Features common to all graphics printers are a page-description language and its interpreter, a raster image processor (or RIP), some kind of marking engine or imaging engine, and a paper-transport mechanism. A **marking engine** is the device (such as a laser) that transfers the digital commands of the image onto the output medium (paper or film).

Laser Printers Certainly one of the most important peripheral devices for digital cartographic production is the one that provides final **hard copy** maps. Hard copy refers to any printed copy of machine output in readable form. In

Figure 17.15 NEC SuperScript 750C ink-jet laser printer. This kind of printer is inexpensive and can be used in many DTM chores, especially for proofing final map designs. (*Illustration courtesy of NEC Technologies, Inc.*)

Figure 17.16 NEC SuperScript 860 black/white laser printer.
Many laser printers today offer medium resolutions (600 dpi) at reasonable prices. Laser printers at these resolutions may be used for final output if budgets do not allow more expensive image setters. (*Illustration courtesy of NEC Technologies, Inc.*)

the present context, this refers to a printed form of the map, and this can be either positive or negative. In the complete, digital cartographic workstation there will probably be at least a low-resolution laser printer (this is usually a black-and-white version). (See Figure 17.16.)

Relatively low-cost laser printers utilize xerography image-transfer technology. A rotating drum coated with photosensitive material is exposed to a light (laser) that is driven by computer control. Areas (dots) on the drum are either charged or not charged by the light source. An electronic pattern is thus composed on the drum. The turning drum picks up toner (plastic particles) of an opposite charge,

---

A PDL has a decisive influence in the quality of the output. A program that outputs a file in a PDL can be output on any device that can interpret the commands contained in the file; your computer doesn't have to issue the standard printer control codes of a character-based printer. It uses the PDL codes instead. A PDL interpreter, which holds the page description code, can be implemented in software running on top of the operating system, or it can be hard-wired in a ROM chip inside the output device. This is the case in printers called "PostScript," "TrueImage," or "PCL." Interpreters not built into printers and other output devices can be included as add-on boards or plug-in cartridges; these implementations are faster than incorporating the interpreter in software.

Source: Barrie Sosinsky, *Beyond the Desktop: Tools and Technology for Computer Publishing,* Bantam ITC Series (New York: Bantam Books, 1991), p. 177.

and the particles adhere to the drum at the dots exposed to the light source. These particles are then transferred to plain paper, heated by rollers, and permanently sealed to the paper. Clear acetate may also be used, and on very expensive printers film negatives can be specified as output.

Resolution, of course, determines the appearance of the final printed product (sharpness of lines or letters on the paper or film negative). **Resolution** of the pattern is specified in dots per inch (dpi) in both directions. A 300-dpi machine, therefore, can produce 300×300, or 90,000 dots per square inch. The cost of a laser printer is usually determined by its resolution. Low-cost machines are 300 dpi, very expensive ones exceed 2,000 dpi. However, most graphics or mapping software programs in use today support printers at both high and low resolutions, without requiring any changes in the CRT image.

Equally important to the cartographer is that many of the low-cost printers also contain graphics languages that pen plotters use, making software designed for plotters that much more useful.

Low-cost low-resolution laser printers may be used in two ways. They may be used to proof maps and the different layers thus incorporated into the design, and to check registration. Or they may be used quite successfully as final output at sizes that will require reduction of the image photographically. Figure 17.17 presents a map compiled in this fashion.

In this illustration, the base map was generated in the MicroCam software program (mentioned earlier), and imported into the Micrografx graphics and drawing program, where the final design was established. The various layers were printed on a 300-dpi laser printer, and registered in the process at sizes that are one-third up from the planned, final printed size. The output of the laser printer was photographed by the photolithographer, reduced to final size, and printed in the usual fashion. In this way the "raggedness" at the edges of letters and lines was virtually eliminated by the photoreduction process, and the final film negatives provided images that are essentially sharp and crisp.

Cartographers have begun earnest experimentation with peripheral devices and their output specifically for the use in mapping. For example, cartographers Jeff Leonard and Barbara Buttenfield have determined the pattern of dots on 300-dpi laser printers that produce an equal-value gray scale.[28] They provide a basic computer program that will generate the pattern of dots to give the different reflectances for approximate perceived gray scales. Any cartographer wishing to control laser output to this level of detail will find their work useful.

Imagesetters   Fundamentally, **imagesetters** are laser printers that have very high resolutions. Originally used only for typesetting operations, they are used today for setting entire pages of both text and graphics. (See Figure 17.18.) Resolutions range from 600 dpi to 2,540 dpi (although some go even higher). Various companies make these products. The Linotype Company (of Hauppauge, New York) has made them for so long that the term linos has almost become a generic term for these devices. The important quality of imagesetters for DTM is that most contain page-description languages (very often PostScript) that make their use very desirable for print-quality output directly from the DTM workstation. Imagesetters are very costly (usually in excess of $20,000), but newly developed DTP-size imagesetters are now available at prices that many small cartography shops may find economical to use.

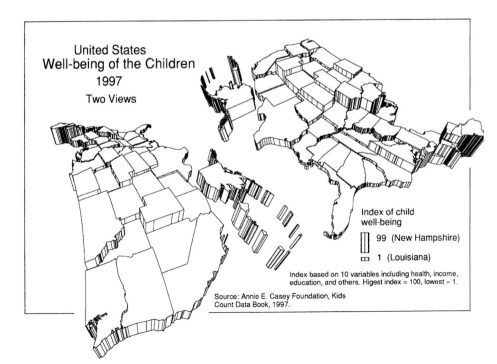

United States
Well-being of the Children
1997
Two Views

Index of child
well-being

99 (New Hampshire)

1 (Louisiana)

Index based on 10 variables including health, income, education, and others. Higest index = 100, lowest = 1.

Source: Annie E. Casey Foundation, Kids Count Data Book, 1997.

Figure 17.17 Digitally produced thematic map.
Attribute data were read into a program called *Map Viewer* (Copyright Golden Software), and the "prism" form of map was selected to view the data. Several views were created, and two were selected to represent the distribution. These views were imported into a graphic design program called *Designer* (Copyright Micrografx, Inc.) for final design work. Although many methods may be selected to print the map, this image was created by first printing it by a 300-dpi printer, scanning by high resolution scanner and importing it into an electronic page compositor. Other methods are available that require fewer steps.

Figure 17.18 A modern high-resolution imagesetter.
Although beyond the means of small DTM workstations, these output devices accept digital map images stored on floppy diskette, in a PDL, and either paper or plate-ready negatives can be processed. (*Photo courtesy of Linotype-Hell Company. Linotronics 230 and PostScript RIP40.*)

Table 17.4 Abbreviated List of Color-Capable Imagesetters at Local Printer Service Bureaus

Manufacturer	Model	Image Area (inches)	Resolution (dpi)	Line Screen (lpi)
Agfa	Accuset	13.6 × 60	3,000	175
DuPont	Crosfield	25 × 27	2,438	391
Linotype-Hell	Chromograph	21.3 × 25.6	4,876	300
Mannesmann	ColorStar	12.9 × 24	3,252	250
Monotype	ImageMaster	14	3,000	175
Optronics	ColorSetter	19.6 × 25	2,000	175
Scitex	Dolev	14 × 19.68	3,556	250
Varityper	ImageRecorder	13.3	2,400	150

Source: Taken and abbreviated from Thad McElroy, "At Your Service," Publish (May 1993), pp. 53–54.

A point has been made that an imagesetter is economical, especially if used in cartography production laboratories that have very high production levels.[29] If this is not possible, as in most cases, the cartographers in charge of production can take the map image, stored on diskette, to a service bureau for preparation of final film negatives. (See Table 17.4.) The cartographer should determine if the service bureau has an imagesetter that will satisfy the requirements of the project.

Thermal Wax and Dye Sublimation Printers

Thermal wax printers work on the principle of laying down bits of hot wax onto the paper surface. Moreover, these types of printers are used only when color is called for. Colored wax is first melted at high temperatures in the printer and then transferred to the paper one dot at a time. *Special paper must be used.* The chief disadvantages of thermal wax printers are their high relative cost and the waxy feel of the final product.

In map reproduction the impact of digital procedures is shattering, as the whole colour separation process is seeing revolutionary changes. Soon the days of the lithographer, of stripping films and of positive or negative procedures will be gone. The images produced will be either plotted and scanned or even transferred directly to printing plates. Much of our current knowledge on map reproduction will soon be as useful as that on copper engraving or on the processing of chalkstone from Solnhofen introduced by Senefelder not far from here in the early 1800s.

Source: Ferjan Ormeling, "Requirements for Computer-Assisted Map Design Courses," *International Yearbook of Cartography* 28 (1988): 265–73.

Dye sublimation printers use the same principle as the thermal wax, except that a transparent dye is used instead of wax. The heated dye is transferred to special paper. In this process each dot, when placed on the special paper, diffuses outward in a feathered way so that, overall, the individual dots appear to blend together. Color is intense and highly saturated. Perhaps because of this, dye sublimation printers are finding their way into prepress to develop color proofs. Neither of these printers produces output of high enough quality for use in producing final art.

CONCLUDING REMARKS

The purpose of this chapter has been to introduce some of the many and varied topics related to the digital production of thematic maps. It is impossible in the space provided here to treat all subjects in detail. Among the many good reference books available, the author encourages the student to read *Beyond the Desktop: Tools and Technology for Computer Publishing.*[30] In that book the thematic cartographer will find a much more thorough coverage of most of the important concepts and organization of the electronic desktop and, by extension, desktop mapping.

Several worthwhile research studies by cartographers dwell on the reading of computer display maps.[31] The results of those studies are important to the design of electronic display maps, but they are not central to the material of this book, and therefore should be found in other cartography texts.

Very important to this summary is that the student must know that electronic map production does not replace the creative design efforts of the cartographer. Logical, thoughtful, and aesthetic judgments are not made by the computer. With desktop mapping the cartographer has only a set of tools at his or her disposal. One cartographer put it this way: "The computer has removed many of the time-consuming processes in map generation and made information much more accessible to researchers, educators, and casual users. However, that does not mean that maps will be better. As always, care will be needed in evaluating them because computers will generate mistakes as quickly as they generate maps."[32]

NOTES

1. Mark Mattson, "Imagesetting in Desktop Mapping," *Cartographic Perspectives* 6 (Summer 1990–91): 13–22.
2. James R. Carter, *Computer Mapping: Progress in the 80s* (Resource Publications in Geography—Washington, DC: Association of American Geographers, 1984).
3. There are two books in computer mapping recommended for a greater in-depth look at the subject: Keith C. Clarke, *Analytical and Computer Cartography* (Englewood Cliffs, NJ: Prentice Hall, 1990); and Mark S. Monmonier, *Computer-Assisted Cartography: Principles and Prospects* (Englewood Cliffs, NJ: Prentice Hall, 1982).
4. Clarke, *Analytical and Computer Cartography,* Chapter 1.
5. J. R. Hunt, "Desktop Mapping in the United Kingdom, 1987–1991: Personal Observations with Special Reference to UK Higher Education," *Cartographic Journal* 28 (1991): 101–4.
6. Clarke, *Analytical and Computer Cartography,* p. 9.
7. *Webster's New World Dictionary of Computer Terms,* 3d. ed. (New York: Prentice Hall, 1988), p. 45.
8. Mattson, "Imagesetting," p. 13.
9. *Webster's New World Dictionary of Computer Terms,* p. 45.
10. Barrie Sosinsky, *Beyond the Desktop: Tools and Technology for Computer Publishing,* Bantam ITC Series (New York: Bantam Books, 1991), p. 31.
11. Harold H. Moellering, "Strategies for Real Time Cartography," *Cartographic Journal* 17 (1980): 12–15.
12. There are many books and articles that describe in detail the CRT display device. This one is recommended for a good introductory treatment: David F. Rogers, *Procedural Elements for Computer Graphics* (New York: McGraw Hill, 1985); see also A. Brown, "Map Design for Screen Display," Cartographic Journal 30 (1993), pp. 129–35.
13. Steven J. Harrington and Robert R. Buckley, *Interpress, The Source Book* (New York: Brady, 1980).
14. Ibid.
15. Herbert Freeman and Goffredo G. Pieroni, *Map Data Processing* (New York: Academic Press, 1980), pp. 4–7.
16. R. W. Anson, ed. *Basic Cartography.* International Cartographic Association (London: Elsevier, 1988), p. 115.
17. Clarke, *Analytical and Computer Cartography,* p. 104.
18. Ibid., p. 103.
19. Refer to documentation for the MicroCAM Microcomputer Automated Mapping Program from MicroConsultants, Mililani, HI, 1991.
20. Robert W. Marx, "The TIGER System, Yesterday, Today, and Tomorrow," *Cartography and Geographic Information Systems* 17 (190): 89–97.
21. Clarke, *Analytical and Computer Cartography,* p. 78.
22. Ibid.
23. *Webster's New World Dictionary of Computer Terms,* p. 357.
24. There are several references that treat the subject of file formats, but especially useful are Sosinsky, *Beyond the Desktop,* various pages; and Steve Rimmer, *The Graphic File Toolkit* (Boston: Addison-Wesley, 1992), all. See also Mark Snoswell and Henry Thomas, "Image File Formats," *Design Graphics,* December 1993, pp. 4–15; and "The Scope of Scanners," *ITC Desktop,* September/October 1989, pp. 42–46.
25. A very interesting article dealing with file formats for the popular CorelDRAW graphics program and the production of maps is Brian Morber and Janet E. Mersey, "Thematic Mapping with Illustration Software: Unraveling the Mystery of Graphic File Formats," *Cartographic Perspectives* 18 (1994), pp. 3–16.

26. Sosinsky, *Beyond the Desktop,* p. 47.
27. Ibid., p. 176.
28. Jeff J. Leonard and Barbara P. Buttenfield, "An Equal Value Gray Scale for Laser Printer Mapping," *American Cartographer* 16 (1989), pp. 97–108.
29. Mattson, "Imagesetting," pp. 13–22.
30. Sosinsky, *Beyond the Desktop,* pp. 346–50.
31. Several research studies dealing with CRT map reading and design include these: Patricia Gilmartin and Elisabeth Shelton, "Choropleth Maps on Higher Resolution CRTs: The Effects of Numbers of Classes and Hue on Communication," *Cartographica* 26 (1989): 40–52; Mark P. Kumber and Richard E. Groop, "Continuous-Tone Mapping of Smooth Surfaces," *Cartography and Geographic Information Systems* 17 (1990): 279–89; Katherine Mak and Michael R. C. Coulson, "Map-Use Response to Computer-Generated Choropleth Maps: Comparative Experiments in Classification and Symbolization," *Cartographic and Geographic Information Systems* 18 (1991): 109–24; Mathew McGranaghan, "Ordering Choropleth Map Symbols: The Effects of Background," *American Cartographer* 16 (1989): 279–85; and Terry A. Slocum, Susan H. Robeson, and Stephen L. Egbert, "Traditional versus Sequenced Choropleth Maps: An Experimental Investigation," *Cartographica* 27 (1990): 67–88.
32. Eugene Turner, "Desktop Mapping Comes of Age," *The American Geographical Society Newsletter* 12, no. 2 (1992), pp. 1–3.

GLOSSARY

applications mapping program   computer program written in a language that instructs the computer to perform certain mapping operations, p. 347

automated mapping   term often used when computers are employed in map development, p. 333

boundary file   computer file that contains information about outlines (polygon shapes) of geographic areas, p. 349

CD-ROM   read-only storage device containing digital information that is read by laser technology, p. 346

central processing unit   also called the CPU; miniaturized electronic component that controls interpretation and execution of instructions, p. 335

character-based printer   computer printer that produces type similar to a typewriter, p. 350

computer-assisted cartography   term often used when computers are employed in the development of maps, p. 333

computer mapping   like computer-assisted cartography and automated mapping, a term often used when computers are employed in the development of maps, p. 333

CRT   cathode-ray tube; type of computer display device, p. 336

cursor   as used here, a handheld device that senses electronic impulses over a digitizer surface to store an image (map) into digital form, p. 341

data entry   entering data into a computer at a keyboard, p. 346

data file   as used in mapping, IDs and attribute data associated with specific geographic areas, p. 349

DEM   refers to digital elevation model; digital file of topographic surface that contains latitude, longitude, and elevation at points, p. 345

desktop mapping   also called DTM; all computer operations necessary to digitally produce a map's component parts; branch of desktop publishing, p. 333

digitizer   peripheral device that converts an analog image into digital format, pp. 340–341

digitizing   the process of using a digitizer to convert an analog image into digital format, p. 340

DIME   refers to Duel Independent Map Encoding georeferencing system; developed by the U.S. Census Bureau for the 1970 census and still used by some agencies, p. 343

DLG   also called digital line graph; digital records of some information from the USGS 7½- and 15-minute topographic quad sheets, p. 345

DRG   digital raster image; digital images of USGS topographic maps, p. 345

DOQ   digital orthophoto quads, p. 345

dye sublimation printers   printer technology that lays down spots of transparent dyes onto paper to create an image, p. 353

EOSAT   quasi-public company that manages, markets, and distributes digital images from the Earth Observation Satellite, pp. 345–346

EPS   encapsulated PostScript file that is a bit-mapped and vector-based blend used in the PostScript PDL, p. 349

EROS   Earth Resources Observational Satellite; source of earth imagery, p. 345

file   a collection of related records that is handled as a basic unit in computer storage, p. 347

format   as in file; specific way information is stored and usually contains codes on how it is to be read, p. 347

format conversion program   program that changes the format of one file into another, p. 347

graphical user interface (GUI)   software part of a computer platform whose purpose is to provide the interface between the computer and the user; uses icons and pointing devices for user control, p. 336

hard copy   printed copy of machine output that is readable in form, p. 350

HSB color system   hue, saturation, and brightness color model; one way to specify color in many color graphics programs, p. 340

image editor   software that can alter the appearance of a scanned (bit-mapped) image, p. 343

imagesetter high-resolution output device capable of handling files in PDLs; can output images on papers and films, pp. 334, 351

ink-jet printer low-resolution printer device that creates an image by forcing ink through small jets onto paper and films, p. 350

interface program on the computer that has the main function of interacting with the user, p. 336

interlaced CRT scanning method that traces every other line on the screen, then goes back and traces the missed lines; as contrasted to noninterlaced, a better method producing less flicker on the screen, p. 338

Macintosh system term generally applied to the graphic interfaces used by the Macintosh family of microcomputers, p. 335

marking engine device that digitally controls the creation of the final image to the output medium (for example, a laser printer), p. 350

microcomputer workstation configuration of computer equipment designed for use by one person at a time, p. 334

MS-DOS refers to Microsoft Disk Operating System; operating system used by the IBM family of computers, p. 335

operating system program of instructions that controls and supervises all other programs running in the computer, p. 335

original map data source of the digital map, p. 340

page description language graphic programming language that places text and graphic objects on a page, p. 349

peripherals associated equipment coupled to the computer (such as printers, plotters, scanners, and others), p. 334

picture element smallest resolvable element on the CRT display; also called pixel or pel, p. 337

pitch spacing of small holes on the shadow mask in the CRT display device, p. 336

platform computer, its operating system, and interface, p. 335

plotter device that develops an image (map) by controlling pen movement over the output medium (paper or films), p. 350

proprietary file privately owned file (may be copyrighted); not in the public domain, p. 345

public domain file files not protected by copyright laws, p. 345

raster format also called a bit-mapped or pixel formatted file; contains descriptions (On/Off and possibly brilliance) of each pixel of the graphic image, p. 347

raster scanner device that optically scans source materials and converts them to digital format, pp. 340

raster scanning as in CRT device, method of creating the image on the display, pp. 336, 340

resolution as in a laser printer, the dots per inch or dpi, in both directions; a 300-dpi printer can place 90,000 dots in each square inch of the image area, p. 351

RGB color model red, green, blue color dots on the color display device, and the intensity of each, determines the colors we see; one way to specify color in many graphics programs, p. 338

RGB color monitor popular color-display device used today at the microcomputer workstation; colors are generated on the display by exciting red, green, and blue phosphor dots, p. 336

scanner see **raster scanner**, pp. 340, 342

shadow mask screen device placed inside a CRT display containing small circular holes that control how the rays strike the screen; some CRTs use small rectangles instead of round holes, p. 336

SPOT refers to System Probatorie d'Observation de la Terra; a European consortium providing and marketing satellite images, p. 346

spreadsheet a computer program that arranges data and formulas in a matrix of cells; useful in business and for organizing large amounts of numerical data, p. 346

thermal transfer printer a nonimpact printer that produces an image onto heat-sensitive media, p. 350

thermal wax printer printer technology that first heats wax particles at high temperatures and then lays them down onto paper to achieve an image, pp. 350, 352

TIFF refers to Tagged Image File Format; pixel-based format containing gray-value information for each pixel; popular format, especially for raster-scanned images, p. 348

TIGER refers to Topologically Integrated Geographic Encoding and Referencing system; georeferencing system developed by the U.S. Bureau of the Census for the 1990 census; contains information by points, lines, and areas, p. 344

vector format also referred to as object-oriented format; contains mathematical instructions for creating graphic images, p. 348

virtual maps digital preliminary maps, not yet completed or permanent, p. 336

READINGS FOR FURTHER UNDERSTANDING

Carter, James R. *Computer Mapping: Progress in the 80s.* Resource Publications in Geography. Washington, DC: Association of American Geographers, 1984.

Clarke, Keith C. *Analytical and Computer Cartography.* Englewood Cliffs, NJ: Prentice Hall, 1990.

Mattson, Mark. "Imagesetting in Desktop Mapping." *Cartographic Perspectives* 6 (Summer 1990–91): 13–22.

Monmonier, Mark S. *Computer-Assisted Cartography: Principles and Prospects.* Englewood Cliffs, NJ: Prentice Hall, 1982.

Sosinsky, Barrie. *Beyond the Desktop: Tools and Technology for Computer Publishing.* Bantam ITC Series. New York: Bantam Books, 1991.

PART V

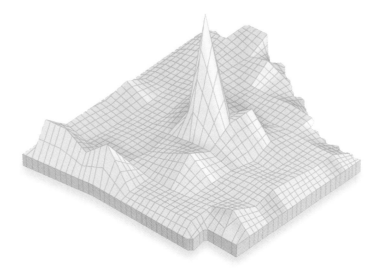

EFFECTIVE GRAPHING FOR CARTOGRAPHERS

The last part in this book contains only one chapter—a chapter that focuses on the development of effective graphs by the cartographer. Because many cartographers in practice are frequently called on to produce graphs, a section on the chief design principles is included, as well as numerous examples. Interestingly, the figure-ground relationship plays an important part in the overall design approach, just as it did in map design. Inasmuch as probability, statistical, and loga-rithmic graphs are so important in the geographical sciences, these are dealt with in some detail. Graphing three-dimensional data sets is likewise included because these data sets are often critical to information display. As in cartography, the grapher needs to develop a spirit of experimentation so that truly creative solutions result. Finally, it is imperative that cartographers adopt ethical standards in graphing just as they do in making maps.

CHAPTER

18

EFFECTIVE GRAPHING FOR CARTOGRAPHERS

CHAPTER PREVIEW

Cartographers often are asked to design effective graphics, especially for books such as atlases and other comprehensive works. There are certain guiding principles of design that can help. Some of these have come from a rather long history of graph design and some come from more recent studies of graph perception tasks. Although several terms are used for the graph, such as charts and diagrams, this chapter uses graphs whenever the display has a mathematical basis. The graphic display of frequency distributions, ogives, and Box-Whisker graphs are common types of graphs and very helpful adjuncts to the map or multimap page display. Showing how much there is of something is a basic display purpose, and the cartog- *rapher frequently utilizes such common forms of display as bar and sector graphs. Trilinear graphs, those that show three components, can also be used. Cartographers often find trend graphs useful, especially the logarithmic plot graph. Logarithmic plots are most useful when one wants to portray rates of change, the fundamental use of these common graphs. The display of multivariate data is very useful today, especially as more and more geographical studies incorporate such forms of analysis. Cartographers need to develop a spirit of exploration, experimentation, and inquiry when designing graphs, as these forms of communication frequently are as important as other methods of communicating information.*

The purpose of this chapter is to provide a brief look at graphing, paying particular attention to several common graphs and design suggestions for their construction. Whole books have been written about graphing, so a complete treatment is not warranted here, and there is not sufficient space to do so.[1] However, as cartographers are often called to render graphs along with their maps, especially in publications such as atlases and other comprehensive works, some guiding principles are useful. The graph types presented here are ones that I consider most useful to many graphing tasks. At the end of the chapter I provide some that are more unique (and innovative), and call for more experimentation in the exciting world of graphing.

At the very beginning, it may be useful to point out what is meant by the word *graph,* at least as it is used in this chapter. There appears to be no formal consensus in the application of the word graph, as both the words chart and diagram are used often in its place. Here I use the term **graph** when there is a mathematical basis for its construction. I prefer to use the terms charts and diagrams for those drawings, pictograms, and organization-based drawings for which no mathematical relations exist between the elements of the drawing. So, then, only graphs that have a mathematical structure, or more specifically a mathematical *grid,* to them are included here.

The underpinning philosophical basis for this chapter rests with two fundamental ideas. First is the idea of graphicacy.[2] **Graphicacy** is a term first used by Balchin and Coleman several years ago referring to the intellectual skill necessary for effective communication of relationships that cannot be communicated well with words or mathematics. It should be pointed out that these authors were discussing graphicacy in the context of communication with other modalities of *literacy, numeracy,* and *articulacy.* To them graphicacy does not stand alone, but works in conjunction with the other forms of communication. Graphicacy also is best suited to show spatial relationship ideas. It is with the idea that graphicacy can share as an equal partner in the communication of ideas that this chapter rests.

Another philosophical foundation upon which I base this chapter is that graphic communication is a part of human communication and is composed of four components: (1) the communicator (the one "sending"), (2) the interpreter (the reader), (3) the communication content (the message), and (4) the communication situation (a perceived need to communicate).[3] As with maps, graphs must have a purpose, and that purpose is to affect change (learning) in some way in the reader. If the desired response is met, then the communication has been successful. I hope that these two foundational pillars support the observations I make in this chapter.

BRIEF HISTORY OF GRAPHING

Contrary to what is thought ordinarily, graphing in some form has had a long history. This history can be summed

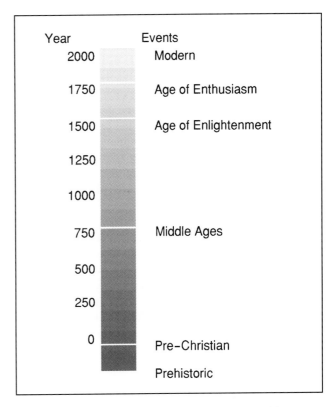

Figure 18.1   Major events in the history of graphing.

into a time line chart showing the main features of several broad periods. (See Figure 18.1.) Although just a thumbnail sketch, this brief introduction will serve to illustrate that graphing, in all forms, runs pretty much parallel to other events and highlights of human history.

During prehistoric times in the western world, cave drawings that exemplified the preoccupation of early humans with hunting and survival were made. Pre-Christian classical times brought about the first attempts to *grid* things. Early attempts at latitude and longitude (the earth's grid) came into being during this time, as did geometry. During the Middle Ages planetary graphs, with time as a variable, came into being. The Age of Enlightenment saw advances made in graphing along with other great achievements. Censuses, analytic geometry with its structure around the grid, and early statistical graphs came on the scene in western Europe. These graphic trends continued into the 1800s—the Age of Enthusiasm (with graphs at least)—and were highlighted by several European statistical conferences. Early nineteenth century saw the first textbooks on graphing, as well as attempts at standards, and statistical mapping. The time line brings us up to date with the "modern" era, a time when graphing can be done effortlessly on microcomputers. A fuller appreciation of graphing and its innovative history comes on the scene with books such as Edward Tufte's *The Visual Display of Quantitative Information, Envisioning Information,* and *Visual Explanation.*

ORGANIZATION OF THIS CHAPTER

This chapter is organized to provide discussions around those graph types that you will most likely be called on to produce. At the very outset, a discussion of general perception tasks when reading graphs and design approaches is given to form a foundation for the remainder of the chapter. Typical graphs include those that show features of statistical distributions, graphs that show how much of something, that show proportions, and that show trends in data arrays; these also include the very typical logarithmic graph. Also, you will often be called on to show how two or more variables are related, with typical examples provided here. In many ways, graphing is an exciting part of graphic design, often calling for new solutions that require you to exercise a great deal of creativity. Although certain standards do exist for standard data arrays, each graph is almost always a new and different challenge. At the very end of the chapter, some unique solutions offered by geographers and others are discussed, and some interesting examples are shown.

PRINCIPLES OF DESIGN APPLIED TO GRAPHS

This brief section of the chapter looks first at the perceptual tasks that graph readers apply when reading graphs, followed by an examination of typical graph components, then followed by a design approach used in map design—planning a graph around a visual hierarchy. It may seem remarkable that this approach can work with graphs, too, but as it relies on sound perceptual principles and rules, it *should* be used with graph design.

HUMAN PERCEPTION AND READING GRAPHS

Just as there are a number of perceptual tasks identified that must be undertaken when reading maps, there are certain perceptual tasks required when reading graphs. The work of William Cleveland can be called on here, as he has assembled and discussed a large number of these tasks.[4] This section of the chapter will present an overview of those most often encountered in graph reading.

Cleveland researched the literature of perceptual tasks and then matched them to those ordinary graphs also found in the literature. The tasks he identified are these (to which I have added the typical graphs that usually apply):

1. Detection (a primitive, applied to all charts)
2. Angle (sector graphs)
3. Area (histograms, triple scatterplots)
4. Color hue (any graph)
5. Color saturation (any graph)
6. Density (scatterplots)
7. Length (distance) (bar graphs)
8. Position along a common scale (dot graphs)
9. Position along identical, nonaligned scales (comparative dot graphs)
10. Slope (rate of change along a data path)
11. Volume (three-dimensional-appearing symbol)

Figure 18.2 presents simple drawings to illustrate these tasks.

According to Cleveland's own research findings, he arrayed the tasks from simple to more complex in this way, and from most accurate to least accurate:

1. Position along a common scale
2. Position along identical, nonaligned scales
3. Length
4. Angle—slope
5. Area
6. Volume
7. Color hue—color saturation—density

What is apparent here is the agreement among cartographic researchers with many of the findings. For example, it has long been known that volume is more difficult to perceive accurately than length or area.[5] Position, as it is a primitive, is least troublesome, and color historically has been known to give perceptual problems.

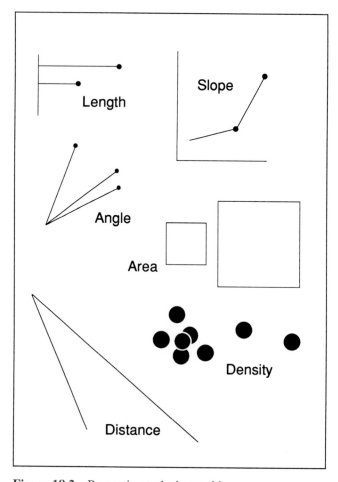

Figure 18.2   Perception tasks in graphing.
These drawings illustrate the principal perception tasks that are required in graph reading.

The order just presented is provided here to caution the graph maker. Certain graphs by their very nature are difficult graphs to read. When coupled with very difficult and complex data, and if not designed incorporating the very best graphic principles, communication failure, or at least unnecessary reader difficulty, results. Figure 18.3 presents two such graphs—one that is difficult for the reasons just mentioned, and one that reflects easier tasks and better design principles.

ELEMENTS OF THE STANDARD GRAPH

It is important in an introduction to graphing that some attention be paid to the typical elements of the standard graph. There are no set uniform labels attached to the elements, so

> There is no single statistical tool that is as powerful as a well-chosen graph. Our eye-brain system is the most sophisticated information processor ever developed, and through graphical displays we can put this system to good use to obtain deep insight into the structure of data.
>
> Source: John M. Chambers, William S. Cleveland, Beat Kleiner, and Paul A. Tukey, *Graphical Methods for Data Analysis* (Boston: Duxbury, 1983), p. 1.

you may encounter variations in the literature. Nevertheless, a brief discussion of these elements will be beneficial.

In the typical graph there is a **data region,** that major part or area of the graph in which the graphic treatment of the data falls. (See Figure 18.4.) The data region is customarily bounded by the **scale lines,** those rulers along which we graph the data. Often the rulers are not included at the top or right of the data region if they are not needed for reader clarity. The scale to the left or right of the graph is called the **vertical scale,** and the scale along the bottom or top is called the **horizontal scale.** Scale lines must have **scale labels,** clearly identifying what the scale is and its units of measure.

In the data region is the graph's grid. The grid may or may not be included, depending on the graph, but its *influence* is always present. The grid is intimately tied to the scale lines at the margins of the data region, and it is upon the grid that the data are plotted. The grid, whether shown or not, should be easily detected by the reader.

A data path may be in the data region, depending on the graph. The **data path** connects data symbols in the graph to show trends. Within the data region there may be a **reference line,** a line identifying a certain place in the data region that is significant to the data presentation (for example, a mean line, or minimal-value line, and so forth). There may also be certain **markers** along either the vertical or horizontal scale. Markers indicate places in the data region, next to a scale, that identify a particular event of importance in the display of this data.

Many graphs require a **key,** or **legend,** which describes any part of the data presentation not immediately clear to the reader—for example, labels identifying the various data in the graph. Labels that mark the symbols on the graph are called **data labels.** These data labels may be in the key or alongside the data in the graph.

Titles are a necessary part of every graph. For stand-alone graphs, the title is usually integral to the graph, typically placed somewhere in the data region but not where it interferes with the graph's message. For graphs that are part of narrative materials (such as in textbooks), the figure caption usually contains the graph title. Titles need to be unambiguous, efficient, and economical. *Titles need as much thought devoted to their design as to the graph itself.*

One last element of the typical graph is the **data source.** This part of the graph is often ignored, but should not be. It is often placed within the data region, or can be found outside the scale lines. Unless the source is implied in the title, it is good practice to include it as part of the graph.

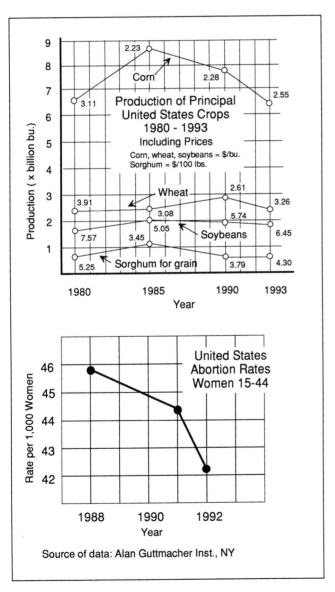

Figure 18.3 Some graphs are easy to read and some are not. Both graph content and graph design play an important part in the overall appearance of graphs. The upper graph suffers from overly complex data and confusing design. The lower graph is unclear, but for opposite reasons.

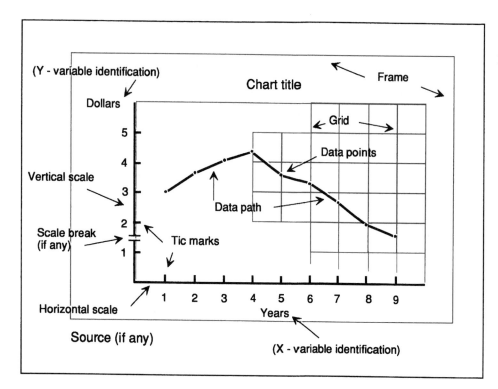

GRAPH PLANNING AND THE VISUAL HIERARCHY

The same principles of design that were the focus of Chapters 12 and 13 are applied to graph design. Two chief design strategies are *contrast* and the *visual hierarchy*. If these are properly integrated into the design, the graph will function successfully.

As with any thematic map, each graph should have a *purpose,* and the type of graph selected must be appropriate for the data it presents. Trend data should not be graphed, for example, with bar graphs. Dot graphs are not the best for showing component parts. Rate of change is best graphed on logarithmic graphs, not linear graphs. Graphs must not be used when a simple table will suffice. And, most emphatically, *graphs should not be applied to maps.* Examples include placing bar graphs or small trend graphs within enumeration units in the map; another is using sector (pie) graphs within similar units. Each graphing project needs to have the designer ask basic questions about its purpose and appropriateness.

Contrast is perhaps the best design tool to use in any graphing design activity. As with maps, there can be contrast of line (weight and character), texture, value (light and dark), detail, lettering, and color. The two that have the most significance are line and texture (and color, too, if color is used for the graph).

Figure 18.5a illustrates a typical graph with little line and texture contrast. This is a rather dull-looking graph, especially when compared to the one in Figure 18.5b, which is identical except for its contrast. With contrast, the image is not as bland, and appears more exciting, unique, and interesting. Also, by using contrast the designer is able

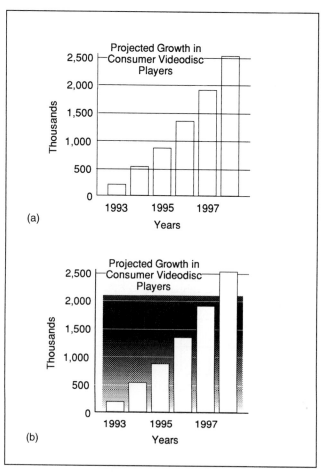

Figure 18.5   Contrast is used in effective graph design. Very little contrast has been employed in (a), but is used effectively in (b). (*Redrawn with changes from* Educator's Tech Exchange, *Winter, 1994, p. 9.*)

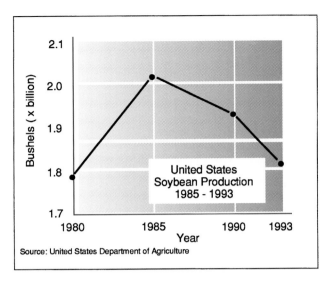

Figure 18.6   The visual hierarchy in graph design.
Data symbols, axis lines, labels, and other key components of the graph are portrayed as strong figures on the graph, with grids and other elements subdued.

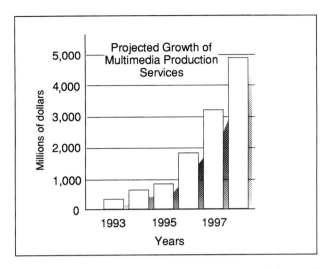

Figure 18.7   Interposition and line screening help in graph design.
See text for explanation. (*Redrawn with changes from* Educator's Tech Exchange, *Winter 1994, p.10.*)

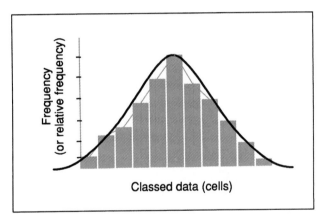

Figure 18.8   A typical histogram with both a frequency polygon and a curve applied.
Histograms are typically used for discrete data, and polygons and smoothed curves for continuous data. Relative frequency is on the vertical axis and data values are arrayed on the horizontal.

to direct the reader's attention to certain parts of the graph, which serves to "scale" the importance of the different parts. Always use the element of contrast when designing graphs.

Data symbols, data paths, titles, and scale rulers are all very important to the graph and should be rendered as figures in the graph's visual hierarchy. Always important—but not needing the visual strength of the aforementioned—the graph's *grid* may be rendered in a subdued fashion. The data symbols and data path should appear to rest *on top* of the grid, and the title should be on top of the grid or other components of the graph. (See Figure 18.6.) Coupled with contrast, developing a visual hierarchy for the graph calls for special attention to design detail, but will *always* yield a better-looking graph. Such a graph will have its data and design well integrated, as they should be.

Screening the grid lines and interposition may be used to effect a desired visual hierarchy, in much the same way that they were used in map design. These two design methods should always be attempted in a graph design activity. (See Figure 18.7.)

THE GRAPHIC DISPLAY OF NUMBERS AND FREQUENCY DISTRIBUTIONS

One of the first displays of numerical data is the simple array of the values in the data set. Often it is adequate to display the numbers in a table of some sort, but when this is not enough, a graphic method may be the solution. Statisticians often use this method and it is a subject frequently dealt with early on in statistics texts. Histograms, frequency distributions, ogives, and normal probability graphs are discussed here because they are fairly common and quite useful.

HISTOGRAMS

Histograms are graphic records of the occurrence of data values in a statistical distribution. These graphs are developed after a researcher has already collected the data—their purpose is to represent the statistical distribution better than can be shown in a table of values. There are certain customs to follow when these graphs are constructed. The observations that follow point out some major rules.

The horizontal scale of the histogram is divided into equal-width segments (classes), each of which represents a range of values and is repeated enough times to accommodate the entire range in the data distribution. (See Figure 18.8.) For reading the histogram, it is especially critical that the bars be of equal width and be drawn contiguous to each other. The vertical scale is marked off to represent either frequencies of occurrence or relative frequencies of occurrence. Relative frequencies are preferred by statisticians. Inside the data region of the graph, bars are erected over each segment on the horizontal scale and their heights scaled to the relative frequencies found in each bar's data range. It is important to understand that the data are represented by *equal units of area.*

It is a good idea to think of area when pondering a histogram. If the width of each bar is considered to be one unit in width, and the heights of the bars are erected proportional to relative frequency, then the total area under all the bars will equal "one." This is a good idea to keep in mind when considering probabilities associated with the normal frequency curve.

The distribution of values found in Table 18.1 was used to construct the graph of Figure 18.9.

One reason for including a discussion here of the typical histogram is that cartographers should begin to display these graphs alongside any map made from the same data. This allows the map reader an opportunity to not only look at a spatial distribution of the data, but also see a nonspatial array of the data. Histograms do this—they permit an easy inspection of how the data values are arrayed along a number line. They allow the map reader to gather a rather detailed picture of overall data characteristics.

Frequency Polygons and Frequency Distributions

If the centers of the tops of the bars in a histogram are joined by straight lines, a **frequency polygon** results. (See Figure 18.8.) If the number of observations in the array of values is *very large* and the class intervals very small, yielding *very small bar widths,* then the straight lines connecting the tops of the bars would appear *smoothed,* giving us a graph called a **frequency curve.** The frequency curve is very special to the statistician because it too serves to enclose an area that accounts for all the values in the data distribution. The area under a frequency polygon and a frequency curve is equal to "one." You may recall from a previous chapter that it is possible to display such distribution characteristics as means and standard deviations directly on the frequency curve.

Again, it is important to remember that when thematic maps contain classed data, it is a good idea to illustrate these classes and the class boundaries on the frequency curve as well as in the map legend. This also provides more information to the map reader about the relationship between the statistical distribution and the map classes.

The data providing the frequency polygon may be reordered in ascending order and, if so, the polygon is called a **Lorenz curve.** These are used in some statistical applications.

Ogives and Probability Graphs

An extension of the relative frequency histogram is the cumulative frequency distribution graph, or cumulative per-

Table 18.1 Random Numbers Frequency Distribution

	Mean = 2.0, $S = 1.0$	
Frequency Class	Frequency	Cumulative Frequency
0–.403	6	.06
.403–.806	5	.11
.806–1.20	9	.20
1.20–1.612	15	.35
1.612–2.015	16	.61
2.015–2.418	9	.70
2.418–2.821	10	.80
2.821–3.224	9	.89
3.224–3.627	8	.97
3.627–4.030	3	1.00

centage graph, more commonly called an **ogive** (pronounced o'-jive). Supposedly this graph was called this by Sir Francis Galton (1822–1911) because of its resemblance to the curved rib of a Gothic vault.[6] The graph is constructed with the horizontal axis ruler laid out to show the upper limits of each class boundary arrayed ascending to the right, and the vertical axis ruler marked to show the cumulative percentage of observations in each class interval. A line graph is then constructed by placing data points at each respective place on the graph along with the cumulative percentage at each point, using straight lines to connect the points. (See Figure 18.10.)

At any point on the graph the relative frequency (percent) for each upper limit of a class interval can show the percent of all observations falling *below* that interval boundary value. It is therefore a quick operation to determine reasonable approximations of the percent of cases falling below a particular observation. **Quantiles,** or the value below which a given percent of the observations falls, can be obtained from the ogive. Certainly one of the uses of the ogive is to see quickly how the data are behaving—for example, is the cumulative percentage going up or down? Ogives are interesting and useful graphs for displaying frequency distributions.

A direct extension of the ogive is the *probability graph.* Here, the ogive is reversed so that the vertical axis becomes the ruler for the variable, and the horizontal axis the cumulative percentage ruler. This causes the curve to be reversed, as well. (See Figure 18.11.) If there were some way to scale the horizontal ruler so that the cumulative frequency curve of a *normal distribution* on the reversed ogive was straight, then the graph could be used to estimate whether other data plotted on the graph were normally distributed. Inasmuch as normality is key to the use of many statistical tests, such a graph would be very useful.

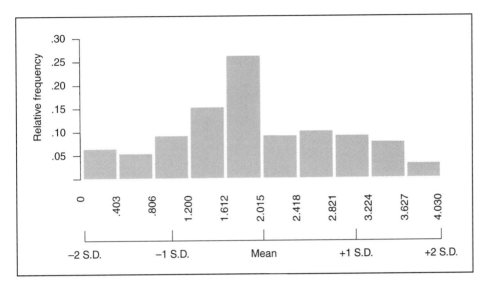

Figure 18.9 A typical frequency polygon.

The graph is drawn from the data shown in Figure 18.8. For a normal frequency distribution, areas under the curve are known, and standard deviations such as the ones shown are commonly displayed.

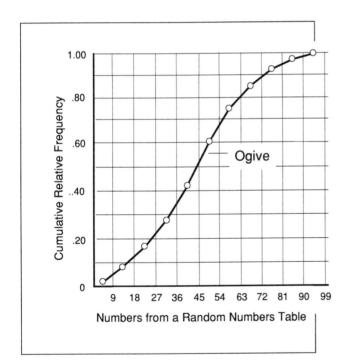

Figure 18.10 An ogive.
Ogives are cumulative frequency graphs, and for normal frequency distributions a characteristic shape results, as in this graph. Data from Table 18.1 were used to create this ogive.

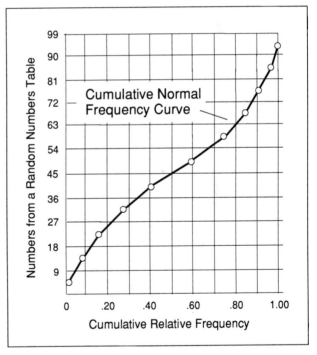

Figure 18.11 A reversed ogive.
When relative frequency is plotted on the horizontal axis, a reversed ogive results, and for normal distributions, a broad S-shape is formed.

Construction of a normal probability grid is rather straightforward. The vertical axis is arithmetically ruled in units of standard deviations (S.D.s) (normally only going from –3.25 to +3.25). The horizontal axis is laid out and its ruler is scaled by examining normal distribution tables. As the distribution is normal, cumulative relative frequencies are known for each standard deviation. (See Table 18.2.)

The lowest cumulative frequency (for –3.25 S.D.) is .0006 and the highest (for +3.25 S.D.) is .9999 and these form the leftmost vertical axis and the rightmost vertical axis, respectively, on the horizontal axis. (See Figure 18.12.) A straight line can be drawn between these two points on the graph. Additional ruler marks on the horizontal axis may be drawn by tracing over other values from the vertical

Table 18.2 Cumulative Relative Frequencies for –3.25 to +3.25 Standard Deviations

Normal Distributions	
Standard Deviation	**Cumulative Relative Frequency**
+3.25	.9999
+3	.9980
+2	.9773
+1	.8406
Mean	.5000
–1	.1587
–2	.0228
–3	.0013
–3.25	.0006

Note: This is an abbreviated table from much larger tables of the normal distribution.

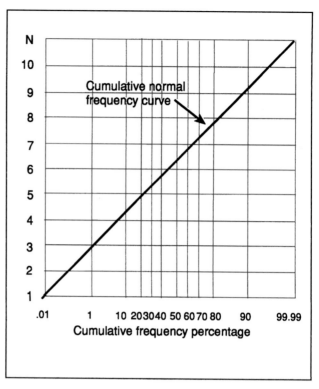

Figure 18.12   A normal probability plot.
If relative frequencies are plotted against z-values in a normal distribution, the reversed ogive becomes a straight line. Any distribution can be checked for normalcy when plotted on a probability graph.

axis to the straight line, and then dropping these to the horizontal axis. These values are then labeled on the horizontal axis and the reverse ogive becomes straight when plotted on this graph.

Often the grapher does not have to construct original grids because probability paper that is already printed is available. This paper is printed with probabilities on the horizontal axis, and ruled on the vertical axis with a convenient rule that can be scaled to accommodate any values of a data set.

BOX-WHISKER PLOTS

A unique graph that shows the percentile summary of a data set is the **box graph,** sometimes called the **Box-Whisker graph.** (See Figure 18.13.) In this graph, a box is drawn so that certain horizontal lines within and at the edges of the box, aligned to the vertical axis of the graph, represent percentile values of the distribution. Customarily the percentiles of 10, 25, 50, 75, and 90 are represented. Individual values in the distribution beyond the 10th and 90th *p*-values are shown with individual dots.

A major use of the box graph is to show two or more data sets in the same data region in order to compare distributions. The compactness of the two distributions around their respective 50th percentiles is immediately seen, as well as the values of the two 50th percentiles. The nature of the tails of the distributions is likewise easily seen. Of course it is possible to show more than two distributions, but showing more than three becomes a bit cumbersome when quick inspection is desired.

Another advantage of the box graph is that many commercial computer graphing programs include these in their gamut of graphs. So the fact that they are relatively easy to produce makes them that much more useful.

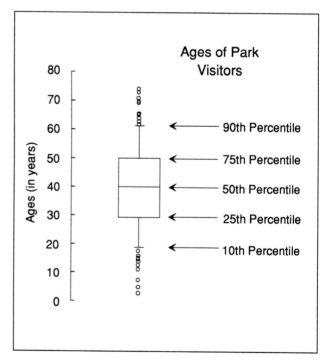

Figure 18.13   A simple Box-Whisker graph.
Source of data: Elifson, Kirk W., Richard P. Runyon, and Audry Haker, Fundamentals of Social Statistics *(McGraw-Hill, 1990, Table 3.2).*

SHOWING HOW MUCH OF SOMETHING

To show how much there is of something accounts for a large percentage of the graphs that are included in scientific writing. In most respects the kinds of graphs designed to show quantity are the easiest and least troublesome to produce, yet they can also incorporate many mistakes. The following discussions look at three commonly used graph types and present some of the basic rules for their design.

THE LINEAR DOT GRAPH

Interestingly, this kind of graph, which employs a newer method, is very easy to construct, but is not used as often as the others in this group. The linear dot graph relies on the reader being able to judge length and, more commonly, comparative length. This kind of graph often displays data that show the *amounts of something* among several compared groups. (See Figure 18.14a.)

For ordinary dot graphing, the different groups are usually displayed along the left of the graph; the amounts of the data for each are displayed along a dot line alongside each and arrayed to the right proportional to each data amount. Small dots are used to show the continuum and connectivity to the labels to the end of the line, which is usually depicted by a data symbol that is larger than the dots. It is important to remember that the data symbol for each group is placed on the grid precisely.

On dot graphs that show fractions of totals (percents) along the horizontal axis, dots may be rendered across the entire length of the line, and the data symbol will therefore show nicely the proportion along the way. (See Figure 18.14b.)

As for the design of these dot graphs, there are only three or four items to consider. The dots must be smaller and appear lighter than the data symbols to ensure that the latter are visually dominant. Although the small dots may be left off, only do so when the horizontal distance to the label is short. Otherwise, the reader will have difficulty seeing the connections between data symbols and group labels. The vertical spacing between the labels of the groups must not be so great that an empty looking graph results. Compactness without congestion is a rule. As with all graphs, all titles and legends must be clear and unambiguous.

THE COMMON BAR GRAPH

The common **bar graph** is one of the oldest forms of data graphing—though simple in concept, it is often not rendered well because of careless mistakes. Numerical values in the data are represented by the length of a series of bars. This form of graph is used to show *comparative amounts* among several groups, facing the reader with the task of making length judgments.

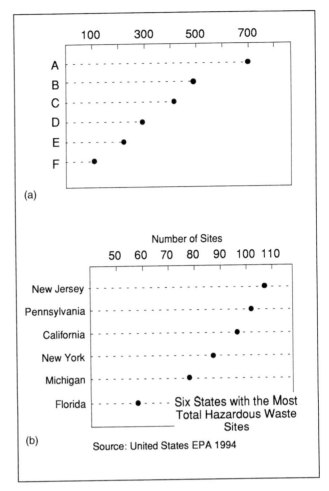

Figure 18.14   Dot graphs.
A simple dot graph uses small dots to establish a visual connection to its base label, and is punctuated at the end by a larger data symbol, as in (a). Another kind of dot graph employs small dots across the data region to the vertical scale on the right, as in (b). In the latter case, the data symbols are placed on the dotted lines in their correct position.

Although the literature abounds with distinctions between column or bar charts, they are the same, differing only in orientation. For our purposes here, they will be treated as the same.

Bar graphs may be used to *trace the rise and fall of amounts* for one group over time, although this is not necessarily the best use of the graph. A better use is simply to show the quantity of a variable among several groups, making this graph type most useful for comparison purposes. (See Figure 18.15a.) A special treatment of this graph is for showing *opposing magnitudes* (such as increase/decrease, rise/fall, and so on). (See Figure 18.15b.)

There are numerous design ideas for bar graphs, some of which have arisen with simple practice, and some that are supported by perception studies. The fundamental rule is that, because these graphs require the reader to make judgments of length, the *bars must all begin on a zero line, and their entire lengths must be shown.* Because total magnitudes are shown,

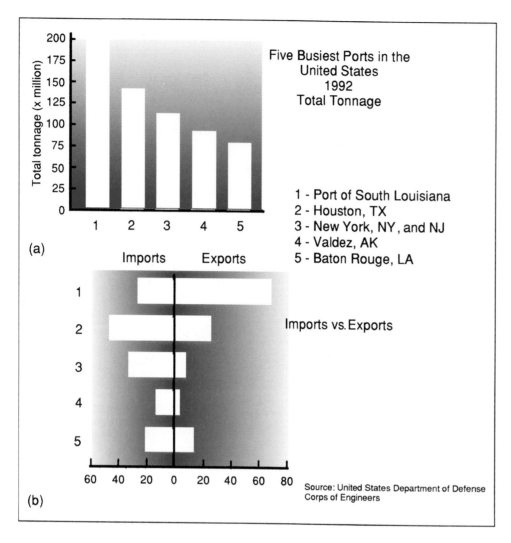

Imports vs. Exports

Figure 18.15 A simple bar graph.

Notice in (a) that the space between bars is one-half the thickness of the bars, and that the bars begin on a zero line and extend unbroken. In (b), opposing bars are used to display directional data.

it is deceptive to eliminate the base of the bars and only show the tops, which can be very misleading. (See Figure 18.16.)

When the horizontal axis represents years, it is always desirable to make the ruler arithmetic and linear, and if there are missing years in the sequence, the scale should not be collapsed to save space. It is important to have the time progression uniform.

The bar graph should be designed with the main axis rulers a bit heavier than the grid of the graph. This provides contrast and improves the look of the graph. And, as explained above, bars should never be broken. The grid should contain enough lines to allow the reader to easily see the magnitudes of the bars but not so many lines as to cause confusion. The ruler should be labeled to represent the data values. If the horizontal axis is years, the earliest year is typically on the left. These design tips will ensure a good-looking bar graph.

The length of bars, of course, should be tailored to the overall size of the graph, but exceptionally long narrow bars look ridiculous and should be avoided. The bars are best arranged by size, or at least by some other noticeable *order*. They should be rendered of uniform width, with spacing between them that is approximately one-half their width.

Typically, bars should be rendered so that they stand out visually above the grid beneath them. They should have strong edge gradients, not open, as this reduces their figural qualities. If they are open their outlines should have good contrast with the grid. (See Figure 18.17a.) Care must be taken when applying patterns to bars because these can cause optical illusions. (See Figure 18.17b.)

Labeling on bar graphs is a designer's choice. If labels are applied to the bars, it is preferable to have them outside the bars so that the labels do not interfere with the figural qualities of the bars. For the same reason, labels should be a bit smaller than the width of the bar. Labels are more often applied to horizontal bars than vertical ones.

Another version of a bar graph is the **clock graph.** On this graph the bars are arranged around a circular time diagram and scaled in length to a concentric grid. (See Figure 18.18.) This kind of graph may be used also for directional data. The wind rose graph used by meteorologists to show prevailing wind directions is a good example. In fact, the clock graph should be explored for any number of unique uses. Design tips include those already mentioned for common bar graphs.

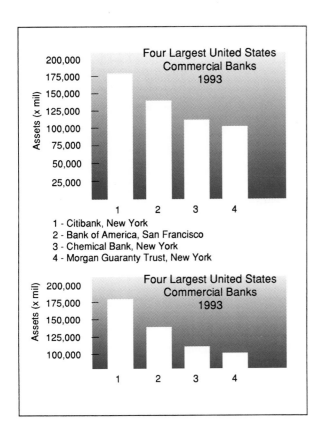

Figure 18.16   Poor practice in bar graphing.
It is always unwise to cut off the beginning, or zero, line of a bar graph. Readers judge total length of bars, and require this length to judge change or comparison in lengths.

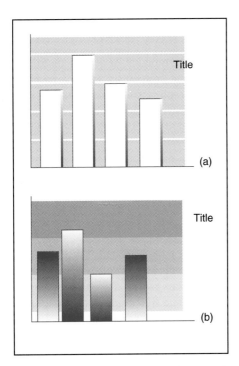

Figure 18.17   Effective and ineffective bar graphs.
(a) Effective bar graphs economically utilize space, contrast, and the visual hierarchy. (b) Some ineffective bar graph designs use annoying optical illusions which should be avoided at all costs.

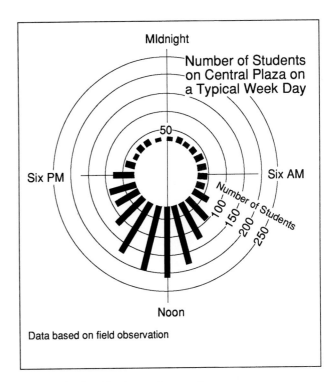

Figure 18.18   Clock-type bar graph.

THE SECTOR GRAPH

The **sector graph** is a commonly used graph, with apparently wide appeal, although not among many scientists. It is often called a pie graph, because it and its parts resemble a pie. In this form of graph, component parts of a total are rendered as wedges of the circle so that the size of each wedge is proportional to its share. Amounts are commonly shown in percentages. The sizes of the wedges are determined by their *angles* at the center, and the graph reader makes size judgments by *comparing angles.*

Certain design rules should be followed carefully when constructing the sector graph. First, always begin by placing the largest wedge so that its left edge is at the 12 o'clock position. (See Figure 18.19.) Second, arrange the sectors by size, with the largest first, then proceed *clockwise* around the graph. Third, never use this graph when there are more than about five or six wedges, especially if some of the wedges are nearly the same size; they are simply too difficult to read for faithful magnitude comparisons.

There is a tendency today for many graphic designers to render sector graphs to look like three-dimensional obliquely viewed coins, such as are seen in weekly news magazines and certain daily papers. (See Figure 18.20.) *Do not render sector graphs as three-dimensional perspective drawings.* Doing so places the wedge angles in perspective, making the perceptual task of the reader that much more difficult. It would be better to use a table of values.

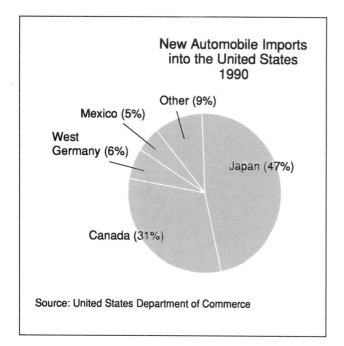

Figure 18.19 A well-designed sector graph.
Well-designed sector graphs are space-efficient, use few sectors, begin at the noon position, and are labeled clearly.

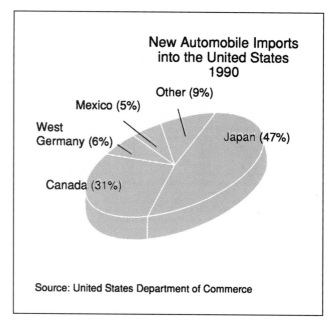

Figure 18.20 A three-dimension perspective sector graph. These graphs should never be used because readers cannot accurately judge angles on them.

Another poor practice is to use several sector graphs for comparative purposes. Once again, it is simply too difficult for the reader to determine comparisons clearly.

Constructing the sector graph is quite simple. The data are transformed into degree parts of a circle. One percent is comparable to 3.6°. The numbers of percentages are multiplied by this number to achieve the sizes of the wedges. I have found that *compactness* will save the design of a sector graph. If labeling is to be used, place the labels in the wedges when possible. If placed outside, anchor the labels to their respective wedges by tag lines for clarity. It is permissible to use different hues for the wedges because they show component parts, and to this extent they are qualitative in nature.

The Trilinear Graph

This type of graph, found in a number of applications in geography, is unique and perhaps could be used more often than it is. The **trilinear graph** is based on the geometric principle that the sum of the perpendiculars from any point within an equilateral triangle to the three sides is equal to the height of the triangle. The trilinear graph is used when there are three components (variables) that always add up to the whole—although each variable may account for a different part of the whole (and may change over time). Each side of the triangle serves as the base line for a variable. Customarily the sides are marked off by a *percentage ruler.*

A common trilinear graph used in geography is the soil-texture graph. (See Figure 18.21.) This illustrates the component parts of a soil sample by noting the percentages

of sand, silt, and clay in the soil. Another example is the percentage of cement, sand, and gravel in a concrete sample. Anything may be represented by the trilinear graph as long as there are three components that make up the sum of something.

SHOWING TRENDS AND TWO-VARIABLE RELATION GRAPHS

After showing geographical or spatial distributions, perhaps the next most frequently applied *numerical technique* of the geographer is to look for trends and relationships among data, and then to display these in some graphic form. There are a number of ways to illustrate trends and relationships for data, as well as guidelines for making these even more effective; this part of the chapter focuses on such graphs.

SIMPLE AND COMPOUND LINE GRAPHS

Perhaps one of the most frequently used graphs is the line graph, sometimes called a *rectilinear line graph.* Structured around the Cartesian grid of Descartes (1596–1650), the purpose of the simple line graph is to show the functional relationship between two variables. Several points are located on the grid, based on their values with respect to the rulers of the two grid variables. (See Figure 18.22.)

There is no set terminology for the different kinds of line graphs. **Simple line graphs** are those that show one variable on the vertical scale, and time on the horizontal scale. These are often referred to as *time series* graphs.

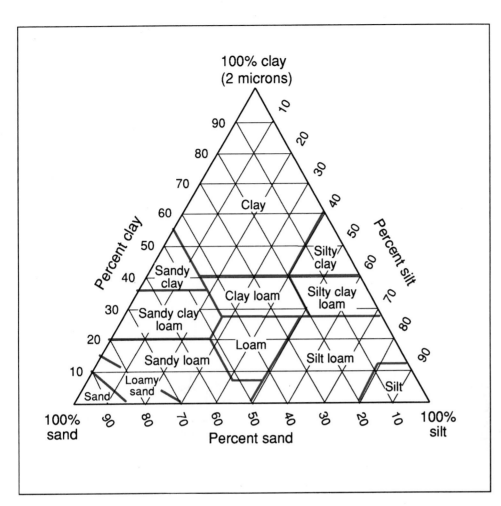

Figure 18.21 A trilinear graph.

These graphs are used to show percentage of the total of the three component parts. This one shows the constituent parts of clay, sand, and silt in soils.

There are **compound line graphs** that show two or more time series on the same graph. *Two-variable* line graphs are those that show one variable on the horizontal axis and another variable on the vertical axis.

Line graphs are easy to construct; they are neat, they attract attention if done well, and provide easy message transfer to the reader. Line graphs are excellent for data analysis, probably because they are uniquely suited for the presentation of *changes in a set of continuous variables.* They are especially good for time series presentations (one of the variables is time), and for observing functional behavior of a pair of variables.

A data point on this graph shows the relationship of the two variables at a given time, and connecting two or more data points on the graph traces out that relationship over the entire set of paired data values. This is the strength of these graphs. The data path tells *where* the data point was and *how* it got to its present location. It is really this "motion" of a point on the plane of the grid that gives the line graph its distinctive functionality and appearance.

The line graph, then, should be used when it is most important to show the overall movement of the relationship of the paired data, and when two or more data paths are used to show comparison on the same graph. Line graphs should *not* be used when there are only a few paired data values in the series, when differences or amounts are more important than movement (trend), or when the movement or change in the series is extremely irregular, violent, or erratic. They are most useful for technical and professional work rather than for casual use for popular audiences.

There are several design suggestions for line graphs that will make them look and function better. The rulers of the two scales should show zero if possible. The grid may be broken, but it is a good idea to show the zero starting point. Scale legends must be clear and succinct, and the grid lines shown to avoid confusion, with enough articulation so that data point values can be seen and interpreted easily. As with all graphs, grid proportions should be selected for balance. If the variable pair are those of a mathematical correlation and regression, then customarily the horizontal scale is the independent variable and the vertical scale is the dependent variable.

There are also some other design considerations. If more than one data path exists on the same graph, then different line *characters* may be used for line differentiation. They absolutely must be seen easily as distinct, and the lines should be labeled to avoid confusion. Do not use multiple data paths on the same graph if confusion results. In addition do not make the reader guess; it would be better to use two graphs or none at all.

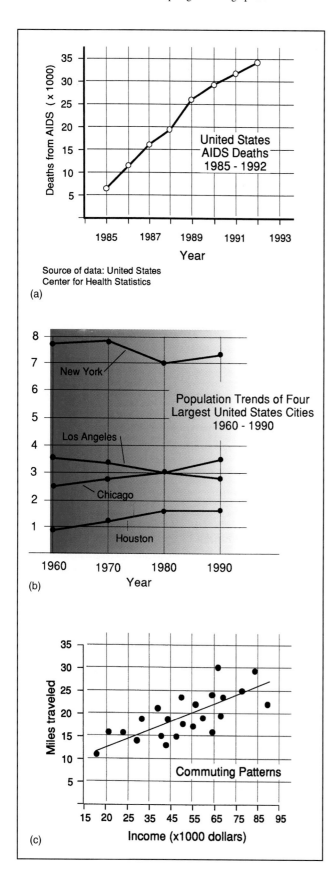

Figure 18.22 Three trend graphs.
These different trend graphs display a single trend of data over time, or two or more data trends over time, and the relationship of two variables.

Generally, try to avoid multiple scales on the same graph. (See Figure 18.23.) If multiple scales are to be used, be certain that they are selected carefully so that data paths do not cross—or if they do cross, that the intersection and its significance, if any, is explained on the graph.

In many instances, when the values of two variables are plotted to show the relationship of the two, a *line of best fit* is fitted to show the relationship rather than an actual data path. There are different strategies to fit these lines—moving averages, weighted moving averages, least-squares lines, or special mathematical curves. To avoid confusion and to add credibility and clarity to the graph, it is of paramount importance to tell the reader how the line of best fit was developed.

THE SCATTERPLOT

Many times geographers study the mathematical relationship between two variables. Most likely in their study they will want to develop a graph to show the relationship visually. If so, they construct a **scatterplot.** Scatterplots have a horizontal scale whose ruler is marked off to accommodate the values of the independent variable in the study, and a vertical scale with a ruler marked off to account for the values in the dependent variable. The researcher will place a dot on the graph that corresponds to each pair of data values, which then represents the paired values. Scatterplots, then, show the *covariation* of two continuous variables. (See Figure 18.24.)

It is important to understand that in a scatterplot the dots are *not* connected in any way. The scatterplot does not show trends—it shows how the two variables are related by their *location patterns* in the graph. Typically a scatterplot is the first step in looking at the statistical correlation between the variables, and is usually followed by a further examination using regression methods. Such studies will lead to the plot of a *regression line,* which predicts the additional values from those actually used in the study. Typically a scatterplot containing a regression line also includes the mathematical formula of the regression line.

Scatterplots can be a useful addition to a map legend when the map in some way depicts both variables in a map study. As with other graph designs, it is important that the data dots and the regression line, if any, be made to appear visually prominent on the graph and appear "above" the data grid. If the data values are classed in some way using the scatter of dots, these can actually be shown on the scatterplot.

THE LOGARITHMIC GRAPH

One of the most prolific graphs used in scientific writing is the logarithmic graph. Although this may be the case, I suspect that the average reader does not fully understand the real advantages of logarithmic graphs, and the producer of these graphs does not fully understand just how difficult it is to read them. I, again, suspect that they are used often

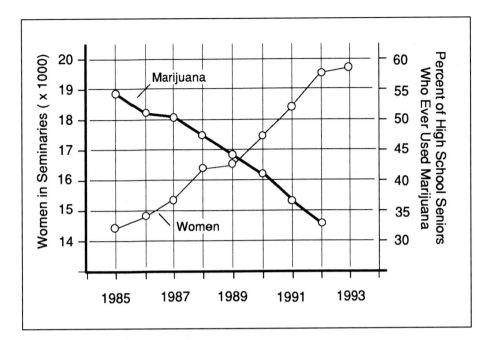

Figure 18.23   Trend graph with multiple scales.

These graphs are to be used with great care. Especially troublesome are those cases where the trend lines intersect, as shown here; readers may see a significance at the intersection when, in fact, there may be none.

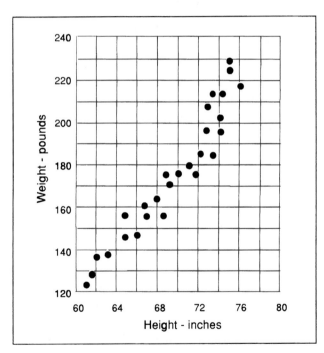

Figure 18.24   Typical scatterplot.

It is important for these graphs that the data symbols (dots) show clearly by making them stand out from the grid.

simply to compress the scales of the data. This is truly not why they *should* be used.

Logarithmic graphs are graphs that contain, at least on one axis, a scale that is marked off with a logarithmic ruler (a ruler marked off in units of a geometric progression). With one scale so marked off, and the other in arithmetic scale, it is referred to as a **semilogarithmic graph.** If both scales con-

tain a logarithmic scale it is called a **log-log graph.** Logarithmic graphs are also sometimes referred to as **ratio graphs.**

The purpose of the logarithmic graph is to show the *rate of change* of one variable against another, or to show the rate of change of one variable over time. Do *not* use these graphs simply to compress the scales of the data. The fundamental idea is that rate of change can be seen by inspecting the slope of the data path. The steeper the slope, the greater the rate of change; the more gradual the slope, the less the rate of change.

A good way to introduce the logarithmic graph is to review two mathematical progressions—the arithmetic and the geometric. The general formula for an arithmetic progression is

$$a, a + d, a + 2d, a + 3d, . . . a + (n - 1)d$$

where

    a = **the first term**
    d = **the common difference**
    n = **the number of terms**
    L = **the last term,** $a + (n - 1)d$

Let's suppose we have this as an example:

$$a = 100, \qquad d = 100, \qquad n = 5$$

which will yield the following series of numbers:

    100   200   300   400   500

If we look at the *rate of change* in the arithmetic progression, we get:

100	**200**	**300**	**400**	**500**
	d = 100	100	100	100
rate of change = 100%	**50%**	**33.3%**	**25%**	

If this progression is graphed on ordinary linear graph paper, the data path shows the *amount* of change as a straight line, but does not show the *rate* of change as decreasing. (See Figure 18.25.)

A review of a geometric progression shows something different. The general formula for a geometric progression is as follows:

$$a, ar, ar^2, ar^3, \ldots ar^{(n-1)}$$

where

a = the first term
r = the common ratio
n = the number of terms
L = the last term, $ar^{(n-1)}$

Now, suppose this is the example:

$a = 100$, $r = 2$, $n = 4$

which will yield the following series of numbers:

100   200   400   800

Looking at the *rate of change* of the geometric progression, we get

100   200   400   800
$r = 2$ 2   2
rate of change =100%   100%   100%

If this progression is graphed on ordinary linear graph paper, the data path shows the *constant rate of change* with a changing slope, and does not show the constant rate of change as a constant slope. (See Figure 18.26.)

If there was some form of graph that showed equal steps on its scales as equal rates of change, we could use the slope of the data path to illustrate identical rates of change. The logarithmic graph does this. First, a quick review of logarithms is necessary.

Logarithms

A common logarithm is the power to which the number 10 must be raised to obtain a given number:

log of 10 = 1(10^1 = 10 × 1 = 10)
log of 100 = 2(10^2 = 10 × 10 = 100)
log of 1,000 = 3(10^3 = 10 × 10 × 10 = 1,000)

In the event the number is not an exact power of 10, its log will be expressed as an approximation containing decimals:

log of 2 = 0.3010
log of 3 = 0.4771

There are two separate components of a logarithm: The *characteristic* component equals the integer of the logarithm, which is to the left of the decimal point. For example, the log of 20 is 1, and 200 = 2. The rule is that the characteristic of a number is always one less than the number of digits in the integer of the number. The *mantissa* equals the decimal part of the logarithm, which is to the

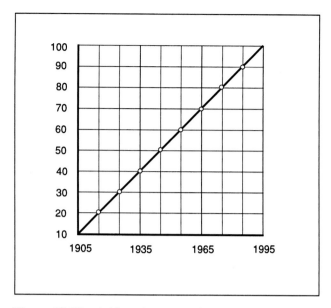

Figure 18.25   Arithmetic progression plotted on an arithmetic grid.
The trend line is straight, in this case rising upward to the right.

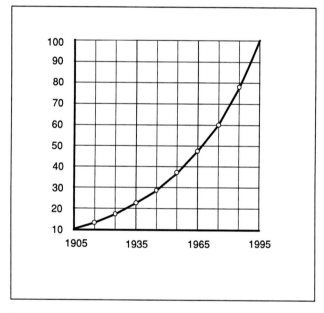

Figure 18.26   Geometric progression plotted on an arithmetic grid.
The trend line plots as a curved line, bending upward to the right.

right of the decimal point. The *antilogarithm* is the number corresponding to a given logarithm.

Certain arithmetic operations on logarithms give them their real power and advantage in mathematical notations. For example, two numbers can be added by *multiplying* their logarithms, and two numbers can be subtracted by *dividing* their logarithms. For purposes here, only adding by multiplication will be discussed. In the multiplication of numbers logarithmically, their logs are *added:*

$$20 \times 30 = 600$$
$$\text{Log of } 20 = 1.3010$$
$$\text{Log of } 30 = \underline{1.4771}$$
$$2.7781 \text{ (log of 600)}$$

(In this example, for quick check, the characteristic 2 suggests immediately that the original number has three digits.)

Let's go back to an example of a geometric progression, but this time logarithms of the numbers will be used:

$a, ar, ar^2, ar^3, \ldots ar^{(n-1)}$, **where $a = 3$, $r = 2$, and $n = 5$**

which yields

3, 6, 12, 24, 48

differences

log of 3 = 0.4771

.3010

log of 6 = 0.7782

.3010

log of 12 = 1.0792

.3010

log of 24 = 1.3802

.3010

log of 48 = 1.6812

Notice that the differences in the logarithms are constant. If the percent change between two or more pairs of values is constant it can be shown that the differences between their logarithms will be the same. Thus, if we could somehow mark a scale logarithmically, we would have a scale that shows rates of change along data paths rather than amounts. This, in fact, can be done so that equal distances on the scale always indicate equal percentages or rates of change, either as an increase or a decrease. This is the logarithmic graph.

Above, only two progressions were considered—an arithmetic increasing by a constant amount, and a ratio increasing by a constant rate. Other progressions are possible, of course. Figure 18.27 shows several series, plotted on both arithmetic and log scale graphs, to show how the slopes of the data paths appear.

Before providing the methods for constructing the log scales, a review of the features of the log graph is important. Assume that you are looking at a graph whose vertical scale is logarithmic and whose horizontal scale is arithmetic (or linear). Then

1. Equal vertical distances on the log scale always indicate equal percentages or rates of change.
2. Rates of change may be different along a data path, and these can be examined by simply looking at the slopes along the data path.
3. It is possible to compare two widely spaced data paths on the same logarithm scale because, regardless of the values, the rates of change (slopes) can be compared.
4. Amounts of change can be read from a log graph, but this is not its chief advantage.

Logarithmic Graph Construction

A logarithmic scale can be constructed simply (assuming there is no computer program available to automatically do the plotting, or that there is not a supply of preprinted log paper). The divisions on the axis ruler are calculated using logarithms. For example, suppose there are 4 inches in which to develop a log scale. The first major division, the second line in the scale, will be placed this way:

log of 2×4 (in) $= (.3010) \times (4) = 1.20$ inches
log of 3×4 (in) $= (.4771) \times (4) = 1.91$ inches

and so on.

Another method is simply to obtain a preprinted log graph and label it accurately to accommodate the range of values in the data array.

Labeling the preprinted log graph is relatively easy. There is *no* zero base line on the scale (the log of 0 is minus infinity). Instead, the bottom line can be labeled with any convenient value—usually slightly less than the smallest value in the data array. The scale should extend to that line that is slightly higher than the largest value in the array. For convenience, the bottom line is usually labeled 10 or some multiple of 10 (.0001, .001, .01, 1, 10, 100, 1,000, and so on). *The value of each successive bold line on the graph is 10 times that of the preceding value.* (See Figure 18.28.)

Each series of multiples of 10 is called a *cycle,* and most preprinted log paper contains multiple cycles.

Even when preparing a graph where the cycle or several cycles must be of a given length, a preprinted sheet of the incorrect size paper may be used. Lay out the preprinted cycle such that its bottom and top lines touch the extremes of the desired final size, then merely project over the intermediate line locations. (See Figure 18.29.)

Measuring Relative Change on the Logarithmic Graph

As stated previously, logarithmic differences between values in the data array in a geometric series will be constant. On a logarithmic graph this means that *equal vertical distances anywhere on the scale represent the same rate or percent change, regardless of their locations on the scale.*

Looked at from another angle, there can be no zero on a ratio chart because, mathematically, such a chart proceeds upward and downward by multiplication and division instead of by addition and subtraction, as in the case of an arithmetic chart, and no matter how many times an amount is divided, there is always something, however small, remaining. Zero may be reached by subtraction, as on an arithmetic chart but never, short of infinity, by division, as on a ratio chart.

Source: Walter E. Weld, *How to Chart: Facts from Figures with Graphs* (Norwood, MA: Codex, 1947), p. 58.

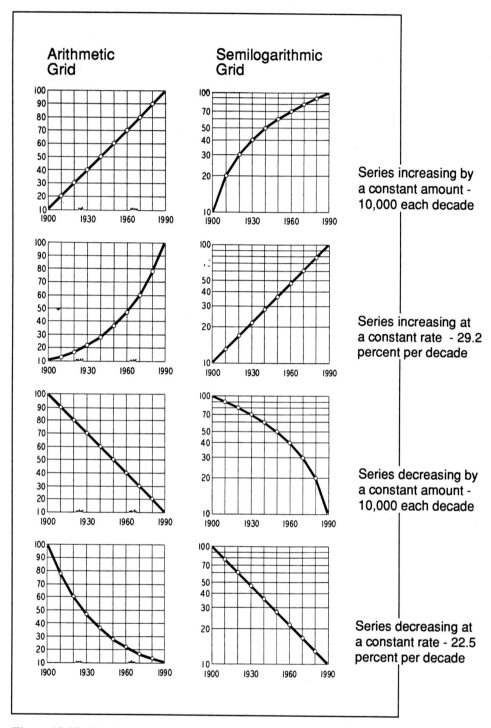

Figure 18.27 Various mathematical progressions plotted on different grids.
Reproduced with changes from Calvin F. Schmid and Stanton E. Schmid. Handbook of Graphic Presentation, *2d ed., John Wiley. (Reprinted by permission of John Wiley and Sons, Inc.)*

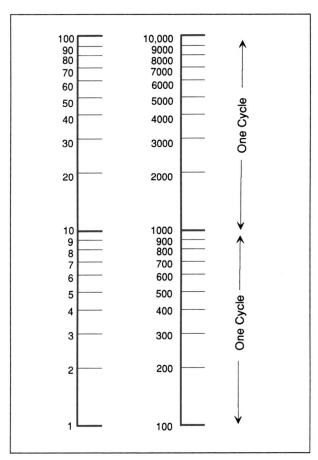

Figure 18.28   Scale labeling on logarithmic grids.
See text for explanation. (*Source: Calvin F. Schmid and Stanton E. Schmid.* Handbook of Graphic Presentation *[New York: Wiley, 1979]. Redrawn by permission of John Wiley and Sons, Inc.)*

On any logarithmic cycle, going from the bottom of a cycle to the first division represents a 100 percent increase; from the bottom to the second, a 200 percent increase, and so on until, from the bottom to the ninth (or end of the cycle), there is a 900 percent increase. Also, going from the bottom of a cycle to the bottom of the next lower cycle represents a 50 percent decrease. A percent change "helper" ruler is convenient here. (See Figure 18.30.)

This feature of logarithmic scales makes them especially useful in *comparing* the rates of change between two or more data paths because the *slope* of the paths provides a measure of the rate of change. In fact, this is the real strength of these graphs and when you want to illustrate rate of change, the logarithmic graph is the best choice. Again, do not select this type of graph simply to compress the graph distance of the vertical scale. (See Figure 18.31.)

It is of course possible to use log-log graphs where a logarithmic ruler is used for both scales. These are used frequently and, again, for perhaps the wrong reasons.

If it is important to illustrate rates of change along both axes of the graph, then this is a good choice. However, remember that such graphs are difficult to read. Their use is not encouraged except in very technical illustrations where it is certain the reader knows the nuances of such graphs.

DESIGNING MULTIVARIATE RELATION GRAPHS

Although much of the work done by geographers looks for relationships between only two related variables, much analytical work is done looking for associations among three or more variables. The graphic display of three or more

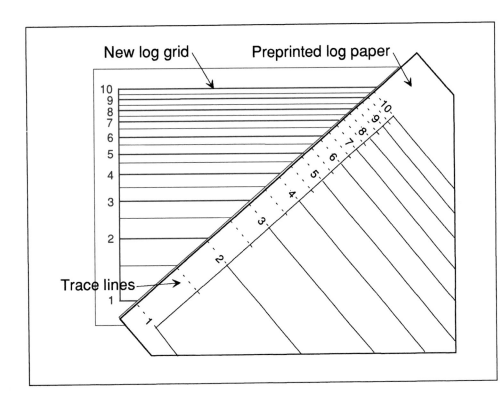

Figure 18.29 Constructing a log scale of any size.
A preprinted log graph is laid out obliquely, connecting the end lines of the new grid. Intermediate grid lines of the preprinted grid are traced over to the new grid. (*Source: Calvin F. Schmid and Stanton E. Schmid,* Handbook of Graphic Presentation *[New York: Wiley, 1979]. Redrawn by permission of John Wiley and Sons, Inc.)*

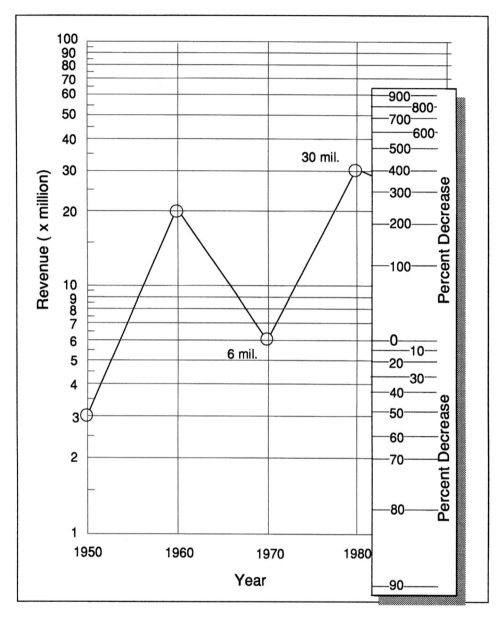

Figure 18.30 Percent change helper.
A portion of the logarithmic grid may be cut out, labeled for percent increase or decrease, and placed on the grid to assist in measuring change.

variables is a bit more difficult, however. Although there are one or two common ways to show these relationships graphically, it is in this arena that geographers frequently employ unique solutions.

THE THREE-VARIABLE GRAPH

Here only the representation of three associated variables is being considered, not those cases where there are any data shown with three-dimensional perspective views. Generally the latter case is to be discouraged because three-dimensional perspectives distort the field of view so that accurate perception tasks are impossible. Thus, the discussion here is to provide one or two ways of displaying three related variables.

Perhaps the most common method is the three-dimensional histogram. A **three-dimensional histogram** illustrates each triad of the data set by first erecting *prisms* proportional to one variable's frequency of occurrence, and

then by locating the prism in a two-dimensional gridded space relative to its proper place, based on the values of the two other variables. (See Figure 18.32a.) In a way similar to the development of a frequency polygon and curve, the three-dimensional histogram may be smoothed out if the data is continuous and it will look like a continuous surface. (See Figure 18.32b.) This diagram has also been called a *block diagram.*[7]

It is also possible to develop a three-variable graph in much the same way as the three-dimensional histogram. Now, however, the third data region is another variable, as opposed to simple frequency. One such example is displayed in Figure 18.33. These histograms must be studied carefully.

The design of any three-dimensional graph requires great care in its treatment of the data regions. There are actually two data regions on these graphs—on the first, two of the variables rest; on the second, the third variable is displayed. It is critical to the successful reading of these

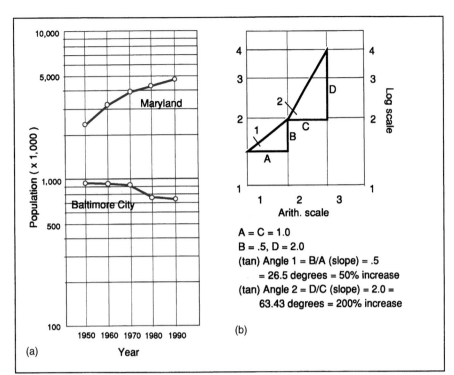

Figure 18.31 Combining two trends on the same logarithmic graph and slope calculation.

Comparison of two trends may be done on the same graph by examining slope (a). Slope determines rate of change on logarithmic graphs (b).

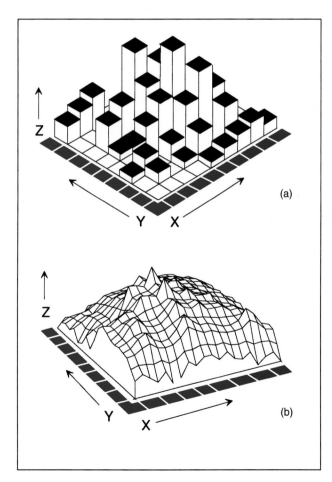

Figure 18.32 Three-dimensional histograms and smoothed frequency surfaces.

The three-dimensional histogram (a), used mostly for discrete variables, may be smoothed when continuous data are being studied (b).

graphs that the reader be able to make a reasonable judgment as to the values of the graphed variables. These tips may be worth following:

1. Avoid perspective views that hide much of the data.
2. Avoid using too many grid lines that obscure the data surface.
3. Make the data surface stand out.
4. Label all scales carefully for easy readability.
5. Label all data regions.
6. Provide generous line and area shading contrast.
7. Be certain the graph easily shows the three-variable association.

THE MULTIVARIABLE GRAPH

Researchers often perform a multivariate statistical test called a multiple correlation or multiple regression. In this instance, several independent variables and their effects on a dependent variable are measured, and the strengths of the associations are noted. Most often, an *intercorrelation table* is displayed to show the overall numerical relations. However, it is possible to graph these, as shown in Figure 18.34. This graph quickly shows how all the variables are correlated, and the existence of any patterns to the associations. Of course it may be necessary to provide the numerical correlations as this graph is not intended to replace the table.

The Triple Scatterplot

The triple scatterplot (sometimes referred to as a **bubble graph**) is an interesting graph, and it probably is not used as often in geographical writing as it should be. It is a common graph found in many computer graphing programs. The **triple scatterplot** depicts the relationship of *three* variables

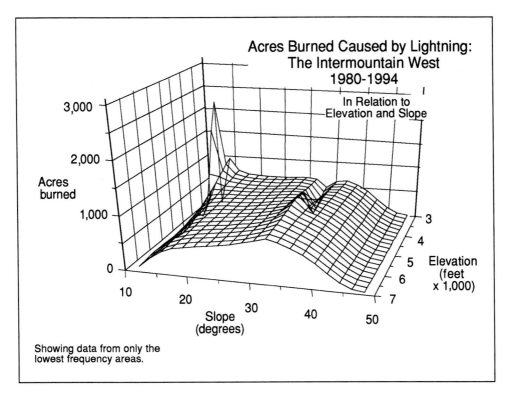

Figure 18.33 Three-dimensional graph.
This graph portrays the relationship of the three variables in an effective manner. Each grid is simply shown and labeled. *(Source of data: From a study by Paul A. Knapp on lightning-caused fires in the American Intermountain West.)*

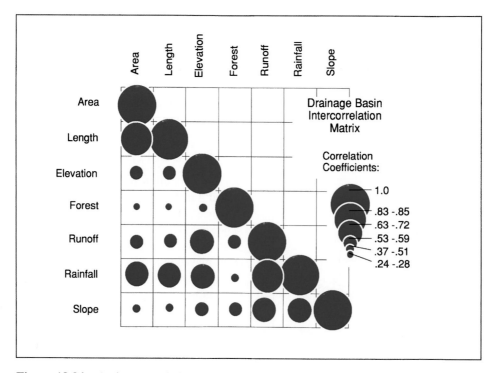

Figure 18.34 An intercorrelation table graph.
Although some data may be lost because of grouping for symbolization, the highly correlated variables are clearly seen.

and is very similar to the graduated symbol map. On this graph, one variable is displayed on the graph's horizontal-axis and one along the vertical-axis; the third is encoded in the sizes of the proportional circles placed at their correct positions within the x–y grid.[8] (See Figure 18.35.)

Position and numeric value along scales must be identified during graph reading, and area judgments must be made when examining the size of the third variable encoded in the circles. In proportional symbol mapping, location on the geographic grid replaces the ruler scales of the x–y grid of the bubble chart. As with any presentation of numerous data points in the x–y space, the reader must make judgments of density as well (i.e., Is there clustering?). Geographers often deal with three-variable data sets, and this type of graph is a unique way of visually presenting the relationships among the variables.

SPECIAL GEOGRAPHIC GRAPHS AND UNIQUE SOLUTIONS

Geographers frequently provide graphic solutions for numerical and relational display that are quite interesting and worth sharing. These are truly unique to specific geography studies and bear mentioning here for one principal reason: they inspire experimentation into seeking graphic solutions that truly excite the mind. Space will not permit an exhaustive list here, but two or three examples make the point.

One interesting graph produced by a geographer is found in Figure 18.36, called by its author, Susan Hanson, a "daily space-time path." On the horizontal axis is relative distance, and on the vertical axis is time during the day. The graph depicts the representative travel pattern of a woman throughout the day. Hanson goes on to say that if the graph is rotated about the home axis, it becomes a space-time prism.[9] It is quite provocative that a person operates in such a solid sphere.

A graph that has been reproduced many times, yet is still quite intriguing, is the one reproduced in Figure 18.37. This graph shows the relationship of three variables—latitude, time of year, and incoming solar radiation. The most interesting aspect of this graph is the *amount of information* it contains. In the spirit of Edward Tufte, this graph has a high data/ink ratio, with little wasted ink.[10]

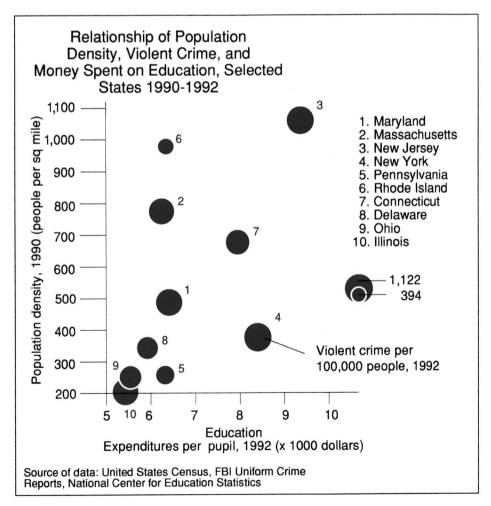

Figure 18.35   A triple scatterplot.

These graphs are often called bubble graphs. Triple scatterplots graph three variables, as illustrated here.

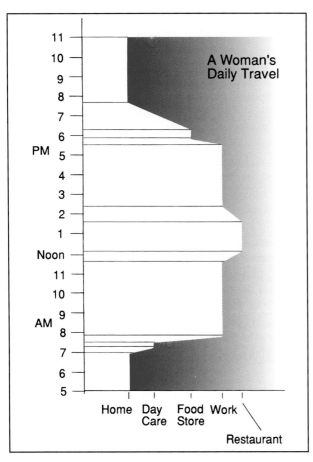

Figure 18.36 Space-time graph.
A geographer's interesting graph of a woman's daily travel. *(After Susan Hanson and Perry Hanson, "The Geography of Everday Life," in* Behavior and Environment, *eds. Tommy Garling and Reginald G. Golledge, [Amsterdam: North Holland, 1993] pp. 249-69.)*

THE CASE FOR EXPERIMENTATION

As with maps and map design, the cartographer involved with graph production should take the posture that exploration into experimental designs will have beneficial results. The works of Tukey, Cleveland, and Tufte are exceptional, and provide excellent guidance in this regard. Every graph provides an *opportunity* to be creative, but can only truly be so if the cartographer takes the time to explore alternatives. This should be a guiding principle.

MISLEADING GRAPHS

As graphing can offer exciting and rewarding prospects, it can also provide opportunities for the unscrupulous worker. Again, just as with maps, graphs can be drawn to mislead. The graph reader, therefore, must be cautious and look out for those graphs that have less than admirable purposes. Ethical practices forbid these activities.

Sometimes, too, graphs can mislead simply because they are poorly designed. A very good example of this is the poorly selected aspect ratio. The ratio of the length of the vertical axis to the horizontal axis (height to width) is the aspect ratio, and this ratio is especially important in the design of the line graph. The impression one gets from the graph can provide different interpretations of the data. (See Figure 18.38.) A higher vertical than horizontal axis makes one see the rise as dramatic compared to how it looks when the vertical axis is shorter than the horizontal, which creates the sense of a nonconsequential rise.[11] Which ratio is correct? There is no definitive one.

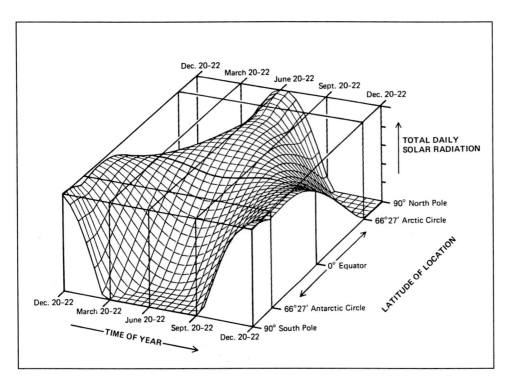

Figure 18.37 Three-dimensional graph of insolation.
A high information graph showing the relationship of three variables. *(Reproduced by permission of John Wiley and Sons, and William M. Marsh.)*

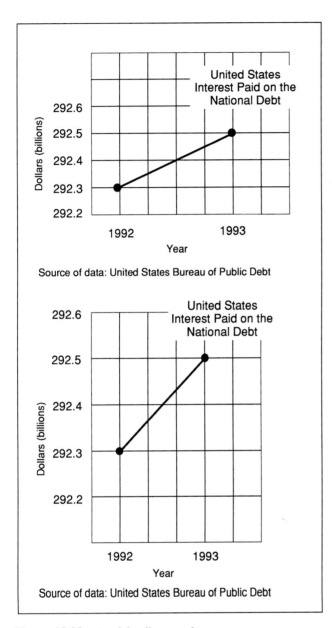

Figure 18.38   A misleading graph.
The aspect ratio of any graph can lead to different visual interpretations of the data. It is better to make the vertical scale length somewhat less than the horizontal scale length.

CHAPTER SUMMARY

This chapter has provided a brief look at graphing, and has reviewed a few graph-reading tasks. A number of design suggestions have been offered that can aid in the development of better graphs. Several graphs have been outlined, and represent the common ones found in the literature. Throughout the chapter several graph examples appeared to illustrate points made in the narrative. Finally, a call for experimentation has gone out, as well as a reminder that graphs can mislead, too.

NOTES

1. For example, see Stephen M. Kosslyn, *Elements of Graph Design* (New York: Freeman, 1994); and Calvin F. Schmid, *Statistical Graphics* (New York: Wiley, 1983).
2. W. G. V. Balchin and Alice M. Coleman, "Graphicacy Should Be the Fourth Ace in the Pack," *Times Educational Supplement,* November 5, 1965. See also W. G. V. Balchin, "Graphicacy," *American Cartographer* 3 (1976): pp. 33–38.
3. Borden D. Dent, "Visual Organization and Thematic Map Communication," *Annals* (Association of American Geographers) 62 (1972):79–93.
4. William S. Cleveland, *The Elements of Graphing* (Monterey, CA: Wadsworth, 1985), pp. 229–94.
5. John I. Clarke, "Statistical Map Reading," *Geography* 44 (1959): 96–104.
6. Harry W. Kershner and Gerald DeMario, *Understanding Basic Statistics* (San Francisco: Holden-Day, 1980), p. 50.
7. Calvin F. Schmid and Stanton E. Schmid, *Handbook of Graphic Presentation,* 2d ed. (New York: Wiley), p. 158.
8. William S. Cleveland and Robert McGill, "Graphical Perception: Theory, Experimentation, and Applications to the Development of Graphical Methods," *Journal of the American Statistical Association* 79 (1984): 531–54; and John M. Chambers, William S. Cleveland, Beat Kleiner, and Paul A. Tukey, *Graphical Methods for Data Analysis* (Boston: Duxbury, 1983), pp. 139–40; see also William S. Cleveland, *The Elements of Graphing* (Monterey, CA: Wadsworth, 1985), pp. 131–33; and John W. Tukey, *Exploratory Data Analysis* (Reading, MA: Addison-Wesley, 1977).
9. Susan Hanson and Perry Hanson, "The Geography of Everyday Life," in *Behavior and Environment,* eds. Tommy Garling and Regionald G. Golledge, (Amsterdam: North-Holland, 1993), pp. 249–69.
10. Edward R. Tufte, *The Visual Display of Quantitative Information* (Cheshire, CT: Graphics Press, 1983), p. 93.
11. Kosslyn, *Elements,* p. 214.

GLOSSARY

bar graph   a series of proportional bars drawn to represent different numerical values in a comparative study, p. 367

Box-Whisker graph   a unique graph that depicts the mean, several percentiles, and extreme values in a statistical distribution; sometimes referred to simply as a box graph, p. 366

bubble graph   another name for a triple scatterplot, pp. 379–380

clock graph   a circular bar graph arranged and scaled around a 24-hour clock, p. 368

compound line graph   a line graph containing the data paths of two or more variables, or components of a compound variable, p. 371

data labels   labels on a graph to identify different data symbols or data paths, p. 361

data path   a line of a graph that portrays the trend of the data, p. 361

data region   the area of a graph containing the grid and in which the graphic treatment of the data falls, p. 361

frequency curve   a smoothed frequency polygon, p. 364

frequency polygon   a graph developed from a frequency histogram by joining the tops of the bars by straight lines, p. 364

graph   a diagram, chart, or drawing, based on a mathematical grid that depicts one or more numerical data sets, p. 359

graphicacy   the intellectual skill necessary to effectively communicate numerical relationships, p. 359

histogram   a graphic record of the occurrence and frequency of data values in a statistical distribution, p. 363

horizontal scale   scale along the top or bottom of a graph, p. 361

key   also called a legend; describes elements of a graph, p. 361

logarithmic graph   a special graph that has at least one axis scaled with a logarithmic ruler; best graph to show rate of change, p. 373

Lorenz curve   a special frequency polygon graph, p. 364

markers   tic marks in a data region that identify important events in the display, p. 361

quantile   value in data array that divides the array into groups with known proportions in each group; deciles, percentiles, quartiles, and quintiles are special quantiles, p. 364

ratio graph   another name for the logarithmic graph, p. 373

reference line   a line on a graph that identifies a certain place in the data region of significance to the data presentation, p. 361

scatterplot   a graph that presents by dots the relationship of two variables, p. 372

scale label   identifiers along the vertical or horizontal scales of a graph, p. 361

scale lines   the rulers that bound the data region on a graph, p. 361

sector graph   a circular graph that shows component parts; commonly called a pie chart, p. 369

semilogarithmic graph   a form of logarithmic graph with a log scale on one axis and an arithmetic scale on the other, p. 373

simple line graph   a line graph that shows time on the horizontal axis and a variable on the vertical axis; sometimes called a time series graph, pp. 370–371

three-dimensional histogram   a special graph that portrays the numeric values of three associated variables, pp. 378–379

trilinear graph   a special triangular graph that shows three component parts relative to the total, p. 370

triple scatterplot   a graph that portrays three variables by using two grids and proportional circles; also called a bubble graph, pp. 379–380

vertical scale   scale along the left or right of a graph, p. 361

READINGS FOR FURTHER UNDERSTANDING

Bohrnstedt, George W., and David Knoke. *Statistics for Social Data Analysis.* 2d ed. Itasca, IL: Peacock, 1988.

Bowman, William J. *Graphic Communication.* New York: Wiley, 1968.

Chambers, John M., William S. Cleveland, Beat Kleiner, and Paul A. Tukey. *Graphical Methods for Data Analysis.* Boston: Duxbury, 1983.

Cleveland, William S. *The Elements of Graphing Data.* Monterey, CA: Wadsworth, 1985.

Dixon, Wilfred J., and Frank J. Massey, Jr. *Introduction to Statistical Analysis.* 3d ed. New York: McGraw-Hill, 1969.

duToit, S. H. C., A. G. W. Steyn, and R. H. Stumpf. *Graphical Exploratory Data Analysis.* New York: Springer-Verlag, 1986.

Everitt, Brian S. *Cluster Analysis.* 3d ed. London: Edward Arnold, 1993.

Kosslyn, Stephen M. *Elements of Graph Design.* New York: Freeman, 1994.

Schmid, Calvin F., and Stanton E. Schmid. *Handbook of Graphic Presentation.* 2d ed. New York: Wiley, 1979.

Spear, Mary Eleanor. *Practical Charting Techniques.* New York: McGraw-Hill, 1969.

Tufte, Edward R. *Data Analysis for Politics and Policy.* Englewood Cliffs, NJ: Prentice Hall, 1974.

———. *The Visual Display of Quantitative Information.* Cheshire, CT: Graphics Press, 1983.

———. *Envisioning Information.* Cheshire, CT: Graphics Press, 1990.

———. *Visual Explanation.* Cheshire, CT: Graphics Press, 1997.

Tukey, John W. *Exploratory Data Analysis.* Reading, MA: Addison-Wesley, 1977.

Wang, Peter C., ed. *Graphical Representation of Multivariate Data.* New York: Academic Press, 1978.

Wildbur, Peter. *Information Graphics: A Survey of Typographic, Diagrammatic, and Cartographic Communication.* New York: Van Nostrand Reinhold, 1989.

APPENDIX A

GEOGRAPHICAL TABLES

The table in this appendix lists the lengths of the degrees of latitude and longitude of the earth's surface, using the Geodetic Reference System (GRS) ellipsoid of 1980.

Lengths of Degrees of Latitude and Longitude on the Earth's Surface

	Length of One Degree of the Meridian				Length of One Degree of the Parallel		
Bottom Latitude (°)	Statute Miles	Nautical Miles	Meters	At Latitude (°)	Statute Miles	Nautical Miles	Meters
0	68.708	59.705	110,574	0	69.171	60.108	111,319
1	68.708	59.706	110,575	1	69.160	60.099	111,303
2	68.709	59.706	110,576	2	69.129	60.071	111,252
3	68.710	59.708	110,578	3	69.077	60.026	111,168
4	68.712	59.709	110,581	4	69.003	59.962	111,050
5	68.714	59.711	110,585	5	68.909	59.881	110,899
6	68.717	59.713	110,589	6	68.794	59.781	110,714
7	68.719	59.716	110,593	7	68.659	59.663	110,495
8	68.723	59.718	110,599	8	68.502	59.527	110,243
9	68.726	59.722	110,605	9	68.325	59.373	109,958
10	68.731	59.725	110,611	10	68.127	59.200	109,639
11	68.735	59.729	110,618	11	67.908	59.011	109,288
12	68.740	59.733	110,626	12	67.669	58.803	108,903
13	68.745	59.738	110,635	13	67.409	58.577	108,485
14	68.751	59.743	110,644	14	67.129	58.334	108,034
15	68.757	59.748	110,654	15	66.829	58.073	107,550
16	68.763	59.754	110,664	16	66.508	57.794	107,034
17	68.770	59.760	110,675	17	66.167	57.498	106,486
18	68.777	59.766	110,686	18	65.806	57.184	105,905
19	68.785	59.772	110,698	19	65.425	56.853	105,292
20	68.792	59.779	110,711	20	65.025	56.505	104,647
21	68.800	59.786	110,724	21	64.604	56.139	103,970
22	68.809	59.793	110,737	22	64.164	55.757	103,262
23	68.818	59.801	110,751	23	63.704	55.358	102,522
24	68.827	59.809	110,766	24	63.225	54.941	101,752
25	68.836	59.817	110,780	25	62.727	54.509	100,950
26	68.845	59.825	110,796	26	62.210	54.059	100,117
27	68.855	59.833	110,811	27	61.674	53.593	99,254.6
28	68.865	59.842	110,828	28	61.119	53.111	98,361.6
29	68.875	59.851	110,844	29	60.546	52.613	97,438.6

Lengths of Degrees of Latitude and Longitude on the Earth's Surface (continued)

	Length of One Degree of the Meridian				**Length of One Degree of the Parallel**		
Bottom Latitude (°)	**Statute Miles**	**Nautical Miles**	**Meters**	**At Latitude (°)**	**Statute Miles**	**Nautical Miles**	**Meters**
30	68.886	59.860	110,861	30	59.954	52.098	96,486.0
31	68.896	59.869	110,878	31	59.343	51.568	95,503.9
32	68.907	59.879	110,896	32	58.715	51.022	94,492.8
33	68.918	59.888	110,913	33	58.069	50.461	93,452.9
34	68.930	59.898	110,931	34	57.405	49.884	92,384.4
35	68.941	59.908	110,950	35	56.724	49.291	91,287.8
36	68.953	59.918	110,968	36	56.025	48.684	90,163.3
37	68.964	59.928	110,987	37	55.309	48.062	89,011.3
38	68.976	59.938	111,006	38	54.576	47.426	87,832.0
39	68.988	59.949	111,025	39	53.827	46.774	86,626.0
40	69.000	59.959	111,044	40	53.061	46.109	85,393.4
41	69.012	59.970	111,064	41	52.279	45.429	84,134.7
42	69.024	59.980	111,083	42	51.481	44.736	82,850.3
43	69.036	59.991	111,102	43	50.667	44.028	81,540.5
44	69.048	60.001	111,122	44	49.838	43.308	80,205.7
45	69.060	60.012	111,142	45	48.993	42.574	78,846.3
46	69.072	60.022	111,161	46	48.133	41.827	77,462.8
47	69.084	60.033	111,181	47	47.259	41.067	76,055.5
48	69.096	60.043	111,200	48	46.370	40.294	74,624.8
49	69.109	60.054	111,219	49	45.467	39.509	73,171.3
50	69.121	60.064	111,239	50	44.549	38.712	71,695.2
51	69.132	60.074	111,258	51	43.618	37.903	70,197.1
52	69.144	60.085	111,277	52	42.674	37.083	68,677.5
53	69.156	60.095	111,296	53	41.717	36.251	67,136.7
54	69.167	60.105	111,314	54	40.747	35.408	65,575.2
55	69.179	60.115	111,333	55	39.764	34.554	63,993.6
56	69.190	60.125	111,351	56	38.769	33.689	62,392.2
57	69.201	60.134	111,369	57	37.762	32.814	60,771.6
58	69.212	60.144	111,386	58	36.743	31.929	59,132.3
59	69.223	60.153	111,404	59	35.713	31.034	57,474.8
60	69.234	60.162	111,421	60	34.672	30.129	55,799.5
61	69.244	60.171	111,437	61	33.621	29.215	54,107.0
62	69.254	60.180	111,454	62	32.558	28.293	52,397.7
63	69.264	60.189	111,470	63	31.486	27.361	50,672.3
64	69.274	60.197	111,485	64	30.404	26.421	48,931.2
65	69.283	60.205	111,500	65	29.313	25.472	47,175.0
66	69.292	60.213	111,515	66	28.213	24.516	45,404.2
67	69.301	60.221	111,529	67	27.104	23.553	43,619.4
68	69.309	60.228	111,542	68	25.986	22.582	41,821.1
69	69.317	60.235	111,556	69	24.861	21.604	40,009.8

Lengths of Degrees of Latitude and Longitude on the Earth's Surface (continued)

	Length of One Degree of the Meridian				Length of One Degree of the Parallel		
Bottom Latitude (°)	Statute Miles	Nautical Miles	Meters	At Latitude (°)	Statute Miles	Nautical Miles	Meters
70	69.325	60.242	111,568	70	23.728	20.619	38,186.1
71	69.333	60.249	111,580	71	22.587	19.628	36,350.6
72	69.340	60.255	111,592	72	21.440	18.631	34,503.8
73	69.347	60.261	111,603	73	20.286	17.628	32,646.4
74	69.353	60.266	111,613	74	19.125	16.619	30,778.8
75	69.359	60.272	111,623	75	17.959	15.606	28,901.7
76	69.365	60.277	111,632	76	16.787	14.587	27,015.6
77	69.371	60.281	111,641	77	15.610	13.564	25,121.1
78	69.376	60.286	111,649	78	14.428	12.537	23,218.8
79	69.380	60.290	111,656	79	13.241	11.506	21,309.3
80	69.384	60.293	111,663	80	12.050	10.472	19,393.2
81	69.388	60.297	111,669	81	10.856	9.434	17,471.1
82	69.391	60.300	111,675	82	9.658	8.393	15,543.6
83	69.394	60.302	111,679	83	8.458	7.349	13,611.2
84	69.397	60.304	111,684	84	7.254	6.304	11,674.6
85	69.399	60.306	111,687	85	6.049	5.256	9,734.4
86	69.401	60.308	111,690	86	4.841	4.207	7,791.2
87	69.402	60.309	111,692	87	3.632	3.156	5,854.5
88	69.403	60.309	111,693	88	2.422	2.105	3,898.0
89	69.403	60.310	111,694	90	0.000	0.000	0.0

The values in this table were obtained by using a program called *The Geographic Calculator*™ from Blue Marble Geographics, Gardner, Maine 04345.

APPENDIX B

DEFINING CONSTANTS FOR THE 1983 STATE PLANE COORDINATE SYSTEM

The table in this appendix lists the zones of the state plane system, and their United States code numbers, by state, the projections used, the central meridians of each zone and scale factor (error), and the latitudes and longitudes of the grid origins, with their desig-nated metric equivalents. Because many thematic cartographers use GIS technology today, this table is a useful reference for anyone requiring this information for grid matching and designation while operating in GIS.

Transverse Mercator (T.M.), Oblique Mercator (O.M.), and Lambert (L.) Projections

State	Zone	Code	Projection	Central Meridian and Scale Factor (T.M.) or Standard Parallels (L.)	Grid Origin Longitude Latitude	Easting Northing (Meters)
Alabama	**AL**					
East	E	0101	T.M.	85° 50'	85° 50'	200,000.
				1:25,000	30° 30'	0
West	W	0102	T.M.	87° 30'	87° 30'	600,000.
				1:15,000	30° 00'	0
Alaska	**AK**					
Zone 1		5001	O.M.	Axis azimuth = arc tan − 3/4	133° 40'	5,000,000.
				1:10,000	57° 00'	-5,000,000.
Zone 2		5002	T.M.	142° 00'	142° 00'	500,000.
				1:10,000	54° 00'	0
Zone 3		5003	T.M.	146° 00'	146° 00'	500,000.
				1:10,000	54° 00'	0
Zone 4		5004	T.M.	150° 00'	150° 00'	500,000.
				1:10,000	54° 00'	0
Zone 5		5005	T.M.	154° 00'	154° 00'	500,000.
				1:10,000	54° 00'	0
Zone 6		5006	T.M.	158° 00'	158° 00'	500,000.
				1:10,000	54° 00'	0
Zone 7		5007	T.M.	162° 00'	162° 00'	0
				1:10,000	54° 00'	0
Zone 8		5008	T.M.	166° 00'	166° 00'	500,000.
				1:10,000	54° 00'	0
Zone 9		5009	T.M.	170° 00'	170° 00'	500,000.
				1:10,000	54° 00'	0
Zone 10		5010	L	51° 50'	176° 00'	1,000,000.
				53° 50'	51° 00'	0

State	Zone	Code	Projection	Central Meridian and Scale Factor (T.M.) or Standard Parallels (L.)	Grid Origin Longitude Latitude	Grid Origin Easting Northing (Meters)
Arizona	**AZ**					
East	E	0201	T.M.	110° 10' 1:10,000	110° 10' 31° 00'	213,360. 0
Central	C	0202	T.M.	111° 55' 1:10,000	111° 55' 31° 00'	213,360. 0
West	W	0203	T.M.	113° 45' 1:15,000	113° 45' 31° 00'	213,360. 0
				(State law defines the origin in International Feet) *(213,360 M. = 700,000 International Feet)*		
Arkansas	**AR**					
North	N	0301	L	34° 56' 36° 14'	92° 00' 34° 20'	400,000. 0
South	S	0302	L	33° 18' 34° 46'	92° 00' 32° 40'	400,000. 400,000.
California	**CA**					
Zone 1		0401	L	40° 00' 41° 40'	122° 00' 39° 20'	2,000,000. 500,000.
Zone 2		0402	L	38° 20' 30° 50'	122° 00' 37° 40'	2,000,000. 500,000.
Zone 3		0403	L	37° 04' 38° 26'	120° 30' 36° 30'	2,000,000. 500,000.
Zone 4		0404	L	36° 00' 37° 15'	119° 00' 35° 20'	2,000,000. 500,000.
Zone 5		0405	L	34° 02' 35° 28'	118° 00' 33° 30'	2,000,000. 500,000.
Zone 6		0406	L	32° 47' 33° 53'	116° 15' 32° 10'	2,000,000. 500,000.
Colorado	**CO**					
North	N	0501	L	39° 43' 40° 47'	105° 30' 39° 20'	914,401.8289 304,800.6096
Central	C	0502	L	38° 27' 39° 45'	105° 30' 37° 50'	914,401.8289 304,800.6096
South	S	0503	L	37° 14' 38° 26'	105° 30' 36° 40'	914,401.8289 304,800.6096
Connecticut	**CT**					
		0600	L	41° 12' 41°52'	72° 45' 40° 50'	304,800.6096 152,400.3048
Delaware	**DE**					
		0700	T.M.	75° 25' 1:200,000	75° 25' 38° 00'	200,000. 0

State	Zone	Code	Projection	Central Meridian and Scale Factor (T.M.) or Standard Parallels (L.)	Grid Origin Longitude Latitude	Easting Northing (Meters)
Florida	**FL**					
East	E	0901	T.M.	81° 00′ 1:17,000	81° 00′ 24° 20′	200,000. 0
West	W	0902	T.M.	82° 00′ 1:17,000	82° 00′ 24° 20′	200,000. 0
North	N	0903	L	29° 35′ 30° 45′	84° 30′ 29° 00′	600,000. 0
Georgia	**GA**					
East	E	1001	T.M.	82° 10′ 1:10,000	82° 10′ 30° 00′	200,000. 0
West	W	1002	T.M.	84° 10′ 1:10,000	84° 10′ 30° 00′	700,000. 0
Hawaii	**HI**					
Zone 1		5101	T.M.	155° 30′ 1:30,000	155° 30′ 18° 50′	500,000. 0
Zone 2		5102	T.M.	156° 40′ 1:30,000	156° 40′ 20° 20′	500,000. 0
Zone 3		5103	T.M.	158° 00′ 1:100,000	158° 00′ 21° 10′	500,000. 0
Zone 4		5104	T.M.	159° 30′ 1:100,000	159° 30′ 21° 50′	500,000. 0
Zone 5		5105	T.M.	160° 10′ 0	160° 10′ 21° 40′	500,000. 0
Idaho	**ID**					
East	E	1101	T.M.	112° 10′ 1:19,000	112° 10′ 41° 40′	200,000. 0
Central	C	1102	T.M.	114° 00′ 1:19,000	114° 00′ 41° 45′	500,000. 0
West	W	1103	T.M.	115° 45′ 1:15,000	115° 45′ 41° 40′	800,000. 0
Illinois	**IL**					
East	E	1201	T.M.	88° 20′ 1:40,000	88° 20′ 36° 40′	300,000. 0
West	W	1202	T.M.	90° 10′ 1:17,000	90° 10′ 36° 40′	700,000. 0
Indiana	**IN**					
East	E	1301	T.M.	85° 40′ 1:30,000	85° 40′ 37° 30′	100,000. 250,000.
West	W	1302	T.M.	87° 05′ 1:30,000	87° 05′ 37° 30′	900,000. 250,000.

State	Zone	Code	Projection	Central Meridian and Scale Factor (T.M.) or Standard Parallels (L.)	Grid Origin Longitude Latitude	Easting Northing (Meters)
Iowa	**IA**					
North	N	1401	L	42° 04′	93° 30′	1,500,000.
				43° 16′	41° 30′	1,000,000.
South	S	1402	L	40° 37′	93° 30′	500,000.
				41° 47′	40° 00′	0
Kansas	**KS**					
North	N	1501	L	38° 43′	98° 00′	400,000.
				39° 47′	38° 20′	0
South	S	1502	L	37° 16′	98° 30′	400,000.
				38° 34′	36° 40′	400,000.
Kentucky	**KY**					
North	N	1601	L	38° 58′	37° 30′	500,000.
				37° 58′	84° 30′	0
South	S	1602	L	36° 44′	85° 45′	500,000.
				37° 56′	36° 20′	500,000.
Louisiana	**LA**					
North	N	1701	L	31° 10′	92° 30′	1,000,000.
				32° 40′	30° 30′*	0
South	S	1702	L	29° 18′	91° 20′	1,000,000.
				30° 42′	28° 30′*	0
Offshore	SH	1703	L	26° 10′	91° 20′	1,000,000.
				27° 50′	25° 30′*	0
Maine	**ME**					
East	E	1801	T.M.	68° 30′	68° 30′	300,000.
				1:10,000	43° 40′*	0
West	W	1802	T.M.	70° 10′	70° 10′	900,000.
				1:30,000	42° 50′	0
Maryland	**MD**					
		1900	L	38° 18′	77° 00′	400,000.
				39° 27′	37° 40′*	0
Massachusetts	**MA**					
Mainland	M	2001	L	41° 43′	71° 30′	200,000.
				42° 41′	41° 00′	750,000.
Island	I	2002	L	41° 17′	70° 30′	500,000.
				41° 29′	41° 00′	0
Michigan	**MI**					
North	N	2111	L	45° 29′	87° 00′	8,000,000.
				47° 05′	44° 47′	0
Central	C	2112	L	44° 11′	84° 22′*	6,000,000.
				45° 42′	43° 19′	0
South	S	2113	L	42° 06′	84° 22′*	4,000,000.
				43° 40′	41° 30′	0

State	Zone	Code	Projection	Central Meridian and Scale Factor (T.M.) or Standard Parallels (L.)	Grid Origin Longitude Latitude	Easting Northing (Meters)
Minnesota	**MN**					
North	N	2201	L	47° 02′	93° 06′	800,000.
				48° 38′	46° 30′	100,000.
Central	C	2202	L	45° 37′	94° 15′	800,000.
				47° 03′	45° 00′	100,000.
South	S	2203	L	43° 47′	94° 00′	800,000.
				45° 13′	43° 00′	100,000.
Mississippi	**MS**					
East	E	2301	T.M.	88° 50′	88° 50′	300,000.
				1:20,000*	29° 30′*	0
West	W	2302	T.M.	90° 20′	90° 20′	700,000.
				1:20,000*	29° 30′*	0
Missouri	**MO**					
East	E	2401	T.M.	90° 30′	90° 30′	250,000.
				1:15,000	35° 50′	0
Central	C	2402	T.M.	92° 30′	92° 30′	500,000.
				1:15,000	35° 50′	0
West	W	2403	T.M.	94° 30′	94° 30′	850,000.
				1:17,000	36° 10′	0
Montana	**MT**					
		2500	L	45° 00′*	109° 30′	600,000.
				49° 00′*	44° 15′*	0
Nebraska	**NE**					
		2600	L	40° 00′*	100° 00′*	500,000.
				43° 00′*	39° 50′*	0
Nevada	**NV**					
East	E	2701	T.M.	115° 35′	115° 35′	200,000.
				1:10,000	34° 45′	8,000,000.
Central	C	2702	T.M.	116° 40′	116° 40′	500,000.
				1:10,000	34° 45′	6,000,000.
West	W	2703	T.M.	118° 35′	118° 35′	800,000.
				1:10,000	34° 34′	4,000,000.
New Hampshire	**NH**					
		2800	T.M.	71° 40′	71° 40′	300,000.
				1:30,000	42° 30′	0
New Jersey (New York East)	**NJ**					
		2900	T.M.	74° 30′*	74° 30′*	150,000.
				1:10,000*	38° 50′	0

State	Zone	Code	Projection	Central Meridian and Scale Factor (T.M.) or Standard Parallels (L.)	Grid Origin Longitude Latitude	Easting Northing (Meters)
New Mexico	**NM**					
East	E	3001	T.M.	104° 20'	104° 20'	165,000.
				1:11,000	31° 00'	0
Central	C	3002	T.M.	106° 15'	106° 15'	500,000.
				1:10,000	31° 00'	0
West	W	3003	T.M.	107° 50'	107° 50'	830,000.
				1:12,000	31° 00'	0
New York	**NY**					
East (New Jersey)	E	3101	T.M.	74° 30'*	74° 30'*	150,000.
				1:10,000*	38° 50'*	0
Central	C	3102	T.M.	76° 35'	76° 35'	250,000.
				1:16,000	40° 00'	0
West	W	3103	T.M.	78° 35'	78° 35'	350,000.
				1:16,000	40° 00'	0
Long Island	L	3104	L	40° 40'	74° 00'	300,000.
				41° 02'	40° 10'*	0
North Carolina	**NC**					
		3200	L	34° 20'	79° 00'	609,601.22
				36° 10'	33° 45'	0
North Dakota	**ND**					
North	N	3301	L	47° 26'	100° 30'	600,000.
				48° 44'	47° 00'	0
South	S	3302	L	46° 11'	100° 30'	600,000.
				47° 29'	45° 40'	0
Ohio	**OH**					
North	N	3401	L	40° 26'	82° 30'	600,000.
				41° 42'	39° 40'	0
South	S	3402	L	38° 44'	82° 30'	600,000.
				40° 02'	38° 00'	0
Oklahoma	**OK**					
North	N	3501	L	35° 34'	98° 00'	600,000.
				36° 46'	35° 00'	0
South	S	3502	L	33° 56'	98° 00'	600,000.
				35° 14'	33° 20'	0
Oregon	**OR**					
North	N	3601	L	44° 20'	120° 30'	2,500,000.
				46° 00'	43° 40'	0
South	S	3602	L	42° 20'	120° 30'	1,500,000.
				44° 00'	41° 40'	0

State	Zone	Code	Projection	Central Meridian and Scale Factor (T.M.) or Standard Parallels (L.)	Grid Origin	
					Longitude Latitude	Easting Northing (Meters)
Pennsylvania	**PA**					
North	N	3701	L	40° 53'	77° 45'	600,000.
				41° 57'	40° 10'	0
South	S	3702	L	39° 56'	77° 45'	600,000.
				40° 58'	39° 20'	0
Rhode Island	**RI**					
		3800	T.M.	71° 30'	71° 30'	100,000.
				1:160,000	41° 05'	0
South Carolina	**SC**					
		3900	L	32° 30'*	81° 00'*	0
				34° 50'*	31° 50'*	0
South Dakota	**SD**					
North	N	4001	L	44° 25'	100° 00'	600,000.
				45° 41'	43° 50'	0
South	S	4002	L	42° 50'	100° 20'	600,000.
				44° 24'	42° 20'	0
Tennessee	**TN**					
		4100	L	35° 15'	86° 00'	600,000.
				36° 25'	34° 20'*	0
Texas	**TX**					
North	N	4201	L	34° 39'	101° 30'	200,000.
				36° 11'	34° 00'	1,000,000.
North Central	NC	4202	L	32° 08'	98° 30'*	600,000.
				33° 58'	31° 40'	2,000,000.
Central	C	4203	L	30° 07'	100° 20'	700,000.
				31° 53'	29° 40'	3,000,000.
South Central	SC	4204	L	28° 23'	99° 00'	600,000.
				30° 17'	27° 50'	4,000,000.
South	S	4205	L	26° 10'	98° 30'	300,000.
				27° 50'	25° 40'	5,000,000.
Utah	**UT**					
North	N	4301	L	40° 43'	111° 30'	500,000.
				41° 47'	40° 20'	1,000,000.
Central	C	4302	L	39° 01'	111° 30'	500,000.
				40° 39'	38° 20'	2,000,000.
South	S	4303	L	37° 13'	111° 30'	500,000.
				38° 21'	36° 40'	3,000,000.
Vermont	**VT**					
		4400	T.M.	72° 30'	72° 30'	500,000.
				1:28,000	42° 30'	0

State	Zone	Code	Projection	Central Meridian and Scale Factor (T.M.) or Standard Parallels (L.)	Grid Origin Longitude Latitude	Grid Origin Easting Northing (Meters)
Virginia	**VA**					
North	N	4501	L	38° 02'	78° 30'	3,500,000.
				39° 12'	37° 40'	2,000,000.
South	S	4502	L	36° 46'	78° 30'	3,500,000.
				37° 58'	36° 20'	1,000,000.
Washington	**WA**					
North	N	4601	L	47° 30'	120° 50'	500,000.
				48° 44'	47° 00'	0
South	S	4602	L	45° 50'	120° 30'	500,000.
				47° 20'	45° 20'	0
West Virginia	**WV**					
North	N	4701	L	39° 00'	79° 30'	600,000.
				40° 15'	38° 30'	0
South	S	4702	L	37° 29'	81° 00'	600,000.
				38° 53'	37° 00'	0
Wisconsin	**WI**					
North	N	4801	L	45° 34'	90° 00'	600,000.
				46° 46'	45° 10'	0
Central	C	4802	L	44° 15'	90° 00'	600,000
				45° 30'	43° 50'	0
South	S	4803	L	42° 44'	90° 00'	600,000.
				44° 04'	42° 00'	0
Wyoming	**WY**					
East	E	4901	T.M.	105° 10'	105° 10'	200,000.
				1:16,000*	40° 30'*	0
East Central	EC	4902	T.M.	107° 20'	107° 20'	400,000.
				1:16,000*	40° 30'*	100,000.
West Central	WC	4903	T.M.	108° 45'	108° 45'	600,000.
				1:16,000*	40 ´ 30'*	0
West	W	4904	T.M.	110° 05'	110° 05'	800,000.
				1:16,000*	40° 30'*	100,000.
Puerto Rico and Virgin Islands	**PR**	5200	L	18° 02'	66° 26'	200,000.
				18° 26'	17° 50'	200,000.

*This represents a change from the defining constant used for the 1927 State Plane Coordinate System. All metric values assigned to the origins also are changes.

Reproduced from National Oceanic and Atmospheric Administration. Manual NOS NGS 5. State Plane Coordinate System of 1983.
Rockville, MD: National Geodetic Information Center, 1989.

APPENDIX C

CENSUS GEOGRAPHY DEFINITIONS AND CENSUS SOURCES*

The cartographer planning to use any of the products of the United States Census Bureau is cautioned to read all introductory materials (especially definitions) of the volumes being used. The current census always contains the most recent definitions used when the census was taken. The material in this appendix describes the Census Bureau area classifications as applied to the 1990 Census of Population, map products, and State Data Centers.

GEOGRAPHIC AREA DEFINITIONS

Alaska Native Regional Corporation (ANRC)

A corporate entity organized "to conduct business for profit," pursuant to the Alaska Native Claims Settlement Act, Public Law (P.L.) 92–203 as amended by P. L. 94–204. It consists of a portion of the State of Alaska inhabited, as far as practicable, by Alaska Natives with a common heritage and common interests. The boundaries of these "regions" have been legally established by the Secretary of the Interior in coordination with Native Alaskans.

Alaska Native Village Statistical Area (ANVSA)

A 1990 census tabulation area approximating jurisdictional areas of an Alaska Native village. These villages are legal entities in Alaska, defined pursuant to the Alaska Native Claims Settlement Act (P.L. 92–203), but they do not have legally defined boundaries.

American Indian Reservation
An American Indian area with boundaries established by treaty, statute, and/or executive or court order. The reservations and their boundaries are identified for the Census Bureau by the Bureau of Indian Affairs (BIA) and State governments.

American Indian Trust Land
An area with boundaries held in trust for American Indians outside the legal boundaries of the American Indian reservations. The trust lands and their boundaries are identified by the BIA.

Block
An area bounded on all sides by visible features such as streets, roads, streams, and railroad tracks, and occasionally by nonvisible boundaries such as city, town, or county limits, property lines, and short imaginary extensions of streets. Blocks do not cross census tract or block numbering area boundaries. A block is the smallest geographic tabulation area from the 1990 census.

Block Group (BG)
A combination of census blocks comprising a subdivision of a census tract or block-numbered area (BNA). For data presentation, BGs are equivalent to and a substitute for the enumeration districts used for reporting data in prior censuses. A BG consists of all blocks whose numbers begin with the same digit in a given census tract or BNA; that is, BG 3 within a tract includes all blocks numbered between 301 and 398. Data are presented separately for each part of a BG that is split by the boundary of a higher-level geographic area.

Block-Numbered Area (BNA)
An area delineated cooperatively by the States and the Census Bureau for grouping and numbering blocks in block-numbered areas where census tracts have not been established. The BNAs do not cross county boundaries.

Census County Division (CCD)
A statistical subdivision of a county that is established cooperatively by the Census Bureau and State and local government authorities, for presenting census data in those States that do not have well-defined minor civil divisions.

Census Designated Place (CDP)
A statistical area comprising a densely settled concentration of population that is not incorporated but which resembles an incorporated place in that local people can identify it with a name.

Census Tract
A small, relatively permanent division of a metropolitan statistical area or selected nonmetropolitan counties, delineated for presenting census data. When census tracts are established, they are designed to be relatively homogeneous for population characteristics, economic status, and

*Reproduced directly, in whole or in part, from United States Bureau of the Census, *1990 Census of Population: Tabulation and Publication Program* (Washington, DC: USGPO, 1989); and United States Bureau of the Census, Data Users Services, *Hidden Treasures: Census Bureau Data and Where to Find It* (Washington, DC: USGPO, 1990).

living conditions, and to contain between 2,500 and 8,000 inhabitants. Census tract boundaries are established cooperatively by local census statistical area's committees and the Census Bureau in accordance with Census Bureau–defined guidelines. Census tracts do not cross county boundaries.

Congressional District (CD)   An area established by State officials or the courts for electing persons to the U.S. House of Representatives.

County   The primary division of a State (except Alaska and Louisiana). Each is a governmental unit with powers defined by State law (except in Connecticut and Rhode Island).

County Equivalent   A geographic entity that is not legally referred to as a county but is treated as such for data tabulation purposes. The statistical equivalents of a county are: independent cities in Virginia; Baltimore City, Maryland; St. Louis City, Missouri; Carson City, Nevada; parishes in Louisiana; boroughs and census areas in Alaska; and the portion of Yellowstone National Park in Montana. The District of Columbia has no subdivisions comparable to counties and its entire area is treated as a county equivalent, for census purposes.

Division (Census Geographic)   A grouping of States within a census geographic region for presentation of census data. The nine divisions, established in 1910, represent relatively homogeneous areas that are subdivisions of the four census geographic regions. See **Region (Census Geographic).**

Geographic Hierarchy   Geographic areas presented in the data products follow a sequence from the highest geographic level to the lowest comparable level. That is, for the National level files and reports, data will be shown for the total United States, followed by regions, divisions, and States. An example of the census hierarchy for State products is State, county, MCD/CCD, place, census tract or BNA, block group, and block. Minor variations of these hierarchies occur in the summary tape files.

Incorporated Place   A governmental unit, incorporated under State law as a city, town (except in New England, New York, and Wisconsin), village, or borough (except in Alaska and New York), having legally prescribed limits, powers, and functions.

Metropolitan Statistical Area (MSA)   A highly populated, economically integrated area defined by the Office of Management and Budget as a Federal statistical standard. An area qualifies for recognition as a MSA in one of two ways: it contains a city of at least 50,000 population, or an urbanized area of at least 50,000 with a total metropolitan population of at least 100,000 (75,000 in New England). The MSAs are defined in terms of counties except in New England, where cities and towns (MCDs) are used.

MSA is also used as a generic term to represent metropolitan statistical areas, primary metropolitan statistical areas, and/or consolidated metropolitan statistical areas in this report.

Minor Civil Division (MCD)   The primary political and administrative subdivision of a county in 28 States. The MCDs are identified by a variety of legal designations such as township, town, borough, magisterial district, or gore. In States where places are or can be independent of any MCD, such places are recorded by the Census Bureau as MCD equivalents, as well as places.

Place   Two types of places are recognized in census products, incorporated places and census designated places, as defined in this appendix.

Region (Census Geographic)   A large grouping of States for presentation of census data. Each of the four regions is subdivided into divisions.

State   A primary governmental division of the United States. For data presentation, the Census Bureau treats the District of Columbia as a statistical State equivalent.

Tribal Designated Statistical Area (TDSA)   An area where Federally and State recognized American Indian tribes without a land base choose to delineate an identifiable land area for the 1990 census.

Tribal Jurisdiction Statistical Area (TJSA)   A 1990 census tabulation area identified by Oklahoma tribal officials containing the American Indian population over which they have jurisdiction.

United States   The 50 States plus the District of Columbia.

Urbanized Area   A central city (or cities) and surrounding closely settled territory ("urban fringe") that together have a minimum population of 50,000. The Census Bureau uses published criteria to determine the eligibility and definition of urbanized areas.

Voting District (VTD)   Any of a variety of types of areas (that is, election districts, precincts, legislative districts, wards, and so forth) defined by States and local governments for elections.

MAP PRODUCTS

The Census Bureau will prepare the map products for the 1990 census tabulation and publication program from the Topologically Integrated Geographic Encoding and

Referencing (TIGER) System. Precensus and postcensus geographic computer tape file extracts from the TIGER database will be available from 1989 through 1992. More detailed information on these files will be provided in later census materials or by contacting Customer Services, Bureau of the Census, Washington, DC 20233, (301/763-4100).

The products for the tabulation and publication program produced from TIGER include various types of paper maps. Each of the following map types will be available for purchase separately. Information on obtaining any/all of the map products can be found in the section, "How to Obtain 1990 Census Data Products." Listed below are descriptions of the map products offered through the 1990 program (NOTE: This list includes only those maps that will be sold separately; it does not include maps that will be prepared and included in the printed reports.)

1990 CENSUS BLOCK-NUMBERED MAPS

The 1990 census block-numbered map series includes county-wide maps prepared on the smallest possible number of map sheets at the maximum practical scale. The 1990 map series will depict each county (or county equivalent) on one or more map sheets—depending on the areal size and shape of the county, the number of blocks in the county, and the density of the block pattern—that will allow displaying all block numbers and feature identifiers as well as show the county boundary and the MCDs/CCDs, places, and census tracts/BNAs in the county. Each county will consist of one or more parent sheets at a single scale, plus insets of densely settled geographic areas as required. As a result, the maps for counties could be at different scales. Insets will be single sheets at a larger scale. In densely developed areas where an inset will not fit on one sheet, multiple-sheet insets will be used. An index showing the map sheet and inset coverage will be included. The standard sheet size planned for all maps is 36″ × 42″ with a maximum 32″ × 32″ map display area. Data users must purchase these maps from the Census Bureau; they will not be printed and sold through the Government Printing Office (GPO) as in 1980. The maps will be produced by map plotting equipment on paper by Census Bureau staff.

COUNTY SUBDIVISION MAPS

These maps will show the names and boundaries of all counties (or county equivalents) and subdivisions (for example, MCDs/CCDs, sub-MCDs) in each State as well as all places for which the Census Bureau tabulates data for the 1990 census. They also will depict American Indian reservations including off-reservation trust lands, tribal jurisdiction statistical areas in Oklahoma, tribal desig-

nated statistical areas, Alaska Native Regional Corporations, and Alaska Native village statistical areas. All boundaries will be as of January 1, 1990. These maps will be available first in electronic plotter version and later in a printed form. The electrostatic plotter paper sheet size will be about 36″ × 42″ and the scale will be 1:500,000. These maps will be produced on a State basis and sold by the Census Bureau upon user request. Later, these maps will be printed and bound in the State reports for the following report series: 1990 CPH-1, 1990 CPH-2, 1990 CPH-5, 1990 CP-1, 1990 CP-2, 1990 CH-1, and 1990 CH-2. In the reports, these maps will be partitioned into multiple, page-size sheets and the scale will vary between States. The printed reports can be purchased through the Superintendent of Documents, Government Printing Office (GPO).

CENSUS TRACT/BLOCK NUMBERING AREA OUTLINE MAPS

These maps will show census tract/block numbering area boundaries and numbers and the features and feature names underlying these boundaries, (for example, the boundaries and names of counties, county subdivisions, and places). The scale of the maps will be determined such that the number of map sheets for each area will be minimal, but will vary by area. For densely settled areas, where the census tract/block numbering area numbers and boundary features cannot be shown, the Census Bureau will issue insets at a larger scale. These maps will be available in both electrostatic plotter version and a printed version. Data users who do not wish to wait for the printed maps can purchase these maps from the Census Bureau for a fee. Data users who want printed maps can purchase the printed maps from the Superintendent of Documents. GPO, beginning in 1992. These maps will be printed and available for purchase at the same time as the *Census Tract/Block Numbering Areas* report series which will be issued by MSA/PMSA.

VOTING-DISTRICT OUTLINE MAPS

The maps in this series will show voting district numbers and boundaries as well as the underlying features such as roads, railroads, and rivers. They also will show the boundaries and names of counties, county subdivisions, and places. The mapping unit will be a county with a variable scale. The maps will not be printed. They will be produced only for counties in those States that participated in Phase 2 of the Voting District Program. These maps will be produced using map plotting equipment on paper by Census Bureau staff.

The following indicates the method of production and the selling agent for each map type described in the section "Geographic Products."

1990 Census Block-Numbered Maps

Method of Production:	Electrostatic (computer-generated) plotter version	Selling agent:	Customer Services Bureau of the Census Washington, DC 20233

County Subdivision Maps

Method of Production:	Electrostatic (computer-generated) plotter version	Selling Agent:	Customer Services Bureau of the Census Washington, DC 20233

Census Tract/Block-Numbered Area Outline Maps

Method of Production:	Electrostatic (computer-generated) plotter version	Selling Agent:	Customer Services Bureau of the Census Washington, DC 20233
Method of Production:	Printed paper copies	Selling Agent:	Superintendent of Documents Government Printing Office Washington, DC 20402 (available for sale on a flow basis beginning in 1992)

Voting District Outline Maps

Method of Production:	Electrostatic (computer-generated) plotter version	Selling Agent:	Customer Services Bureau of the Census Washington, DC 20233

STATE DATA CENTERS

Probably the single most valuable treasure trove you can find on the local scene is your State Data Center (SDC). Begun in 1978, this program now numbers over 1,400 organizations.

You can find SDCs in every State, the District of Columbia, Puerto Rico, Guam, and the U.S. Virgin Islands.

These organizations are not part of the Census Bureau, but they are vital partners in helping us disseminate data. They are generally State or local government agencies, libraries, and academic centers.

The organization of each SDC varies from State to State, but usually involves a major State executive or planning agency, a major State university, or the State library. In addition, each State has a network of affiliated data centers.

SDCs receive Census Bureau data for their States and surrounding areas and make the data available to the public, often at or below cost. Many provide special services, some of which are not available from the Census Bureau.

Here are just a few of the services SDCs may be able to offer:

Downloading Computer Tape Files onto Microcomputer Diskettes
SDCs have given the microcomputer revolution a real push.

State Profiles with Information from the Census Bureau and Other Government and Private Sources
Some publish annual State statistical abstracts similar to the Census Bureau's best-selling *Statistical Abstract of the United States*. Often they prepare these reports using data otherwise available only on tape from the Census. In addition, the reports often have data collected by the State itself.

Special Extracts from Large Census Bureau Tape Files.

On-line Data Sources
If you've got a microcomputer and a modem, you're in business!

Marketing Research
A few SDCs help clients in profiling the local market. Some even conduct telemarketing surveys or studies to fit a client's specifications. Do you need a market analysis? An SDC may be able to help.

Guides to Local Data Sources
SDCs have newsletters and brochures that steer you to the information you need. Ask about getting on the mailing list.

Small-Area Profiles
Many SDCs have prepared county, ZIP code, neighborhood, or census tract profiles so you can zoom in for a close look at the area you're studying.

Studies on Issues of Local Importance
SDCs are occasionally called upon to provide research for bills pending in their legislature. They'll share this research.

Maps
If you need maps for special areas—school districts, neighborhoods, census tracts—call an SDC. The newly released TIGER/Line™ geographic reference files allow SDCs to be more creative than ever in doing maps.

Reference Libraries
SDCs are often libraries or include libraries where you can do your own research.

Services vary from center to center, so call and get a rundown on how your center can help you. For information about the special services of your State Data Center, you should contact the lead agency in your State.

LEAD AGENCIES IN THE SDC PROGRAM

Alabama
Center for Business and Economic Research
University of Alabama
Tuscaloosa, AL 35487-0221
(205/348-6191)

Alaska
Research and Analysis
Department of Labor
Juneau, AK 99802-5504
(907/465-4500)

Arizona
Arizona Department of Economic Security
Phoenix, AZ 85005
(602/542-5984)

Arkansas
University of Arkansas–Little Rock
Little Rock, AR 72204
(501/569-8530)

California
Department of Finance
Sacramento, CA 95814
(916/322-4651)

Colorado
Colorado Department of Local Affairs
Denver, CO 80203
(303/866-2156)

Connecticut
Connecticut Office of Policy and Management
Hartford, CT 06106
(203/566-8285)

Delaware
Delaware Development Office
Dover, DE 19903
(302/736-4271)

District of Columbia
Mayor's Office of Planning
Washington, DC 20004
(202/727-6533)

Florida
Executive Office of the Governor
Tallahassee, FL 32399-0001
(850/487-2814)

Georgia
Georgia Office of Planning and Budget
Atlanta, GA 30334
(404/656-0911)

Guam
Guam Department of Commerce
Tamuning, Guam 96911
(671/646-5841)

Hawaii
State Department of Business and Economic Development
Honolulu, HI 96804
(808/548-3082)

Idaho
Idaho Department of Commerce
Boise, ID 83720
(208/334-2470)

Illinois
Illinois Bureau of the Budget
Springfield, IL 62706
(217/782-1381)

Indiana
Indiana State Data Center
Indiana State Library
Indianapolis, IN 46204
(317/232-3733)

Iowa
State Library of Iowa
Des Moines, IA 50319
(515/281-4105)

Kansas
State Library
Topeka, KS 66612
(913/432-3919)

Kentucky
Urban Studies Center
University of Louisville
Louisville, KY 40292
(502/588-7990)

Louisiana
Office of Planning and Budget
Division of Administration
Baton Rouge, LA 70804
(504/342-7410)

Maine
Maine Department of Labor
Augusta, ME 04330
(207/289-2271)

Maryland
Maryland Department of State Planning
Baltimore, MD 21201
(410/255-4450)

Massachusetts
Massachusetts Institute for Social and Economic Research
University of Massachusetts
Amherst, MA 01003
(413/545-0176)

Michigan
Department of Management and Budget
Lansing, MI 48909
(517/373-7910)

Minnesota
State Demographer's Office
Minnesota State Planning Agency
St. Paul, MN 55155
(612/297-2360)

Mississippi
Center for Population Studies
University of Mississippi
University, MS 38677
(601/232-7288)

Missouri
Missouri State Library
Jefferson City, MO 65102
(314/751-3615)

Montana
Montana Department of Commerce
Helena, MT 59620-0401
(406/444-2896)

Nebraska
Center for Applied Urban Research
University of Nebraska–Omaha
Omaha, NE 68182
(402/595-2311)

Nevada
Nevada State Library
Carson City, NV 89710
(702/885-5160)

New Hampshire
Office of State Planning
Concord, NH 03301
(603/271-2155)

New Jersey
New Jersey Department of Labor
Trenton, NJ 08625-0388
(609/984-2593)

New Mexico
Economic Development and Tourism Department
Santa Fe, NM 87503
(505/827-0276)

New York
New York Department of Economic Development
Albany, NY 12245
(518/474-6005)

North Carolina
North Carolina Office of State Budget and Management
Raleigh, NC 27603-8005
(919/733-7061)

North Dakota
Department of Agricultural Economics
North Dakota State University
Fargo, ND 58105
(701/237-8621)

Ohio
Ohio Department of Development
Columbus, OH 43266-0101
(614/466-2115)

Oklahoma
Oklahoma Department of Commerce
Oklahoma City, OK 73126-0980
(405/841-5184)

Oregon
Center for Population Research and Census
Portland State University
Portland, OR 97207-0751
(503/725-3922)

Pennsylvania
Institute of State and Regional Affairs
Pennsylvania State University
Middletown, PA 17057-4898
(717/948-6336)

Puerto Rico
Puerto Rico Planning Board
San Juan, PR 00940-9985
(787/728-4430)

Rhode Island
Office of Municipal Affairs
Providence, RI 02908-5873
(401/277-6493)

South Carolina
South Carolina Budget and Control Board
Columbia, SC 29201
(803/734-3780)

South Dakota
Business Research Bureau
University of South Dakota
Vermillion, SD 57069
(605/677-5287)

Tennessee
State Planning Office
Nashville, TN 37219
(615/741-1676)

Texas
Texas Department of Commerce
Capitol Station, Austin, TX 78711
(512/472-5059)

Utah
Office of Planning and Budget
Salt Lake City, UT 84114
(801/538-1036)

Vermont
Office of Policy Research and Coordination
Montpelier, VT 05602
(802/828-3326)

Virgin Islands
Caribbean Research Institute
University of the Virgin Islands
Charlotte Amalie,
St. Thomas, VI 00802
(340/776-9200)

Virginia
Virginia Employment Commission
Richmond, VA 23219
(804/786-8624)

Washington
Office of Financial Management
Olympia, WA 98504-0202
(206/586-2504)

West Virginia
Governor's Office of Community
and Industrial Development
Charleston, WV 25305
(304/348-4010)

Wisconsin
Demographic Services Center
Madison, WI 53707-7868
(608/266-1927)

Wyoming
Department of Administration and Fiscal Control
Cheyenne, WY 82002-0060
(307/777-7505)

APPENDIX D

WORKED PROBLEMS

This appendix presents several worked problems in the areas of map scale, classing, proportional symbols, and value-by-area cartograms. The intention is *to give you additional experience with these problems to strengthen your understanding of these topics.*

MAP SCALE

Map scale problems occur routinely when you are dealing with maps. These problems are not "made up" in the fictitious sense; they are all encountered in the normal day-to-day world of handling manuscript maps as resource maps for scanning, when you are assembling many maps to produce one source map for compilation, or when you are figuring the map scale for a new computer output map, or when you are faced with changing an RF scale to a verbal or graphic scale. Knowing how to easily deal with map scale will smooth out your tasks.

The first thing to do in *any* map scale problem is to write down the fundamental relationship: map distance divided by earth distance. Doing this first will usually solve most map scale problems. Let's look at our first problem.

Map Scale Problem 1

A straight stretch of road is known to be 500 yards long. The same road segment measured on an aerial photograph is measured to be 3/4 of an inch long. What is the RF of the photograph?

1. First, put down the basic relationship:

$$\frac{\text{map distance}}{\text{earth distance}}$$

2. Now substitute what you know to be true. Map distance in this case is .75 inches, and the earth distance is 500 yards. You know that you must always deal in the same units, so convert the 500 yards into inches:

500 yards × 3 feet × 12 inches = 18,000 inches

3. Simply place the number of units into the basic relationship:

$$\frac{.75 \text{ inches (map distance)}}{18,000 \text{ inches (earth distance)}}$$

Now, you know that customarily the numerator in any RF scale is always expressed as unity, so you must convert the fraction into one with unity in the numerator:

4. To form a proportional relationship you must place two fractions equal to one another, but with one element unknown, and solve the equation for the unknown (*X*):

$$\frac{.75}{18,000} = \frac{1}{X}$$

Solve for *X*.

$$(.75)(X) = (1)(18,000)$$
$$(X) = 24,000$$

So, then, the answer to the problem is **1:24,000.**

Map Scale Problem 2

In this problem you have a map whose RF scale is 1:29,500; you are trying to find the distance between two cellular telephone towers which are 6.7 inches apart on the map. The first thing you do is to set down the fundamental relationship:

1. $$\frac{\text{map distance}}{\text{earth distance}}$$

2. Now you substitute into the basic fraction those parts of the problem you already know:

$$\frac{6.7 \text{ inches (map distance)}}{\text{earth distance}}$$

Because you know their relationship to one another (e.g., the map scale), you can write:

$$\frac{6.7 \text{ inches}}{X} = \frac{1}{29,500}$$

Solve for *X*.

$$(1)(X) = (6.7 \text{ inches})(29,5000)$$
$$X = 197,650$$

3. It is unlikely that you will want the distance between the two towers in inches, so you convert the inches into miles by dividing the total inches by 63,360 (the number of inches in 1 mile):

$$\frac{197,650}{63,360} = 3.12 \text{ miles}$$

Map Scale Problem 3

Many times in mapping you are faced with the problem of expressing scale in terms that are different from what you have, and having to convert to different units. Imagine that a European friend of yours has a map of Europe, made in the United States, whose scale is expressed in inches to the mile, and she wishes you to convert it to centimeters to the kilometer. Her map has a verbal scale of 2 inches, representing 5 miles. There are several ways you could do this, but you decide to make this an educational lesson for your friend, so you first compute an RF.

1. First, you put down the basic relationship:

$$\frac{\textbf{map distance}}{\textbf{earth distance}}$$

2. You already have these values, so you substitute

$$\frac{\textbf{2 inches}}{\textbf{5 miles}}$$

3. But, you want to find the RF, so you set two fractions equal, and solve for the denominator (X):

$$\frac{\textbf{2 inches}}{\textbf{5 miles}} = \frac{1}{X}$$

then

$$(\textbf{2 inches})(X) = (\textbf{5 miles})(1)$$

and converting miles to inches:

$$(2)(X) = 316,000$$

then

$$X = 158,000$$

so the RF is **1:158,000**.

4. Because any RF scale shows only a relationship, it can be expressed in any units; therefore, to answer your friend's question, you can do this:

$$\frac{1}{158,000} = \frac{1}{X} \quad \textbf{(distance on map in centimeters)}$$

and

$$X = 158,000$$

To get kilometers, simply divide 158,000 by the number of centimeters in 1 kilometer (100,000), and the answer becomes **1.58 kilometers**.

Map Scale Problem 4

Many times in small-scale mapping you are faced with determining different scales on the same projection in order to display these relationships on a map. In this particular problem you have a hemispheric Bonne projection, and you know that there is excessive linear scale distortion along the outer meridians of this plot (you know this from previous experience). (See Figure D.1.) The task in this problem is to find how much greater in length the 60-degree east meridian is than the central meridian (but just from the equator to latitude 80-degrees north). You therefore divide the problem into three components: the RF scale of the central meridian; the RF scale along the 60-degree east meridian (or west, as it is a symmetrical projection), but only from the equator to 80-degrees north;

and how much longer it is than it should be when compared to the projection's central meridian.

You begin tackling this problem by putting down what you know:

1.
$$\frac{\textbf{map distance}}{\textbf{earth distance}}$$

2. You also know that the earth is approximately 24,900 miles around the equator and, for all practical purposes, the length around all meridians is the same (you remember these distances from a cartography course you took).

You first decide to figure the earth distance in the scale problem. From the equator to 80-degrees north is 8/9 of the way (.88). The whole way from the equator to the pole is 24,900 miles divided by 4, which yields 6,225 miles. So the distance from the equator to 80-degrees north is

$$\textbf{6,225 miles} \times \textbf{.88} = \textbf{5,478 miles}$$

3. Now you decide to tackle the map distance. This can be done by *measuring* with a scale as closely as you can the distance in inches between the equator to 80-degrees north. You determine that it is $8 \times .3$ inches, yielding 2.4 inches (one 10-degree segment along the meridian is .3 inches; all segments are very nearly the same length, and you have eight segments).

4. Coming back to the fundamental relationship, you have

$$\frac{\textbf{2.4 inches}}{\textbf{5,478 miles}} = \frac{1}{X}$$

so

$$2.4X = 5,478$$

and converting:

$$2.4X = 347,086,080$$

Reducing to unity and rounding, the RF is **1:145,000,000**.

But as this is only part of the problem, you next need to determine the scale along the central meridian. *Using the same procedures as before,* you determine the linear scale of the central meridian to be 1:153,000,000 by measuring its length from the equator to 80-degrees north to be 2.27 inches and knowing that this represents 5,478 miles.

All of the meridians should be the same scale (they are on the globe), but as the 60-degree east is *longer* than it should be on the projection, its linear scale is exaggerated; thus, it has a scale factor greater than one. The 60-degree meridian length is divided by the length of the central meridian to find out *how much longer it is than it should be,* which turns out to be 1.05. This concludes your work on this scale problem.

All scale problems can be approached and solved the same way: first, put down the basic relationship of map distance/earth distance; second, substitute all values you have into this relationship; and third, draw on the knowledge you have at hand, and solve the problem.

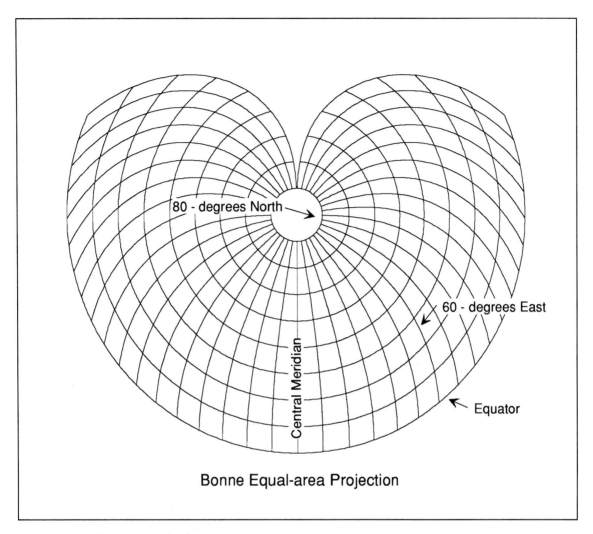

Figure D.1   The Bonne projection.
The Bonne equal-area projection centered on 0° and extends from the equator to 80° N, and from 180° W to 180° E.

Class-Limit Problems

There are different ways of calculating class limits for quantitative maps; some were presented in Chapter 7. Four examples of worked problems are discussed here to extend the earlier discussions.

The first worked problem is for calculating class limits using the arithmetic progression—the method you want to choose if you desire constant intervals. The first thing you do is to calculate the range of the data:

1. **Range (R) = highest value (H) − lowest value (L)**

In this example, suppose you wish to have seven classes, and your data range is from 7 to 111. Now, calculate the common difference (CD), which is also the class range for each class:

2. **Class range (CD) = $\dfrac{R}{\text{Number of classes (7)}}$**

which in this case is $\dfrac{111-7}{7} = 14.85$

To obtain each upper class limit, you begin by adding the CD to the lowest value (L), and then to each successive upper class limit, like this:

3. **L + (1 × CD)   = first class limit**
4. **7 + (1 × 14.85) = 21.85**

The next class limit would be

5. **21.85 + 14.85 = 36.7**

and so on until you reach the highest value, which in this case will be these upper class limits:

21.85, 36.76, 51.5, 66.40, 81.25, 96.10

Notice that although you specified seven classes, you only have six upper class limits. You will always have one fewer upper class limit than you have number of classes.

Although this helps you compute the class limits, it does not tell you how to develop the *legend*. With this first example, the legend classes should be 7–21.85, 21.86–36.76, 36.77–51.50, 51.51–66.40, 66.41–81.25, 81.26–96.10, 96.11–111.00. You may also wish to round off the numbers to whole numbers.

Our next worked problem is for the times that you wish to calculate class limits when you are *multiplying* each class interval by a common multiplier, which is the case when you know your data are approximating a geometrical rate of increase. With this form of class limit determination, you will be using a geometric progression:

$$a, ar, ar^2, ar^3, \ldots ar^{n-1}$$

where

a = the first term (the lowest value in the data array)
r = the common ratio (the multiplier)
n = the number of classes
$l = ar^{n-1}$ (the highest value in the data array).

In this example, suppose that your lowest value (L) in the data set is 15, and the highest value (H) is 240, and you note (perhaps by plotting the values into a data array) that the array appears to follow the pattern of a geometric progression. You want to have five classes, which means of course that you will have only four upper class limits. The formula (derived from the progression series above) is

1. **$L \times X^{n-1} = H$**

where L is the lowest value in the data set, H is the highest value, X is the multiplier, and n is the number of classes. For our example here:

2. **$15 \times X^4 = 240$**
3. **$X^4 = \dfrac{240}{15}$**
4. **$X = 2$**

Knowing this, it is possible to determine the upper class limits using the progression series:

lowest value = 15
1st class upper limit = 15×2 = 30
2nd class upper limit = $15 \times 2^2 = 60$
3rd class upper limit = $15 \times 2^3 = 120$
4th class upper limit = $15 \times 2^4 = 240$

The legend would appear like this: 15–30, 31–60, 61–120, and 121–240 (four classes). Determining the terms in a geometric progression is relatively easy with handheld calculators that extract roots easily.

A third example is when you wish to divide the data set into a number of classes so that there are an equal number of area units in each class. This is quantile classing, and yields a very uneven or irregular class interval system. Suppose for this example that you have the values found in the following array:

2, 2, 4, 4, 5, 7, 8, 8, 9, 9, 10, 11, 12, 14, 16, 16, 17, 19, 19, 25

and you wish to have five classes. You begin, as we have here, by arraying the data from lowest to highest (easy with spreadsheet programs). Then use this simple formula:

$$K = \frac{N}{n}$$

where K is the number of values in each class, N is the number of areal units, and n is the number of classes.

In this example,

1. $K = \dfrac{20}{5}$
2. $K = 4$

Beginning with the lowest value, count up four values. Normally, the mean between the fourth and the fifth values is the class limit, and so on, like this:

4.5, 8.5, 11.5, and 16.5

The values placed in the legend could be 2–4.5, 4.6–8.5, 8.6–11.5, 11.6–16.5, and 16.6–25.0.

Data arrays were explained and used in Chapter 7 where it was pointed out that class breaks are placed at points on the array where slope changes markedly. In that example, ordinary linear graph paper was used. It is also possible to place the array on log paper, which dramatizes slope changes even more (the steeper the slope, the greater the rate of change).

Proportional Symbol Problems

A very common computational problem in quantitative mapping is the development of the sizes for proportional symbols. As the circle is the most common, it will be the one treated here. There are two ordinary ways of scaling circles, as explained earlier in the text: linearly, in which circle sizes are computed by calculating square roots of the values (where the power in the equation is .5); and the Flannery method, in which the circle values are raised to the .57 power in the equation.

The first example has a number of values between the lowest and the highest, 76 and 284 respectively. Because deciding the sizes of the end values is most critical, this example will skip the computations necessary for the intermediate values (see Chapter 9), and concentrate only on the extremes. This first example will compare both conventional circles and Flannery circles.

A general strategy is first to calculate the size of the small circle. Remember with circle size scaling, it is the *area* of the circle that is scaled. A radius is set for the small circle by looking at the map and judging a size that looks appropriate. After a radius is chosen, then the proportional size for the largest is calculated to see if it is too large or too small for the map.

For the first attempt, suppose a radius of .07 inches is selected for the small circle, then [as the area of a circle is $(\pi)(r^2)$] the largest circle is calculated this way:

1.
$$\frac{(\pi)(r_S)^2}{(\pi)(r_L)^2} = \frac{76}{284}$$

Because r_s (smallest circle radius) is .07

2.
$$\frac{(.07)^2}{(6r_L)^2} = \frac{76}{284}$$

3.
$$\frac{(.07)}{r_L} = \frac{\sqrt{76}}{\sqrt{284}}$$

4.
$$\frac{.07}{r_L} = \frac{8.71}{16.85}$$

5.
$$8.71\ (r_L) = 1.179$$

6.
$$r_L = .135\ \text{inches}$$

Now you have the proportional size of the largest circle and you need to examine it for suitability on the map. If it is appropriate, then all intermediate circle sizes can be calculated; but if it is not, a new radius of the small circle must be selected, and the process repeated.

Fortunately, today there are two companions that can help you do these problems quickly—the computer spreadsheet and/or the handheld calculator.

The square root of a number is the same as raising it to the power of .5—an easy task for most handheld calculators. Flannery introduced the idea of using .57 instead of .5, again a simple task for the calculator and for most spreadsheet programs. Spreadsheet programs will be especially helpful when calculating all the intermediate circle sizes. Table D.1 shows the differences of the two largest circles when computed for both .5 and .57.

There are two considerations regarding which method to use. In one, research has shown that some map readers will be able to estimate circle sizes better if drawn by the Flannery methods. The other consideration is that Flannery circles take up far more room on the map than conventional circles. Which is best? Each cartographer will have to make a judgment based on map purpose and sophistication of the map reader.

Value-by-Area Cartogram Problem

Table 11.1 of Chapter 11 provided one example of a value-by-area scaling problem. Here we will resolve yet another problem to give you more experience with them.

In the present example, there are a few constraints given for the mapping task at hand, which you need to know at the outset. First, you can only have 20 square inches in which to draw the cartogram. Second, you decide that you will use a .01-square-inch (100 per square inch) grid paper because you like to work with this size. This means that there will be a total of 2,000 grid units, or *counting units* (20 × 100).

Table D.1 Comparison of the Largest Circle Radius by Conventional and Flannery Methods—Proportional Symbol Problem #1

Small Circle (r)	Large Circle (r)	
	Conventional	Flannery
.07 inches	16.85 inches	25.0 inches

Table D.2 Per Capita Spending in Mental Health Programs in Selected Mid-Atlantic and New England States and Washington, DC

State	Per Capita Spending (Dollars)	Number of Counting Units
Maine	53	147
New Hampshire	51	140
Vermont	52	143
Massachusetts	62	171
Rhode Island	52	143
Connecticut	68	187
New York	140	386
New Jersey	51	140
Pennsylvania	68	187
Washington, DC	129	355
Total	**726**	**1,999+ (2,000)**

For this example, you are doing a cartogram of the per capita spending (dollars) on mental health programs in Washington, D.C., and the 10 New England and Middle Atlantic states. You have inspected the data and are certain that they qualify for a reasonable cartogram. The data are provided in Table D.2.

The first step is to add all the values in the per capita spending column; the result is 726. As there are 2,000 counting units that will be used to portray 726 dollars, then each counting unit will be 726 ÷ 2,000, or .363 dollars each. The next step is to divide each state's dollars spent by .363 to determine how many counting units you should assign to each state for mapping. These values are shown also in the table. The total of this column should equal the total of all counting units (2,000), or nearly so because of rounding.

The remainder of the cartogram mapping problem involves the mapping activity, and need not be discussed here.

APPENDIX E

IMPORTANT WORLD WIDE WEB SITES FOR CARTOGRAPHERS

Data and Map Sources

Note: All sites listed have the http:// protocol unless otherwise listed. In some cases entire pathnames are listed if particularly important; otherwise, only addresses are given. Most sites with only addresses have numerous internal sites.

1. Academic research centers
   www.cast.uark.edu Center for Advanced Spatial Technologies, University of Arkansas
   www.cfm.ohio-state.edu Center for Mapping, Ohio State University
   www.ncqia.ucsb.edu National Center for Geographic Information and Analysis
   www.utexas.edu/dept/grg/virtdept/sources/contents.html University of Texas, Geography and Cartography Resources
2. United States federal agencies
   www.census.gov Bureau of the Census
   www.odci.gov/cia/publications/pub.html Central Intelligence Agency
   www.usda.gov Department of Agriculture
   www.fs.fed.us Forest Service

   Department of the Interior:
   www.blm.gov Bureau of Land Management
   www.fws.gov Fish and Wildlife Service
   www.blm.gov/gis/nsdi.html National Spatial Data Infrastructure MetaData and www Mapping Sites
   www.bts.gov Department of Transportation, national transportation data
   www.epa.gov Environmental Protection Agency
   www.nasa.gov National Aeronautics and Space Administration
   www.nima.mil National Imagery and Mapping Agency
   www.usno.navy.mil Naval Observatory

   National Oceanographic and Atmospheric Agency:
   www.ncdc.noaa.gov National Climate Data Center
   www.ngs.noaa.gov National Geodetic Survey
   www.ngdc.noaa.gov National Geophysical Data Center
   www.nodc.noaa.gov National Oceanographic Data Center
   www.nws.noaa.gov National Weather Service
   www.saa.noaa.gov Satellite Active Archives
   www-mel.nrlmry.navy.mil Department of Defense, Master Environmental Library
   www.usno.navy.mil Naval Observatory

   United States Geological Survey:
   www.usgs.gov USGS Home Page: biology, earthquakes, geology, mapping, water topics, resources
   nsdi.usgs.gov USGS, National Geospatial Data Clearinghouse
   edcwww.cr.usga.gov USGS, EROS Data Center
   edcwww.cr.usgs.gov USGS, Global Land Information Center (GLIS)
   edcwww.cr.usgs.gov/land/daac/gtopo30/gtopo30.html global 3arc second elevation data
   edcwww.cr.usgs.gov/doc/edchome/ndcdb/ndcdb.html USGS DEMs, DLGs LULCs
   H2o.er.usgs.gov/nsdi/dcw/dcwindex.html United States Index, digital chart of the world
3. Canada federal agencies
   www.geod.emr.ca Geomatics Canada: Geodetic Survey of Canada
   www.ellesmere.ccm.emr.ca/sales National Atlas of Canada
   www.ellesmere.ccm.emr.ca/naismap National Atlas of Canada, interactive map development
   www.statcan.ca/start.html Statistics Canada
4. Sources and data for outside the United States and Canada
   www.geo.ed.ac.uk/home/ded.html Digital Elevation Catalog
   www.un.org United Nations
   www.library.yale.edu/un/unhome.htm United Nations Scholars Workstation
   www.grid.unep.ch UN Environmental Program, Global Data Set Catalog
   www.geo.ed.ac.uk/home/ded.html Digital Elevation Catalog
   ftp:sepftp.stanford.edu/pub/world-map World Map Directory

Newsgroups

An Internet newsgroup is analogous to a bulletin board where messages, or articles as they are sometimes called, are posted for review by any Internet user. There are a large number of newgroups encompassing a wide variety of topics. News servers, operated by Internet service providers, educational institutions, companies, and other organizations, distribute these messages to a worldwide audience. Most Internet browsers provide a news reader program that allows users to read or post messages. Internet users may reply to postings with the option of responding to an individual or the entire group. Some non–software specific cartography-related newsgroups are listed below:

bit.listserv.geograph   general geography discussion

comp.infosystems.gis   discussion of geographic information systems (GIS)

comp.graphics.visualization   visualization issues

sci.geo.satellite-nav   discussion of global positioning systems (GPS)

sci.image.processing   image-processing issues

Mailing Lists

An Internet mailing list is an E-mail service, generally free, that sends messages to a group of subscribers. There are two addresses associated with every mailing list; the server address and the list address. The server address is where users send an E-mail message to subscribe or cancel. The list address is the address to which subscribers post E-mail queries, responses, or comments related to the list's topic. To subscribe to a mailing list, send an E-mail to the server address. In the body of the message write Subscribe < list name >. For example, if subscribing to MAPS-L, send an E-mail to listserv@ugs.cc.ugs.edu and in the body of the message, type Subscribe MAPS-L. A selection of non–software specific mailing lists related to cartography and GIS follows:

CARTO-SOC
Society of Cartographers
Server: listproc@sheffield.ac.uk
List Address: CARTO-SOC@sheffield.ac.uk

(**Note:** When subscribing, type subscribe CARTO-SOC <your full name>)

GEOGRAPH
General Geography Discussion
Server: listserv@segate.sunet.se
List Address: GEOGRAPH@segate.sunet.se

GEONET-L
Geoscience Librarians and Information Specialists
Server: listserv@iubvm.ucs.indiana.edu
List Address: GEONET-L@lubum.ucs.indiana.edu

GEOWEB
GIS Resource Discussion
Server: majordomo@census.gov
List Address: GEOWEB@census.gov

GIS-L
General GIS Discussion
Server: listserver@geoint.com
List Address: GIS-L@geoint.com

MAPS-L
Maps and Air Photos Systems Forum
Server: listserv@uga.cc.uga.edu
List Address: MAPS-L@uga.cc.uga.edu

MAPHIST
Historical Maps Topics
Server: listserv@harvarda.harvard.edu
List Address: MAPHIST@harvarda.harvard.edu

NSDI-L
National Spatial Data Infrastructure
Server: listproc@fgdc.er.usgs.gov
List Address: NSDI-L@fgdc.er.usgs.gov

INDEX